Power Systems

Ryszard Strzelecki • Grzegorz Benysek
Editors

Power Electronics in Smart Electrical Energy Networks

Springer

Ryszard Strzelecki, DSc, PhD
Department of Electrical Engineering
Gdynia Maritime University
81-87 Morska street
81-225 Gdynia
Poland

Grzegorz Benysek, DSc, PhD
Institute of Electrical Engineering
University of Zielona Góra
50 Podgórna street
65-246 Zielona Góra
Poland

ISBN 978-1-84800-317-0 e-ISBN 978-1-84800-318-7

DOI 10.1007/978-1-84800-318-7

Power Systems Series ISSN 1612-1287

British Library Cataloguing in Publication Data
Power electronics in smart electrical energy networks. -
(Power systems)
1. Power electronics 2. Electric networks
I. Strzelecki, Ryszard II. Benysek, Grzegorz
621.3'17
ISBN-13: 9781848003170

Library of Congress Control Number: 2008926908

Cover design: deblik, Berlin, Germany

Printed on acid-free paper

9 8 7 6 5 4 3 2 1

springer.com

Dedicated to our loving families

Preface

The book arises from the conviction that it is necessary to re-think the basic philosophy governing the electricity distribution systems. In the authors' opinion there is a need to exploit fully the potential advantages of renewable energy sources, distributed generation, energy storage and other factors which should not only be connected but also fully integrated into the system to increase the efficiency, flexibility, safety, reliability and quality of the electricity and networks. Transformation of the current electricity grids into a smart (resilient, interactive *etc.*) network necessitates the development, propagation and demonstration of key cost effective technologies enabling (*e.g.*, innovative interconnection solutions, storage technologies for renewable energy sources, power electronics, communications *etc.*). On the basis of the above, the major aim of this book is to present the features, solutions and applications of the power electronics arrangements likely to be useful in future smart electrical energy networks.

The first part of this book introduces the structure and fundamental problems of the current electricity grids together with the concept of smart electrical energy networks.

Next there is a critical overview of power theories, mainly under non-sinusoidal conditions in single-phase and three-phase systems, in both time and frequency domains. The basic criterion for the choice of the discussed theories is historical development of knowledge in this field and the usefulness of power theory in solving practical problems: reactive power compensation, balancing the supply network load and mitigation of voltage and current distortion. Particular attention is given to the theories defining the current components in the time domain as the basis for present-day interconnection, active compensation and filtering systems. The content of this part is essential for understanding both the principle of operation and the control algorithms of the majority of the currently used power quality improvement and interconnecting systems.

Additionally, in this part an overview of control methods in power systems with the focus on damping of electromechanical oscillations and mitigation of power quality problems is presented. The focus is on power systems with increased levels of uncertainty resulting from deregulation of theelectrical power industry and the presence of non-conventional types of generation (renewable energy sources and distributed generation). The issue of finding the best techno-economical solution for the problems is also briefly mentioned. The focus in thepower quality section is on probabilistic modelling of disturbances and their consequences.

In the next part of the book the main emphasis is on low, medium, and high power conversion issues and the power electronic converters that process power for a variety of applications in smart grids. Following recent trends in power electronics technology, greater stress is placed on modern power electronic converters, such as resonant and multi-level inverters or matrix converters, and these are thoroughly covered. Special features include in-depth discussions of all power conversion types: AC/DC, AC/AC, DC/DC, and DC/AC.

After that, both the relationships and the differences between electrical power quality and electromagnetic compatibility are explained and definitions of these notions are provided. The principles of standardization in both fields are also be discussed. The power quality survey is a useful procedure for identifying and resolving power-related equipment or facility problems. It is an organized, systematic approach to problem solving. If all the steps for a power quality survey are completed, information is obtained that either identifies a solution to a power-related problem or reveals that the problem is not related to the electrical power system.

After that, EMC related problems in smart electrical power systems as well as some EMC regulations are overviewed. Special attention is paid to the origin and the spreading of the conducted EMI over power systems containing power converters. This is true because the diversity of power converters makes difficult the general analysis of the EMI spectra. However, there are some common features which can be derived from typical applications and layouts of the systems with power converters. Specific key aspects of electromagnetic compatibility in power electronics are presented, such as a typical role of power converters and their place in the smart power system, a typical frequency range of generated EMI noises, specific features of the common mode source in three-phase power converter systems and traveling wave phenomena. This part gives a detailed analysis based on the authors' own experimental results in the systems with converters that are common in smart power systems.

The next part of the book introduces high frequency AC power distribution systems as relatively new and promising developments in the field of electric power. Compared with low frequency or DC link power systems, the high frequency system offers many key advantages including system compactness due to small filtering and transforming components, better power quality, freedom from acoustic noise and mechanical resonance. In addition, it is particularly conducive to the distributed and amalgamated structures of future power systems, which are likely to converge with the information superhighways. Also described are the motivations and performances of the earliest high frequency systems used in telecommunications and NASA's Space Station, and to those more recently introduced in the fields of electric vehicles, micro-grids and renewable energies. Additionally there is discussion of the many potential benefits these systems can offer in shaping the future electric power infrastructure, and also the challenges that need to be overcome.

Next addresed are the technical considerations for interconnecting distributed generation equipment with conventional electric utility systems. This discussion arises from the fact that most electric distribution systems are designed, protected, and operated on the premise of being a single source of electric potential on each distribution feeder at any given time. Distributed generation violates this fundamental assumption, and therefore special requirements for connecting to the utility distribution grid are critical to ensure safe and reliable operation. Manufacturers, vendors, and end-users often see distributeed generation interconnection requirements as a huge market barrier, whereas utility engineers consider them to be absolutely necessary. Thus tools to help assess practical interconnection for specific projects and equipment are provided; we also create a clearinghouse for the many ongoing domestic and international efforts to develop uniform standards for interconnection.

After that, the next part of this book is targeted at known electric energy storage systems as well as development of methodologies and tools for assessing the economic value and the strategic aspects of storage systems integrated into electricity grids. Such tools should be ble to evaluate and analyse energy storage solutions in a variety of applications, such as integration of distributed/renewable energy resources, reduction of peak loading, improvement of transmission grid stability and reliability. Additionally, electricity storage is presented as a strategic enabling technology which not only reduces costs and increases the efficient use of grid assets, but is key for accelerating the integration of distributed generation and renewable sources of energy.

The next part of our book deals with grid integration of wind energy systems. The focus of this topic is on the electrical side of wind conversion systems. After a short description of the basics, such as energy conversion, power limitation and speed control ranges, the existing generator types in wind energy conversion system are described. Because of the practical problems arising with wind turbine installations, their grid integration is an interesting field, whereas the characteristics of wind energy conversion itself, the common types of grid coupling and resulting wind park designs are discussed. On the point of common coupling, wind energy generation may produce distortions of the grid, *e.g.*, flicker effects and harmonics. The causes of their generation, superposition and mitigation are described in detail. Existing standards and the requirements of the transmission system operators are also discussed from the point of view of the conversion system.

Because of limited onshore areas for wind energy systems in Europe, powerful wind parks can be installed only at selected places. A solution of this problem is offshore technology which, due to better wind conditions, brings higher energy yields, but also a lot of additional requirements for the installation and operation of the wind turbines. This includes a special generator design necessitated by the salty environment and different possibilities for the wind park structure, which has internal fixed or free adjustable parameters such as frequency, voltage range and transmission type. The external energy transmission to the onshore substation can be realized with different system configurations. Their advantages and disadvantages are explained.

The next part of the book describes grid integration of photovoltaic systems and fuel cell systems. First the cell types and their efficiency and place requirement are explained. The focus lies on grid-connected photovoltaics, mainly their plant design and grid interfacing of systems depending on isolation conditions, and the possible use of different components is a topic of current interest. Power quality becomes an important issue if higher unit powers are installed. Special problems arising from common connection at the low voltage level are discussed. Derived from the existing devices and their assigned problems in the grid, possibilities for future development are presented.

Fuel cells, photovoltaic systems, generate DC voltage and need a power electronic conversion unit for their grid connection. The different types of fuel cells and their typical applications are described. But the focus lies on plant design, grid interfacing and future development. At the moment only a few fuel cell applications exist. The big potential of this technology may lead to large installation numbers within the next five years. Existing standards of this technology are listed to assist the understanding of this technology.

Gdynia, Poland *Ryszard Strzelecki*
Zielona Góra, Poland *Grzegorz Benysek*
January 2008

Contents

List of Contributors

Grzegorz Benysek
University of Zielona Gora
Institute of Electrical Engineering
Podgorna 50 Street
65-246 Zielona Góra, Poland
G.Benysek@iee.zu.zgora.pl

Piotr Biczel
Warsaw University of Technology
Institute of Electrical Power Engineering
Koszykowa 75 Street
00-662 Warszawa, Poland
Biczelp@ee.pw.edu.pl

Zbigniew Hanzelka
AGH University of Science
and Technology
Department of Electrical Drive
and Industrial Equipment
al. Mickiewicza 30
30-059 Krakow, Poland
Hanzel@agh.edu.pl

Matthias Jahn
Fraunhofer Institut für Keramische
Technologien und Systeme
Winterbergstraße 28, 01277
Dresden, Germany
Matthias.Jahn@ikts.fraunhofer.de

Adam Kempski
University of Zielona Góra
Institute of Electrical Engineering
Podgorna 50 Street
65-246 Zielona Góra, Poland
A.Kempski@iee.uz.zgora.pl

Włodzimierz Koczara
Warsaw University of Technology
Institute of Control and Industrial
Electronics
Koszykowa 75 Street
00-662 Warszawa, Poland
Koczara@isep.pw.edu.pl

Patrick Chi-Kwong Luk
Cranfield University
Defence College of Management
and Technology
Shrivenham Wiltshire SN6 8LA, UK
P.C.K.Luk@cranfield.ac.uk

Jovica V. Milanović
University of Manchester
School of Electrical and Electronic
Engineering
B11 P.O. Box 88, Sackville Street,
Manchester M60 1QD, UK
Milanovic@manchester.ac.uk

Andy Seng Yim Ng
Cranfield University
Defence College of Management
and Technology
Shrivenham Wiltshire SN6 8LA, UK
S.Ng@cranfield.ac.uk

Khaled Nigim
Conetsoga College
Institute of Technology and Advanced
Learning
Doon Campus
Doon Valley Drive Kitchener
Ontario, N2G 4M4, Canada
KNigim@conestogac.on.ca

Thomas Pfeifer
Fraunhofer Institut für Keramische
Technologien und Systeme
Winterbergstraße 28, 01277
Dresden, Germany
Thomas.Pfeifer@ikts.fraunhofer.de

Detlef Schulz
Helmut Schmidt University
Department of Electrical Engineering
Electrical Power Engineering
Holstenhofweg 85, 22043
Hamburg, Germany
Detlef.Schulz@hsu-hh.de

Robert Smoleński
University of Zielona Góra
Institute of Electrical Engineering
Podgorna 50 Street
65-246 Zielona Gora, Poland
R.Smolenski@iee.uz.zgora.pl

Ryszard Strzelecki
Gdynia Maritime University
Department of Electrical Engineering
81-87 Morska Street
81-225 Gdynia, Poland
Rstrzele@am.gdynia.pl

Genady S. Zinoviev
Novosibirsk State Technical University
Department of Industrial Electronics
20 Karla Marksa Prospect
Novosibirsk, Russia
Genstep@mail.ru

1

Introduction

Ryszard Strzelecki[1] and Grzegorz Benysek[2]

[1]Department of Electrical Engineering,
Gdynia Maritime University,
81-87 Morska Street, 81-225 Gdynia, Poland.
Email: Rstrzele@am.gdynia.pl

[2]Institute of Electrical Engineering,
University of Zielona Góra,
50 Podgórna Street, 65-246 Zielona Góra, Poland.
Email: G.Benysek@iee.uz.zgora.pl

1.1 Structure and Fundamental Problems of Electrical Power Systems

Today's grids are primarily based on large power stations connected to transmission lines which supply power to distribution systems thus the overall image is still the same: one way power flow from the power stations, *via* the transmission and distribution systems, to the final customer (end-user).

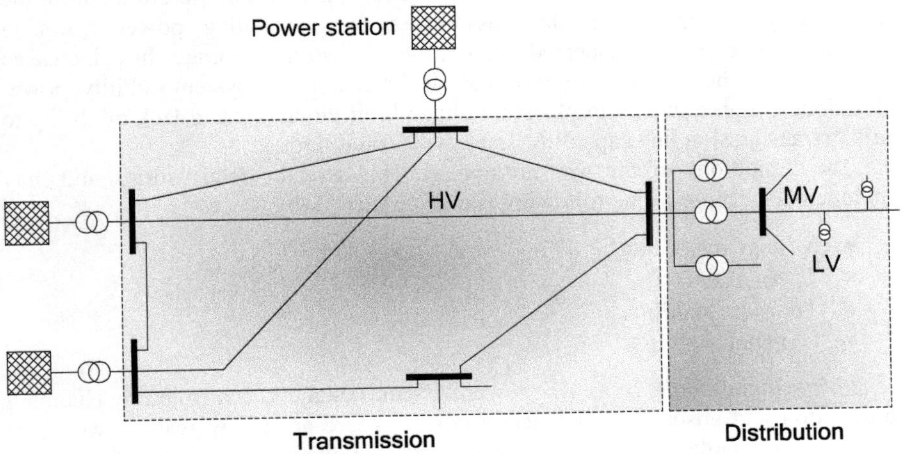

Figure 1.1. A simplified one-line diagram of the power system

The one-line diagram shown in Figure 1.1 illustrates today's Electrical Power System (EPS) and its major components: generation, transmission and distribution. Electric power is generated at power stations predominantly by synchronous generators that are mostly driven by steam or hydro turbines. Hence, the electric power generated at any such station usually has to be transmitted over a great distance, through transmission systems to distribution systems. The distribution networks distribute the energy from the transmission grid or small/local Distributed Resources (DR) to customers.

The three mentioned components – generation, transmission and distribution – have different influences, individual and sometimes common, on the level of the quality of delivered electrical energy. There are many issues involved, such as the maintenance of power apparatus and system, the stability of the operation system, faults, distortions, loads non-linearities *etc*.

One must understand the potential impact of failure within one component on the performance of the whole. For example, a failure in the generation component may lead to failure in the transmission system and in a consequent loss of load in the distribution system, while a failure in the transmission component may lead to failure in the generation component and subsequent loss of customer load in distribution. A failure in the distribution system rarely leads to failure in the other two components and causes very minimal, local losses of customer load. Some of these problems are related to power transmission systems and some of them to power distribution systems, but all are fundamental from the point of view of quality of delivered power.

Transmission Systems

As noted earlier, transmission systems are being pushed ever closer to their stability and thermal limits and if the facilities are not suitably upgraded the power system becomes vulnerable to steady-state and transient stability problems [1, 2]. In such environments, transmission capacity becomes a virtue; therefore during the last few years interest in the possibilities for controlling power flows in transmission systems, dispersed generation or energy storage has increased significantly. There are a number of reasons for this: loss of system stability, power flow loops, high transmission losses, voltage limit violations and lack of ability to utilize transmission line capability up to the thermal limit.

The limitations of the transmission system can take many forms and may include one or more of the following characteristics [3–6]:

- Voltage magnitude;
- Thermal limits;
- Transient stability;
- Dynamic stability.

Voltage magnitude. In an AC power system, voltage is controlled by changing production and absorption of reactive power. There are a few reasons why it is necessary to handle reactive power and to control voltage. First, both customer and EPS equipment are designed to operate within a range of voltages. At low voltages, many types of equipment perform poorly; induction motors can overheat and be damaged, and some electronic equipment will not operate at all. High voltages can damage equipment and shorten its working life. Second, to maximize the amount of real power that can be transferred across a transmission system, reactive power flows must be minimized. Third, reactive power on transmission system causes real-power losses. Both capacity and energy must be supplied to replace these losses. Voltage control is complicated by two additional factors. First, the transmission system itself is a non-linear consumer of reactive power, depending on system loading. At very low levels of system load, transmission lines act as capacitors and increase voltages (the system consumes reactive power that must be generated). At high load levels, transmission lines absorb reactive power and thereby lower voltages (the system consumes a large amount of reactive power that must be replaced). The system's reactive power requirements also depend on the generation and transmission configuration. Consequently, system reactive requirements vary in time as load levels and load and generation patterns change.

The EPS operator has several devices available that can be used to control voltages: for example, generators which inject reactive power into the power system, tending to raise system voltage, or which absorb reactive power, tending to lower system voltage. Additionally transformer tap changers can be used for voltage control. These arrangements can force voltage up (or down) on one side of a transformer, but it is at the expense of reducing (or raising) the voltage on the other side. The reactive power required to raise (or lower) voltage on a bus is forced to flow through the transformer from the bus on the other side. Fixed or variable taps often provide ±10% voltage selection.

Thermal limits. If the transmission line has not been loaded to its thermal limit (the thermal rating of normally designed transmission lines depends mainly on the voltage level at which they operate and the reactance) the power transfer capability can be increased by the use of, *e.g.*, switchable capacitors and controlled reactors. Such devices can supply or absorb reactive power, respectively raising or lowering the voltage of the transmission line. Also series compensation is used to increase the capability of power transfer by reducing the reactance of the transmission line.

Transient and dynamic stability. Transient stability refers to the ability of the power system to survive after a major disturbance, while dynamic stability refers to sustained or growing power swing oscillations between generators or a group of generators initiated by a disturbance (fault, major load changes *etc.*). The mitigation of these oscillations is commonly performed with Power System Stabilizers (PSSs) and sometimes in conjunction with Automatic Voltage Regulators (AVRs).

Traditional solutions for upgrading electrical transmission system infrastructure have been primarily in the form of new power plants, new transmission lines, substations, and associated equipment. However, as experience has proven, the process of authorizing, locating, and constructing new transmission lines has become extremely difficult, expensive and time-consuming. It is envisaged that, alternatively, Flexible Alternating Current Transmission System (FACTS), Energy Storage Systems (ESS) and Distributed Generation (DG) can enable the same objectives to be met.

The potential benefits of employing the above-mentioned solutions include reduction of operation and transmission investment costs and implementation time compared to the construction of new transmission lines, increased system security and reliability, increased power transfer capabilities, and an overall enhancement of the quality of the electric energy delivered to customers [7, 8].

Distribution Systems
With the emergence of computers, sensitive loads and modern communications, a reliable electricity supply with high quality voltage has become a necessity.

A few years back, the main concern of consumers of electricity was reliability of supply *per se*. It is however not only simple supply reliability that consumers want today — they want an ideal AC line supply, that is, a pure sine wave of fundamental frequency and, in addition, a rated peak voltage value. Unfortunately the actual AC line supply that we receive differs from this ideal. There are many ways in which the lack of quality power affects customers.

Voltage sags and dips can cause loss of production in automated processes, and can also force a computer system or data processing system to crash. To prevent such events an Uninterrupted Power Supply (UPS) is often used, which in turn may generate harmonics. A consumer that is connected to the same bus that supplies a large motor load may have to face a critical dip in supply voltage every time the motor load is switched on. This may be quite unacceptable to many consumers. There are also very sensitive consumers, such as hospitals, air traffic control and financial institutions, that require clean and uninterrupted power.

A sustained overvoltage can cause damage to household appliances. An undervoltage has the same effect as that of voltage sag. Voltage imbalance can cause temperature rises in motors. Harmonics, DC offset, can cause waveform distortions. Unwanted current harmonics flowing across the distribution network can cause losses and heating in transformers and Electromagnetic Interference (EMI) [9–11]. Interharmonics voltages can upset the operation of fluorescent lamps and television receivers. They can also produce acoustic noise. It can be concluded that the lack of quality power can cause loss of production, and damage to equipment.

As with FACTS and other players in transmission systems, power electronics devices called Custom Power Systems (CUPS) together with ESS, DG and smart end-user appliances can be applied to the power distribution systems to increase reliability and quality of power supplied to customers [12–15]. Through those technologies the reliability and quality of the power delivered can be improved in terms of reduced interruptions, reduced voltage, current variations and distortions. The proper use of these technologies will benefit all industrial, commercial and domestic customers.

1.2 Power Flow Control, Distributed Generation and Energy Storage Benefits to Grids

Today, most users are inactive receivers of electricity without further involvement in the management of the sources and the grid; each user is simply a hole for electrical energy. Additionally, much of the equipment of today's EPS was installed with a nominal design life of about 30 – 40 years; meanwhile, the amount of load has grown above what was predicted when the grid was designed. In response to the above and the climate change challenge, forcing reduction of greenhouse gases emissions, many countries have started the process of liberalization of their electrical power systems, opening access to grids and encouraging renewable energy sources. In this situation, there is a great opportunity for DG and other new players to provide better network capability, flexibility and functionality. As a result, an amount of the electrical energy produced by large conventional plants will be displaced by DGs, demand response and energy storage. Additionally, integration of the new above-mentioned players, will be made without changes to transmission and distribution systems, planning and operating procedures. On the other hand, DGs and other technologies could change the present radial architecture of distribution systems and new operations in closed loops could appear. These new operating points would require an increase in smart distribution and synchronization between components.

Power Flow Control

During the last several years interest in the control of (active) power flows in transmission systems has increased significantly. There are a number of reasons for this:

- Thermal issues are generally related to thermal limits caused, for example, by a change in the network configuration. Additionally, in a meshed power system, there can occur a situation where a low impedance line carries much more power than originally designed for, while parallel paths are underutilized. With power flow control, the stressed line can be relieved, resulting in better overall utilization of the network;
- In the future when, among others, private companies will operate transmission lines and sell energy to interested parties, the load flow will have to be controlled. One possibility is to use HVDC lines; another possibility is load flow control using FACTS devices in an AC network;
- Voltage and reactive power control issues—low voltage at heavily loaded transmission lines as well high voltage at lightly loaded lines are undesirable occurrences in transmission lines. The first can be a limiting factor responsible for reduced value of the transmitted power and the second can cause equipment damage. Both low voltage as well as high voltage can exceed the voltage limits and therefore corrective actions have to be taken. The corrective actions with utilization of selected FACTS devices include correcting the power factor and compensating reactive losses in lines by supplying reactive power;
- Loss reduction—generally, total losses in a system cannot be reduced to such an extent that the installation of power flow controllers is justified. Only the losses due to reactive power flow, which usually are quite small, are easily avoidable. A reduction of the losses due to active power flow would require a decrease of the line resistances. However, loss reduction in a particular area of the system is a relevant issue. Power transfers from one point to another will physically flow on a number of parallel paths and thereby impel losses on lines that might belong to another utility, thus

causing increased costs for that company. If the latter utility cannot accept these losses, power flow control can be a solution;

- Transient and dynamic stability control issues—transient stability describes the ability of the power system to survive after a major disturbance, while dynamic stability describes sustained or growing power swing oscillations between generators or a group of generators initiated by a disturbance (fault, major load changes *etc.*). The first phenomenon can be improved by synchronizing power flow between sending and receiving ends. A solution for the second phenomenon lies in the use of equipment that permits dynamic damping of such oscillations. In the first as well as the second situation, active power flow control can be a solution.

Summarizing, power flow control technologies have the abilities to solve both steady-state (better utilization of the transmission assets, minimization of losses, limit flows to contract paths *etc.*) and dynamic issues (dynamic dumping of the oscillations) of transmission systems.

Distributed Generation
DG can be some relief for Transmission and Distribution (T&D) capacity and these new energy injections can have some benefits in the operation on the whole EPS [16, 17]:

- The production of energy near the load reduces the losses of the grids, because energy is generated where it is consumed;
- Normally voltage control is carried out by means of manually operated or automatic tap changers, or by utilization of capacitor banks. In both cases, the existence of DG units could be an important way to increase the voltage; the insertion of a DG in a bus raises its voltage;
- End-users who place DG can benefit by having backup generation to improve reliability; they may also receive compensation for making their generation capacity available to the grid in areas where power outages are possible;
- Traditionally, grid operators have built sufficient delivery capacity to serve the peak load, assuming one major malfunction. At the distribution level, this forces implementation of extra ties to other feeders; thus the load can be switched to an alternate feeder when a fault occurs. As a result, in a substation there must be enough capacity to serve the normal load and the extra load expected during failure; this results in overcapacity when the grid is in its normal state. Thus one good implicated DG could be support for feeders when needed to switch to an alternate during repairs.

One must keep in mind that the value of DG to the power system depends among other things on location and the understandable choice for location is a substation with sufficient space and communications [18, 19]. Of course this is all right if the goal is T&D capacity relief (DG generation is usually customer-owned and grid operators don't want to rely on it for capacity). To provide support for distribution feeders, the DG must be sited away from the substation. For example, if the load is uniformly distributed along the feeder, the optimal point for loss reduction and capacity relief is approximately two-thirds of the way down the main feeder [19].

The optimal DG siting problem is similar to the optimal siting problem for ESS, shunt FACTS or shunt capacitor banks and some of the same rules could be applied. However for customer-owned generation the utility generally does not have a choice in the DG location; the siting is given, and the only dilemma is to determine whatever the location has influence on T&D capacity relief and losses, or not? In most cases the answer is positive, because in very complex transmission and distribution grids, often one small area affects a large area; thus a relatively small load decrease in the controlled area allows several times that amount of load to be served by the system.

Energy Storage
In a very similar way to that described earlier in this chapter for DGs, the ESS also could have some benefits in the operation on the whole EPS. In order to realize as many ESS benefits as possible, T&D operators require a second, minute, hour and sometimes even multi-hour energy storage system. Benefits of well-penetrated ESS include these listed below.

Electricity pricing. Cheap electricity, accessible during periods when demand for electricity is low (low priced energy), stored in ESS can be sold at a later time when the price for energy is higher.

Generation capacity. For grids where electric generation capacity is tight, ESS could be used to compensate for the need to install new generation equipment. In other words, storage is used instead of adding central generation capacity.

Transmission support, power quality and capacity reduction. ESS may be used to support T&D systems by compensating for such anomalies as outages, voltage sags (storage provides a more reliable service). Additionally energy storage can be involved to protect against events which affect the quality of delivered energy (voltage and frequency variations, harmonics *etc.*). Finally, reducing capacity needs by storing cheap off-peak electric energy and then discharging it during peak demand periods can reduce the load on the grids and delay utility investments.

Increased DG profits. In many applications there is a need to provide for steady distributed generation; thus irregular DGs cannot be used to serve loads. In these cases, energy is stored when demand and prices for electricity are low so that energy can be used when demand and price is high, and/or when output from the normal source is low.

Below are described five options among energy storage devices which could have some benefits for the operation as a whole EPS [20–23]:

- Available now, sodium sulfur, lead acid or vanadium redox batteries provide up to 9 h of energy and can be used for peak shaving of loads, voltage control and dynamic stability improvement;
- Available beacon flywheels can be used for voltage and transient stability support or frequency regulation;
- Superconducting magnetic energy storage systems store energy in the magnetic field created by the flow of direct current in a superconducting coil, cooled to a temperature below its superconducting temperature. These

systems can provide voltage support and have enough real power to improve system dynamic properties;

- Compressed air energy storage provides GW of energy and can be used for frequency regulation;
- Ultra-capacitors are high energy and power density electrochemical devices. They are able to store energy like batteries for hours, but can quickly discharge the energy like capacitors.

1.3 Smart Electrical Energy Networks Concept

Over the last few years, electrical energy consumption has continually grown and, at the same time, investment in the T&D infrastructure has correspondingly declined. Traditional solutions for upgrading the electrical system infrastructure have been primarily in the form of new power plants, new transmission lines, substations, and associated equipment. However, as experience has proven, the process of authorizing, locating, and constructing new transmission lines has become extremely difficult, expensive and time-consuming. As a result, the power grid is under stress, resulting in compromised reliability and higher energy costs [15].

Despite the above problems, system reliability is vital and cannot be compromised. To overcome this problem, grid operators are moving away from radial systems towards networked; however this degrades controllability of the network because current flows along particular lines which cannot easily be controlled. The situation is even worse if an incident such as loss of a line results in overload, increasing the possibility of a blackout. Additionally, rapid load growth leads to jamming on key lines which, in consequence, leads to an inefficient operation of energy markets.

The answer seems to lie in transforming the current EPS into Smart Electrical Energy Network (SEEN). Future SEENs will be strong, more flexible, reliable, self-healing, fully controllable, asset efficient and will be a platform to make possible the coexistence of smart-self-controlling grids with great numbers of DGs and large-scale centralized power plants [24–26]. The need for modifications, demands to remove the barriers to the large-scale exploitation and integration of DGs and other players, will necessitate research and development new innovative technologies from generation, transmission and distribution to communication tools, with far more sensors than at present. Thus it is envisaged that FACTS, CUPS, ESS, DG, smart end-user appliances together with communications will be at the heart of the future SEENs; see Figure 1.2 [27, 28].

SEEN will allow the customer to take an active role in the supply of electricity, which can help the electricity system respond to equipment failures, weather-related emergencies, and other conditions. At present, the system operator must maintain enough excess generating capacity online or quickly available to continue supplying system load if a large generating unit or transmission line fails. In Smart Electrical Energy Networks, much of that reserve could be provided by EPS or small DG, ESS units located near end-user sites.

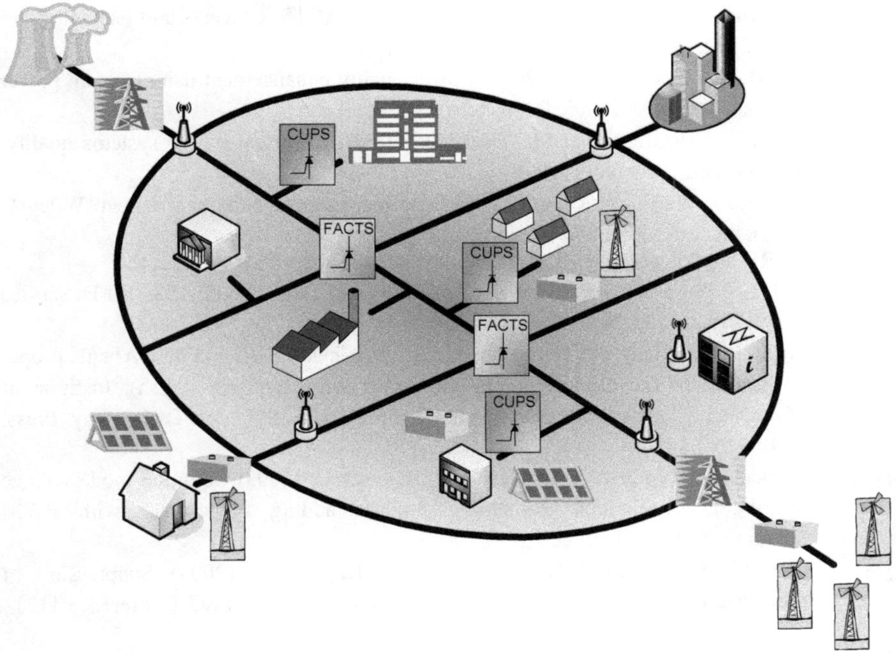

Figure 1.2. Smart Electrical Energy Network concept

Summarizing, a modernized smart grid would create EPS that:

- Will reduce peak loads and generate reserve margins;
- Will delete capital costs of new T&D infrastructure as well as generating plants;
- Will lower T&D line losses together with operation and maintenance costs;
- Will redirect power flows, change load patterns, improve voltage profiles and stability;
- Will enable loads ESS and DG to participate in system operations;
- Through extensive monitoring, quick communications, and feedback control of operations, will have much more information about system rising problems before they affect service;
- Provide system utilities with advanced visualization tools to enhance their ability to oversee the system.

References

[1] Kundur P, (1994) Power system stability and control. McGraw-Hill, New York

[2] Machowski J, Białek J, Bumby J, (1997) Power system dynamics and stability. John Wiley & Sons, Chichester, New York

[3] Hingorani N, Gyugyi L, (2000) Understanding FACTS: concepts and technology of flexible ac transmission systems. IEEE, New York
[4] Ghosh A, Ledwich G, (2002) Power quality enhancement using custom power devices. Kluwer Academic Publishers, Boston
[5] Dugan R, McGranaghan M, Beaty W, (1996) Electrical power systems quality. McGraw-Hill, New York, 1996
[6] Arrillaga J, Watson N, Chan S, (2000) Power system quality assessment. Wiley & Sons, Chichester, England
[7] CIGRE Working Group 14.31, (1999) Custom power – state of the art
[8] IEEE/CIGRE, (1995) FACTS overview. Special issue 95-TP-108, IEEE Service Center, Piscataway, New York
[9] Kurowski T, Benysek G, Kempski A, Smoleński R, (2000) About proper cooperation of the electric drives and static converters. (in Polish), Institute of Drives and Electrical Measurements, Wrocław University of Technology Press, vol.48:326–334
[10] Strzelecki R, Kempski A, Smoleński R, Benysek G, (2003) Common mode voltage cancellation in systems containing 3-phase adding transformer with PWM excitation. EPE Conference
[11] Kempski A, Strzelecki R, Smoleński R, Benysek G, (2003) Suppression of conducted EMI in four-quadrant ac drive system. IEEE-PESC Conference:1121–1126
[12] Thomsen P, (1999) Application and control of CUPS in the distribution grid. Institute of Energy Technology, Aalborg University, vol.3:2–11
[13] Strzelecki R, Benysek G, (2004) Conceptions and properties of the arrangements in distributed electrical power systems. MITEL Conference:241–248
[14] Strzelecki R, Jarnut M, Benysek G, (2003) Active electrical energy conditioners for individual customers. PES Conference, Warsaw University of Technology Press, vol.1:27–34
[15] Benysek G, (2007) Improvement in the quality of delivery of electrical energy using power electronics systems. Springer-Verlag, London
[16] Feinstein C, Chapel S, (2000) The strategic role of distributed resources in distribution systems. Energy 2000 Proceedings, CRC Press
[17] Ford A, (2002) Selected benefits of distributed generation in a restructured electricity system. Report to Pacific Northwest National Laboratory
[18] Schienbein L, DeSteese J, (2002) Distributed energy resources, power quality and reliability. Pacific Northwest National Laboratory, PNNL-13779
[19] Willis H, Walter G, (2000) Distributed power generation: planning and evaluation. Marcel Dekker, New York
[20] Iannucci J, Eyer J, (2005) Innovative applications of energy storage in a restructured electricity marketplace. SAND2003-2546
[21] Eyer J, Iannucci J, Garth P, (2004) Energy storage benefits and market analysis handbook. A study for the DOE energy storage systems program, SAND2004-6177
[22] Hadley S, Van Dyke J, Poore W, (2003) Quantitative assessment of distributed energy resource benefits. ORNL/TM-2003/20:35
[23] Nourai A, (2002) Large-scale electricity storage technologies for energy management. IEEE Power Engineering Society Summer Meeting, vol.1:310–315
[24] Gellings C, (2003) Smart power delivery: a vision for the future. EPRI Journal, June

[25] GridWise Alliance, (2003) Rethinking energy from generation to consumption. Brochure
[26] Kannberg L, (2003) GridWise: transforming the energy system. Pacific Northwest National Laboratory, Conference Presentation
[27] Mazza P, (2003) The smart energy network: electricity's third great revolution. Http://www.climatesolutions.org
[28] Massoud S, Wollenberg B, (2005) Toward a smart grid: power delivery for the 21st century. IEEE Power and Energy Magazine, vol.3:34–41

2

Principles of Electrical Power Control

Zbigniew Hanzelka[1] and Jovica V. Milanović[2]

[1]Department of Electrical Drive and Industrial Equipment,
AGH University of Science and Technology,
al. Mickiewicza 30, 30-059 Krakow, Poland.
Email: Hanzel@agh.edu.pl

[2]School of Electrical and Electronic Engineering
The University of Manchester,
B11 P.O. Box 88, Sackville Street, Manchester M60 1QD, UK.
Email: Milanovic@manchester.ac.uk

2.1 Power Theory

The term power theory of circuits can be understood as the state of knowledge on their power properties. In that sense it is a set of true statements, interpretations, definitions and equations describing these properties [1]. The theory of power, understood that way, is a collective product of those who seek an answer to the question why a load with the active power P usually demands a power source with an apparent power S greater than its active power [2]. This question is closely related to the need for interpretation of power phenomena in electric circuits. Another factor is of a practical nature – power theory attempts to answer the question how the apparent power of the source can be reduced without the reduction in the load active power.

2.1.1 Critical Review of Classical Power Theory

In the 1920s two major trends in power theory had developed. The first uses Fourier series expansion to describe power properties of a circuit. Since electric quantities are regarded as sums of components with different frequencies, the electric circuit properties are defined in the frequency domain. Budeanu's power theory [3] is the most widespread theory in the frequency domain. Almost simultaneously, another trend has emerged which does not employ Fourier series and emphasizes defining the circuit power properties in the time domain. Since its conception the power theory in time domain is associated with Fryze's name [4]. The existence of these two trends has been evident in power theory development in the past as well as the present.

2.1.1.1 Single-phase Circuits Under Sinusoidal Conditions

This case, the simplest in terms of interpretation, is a starting point for all further considerations. If a linear load is supplied with a sinusoidal voltage (see Figure 2.1)

$$v = \sqrt{2} V_{RMS} \sin(\omega_{(1)} t - \varphi_u) \tag{2.1}$$

and the source current value is

$$i = \sqrt{2} I_{RMS} \sin(\omega_{(1)} t - \varphi_i) \tag{2.2}$$

where V_{RMS} and I_{RMS} are the RMS values of supply voltage v and load current I; $\varphi=\varphi_u-\varphi_i$ – phase angle between voltage v and current i.

The source's instantaneous power p, which is a measure of energy flow rate w from the source to the receiver, is (for $\varphi_u=0$, $\varphi=\varphi_i$)

$$p = v \cdot i = 2 V_{RMS} I_{RMS} \sin \omega_{(1)} t \sin(\omega_{(1)} t - \varphi) = p_a + p_b \tag{2.3}$$

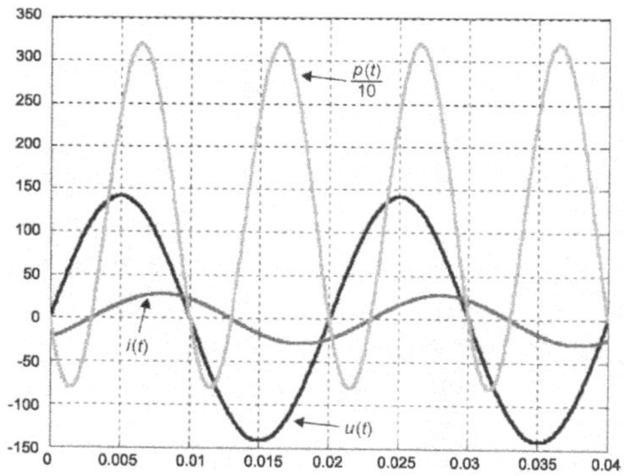

Figure 2.1. The waveforms of supply voltage u, load current i and instantaneous power p for an AC circuit with a linear (resistance-inductive) load under a steady-state operation ($\varphi_u=0$) [5]

Thus the instantaneous power p can be decomposed into the nonnegative component p_a and the oscillatory component p_b. The waveforms of separated components of instantaneous power p are shown in Figure 2.2.

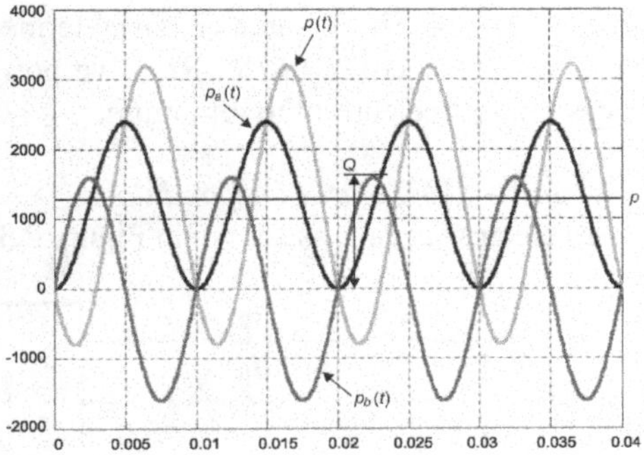

Figure 2.2. The waveforms of separated components of instantaneous power p [5]

The reactive power $Q = V_{RMS} I_{RMS} \sin\varphi$, in a circuit under sinusoidal conditions, is interpreted as the amplitude of the p_b component of the instantaneous power. According to the definition, the active power P equals the average value (DC component) over one period T of the instantaneous power p

$$P = \frac{1}{T}\int_0^T p(t)dt = V_{RMS} I_{RMS} \cos\varphi = S\cos\varphi \tag{2.4}$$

where S is the apparent power.

The source current can be expressed in the form

$$i = \sqrt{2} I_{RMS} \cos\varphi \sin\omega_{(1)} t + \sqrt{2} I_{RMS} \sin\varphi \cos\omega_{(1)} t = \sqrt{2}\frac{P}{V_{RMS}}\sin\omega_{(1)} t + \\ + \sqrt{2}\frac{Q}{V_{RMS}}\cos\omega_{(1)} t \tag{2.5}$$

and, thus, considering the orthogonality of both components, the current RMS value is

$$I_{RMS} = \sqrt{\left(\frac{P}{V_{RMS}}\right)^2 + \left(\frac{Q}{V_{RMS}}\right)^2} \tag{2.6}$$

and the equation which determines the apparent power value $S^2 = P^2 + Q^2$ is satisfied. The magnitude of the complex power

$$\underline{S} = \underline{V}\underline{I}^* = Ve^{j\varphi_u}(Ie^{j\varphi_i})^* = Se^{j(\varphi_u - \varphi_i)} = Se^{j\varphi} = S\cos\varphi + jS\sin\varphi = P + jQ \tag{2.7}$$

is equal to the apparent power S. The parameter $\cos\varphi$, termed the displacement power factor, is defined as $DPF = \cos\varphi = P/S$. Reactive power compensation to the value of Q=0 reduces the source current to minimum.

2.1.1.2 Three-phase Circuits Under Sinusoidal Conditions

A three-phase, three-wire electric circuit, as shown in Figure 2.3, is considered.

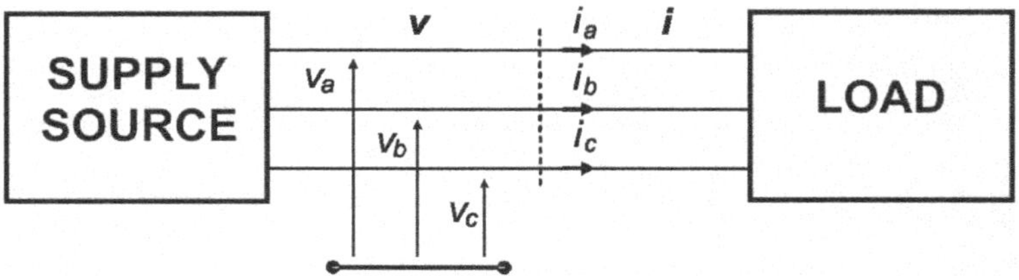

Figure 2.3. Phase voltages and currents at the cross-section *a-b-c*

Phase voltages and currents are expressed in the form of column vectors: $v = v = [v_a, v_b, v_c]^T$ and $i = i = [i_a, i_b, i_c]^T$. Figure 2.4 shows the sinusoidal symmetrical waveforms of supply voltage v and load current i in the three-phase electric circuit with linear load. This is the simplest case from the point of view of power phenomena that take place in three-phase electrical circuits.

Instantaneous power p_{3f} of a three-phase circuit takes the form of

$$p_{3f} = p_a + p_b + p_c = v_a \cdot i_a + v_b \cdot i_b + v_c \cdot i_c \tag{2.8}$$

In this case the waveform of instantaneous power p_{3f} is constant, in contrast to the single-phase circuit (Figures 2.1 and 2.2). The relations of powers of three-phase circuit take the form of

$$P_a = V_{RMS,a} I_{RMS,a} \cos\varphi = P_b = P_c = P \Rightarrow P_{3f} = 3P \tag{2.9}$$

$$Q_a = V_{RMS,a} I_{RMS,a} \sin\varphi = Q_b = Q_c = Q \Rightarrow Q_{3f} = 3Q \tag{2.10}$$

$$S_a = S_{a(1)} = V_{RMS,a} I_{RMS,a} = S_b = S_c = S \Rightarrow S_{3f} = S_{3f(1)} = 3S \tag{2.11}$$

and power factors PF_{3f} and DPF_{3f} take the form of

$$PF_{3f} = DPF_{3f} = P_{3f} / S_{3f} \tag{2.12}$$

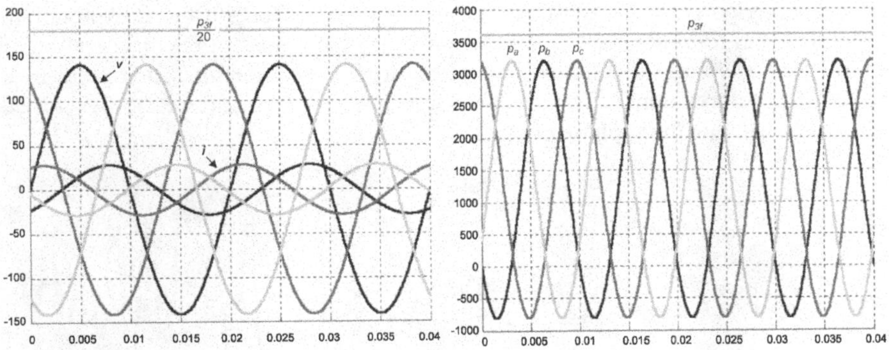

a b

Figure 2.4. The waveforms of: **a** supply voltage v, load current i; **b** phase instantaneous powers p_a, p_b, p_c and instantaneous power p_{3f} (balanced resistive load) [5]

In the latter case the interpretation, definitions and equations describing these properties for single-phase circuits with sinusoidal supply voltage and current were applied.

In three-phase systems it is impossible, on the basis of the instantaneous power, to separate the reactive power. Reactive power interpretation according to what exists in one-phase circuits cannot be applied, so it loses its general physical sense.

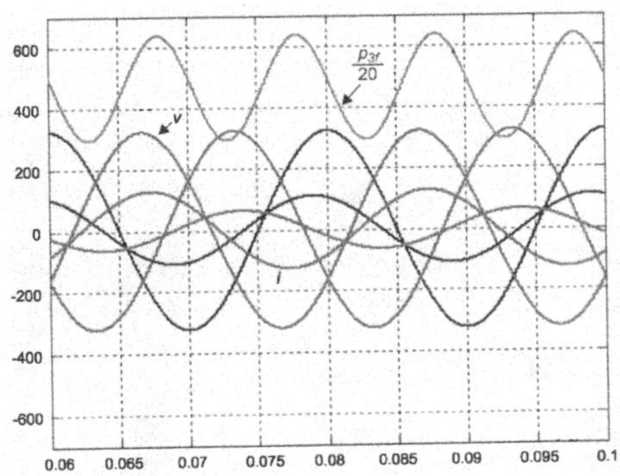

Figure 2.5. The waveforms of supply voltage v, load current i and instantaneous power p_{3f} in the three-phase thee-wire electric circuit with and unbalanced resistive load [5]

In three-phase circuits, however, a phenomenon occurs that was not present in single-phase circuits, namely the asymmetry of supply voltages v and/or load currents i waveforms. It is assumed that the asymmetry of resistive load currents is the only cause for the phase shift between the voltages and currents in three-phase three-wire electric circuits (despite the lack of passive elements). Therefore, the displacement power factor DPF_{3f} value is less than 1, $DPF_{3f} < 1$.

Figure 2.5 shows the waveforms of voltages v and currents i in three-phase three-wire electric circuits with unbalanced resistive load. One can see that the instantaneous power p_{3f} is no longer constant. In this situation the interpretation, definitions and formulas for single-phase circuits do not hold. Any consideration of a three-phase circuit as composed of three single-phase circuits could lead to a major misinterpretation of power phenomena in such a circuit. It has to be considered a single three-phase circuit.

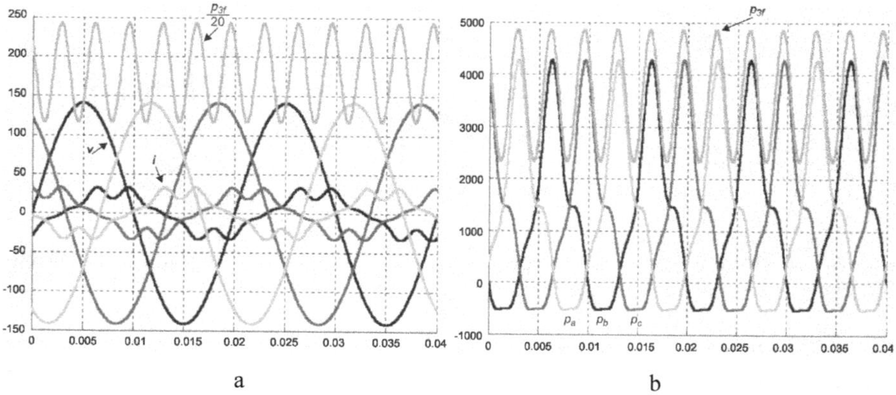

Figure 2.6. The waveforms of: **a** supply voltage v, load current i; **b** phase instantaneous powers p_a, p_b, p_c and instantaneous power p_{3f}

Figure 2.6 shows sinusoidal symmetrical waveforms of the supply voltages v and load current i in the three-phase electric circuit with symmetrical non-linear load. The presence of distorted phase load currents results in the fact that instantaneous power p_{3f} is no longer constant. However, in this case it is possible to use the interpretation, definitions and formulas for single-phase circuits. It is not possible if the waveforms of supply voltages v and/or load currents i are asymmetrical.

The extension of the apparent power concept to multi-phase systems has lead to many controversies. Essentially, three definitions are in use:

- Arithmetic apparent power $S_A = V_{RMS,a} I_{RMS,a} + V_{RMS,b} I_{RMS,b} + V_{RMS,c} I_{RMS,c}$;
- Geometric apparent power [6] $S_G = \sqrt{P_{3f}^2 + Q_{3f}^2}$;
- Buchholz's apparent power [3] $S_B = \sqrt{V_a^2 + V_b^2 + V_c^2} \cdot \sqrt{I_a^2 + I_b^2 + I_c^2}$.

As long as the supply voltage is sinusoidal and symmetrical, and the load is balanced, these relations give the same correct result. If one of the above-mentioned conditions is not met the obtained results will differ. The consequence of this will be that we get different values of power factors for the same electric circuit. For obvious reasons such a situation is neither desired nor admissible. In publications [7] it has been proved that in such a case only the Buchholz's definition of apparent power S_B allows for the correct calculation of apparent power, and therefore of the power factor value. Moreover, it can be proven that the Buchholz's apparent power can be extended to circuits with non-sinusoidal voltages and currents.

2.1.1.3 Powers in Circuits with Non-sinusoidal Voltages and Currents
The choice of the presented power theories is subjective; only those theories which in the author's opinion are most widely used will be discussed.

The works of Steinmetz [2] should certainly be mentioned here – he was the first who, examining the example of a mercury rectifier, found that the apparent power is greater than the active power due to the distortion of a current with respect to a sinusoidal voltage. This finding worked the inception of power theories of electric circuits with non-sinusoidal voltages

$$v = \sum_{n=0}^{\infty} v_{(n)} \tag{2.13}$$

and currents

$$i = \sum_{n=0}^{\infty} i_{(n)} \tag{2.14}$$

where n is the harmonic order.

Budeanu Theory (Frequency Domain)[8]
The active and reactive powers are defined as superposition of active and reactive powers of all harmonics, respectively

$$P = \frac{1}{T}\int_0^T v \cdot i dt = V_0 I_0 + \sum_{n=1}^{\infty} V_{RMS(n)} I_{RMS(n)} \cos\varphi_{(n)} = \sum_{n=0}^{\infty} P_{(n)} \tag{2.15}$$

$$Q_B = \sum_{n=1}^{\infty} V_{RMS(n)} I_{RMS(n)} \sin\varphi_{(n)} = \sum_{n=1}^{\infty} Q_{(n)} \tag{2.16}$$

(satisfying the principle of energy balance) and the distortion power

$$D = \sqrt{S^2 - (P^2 + Q_B^2)} \tag{2.17}$$

where $I_{RMS(n)}$, $V_{RMS(n)}$ are RMS values of n-th harmonic of current and voltage; $\varphi_{(n)}$ – phase angle between voltage v and current i n-th harmonics; V_0, I_0 are the DC components.

According to this theory the apparent power is determined as

$$S = V_{RMS} I_{RMS} = \sqrt{\sum_{n=0}^{\infty} V_{RMS(n)}^2 \sum_{n=0}^{\infty} I_{RMS(n)}^2} = \sqrt{P^2 + Q_B^2 + D^2} \tag{2.18}$$

Errors in this theory have been demonstrated for the first time in the work [9] proving that:

- It suggests a misinterpretation of power phenomena in circuits with periodic distorted voltages and currents. The defining formula introduces the distortion power D as a completion of existing powers to the apparent power $S=V_{RMS}I_{RMS}$ (Schwartz's inequality), and does not link it directly with the load properties. The distortion power is not related to the current distortion against the voltage and has absolutely no physical meaning. A necessary condition for the distortion power zero value is the load admittance having the same value $Y_{(n)}$ for each voltage harmonic, *i.e.*, $Y_{(n)} = Y_{(n)} \exp(j\varphi_{(n)}) = const$. At the same time the necessary condition for the lack of current waveform distortion against the voltage is that the load admittance for each harmonic takes the value $Y_{(n)} = Y \exp(-jn\omega_{(1)}t) = Y \exp(-jn\varphi)$. These conditions are mutually exclusive. If, under non-sinusoidal supply voltage, the distortion power $D=0$, the load current must be distorted against the voltage. When, under the same conditions, the load current is not distorted against the voltage, the distortion power D cannot be zero. The properties of distortion power are therefore quite the opposite to what its name suggests;
- The reactive power according to Budeanu's definition is not a measure of energy oscillation. This quantity can be equal to zero though one-directional power components may have significant amplitudes and cause power oscillation;
- Moreover, powers Q_B and D are of no significance for the power factor correction. The reactive power cannot be minimized employing this theory and therefore the power factor cannot be improved. This theory does not allow one to determine the compensation capacitance at which the power factor attains its highest value. Full compensation of reactive power Q_B in some situations may even worsen the power factor;
- There is no direct relationship between the RMS current value and distortion power. The RMS value of a source current harmonic is

$$I_{RMS(n)} = \sqrt{\left(\frac{P_{(n)}}{V_{RMS(n)}}\right)^2 + \left(\frac{Q_{(n)}}{V_{RMS(n)}}\right)^2}$$

Since the current harmonics, as the waveforms with different frequencies, are mutually orthogonal, the RMS source current value can be expressed in the form

$$I_{RMS} = \sqrt{\sum_{n=0}^{N} I_{RMS(n)}^2} = \sqrt{\sum_{n=0}^{N}\left(\left(\frac{P_{(n)}}{V_{RMS(n)}}\right)^2 + \sum_{n=1}^{N}\left(\frac{Q_{(n)}}{V_{RMS(n)}}\right)^2\right)}$$

It is therefore evident that the RMS source current value does not depend on the sum of reactive powers of individual harmonics $Q_{(n)}$, *i.e.*, the reactive power according to Budeanu, but on the sum of their squares. For unchanged values of $P_{(n)}$ the RMS source current value is minimum if the reactive power of each harmonic $Q_{(n)}$ equals zero, and not when Budeanu reactive power is zero. Thus the zero value of Budeanu reactive power is not the necessary condition for the minimum RMS current value – it is only a sufficient condition.

A special case of considerations regarding this theory is the analysis of a circuit with the sinusoidal supply voltage and non-linear load. Therefore, the equation on the load current is as follows (see Figure 2.7)

$$i = \sum_{n=1}^{\infty} \sqrt{2} I_{RMS(n)} \sin(n\omega_{(1)}t - \varphi_{(n)}) \tag{2.19}$$

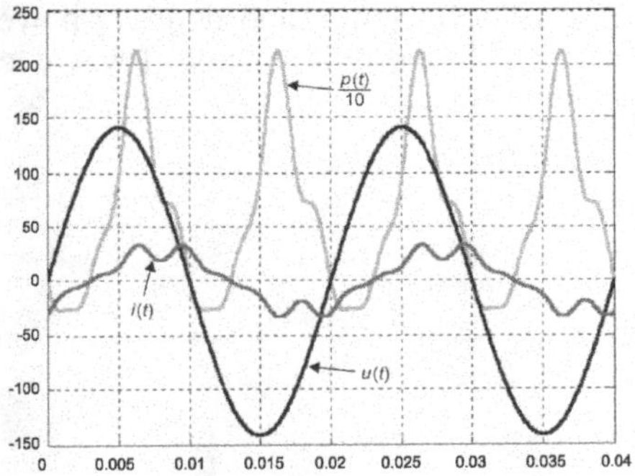

Figure 2.7. The waveforms of voltage *v*, current *i* and instantaneous power *p* for an AC circuit with a non-linear load under a steady-state operation [5]

In that case the equation of instantaneous power *p* is

$$\begin{aligned} p = & V_{RMS} I_{RMS(1)} \cos\varphi_{(1)} (1 - \cos 2\omega_{(1)}t) - \\ & - V_{RMS} I_{RMS(1)} \sin\varphi_{(1)} \sin 2\omega_{(1)}t + \\ & + \sum_{n=2}^{\infty} 2 V_{RMS} I_{RMS(n)} \sin\omega_{(1)}t \sin(n\omega_{(1)}t - \varphi_{(n)}) \end{aligned} \tag{2.20}$$

The definitions of active power $P_{(1)} = V_{RMS} I_{RMS(1)} \cos\varphi_{(1)} = P$ and reactive power $Q_{(1)} = V_{RMS} I_{RMS(1)} \sin\varphi_{(1)} = Q$ were specified for the first harmonic. The equation of apparent power *S* changes, as follows

$$S = V_{RMS} I_{RMS} = V_{RMS} \sqrt{I_{RMS(1)}^2 + \sum_{n=2}^{\infty} I_{RMS(n)}^2} = \sqrt{P_{(1)}^2 + Q_{(1)}^2 + D^2} \tag{2.21}$$

In this way the apparent power in the first harmonic domain $S_{(1)}$ and the distortion power *D* were separated.

Therefore, in this case of non-linear load the power factor denoted as *PF* was defined as

$$PF = \frac{P_{(1)}}{S} = \frac{P_{(1)}}{\sqrt{S_{(1)}^2 + D^2}} = \frac{1}{\sqrt{1+THD_I^2}}\cos\varphi_{(1)} = \frac{1}{\sqrt{1+THD_I^2}}DPF \tag{2.22}$$

where THD_I is the so called - total harmonic distortion of the current i (the same form for the voltage).

Fryze Theory (Time Domain)[4]
Fryze introduced a number of important and lasting elements to the theory of power. The most important of them are decomposition of a source current into orthogonal components related to the electric energy phenomena, identification of the active current as a separate quantity, and defining power quantities without employing Fourier series, *i.e.*, in the time domain. Fryze decomposed the source current into the active component

$$i_P = \frac{P}{V_{RMS}^2}\cdot v = G\cdot v \tag{2.23}$$

and the reactive component

$$i_{QF} = i - i_P \tag{2.24}$$

The active current is the current of a resistive load with conductance G (termed equivalent conductance) having the same active power P at the same voltage v. The portion of the source current remaining after subtracting the active current is the Fryze's reactive current i_{QF}. The scalar product of such defined currents is zero, *i.e.*, $(i_P, i_{QF}) = 0$, so they are mutually orthogonal and their RMS values satisfy the equitation

$$I_{RMS}^2 = I_P^2 + I_{QF}^2 \tag{2.25}$$

Multiplying both sides of this equation by the square of the RMS voltage value V_{RMS}, we obtain the equation of power $S^2 = P^2 + Q_F^2$ with the reactive power defined by the formula $Q_F = V_{RMS} I_{QF}$. This power does not satisfy the principle of energy conservation and is always positive, which precludes determining whether the power factor is leading or lagging.

The Fryze power theory does not explain the physical sense of the reactive current i_{QF}, except by concluding that it is a needless load for the source and causes extra transmission losses. The Fryze theory, moreover, does not explain either what phenomena in a circuit give rise to this current or the load parameter's influence on its RMS value.

This theory does not allow determination parameters of LC passive compensators, whereas a current source connected in parallel with the load and generating the current ($-i_{QF}$) corrects power factor to unity. Such a source, with its control circuits, is an active power filter. Its principle of operation is an immediate conclusion inferred from the Fryze's power theory.

Orthogonal decompositions are valid for circuits supplied from an ideal source, but for circuits supplied from sources with nonzero internal impedance they do not hold. There are, however, proposals for solution of the optimisation task that minimizes given quality index, *e.g.*, the RMS value of the source current. Such a solution allows one to obtain, under given constraints, the energy-optimal state of an electric system, by using compensators with parameters determined using this method [10, 11].

In their original form, Budeanu's and Fryze's theories describe power properties exclusive of single-phase circuits. They can, however, be easily extended to balanced three-phase circuits.

Shepherd and Zakikhani Power Theory (Frequency Domain) [12]

For a load with the current

$$i = \sqrt{2}\sum_{n=1}^{\infty} I_{RMS(n)} \cos(n\omega_{(1)}t - \beta_{(n)}) \tag{2.26}$$

supplied with the voltage

$$v = \sqrt{2}\sum_{n=1}^{\infty} V_{RMS(n)} \cos(n\omega_{(1)}t - \alpha_{(n)}) \tag{2.27}$$

the resistive current

$$i_R = \sqrt{2}\sum_{n=1}^{\infty} I_{RMS(n)} \cos\varphi_{(n)} \cos(n\omega_{(1)}t - \alpha_{(n)}) \tag{2.28}$$

and reactive current have been defined as

$$i_r = \sqrt{2}\sum_{n=1}^{\infty} I_{RMS(n)} \sin\varphi_{(n)} \sin(n\omega_{(1)}t - \alpha_{(n)}) \tag{2.29}$$

where $\varphi_{(n)} = \alpha_{(n)} - \beta_{(n)}$.

The scalar product of these currents is zero, so they are mutually orthogonal and their RMS values satisfy the relation

$$I_{RMS}^2 = I_{RMS,R}^2 + I_{RMS,r}^2 \tag{2.30}$$

Multiplying both sides of this equation by the square of RMS voltage value we obtain an equation of the source power

$$S = V_{RMS} I_{RMS} = \sqrt{V_{RMS}^2 I_{RMS,R}^2 + V_{RMS}^2 I_{RMS,r}^2} = \sqrt{S_R^2 + Q_{SZ}^2} \quad (2.31)$$

Powers S_R and Q_{SZ} do not satisfy the principle of power balance. The current i_r can be compensated for a finite number of harmonics by means of a two terminal LC network connected in parallel with the load [13].

The Shepherd's and Zakihani's power theory for the first time allowed are to obtain a result of practical significance since it enables one to calculate such compensating capacitance that ensures a maximum power factor value of the supply source (assuming the supply source is an ideal voltage source. *i.e.*, RMS values of the voltage harmonics at the load are independent of the compensating capacitance, which is not satisfied in real power supply systems). This capacitance is termed the optimal capacitance (in the sense of minimization of the source current RMS value)

$$C_{opt} = \frac{\sum_{n=1}^{N} n V_{(n)} I_{(n)} \sin \varphi_{(n)}}{\omega_{(1)} \sum_{n=1}^{N} n^2 V_{(n)}^2} \quad (2.32)$$

The decomposition of power lacks a single, well defined power quantity, namely active power. Sharon attempted to eliminate this inconvenience separating from the power S_R a quantity which he termed associated reactive power

$$S_C = \sqrt{S_R^2 - P^2} \quad (2.33)$$

Thus, the equation of power according to Sharon [14] takes the form

$$S = \sqrt{P^2 + Q_{SZ}^2 + S_C^2} \quad (2.34)$$

This equation, however, does not explain the nature of power phenomena in a circuit and their relationship with the load properties. The Shepherd and Zakichani theory has never been extended to polyphase circuits.

Kusters and Moore Theory of Power (Time Domain)[15]

The theory is based on the source current decomposition, for a resistive-inductive load, into an active current, capacitive reactive current i_{qC} and the residual reactive current i_{qCr}. These currents are defined by the formulas

$$i_{qC} = \frac{\frac{1}{T}\int_0^T \frac{dv}{dt} i dt}{\left(\frac{dv}{dt}\right)_{RMS}^2} \frac{dv}{dt} = \frac{\left(\frac{dv}{dt}, i\right)}{\left(\frac{dv}{dt}\right)_{RMS}^2} \frac{dv}{dt} \tag{2.35}$$

and

$$i_{qCr} = i - i_P - i_{qC} \tag{2.36}$$

The equation of power according to Kusters and Moore has the form

$$S = \sqrt{P^2 + Q_C^2 + Q_{Cr}^2} \tag{2.37}$$

where $Q_C = V_{RMS} I_{RMS,qC} \operatorname{sgn}(dv/dt, i)$ (“*sgn*” denotes the sign of a scalar product); $Q_{Cr} = V_{RMS} I_{RMS,qCr}$.

According to the authors of this theory, the capacitive reactive power Q_C can be fully compensated by means of a capacitor with capacitance

$$C_{opt} = -Q_C / \left[V_{RMS} \left(\frac{dv}{dt}\right)_{RMS}^2 \right] \tag{2.38}$$

connected in parallel to the load. The source power factor attains its highest value for that capacitance. It is the same optimal capacitance as determined from the Shepherd and Zakikhani theory. Calculating the optimal capacitance from this formula requires the knowledge of the RMS values of current and voltage harmonics and their phase shifts. Kusters and Moore theory allows one to calculate this capacitance directly in the time domain. On the basis of this theory, Page [16] proposed a method for computing the parameters of a shunt LC compensator.

Czarnecki Power Theory (Frequency Domain)[17]

Czarnecki, taking the concept of active current i_p from Fryze's theory, the concept of reactive current i_r from the Shepherd and Zakikhani theory and introducing the new component i_s termed scattered current, proposed orthogonal decomposition of the current of a source feeding a lineal load in the form $i = i_p + i_r + i_s$. In this case the reactive current according to Fryze is expressed as the sum of currents i_r and i_s, whereas the resistive current according to Shepherd and Zakikhani is the sum of currents i_p and i_s. If N is the set of supply voltage harmonic orders n

$$v = \sqrt{2}\, \mathrm{Re} \sum_{n \in N} V_{RMS(n)} \exp(jn\omega_{(1)}t) \tag{2.39}$$

and then the current components i_s and i_r are defined by formulas

$$i_S = \sqrt{2}\,\mathrm{Re}\sum_{n\in N}\left(G_{(n)} - G_e\right)V_{RMS(n)}\exp(jn\omega_{(1)}t) \tag{2.40}$$

$$i_r = \sqrt{2}\,\mathrm{Re}\sum_{n\in N} jB_{(n)}V_{RMS(n)}\exp(jn\omega_{(1)}t) \tag{2.41}$$

where $G_{(n)} + jB_{(n)}$ is the load admittance for the n-th harmonic and $G_e\,(= G)$ is the equivalent conductance defined according to Fryze's theory.

Currents i_p, i_s and i_r are mutually orthogonal, thus

$$I_{RMS} = \sqrt{I^2_{RMS,P} + I^2_{RMS,S} + I^2_{RMS,r}} \tag{2.42}$$

According to this equation the RMS value of the source current I_{RMS} in a linear circuit with non-sinusoidal voltage is greater than the active current RMS value $I_{RMS,P}$, not only when the load susceptance for voltage harmonics $B_{(n)}$ is nonzero, but also when the load conductance $G_{(n)}$ varies with frequency.

Decomposition of the source current in a linear circuit into the active, scattered and reactive currents was extended to non-linear circuits by means of separation of the generated current i_g, which consists of harmonics generated in a load due to its non-linearity or the circuit time-variant parameters. Thus, in the general case, the current was finally decomposed into four orthogonal components: $i = i_p + i_r + i_s + i_g$. The proposed components of current, even if they are only mathematical quantities (only the current is a physical quantity), exhibit properties which distinguish them from (infinitely many) other components. They can be associated with various physical phenomena (hence the theory is referred to as the theory of the current's physical components) and, consequently, with various components of reactive power.

2.1.1.4 Three-phase Circuits

A number of power theories in three-phase circuits with periodic non-sinusoidal voltages and currents have been proposed. This issue still provokes discussion and controversy. The most popular power theory, commonly employed in active filters control, has been proposed by Akagi and co-authors; it is known as the *p-q* power theory.

One of the most interesting power theories (known as the power theory based on the current's physical components) was presented by Czarnecki. This theory is a proposal of the physical interpretation of power phenomena occurring in electric circuits under unbalanced conditions and in the presence of non-sinusoidal waveforms. The complete Czarnecki theory for three-phase unbalanced circuits with periodic non-sinusoidal source voltage was presented in his work in 1994 [18].

Both theories are of particular importance for the development of power theory.

2.1.2 Instantaneous Power Theory

Nabae and Akagi instantaneous reactive power *p-q* theory [19–23] not only provides theoretical basis for control algorithms for switching compensators but also became the method of describing power properties of three-phase circuits. It is popular although it can exclusively be employed to analyze three-phase circuits and therefore is not a general power theory. The description of power properties of electric circuits, using instantaneous voltage and current values, without the use of Fourier series *i.e.*, in time domain, to a great extent explains the interest in this concept. Since a compensator control algorithm based on the *p-q* theory involves no harmonic analysis, the number of necessary mathematical operations is reduced with respect to frequency-based methods.

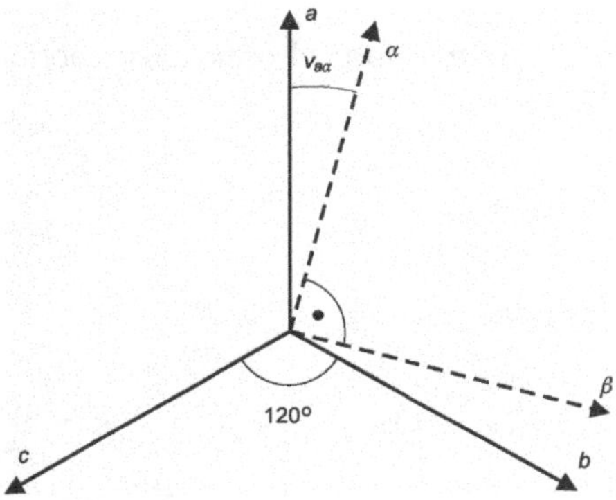

Figure 2.8. Transformation of *a-b-c* coordinates into *α-β-0* coordinates

The instantaneous reactive power theory employs the Clark transformation of three-phase voltages and currents in natural phase coordinates (Figure 2.8), into orthogonal *α-β-0* coordinates. The transformation is performed computing the instantaneous values according to Equation 2.43, where υ_{ax} is the phase shift angle between the *x* axis of the natural three-phase coordinate system and the *α* axis of the orthogonal coordinate system; see Figure 2.8.

$$\begin{bmatrix} F_\alpha \\ F_\beta \\ F_0 \end{bmatrix} = \sqrt{\frac{2}{3}} \begin{bmatrix} \cos\upsilon_{a\alpha} & \cos\upsilon_{ab} & \cos\upsilon_{ac} \\ -\sin\upsilon_{a\alpha} & -\sin\upsilon_{ab} & -\sin\upsilon_{ac} \\ \frac{1}{\sqrt{2}} & \frac{1}{\sqrt{2}} & \frac{1}{\sqrt{2}} \end{bmatrix} \begin{bmatrix} F_a \\ F_b \\ F_c \end{bmatrix} \tag{2.43}$$

Due to the fact that energy is mostly transferred by means of three-wire systems, the zero-sequence component in the transformation matrix can be omitted and considerations are confined to three-wire circuits supplied from balanced sources of three-phase sinusoidal voltages, where $i_a + i_b + i_c = 0$ and $v_a + v_b + v_c = 0$. Assuming additionally that axes *a* and α coincide, *i.e.*, $\upsilon_{a\alpha} = 0$, the Clark transformation of phase voltages and currents simplifies to

$$\begin{bmatrix} e_\alpha \\ e_\beta \end{bmatrix} = \sqrt{\frac{2}{3}} \cdot \begin{bmatrix} 1 & -\frac{1}{2} & -\frac{1}{2} \\ 0 & \frac{\sqrt{3}}{2} & -\frac{\sqrt{3}}{2} \end{bmatrix} \cdot \begin{bmatrix} v_a \\ v_b \\ v_c \end{bmatrix} \tag{2.44}$$

and

$$\begin{bmatrix} i_\alpha \\ i_\beta \end{bmatrix} = \sqrt{\frac{2}{3}} \cdot \begin{bmatrix} 1 & -\frac{1}{2} & -\frac{1}{2} \\ 0 & \frac{\sqrt{3}}{2} & -\frac{\sqrt{3}}{2} \end{bmatrix} \cdot \begin{bmatrix} i_a \\ i_b \\ i_c \end{bmatrix} \tag{2.45}$$

The currents and voltages transformed orthogonal coordinates α-β are used to define the instantaneous real power

$$p = e_\alpha i_\alpha + e_\beta i_\beta = v_a i_a + v_b i_b + v_c i_c = \frac{dW}{dt} \tag{2.46}$$

and the instantaneous imaginary power

$$q = e_\alpha i_\beta - e_\beta i_\alpha = \frac{1}{\sqrt{3}}\left[i_a(v_c - v_b) + i_b(v_a - v_c) + i_c(v_b - v_a)\right] \tag{2.47}$$

Powers p and q are determined with a delay that results only from current and voltage sampling and the necessary computing time, *i.e.*, almost instantaneously.

The instantaneous real power p is the well-known instantaneous power of a load being the measure of the energy flow rate to the load, whereas there is no physical interpretation of the instantaneous imaginary power q.

Denoting instantaneous powers in axes α and β as p_α and p_β the instantaneous power p can be expressed as

$$\begin{aligned} p &= p_\alpha + p_\beta = e_\alpha i_{\alpha p} + e_\alpha i_{\alpha q} + e_\beta i_{\beta p} + e_\beta i_{\beta q} = \\ &= e_\alpha \frac{e_\alpha}{e_\alpha^2 + e_\beta^2} p - e_\alpha \frac{e_\beta}{e_\alpha^2 + e_\beta^2} q + e_\beta \frac{e_\beta}{e_\alpha^2 + e_\beta^2} p + e_\beta \frac{e_\alpha}{e_\alpha^2 + e_\beta^2} q = \\ &= p_{\alpha p} + p_{\alpha q} + p_{\beta p} + p_{\beta q} \end{aligned} \tag{2.48}$$

where $i_{\alpha p}$ – instantaneous active current in α axis; $i_{\beta p}$ – instantaneous active current in β axis; $i_{\alpha q}$ – instantaneous reactive current in α axis; $i_{\beta q}$ – instantaneous reactive current in β axis; $p_{\alpha p}$ – instantaneous active power in α axis; $p_{\alpha q}$ – instantaneous reactive power in α axis; $p_{\beta p}$ – instantaneous active power in β axis; $p_{\beta q}$ – instantaneous reactive power in β axis.

These components have no physical interpretations and do not participate in the energy transfer from the source to load.

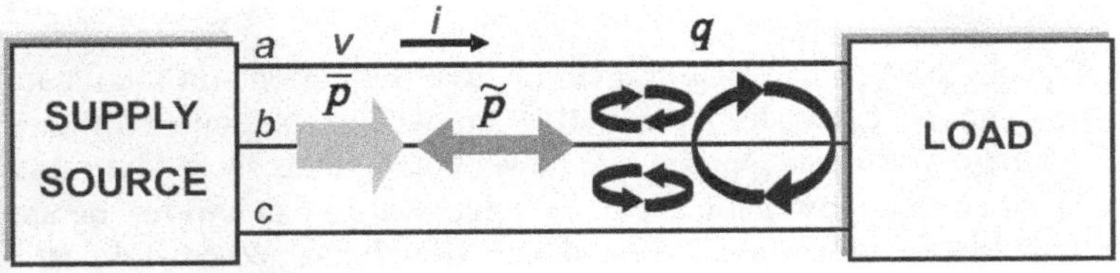

Figure 2.9. The *p-q* power theory – the graphical representation of defined power components in a three-phase circuit

The sum of the two remaining components (instantaneous active powers)

$$p_{\alpha p} + p_{\beta p} = p \tag{2.49}$$

is consistent with the common interpretation of an instantaneous power applied to three-phase circuits. Figure 2.9 shows the graphical representation of defined power components in a three-phase circuit.

When the load is non-linear and unbalanced the real and imaginary powers can be split into average and oscillating components, as follows

$$p = \overline{p} + \widetilde{p} = \overline{p} + \widetilde{p}_h + \widetilde{p}_{2f_{(1)}} \tag{2.50}$$

and

$$q = \overline{q} + \widetilde{q} = \overline{q} + \widetilde{q}_h + \widetilde{q}_{2f_{(1)}} \tag{2.51}$$

where $\overline{p}, \overline{q}$ – average components; $\widetilde{p}_h, \widetilde{q}_h$ – oscillating components (*h* – "*harmonic*"); $\widetilde{p}_{2f_{(1)}}, \widetilde{q}_{2f_{(1)}}$ – oscillating components ($2f_{(1)}$ – with double frequency of the fundamental component frequency).

According to the *p-q* power theory, the current of three-phase unbalanced non-linear load has been decomposed into four components

$$i = i_{\overline{p}} + i_{\overline{q}} + i_h + i_{2f_{(1)}} \tag{2.52}$$

where $i_{\overline{p}}$ – is associated with $\overline{p}$; $i_{\overline{q}}$ is associated with $\overline{q}$; i_h is associated with $\widetilde{p}_h$ and $\widetilde{q}_h$, *i.e.*, with the presence of harmonics in voltage and current waveforms; $i_{2f_{(1)}}$ is associated with $\widetilde{p}_{2f_{(1)}}$ and $\widetilde{q}_{2f_{(1)}}$, *i.e.*, the unbalance load currents.

From these power components the current components in the α-β coordinates can be calculated

$$\begin{bmatrix} i_\alpha \\ i_\beta \end{bmatrix} = \begin{bmatrix} e_\alpha & e_\beta \\ -e_\beta & e_\alpha \end{bmatrix}^{-1} \cdot \begin{bmatrix} p \\ q \end{bmatrix} = \frac{1}{e_\alpha^2 + e_\beta^2} \begin{bmatrix} e_\alpha & -e_\beta \\ e_\beta & e_\alpha \end{bmatrix} \cdot \begin{bmatrix} p \\ q \end{bmatrix} \tag{2.53}$$

Then, using the Clark inverse transformation, the currents in the natural *a-b-c* coordinates can be calculated. Depending on which component of the source current has to be eliminated, appropriate power components should be substituted.

The instantaneous power theory, in its original form, should not be applied to circuits unbalanced or supplied with distorted voltage. When a linear load is supplied with distorted periodic voltage, the distortions caused by the supply voltage harmonics are still present in the source current after the compensation.

The author of [24] has demonstrated that the instantaneous power *p-q* theory suggests an erroneous interpretation of power properties of three-phase circuits. According to this theory, the instantaneous imaginary current may occur in the current of a load with zero reactive power whereas the instantaneous active current may occur in the current of a load with zero active power. Moreover, these currents are non-sinusoidal, even if the supply voltage is sinusoidal and the load contains no harmonic sources.

If linear, time-invariant three-phase loads in a three-wire system supplied with balanced sinusoidal voltage are considered, the following three features are decisive for their power properties under the circuit steady state conditions. These are:

- Irreversible conversion of electric energy into other forms of energy determined by the load active power P;
- Reversible accumulation of energy in the load reactive elements, determined by the load reactive power Q, causing the current phase shift with respect to the voltage;
- The load unbalance resulting in the asymmetry of supply currents, determined by the distortion power D of a load.

The instantaneous power *p-q* theory employs only two power components: p and q. Thus, the pair of powers, measured at any given instant of time, does not allow determining three powers P, Q and D, and therefore does not allow determining the power properties of a three-phase load. The knowledge of instantaneous powers p and q at any given instant of time does not allow the drawing of any conclusions regarding the load properties. For this purpose the instantaneous powers p and q must be monitored over the entire period of their variability. Power properties of three-phase three-wire loads with sinusoidal currents and voltages are specified in terms of the active, reactive and unbalanced power. The active power P equals the average value of instantaneous active power p: $P = \bar{p}$. The reactive power Q equals the negative value of average instantaneous imaginary power q: $Q = -\bar{q}$. Thus, the active and reactive powers of a load can only be known after computing these values. Still the unbalanced power D remains unknown, what is the main drawback of the discussed theory. This concept lacks any power component uniquely associated with the load unbalance.

Similarly, since the sum of the squares of powers p and q does not define the apparent power S, the power factor $\lambda=P/S$ cannot be instantaneously determined and therefore cannot be determined whether or not the load requires reactive power compensation.

The p-q power theory turned out to be a very useful tool for developing the control algorithms of active power filters.

2.2 General Problems and Solutions of Control in Smart Power Systems

2.2.1 Control in Smart Power Systems

A substantial growth in the use of DR has been experienced in power systems in many countries. This growth is influenced by several factors including pressure to reduce CO^2 emissions, re-regulation of electric power industry, progress in generation technologies, cost reduction in materials, economic incentives (*e.g.*, special purchase tariffs for electric energy produced by Renewable Energy Sources (RES) and Combined Heat and Power (CHP) systems or plants making use of waste). The exploitation of DR leads to the connection of DG to low voltage and medium voltage grids and has given rise to new and often challenging problems. The distribution networks were not initially designed to host generation and they were usually operated with energy flowing in only one direction, namely from the substation to the customers, which is no longer true with the advent of DR generating units. DR deployment has often led system operators, electric utilities, governments or regulatory boards to define technical specifications for the grid connection and operation of DG units. The issues that came into the spotlight with the advent of DR generating units on distribution networks include: steady-state and short-circuit current constraints; power quality; voltage profile, reactive power and voltage control; contribution to ancillary services; stability and capability of DR generating units to withstand disturbances; protection aspects and islanding and islanded operation.

Connection of large RES plants whose location is largely influenced by availability of primary energy source (*e.g.*, wind) poses problems of bulk transfer of large amounts of power from remote (previously not considered) locations to load centers. The existing transmission systems often lack the capacity for transferring such amounts of power due unavailability of transmission lines or limited capacities of existing transmission corridors. The transmission system therefore needs to be strengthened in order to ensure security of supply and to fulfil its role as a marketplace where electrical energy is freely bought and sold following the rules of fair competition. There are several options to increase the transfer capacity, ranging from the building of new transmission lines (which in many countries will be very difficult due to environmental and legislative constraints), increased deployment of smart transmission technologies, such as

HVDC lines and FACTS devices, to a better utilization of the existing network due to a more efficient operation and improved control and regulation.

The role of electrical power system, *i.e.*, to be able to meet the continually changing load demand for active and reactive power, to supply energy at minimum cost and with minimum ecological impact and to ensure that the quality of power supply meets minimum standards with regard to constancy of frequency, constancy of voltage and level of reliability, will remain basically the same in spite of the changed composition and characteristics of its constituting elements, be those primary sources of electrical energy (generators) or power electronics based control devices that facilitate required transfer and control of electrical power. They will be required to provide reliable, flexible and cost-effective power supply, fully exploiting the use of both large centralized generators and DR, ensuring at the same time that they are more than just massive integration of DR generating units into the grids. Adoption of more ambitious concepts related to active management of the distribution grids in particular is required, where responsive loads, storage devices and DG can be used together to improve the overall system efficiency, quality of electricity supply and operating conditions, leading to a fully active distribution network. Exploiting the active networks infrastructure (SmartGrids) requires intelligent tools that will help decision makers to assess the network impacts and benefits resulting from a wide deployment of DR (including RES) and optimize the operating performance of the system with a high penetration of RES and DR. The evolution of electrical power networks towards the fully active structure requires several intermediate steps including: definition of connection standards and operational rules (grid codes); identification of new protection schemes and new relay settings; definition of new control procedures and managing approaches, integrating DR generation, storage devices and responsive loads; feasibility evaluation regarding islanding operation and ancillary services provision from DR generation units, as well as the quantification of the required volume of these services; regulatory studies to identify policies that may promote the deployment of DR.

Following the changes discussed above, power systems, are already considered to be one of the most complex systems built by humans, will become even more complex and more difficult to control and operate in the future. They will include a significant amount of non-conventional, generation (*i.e.*, non-synchronous generators) whose power output will generally be stochastic and greatly dependent on environmental conditions. This is particularly the case in wind, photovoltaic and combined heat and power generation. In addition to those new types of generators, proliferation of power electronic devices, either as an essential, constitutive part of new generator types, power electronic interfaces used to connect variable output (*e.g.*, doubly fed induction generators, full converter connected synchronous or induction wind generators) or essentially DC generation (*e.g.*, photovoltaics) to AC grid, or in the form of stand alone control devices, will further increase the complexity and versatility of the system.

One of the basic challenges facing control and operation of future power systems is development of non-deterministic, *i.e.*, stochastic/probabilistic, methodologies for system analysis to avoid either over conservative or too risky solutions. Deterministic approaches, based on maximum, minimum or selected subset of conditions only, cannot deal successfully with the intermittent, stochastic nature of RES and the requirement to process quickly and efficiently thousands of possible scenarios that may occur in a highly complex system in order to select the most appropriate control action.

In the rest of this chapter some basic requirements with regard to control of power system oscillations and improvement of power quality will be reviewed.

2.2.2 Damping of the System Oscillations

Power system stability is defined as "*the property of a power system that enables it to remain in a state of operating equilibrium under normal conditions and to regain an acceptable state of equilibrium after being subjected to a disturbance*" [25]. It is essential to the performance of a power network. As the complexity of the network increases, the task of maintaining system stability increases as well. In the re-regulated framework with an increased number of non-conventional generators, high proliferation of power electronic devices and demands for power transfers dictated by market rules, the system is foreseen to become more stressed and pushed to operate closer to the stability limits.

Power system stability is broadly sub-divided into rotor angle stability and voltage stability. The former is (or at least used to be) more, if not exclusively, an issue for transmission network operation while the latter was predominantly a "*local issue*", *i.e.*, typically confined to distribution networks. With new structure and operational requirements of power networks of the future, the above distinction will start to become more and more blurred and rotor angle stability studies will proliferate to lover voltage levels, *i.e.*, to distribution networks. Rotor angle stability involves the ability of the system to remain in synchronism as a whole, even after experiencing a disturbance. As power systems rely largely, and will continue to do so for the foreseeable future, on synchronous machines for electrical power generation, there is an issue of maintaining synchronism of interconnected generators. Voltage stability refers to the ability of the system to maintain voltage of all buses within statutory limits (usually within ±10% of the nominal) under both normal operating conditions and following a disturbance. The essential requirement for preventing voltage instability (and ultimately voltage collapse and wide-spread system blackouts) is to provide adequate reactive power support in the network and proper coordination between voltage regulating devices (*e.g.*, tap changing transformers, active and passive shunt or series connected voltage support devices, *etc.*). For low voltage networks with more resistive transmission lines and cables (high resistance to reactance ratios) a reactive support alone may not be sufficient and real power support is required as well. Networks with stochastically varying DR and large numbers of power electronics based control devices (or generator interfaces) will impose more demanding voltage control methodologies.

The focus of this chapter is on rotor angle stability which is further classified into transient or large disturbance stability, and small disturbance stability. Transient stability is the ability of the system to remain in synchronism after being subjected to a large disturbance (*e.g.*, a three-phase fault at any bus) while small disturbance stability is the ability of the system to remain in synchronism following a small disturbances (*e.g.*, naturally occurring, scheduled, load or generation changes, *etc.*). For both of those types of stability the damping of electromechanical oscillations in the system is essential. The physical insight in to the phenomenon of electromechanical oscillations is given in [26] through the concept of damping and synchronizing torque.

As the name suggests, these oscillations occurs as a result of interactions between the electrical and mechanical processes in the power system, namely the conversion between mechanical power (governor and turbine) into electrical power (generator). They are characterized by changes in generator rotor speed on one hand (hence "*mechanical*" in their name) and consequent fluctuations in generated electrical power (hence "*electro*" in their name) on the other. These oscillations are inherent to the synchronous machine. The equation of motion of the rotor of synchronous machine, commonly referred to as the swing equation, is given by [25, 27]

$$T_m - T_e = 2H\frac{d^2\delta}{dt^2} + D\frac{d\delta}{dt} \tag{2.54}$$

where $T_e = T(\delta, V)$.

Parameters T_m and T_e refer to the mechanical and electrical torque respectively, H is the moment of inertia of the machine, D is the damping constant, V is voltage, K is a constant and δ is the rotor angle.

Considering only small disturbances (see Equation 2.54), these can be literalized about a given equilibrium point yielding

$$\Delta T_m = 2H\frac{d^2\Delta\delta}{dt^2} + D\frac{d\Delta\delta}{dt} + K\cos\delta_o\Delta\delta \tag{2.55}$$

Solving the differential Equation 2.55 for rotor angle gives

$$\Delta\delta = \frac{\Delta T_m}{K\cos\delta_0}\left[1 - e^{-\left(\frac{D}{4H}\right)t}\sin(\omega t + \phi)\right] \tag{2.56}$$

where the frequency of the oscillation, ω is given by

$$\omega = \sqrt{\frac{K\cos\delta_o}{2H} - \left(\frac{D}{4H}\right)^2} \tag{2.57}$$

From the equations above it is obvious that, following a disturbance, the rotor angle will experience oscillatory motion with a frequency given by Equation 2.57. Taking into account generic parameters of generators and substituting the values into the equations above, the frequency range of the power system electromechanical oscillations will typically be within the range of 0.1 – 2.5 Hz [26, 27].

Electromechanical oscillations can be initiated by small disturbances occurring almost continually in the system. A system with unstable electromechanical oscillations would not be possible to operate as the response to any disturbance would result in oscillations with increasing magnitude that would lead to triggering protection and disconnection of the generator from the system. Oscillatory modes that are localized at the individual machines are referred to as local modes. They tend to occupy the higher end of the frequency range for electromechanical modes; *i.e.*, the range from 0.7 to 2.0 Hz [25, 28]. These modes tend to involve only a small part of the system and are usually associated with the angle oscillations of a single machine or oscillations of the single plant against the rest of the system [25]. Modes that are typically far less damped typically involve large groups of generators or plants and have the typical frequency range of 0.2 to 0.8 Hz [29]. They result from the interchange of power across transmission corridors (lines) between generating units in different parts of the system and involve groups of machines in one part of the network oscillating against groups of machines in other parts of the network. Inter-area oscillations have been observed in many power systems around the world over the last few decades and were typically associated with large power transfers across week transmission lines. The unstable electromechanical oscillations can cause massive disruptions to the network and ultimately blackouts [30]. They can also can cause fatigue to the machine shafts, added wear and tear of mechanical actuators of the machine controllers leading to early replacements or breakdowns the costly components. Oscillatory behavior in the network also limits power transfer capacity of tie-lines [29, 31]. To summarize, electromechanical oscillations have significant impacts on the system performance and they need to be well damped in order to insure proper system operation. Appropriate damping of electromechanical oscillations is typically achieved by installing damping controllers in the system. They can be designed and connected as local controllers, *i.e.*, as part of the generator excitation system (the most common of those is the Power System Stabilizer) or as additional damping controllers in control loops of FACTS devices connected to non-generation buses in the network. The latter are typically used for damping of system-wide, *i.e.*, inter-area oscillations.

Locally Installed Damping Controllers

By far the most widely used damping controller in the power system is the Power System Stabilizer. It uses local signals (typically speed, electrical power or frequency) available at generator terminals and injects damping signals (derived based on principle of damping and synchronizing torque [26]) directly into the generator excitation loop. The damping torque provided by PSS is in phase with generator rotor speed deviation and as such directly adds to the system damping torque [25, 26, 32]. PSS usually consists of a series (up to three or four) of tunable cascaded lead-lag compensators (blocks), and low and high frequency filters (to prevent negative interactions outside the target frequency range, *i.e.*, 0.1 to 2.5 Hz) with pre set parameters, though other types of PSSs have been developed over the years [33, 34].

Damping Controllers in Control Loops of FACTS Devices

FACTS devices were originally designed in the 1980s [35] to provide flexible control and operation of the transmission system, *i.e.*, to improve power transfer capability and to restrict or redirect the power flow in the system to designated transmission corridors [36]. They can be split broadly into two main groups, one that redirects the power flow through the control of reactances in the network and the other that uses static converters as voltage sources to inject or absorb power as appropriate. The first group includes devices such as Static Var Compensator (SVC), Thyristor Controlled Series Capacitor (TCSC) and Thyristor Controlled Phase-Shifter (TCPS). The second group includes Static Synchronous Compensator (STATCOM), Static Synchronous Series Compensator (SSSC), Unified Power Flow Controller (UPFC) and the Interline Power Flow Controller (IPFC). More details about FACTS devices can be found in [35].

At the later stages it was realized that FACTS devices also have a positive impact on damping of electromechanical oscillations (inter-area modes in particular [31]) and ever since there has been huge interest in designing supplementary damping controllers for FACTS devices [35–54].

Design of Damping Controllers

Design and tuning of power system controllers has been a research topic for many years. Many different methods have been proposed [32, 39, 40, 55–65] ranging from classical linear control system theories based on residues and frequency responses to more complex theories such as, Linear Matrix Inequalities (LMI), Multivariable Control and Linear Optimal Control (LOC). While classical methods generally suffer from lack of robustness of the solution, the advanced methods usually require either the oversimplification of the power system model or result in very complex structures of the proposed controllers. Power system engineers, however, overwhelmingly favor classical tuning methods due to their simplicity. Classical tuning methods of damping controllers (*i.e.*, PSSs) can be generally subdivided into three categories: tuning based on the torque-angle loop [32, 66, 67], tuning based on the power-voltage reference loop [45, 55, 68] and tuning based on the exciter control loop [56, 57, 69].

Even though in the vast majority of cases in realistic power systems, locally tuned damping controllers (PSSs) achieved the initially set objective of successful damping of electromechanical oscillations, it has been noted that in order to achieve optimal performance of the system the interactions between different controllers have to be taken into account [70]. This resulted in requirement for coordinated tuning of different controllers in order to enhance the operation of the system as a whole. Coordinated tuning of different controllers is typically performed using optimization methods and techniques. Methods ranging from linear programming to non-linear constrained optimization have been developed in the past [71, 72]. Depending on the formulation of the objective function to be optimized, different optimization routines should be used [37, 73–77]. In most cases the objective function is based on frequency domain information such as eigenvalues and damping factors. Other forms of objective functions, *e.g.*, using controller gain [78] or the phase and gain margin of a specific control loop [79] have also been effective. In addition to these more conventional, analytical optimization methods, more recently optimization has also been performed using evolutionary programming methods such as Genetic Algorithm, Tabu search, Simulated Annealing and Particle Swarm Optimization [80]. These novel optimization methods are particularly suited to complex non-linear systems and have been successfully applied to coordinate multiple controllers in the power system [27, 81–87].

2.2.3 Power Quality Control

The phrase Power Quality appeared as such only in the early to mid- 1980s. Prior to that, all individual phenomena coming under this generic name, voltage sags (dips), harmonics, voltage transients, voltage regulation, voltage flicker and reliability, were studied and referred to separately. Today, Power Quality is an expression used to describe broadly the entire scope of the complex interaction among electrical producers and suppliers, the environment, the systems and devices supplied by the electrical energy and the users of those systems and devices. It generally involves the maintainability of the power delivered, the design, selection, and the installation of every piece of equipment, whether hardware or software in the electrical energy system. It covers all areas from the generation plant to the last customer in the chain of electricity supply and is a measure of how the elements affect the system as a whole.

From the very beginning of generation transmission, distribution and use of electricity, the requirements for continuous delivery of the "*clean*" electrical energy and the efficient and comfortable use of all electrical equipment were, and still are, primary objectives of all electrical power systems [88]. The supply voltage, as a system-wide characteristic, and current as a more local property of the system were required to have: constant sine wave shape with fundamental frequency only, that they are supplied at constant frequency, that they form a symmetrical three-phase power system and have constant RMS value, unchanged over time, that the voltage is unaffected by load changes and that the supply is reliable, *i.e.*, the energy is available when required. Only under these conditions is the quality of electricity supply considered to be good enough. The main types of equipment served by the power supply did not changed dramatically. From the very beginning of the use of electricity, these were, and still are, light sources, motors and heaters/coolers. Further, almost all power quality disturbances that are of interest now have been an intrinsic feature of power supply systems since the earliest times. Finally, electrical systems, devices and controls for generation, transmission and distribution of electricity have evolved over the years and they are certainly more efficient and reliable today than they were in the past. What has changed over the years though, is that equipment characteristics and ways of equipment utilization have changed and influenced increase in the awareness about the power quality issues. Once almost harmless and hardly noticeable, power quality disturbances today have become an increasingly trouble some phenomenon, giving rise to intolerable inconvenience and, more importantly, considerable economic losses. Modern high-efficiency and high-intensity discharge light sources are increasingly used instead of simple incandescent lamps. Sophisticated AC and DC adjustable speed drives are now widely used for more precise control of the motors, and complex air-conditioning systems that have superseded traditional resistive heaters.

These modern "*substitutes*" are much more sensitive to various power quality disturbances than their almost insensitive predecessors. Almost all contemporary electronic equipment is sensitive to various power quality disturbances, either due to its own design features or because of the incorporated control and communication features. This step change in the equipment sensitivity to power quality disturbances was initiated by the introduction of semiconductor components in the 1960s and further strengthened by the spread of personal computers in the 1980s. Since that time, heavy dependence of customers on the comfort provided by the electronic and microprocessor-based systems grows at an exponential rate. Introduction of computer networking, mobile and other communication systems in the 1990s also increased both the range and level of exposure and interference of various commercial and industrial processes. Praised for its high efficiency, effects on energy saving and provision of accurate and comfortable control, modern electronic/microprocessor equipment was promoted and installed at a wholesale rate, demonstrating extraordinary diversity of its usage and affecting almost all aspects of life. At the same time, high sensitivity of this equipment to power quality disturbances (especially to short voltage reduction events), and adverse effects resulting from its nature of operation (*e.g.*, harmonic emission, electromagnetic interference, overheating effects, *etc.*) have resulted in increased interest in power quality.

Besides the use of the more susceptible electrical and electronic equipment, changes in modern production concepts also helped in putting power quality at the focus of interest. Modern manufacturing and service industries are characterized by highly automated processes which are more complex and more sensitive to supply disturbances, and production/operation philosophy governed by "*time-is-money*" approach, "*near the peak*" production plant operating conditions and by "*just-in-time*" manufacturing/delivery concepts. Any disruption of operation, service or production is therefore related to substantially higher costs and losses than before. The best indicator of the severity of power quality problems is probably the volume of the related investments. Industrial and commercial customers with continuous production processes or services are nowadays much more willing to invest capital in order to improve the quality of supply and to reduce the costs of power quality disturbances [89–94]. The interest in maintaining "*good*" power quality is even higher in the financial market services (stock exchanges, credit card transactions and on-line banking) and Internet data centres, where loss of information due to service interruption often exceed several thousands of dollars, pounds, euros, yens *etc.* per second of downtime [95]. Semiconductor wafer plants are probably exposed to the highest risk of potential financial losses due to power quality disturbances, as they concentrate enormously expensive facilities and equipment in a small geographic area [96].

The introduction of the electricity market in recent years also contributed very strongly to the increased interest in power quality. Electric power utilities (network operators, transmission and distribution companies, and power suppliers) are faced with re-regulation and opening of the electricity market. Distribution companies are under considerable pressure to improve and guarantee the quality of the service they provide. A failure to deliver electrical energy of an adequate quality to the customers is likely to result in severe financial penalties and affect their ability to gain new and keep existing customers. Their ability to provide the expected level of quality of supplied electricity is further challenged by proliferation of non-conventional, stochastic, renewable generation and power electronic devices in their system. New types of generators typically have variable voltage and power output requiring more sophisticated control. They are often connected to the system through power electronic devices which are both more sensitive to power quality disturbances, and sources of those disturbances (harmonics) themselves. Power electronic devices are not only used for connection and control of renewable generation but also as a constitutive part of other system controllers (FACTS devices, HVDC systems, *etc.*). It is thus essential for the future of these companies to target investments towards the service and network improvements that will be the most effective. Reliability, availability and good overall quality of delivered electrical energy are still the most important aims of power suppliers. The meaning of these supply attributes, however, has changed somewhat over the past decades taking more into account the situations on the customer's side. The standard assessment of "*customer minutes lost*" has a completely different meaning for customers whose processes and equipment are sensitive to power quality disturbances. Thus, reliability and other traditional concepts used for description of supply system performance have to be reformulated and generalized in order to include at least the most important power quality concepts, as they are becoming more and more appreciated as the standard performance criteria of power supply services.

The manufacturers of (sensitive) electrical equipment and power quality mitigation devices have also shown strong interest in understanding, characterization and quantification of power quality problems. They need to know the basic characteristics of the most frequent power quality disturbances in order to design their products to withstand/ride through those disturbances or to mitigate them. Having in mind the costs of mitigation of power quality problems, improvement of equipment immunity and ride-through capabilities at the equipment manufacturing stage is the most cost-effective approach. The developers of the control software for the microprocessor-based equipment too, are also interested in understanding power quality disturbances in order to prevent software-driven equipment disconnections and mal-operation due to the rigid control or protection setting [97].

Power quality disturbances shut down industrial processes and computer systems, causing losses in productivity and materials at a level never encountered before. These disturbances also annoy and inconvenience residential customers who own an increasing array of modern electronic equipment and find such problems less acceptable in the electricity supply industry that is privately owned and operates in a competitive environment. Regarding the frequency of occurrences and associated costs, voltage sags and short interruptions and harmonics are the two most detrimental power quality disturbances (*e.g.*, [98]).

Power system harmonics are another important power quality phenomenon, which attracts significant attention due their high potential impact on system and equipment operation and performance.

Voltage Sags and Harmonics

Voltage sag (dip) is defined as a decrease in the Root-Mean-Square (RMS) value of an AC voltage between 0.1 p.u. and 0.9 p.u. at power frequency for a duration from 0.5 cycles (10 ms in 50 Hz supply system) to 1 min [99–101]. The magnitude and the duration of a sag have been extensively used for development of equipment compatibility charts and indices and characterization of system sag performance. They are identified as the main characteristics of a voltage sag [102]. Other voltage sag characteristics, *i.e.*, phase angle jump, three-phase unbalance and point on wave of sag initiation and recovery, though very important for assessing equipment sensitivity to voltage sags, received much less attention. Voltage sags are mainly caused by power system faults on transmission or distribution systems, though the faults within an industrial facility, or starting of large induction motors, can also cause voltage sags. As a result of voltage sags at the point of connection, the equipment may trip or fail to perform the intended operation. Examples include: unwanted tripping of sensitive controls; dropping out of AC contactors, personal computers, inverters, variable speed drives, programmable logic controllers; slowing down of induction motors; disconnection of high-intensity discharge lighting, *etc.* As a consequence of individual equipment failure, the whole production process (or service) controlled by that equipment may be interrupted and thus cause significant financial losses to the owner.

Power system harmonics are sinusoidal voltages or currents with a frequencies which are the integer multiple of the frequency at which the supply system is designed to operate (50 or 60 Hz). Harmonics are generated by the whole range of non-linear/harmonic loads where the relationship between voltage and current at every instant of time is not constant, *i.e.*, the load is non-linear. The non-linear currents flowing through the system impedance result in harmonic voltages at the load. Harmonic sources can be broadly classified as:

- Saturable devices where the source of non-linearities is the physical characteristics of the iron core. They include transformers, rotating machines, non-linear reactors, *etc.*;
- Arcing devices where the source of non-linearities is the physical characteristics of the electric arc. They include arc furnaces, arc welders and fluorescent lighting;
- Power electronics where the source of non-linearities is semiconductor device switching which occurs within a single cycle of the power system fundamental frequency. They include Variable Speed Drives (VSD), DC motor drives, electronic power supplies, rectifiers, inverters, FACTS devices, HVDC transmission systems, *etc.*

The principle measure of the effective (RMS) value of harmonic distortion is defined as Total Harmonic Distortion (THD) and is given by

$$THD_M = \frac{\sqrt{\sum_{h=2}^{h_{max}} M_h^2}}{M_1} \tag{2.58}$$

where h is harmonic order and M is RMS value of the individual current or voltage harmonic.

However, current distortion levels characterized by THDI can be misleading when the fundamental current is low. The impact of a high THDI value on the system for light load is insignificant, since the magnitude of harmonic currents is low even though its relative distortion to the fundamental frequency is high. Hence, another distortion index known as the Total Demand Distortion (TDD) [103–105] is often used to indicate a more realistic impact of harmonic current distortion on the system

$$TDD_I = \frac{\sqrt{\sum_{h=2}^{h_{max}} I_h^2}}{I_L} \tag{2.59}$$

where I_L is the maximum demand load current (15 or 30 min demand) at fundamental frequency at the Point of Common Coupling (PCC), calculated as the average current of the maximum demands for the previous 12 months [104].

Besides THD and TDD, several other indices specific to the type of equipment affected by harmonics, particularly telecommunication and audio system are available [106, 107].

The overall magnitude of harmonic currents in a distribution network typically follows the trend of fundamental current (*i.e.*, power demand), as harmonic (non-linear) and linear loads tend to be simultaneously present in the network, particularly commercial loads, which operate over distinct periods of time. In most cases, in the short term, an upward trend in power demand is likely to be followed by a similar upward trend in magnitude of harmonic currents [108–111]. In addition, harmonic distortion in distribution networks is influenced by load composition. For example, a distribution feeder that supplies 100% industrial load is likely to have different harmonic distortion levels from a feeder that supplies 40% commercial and 60% residential load [108]. The difference arises due to the characteristics of harmonic loads in each case which, besides different THD level, also produce different characteristic harmonics. Large power VSD and converters used in industrial environment generate predominantly 5th and 7th harmonics, due to the use of six pulse converters (dominant harmonics produced by n-pulse converter are typically $n\pm1$). Single-phase power electronic devices, used in commercial and residential loads, on the other hand generate predominantly 3rd, 5th and 7th harmonics. As far as variation in THD with time is concerned, distribution feeders dominated by residential loads are likely to experience an increase in magnitude of harmonic currents during the night when television and lights are switched on. Feeders supplying predominantly commercial loads typically experience decrease in magnitude of harmonic currents during late night periods when premises are closed for business.

With the rapid development of electronic, information and communication technology, usage of non-linear/harmonic loads is expected to increase at both system and end user level, and so does the level of voltage distortion in power systems. In commercial and residential buildings, single phase non-linear/harmonic loads are already being used in large quantities (personal computers, fluorescent lighting, office equipment, *etc.*). It is envisaged that higher power non-linear devices such as central airconditioning with variable speed drives will increasingly find their way into customer loads in the near future. At the system level, power electronic interfaces employed for connection and control of renewable generation, FACTS devices that are increasingly employed for efficient control of power networks and HVDC transmission systems used for bulk power transfer across long distances may contribute further to the overall harmonic pollution of the network. From the utility viewpoint, high harmonic distortion increases the risk of customer complaints as a result of equipment malfunction, and in extreme cases power shut down. In addition, harmonic distortion increases the volt-ampere demand on transformers and lines/cables through additional heat loss, which decreases the margin on the entire power distribution system and the opportunity to add more customers on existing systems. Customers, on the other hand, are inclined to use non-linear loads due to their state-of-the-art technology, but are reluctant to invest in harmonic mitigation devices, particularly those from the commercial and residential sectors. The major effects of voltage and current harmonic distortion can be classified as follows:

- Thermal stress, through increasing copper, iron and dielectric losses;
- Insulation stress, through the increase of peak voltage, *i.e.*, voltage crest factor;
- Load disruption.

Since the major effects of harmonic distortion typically have cumulative effect, *i.e.*, overheating and insulation fatigue, they depend on the level of THD and duration of the state in which this increased THD exists. Because of that, harmonic resonance [104], either parallel or series, contributes significantly to the ultimate effect of harmonics on equipment and network. During the resonant condition in the network, critical harmonic is particularly magnified, resulting in high voltages and currents which can contribute to excessive heating of the device, insulation breakdown or load/process interruption. A major cause of resonant condition in the network is the installation of inadequately sized Power Factor Correction Capacitors (PFCC) which brings natural resonant frequency down to the level of 5th (250 Hz) or 7th (350 Hz) harmonic.

Realizing the potential magnitude of losses due to power quality disturbances [112–126], more and more industrial and commercial customers have started to seek protection, both technical and contractual, from the impact of power quality disturbances. The most common approach is to install mitigation devices (custom power) to reduce the number of disruptive events. Others go for higher quality of supply through power quality contracts with the electrical utility. However, regardless of which option is preferred, the investment must be economically justifiable.

From the electrical utilities' point of view, reducing the number of voltage sags and short interruptions and controlling harmonic levels in the network would boost customer satisfaction and improve future business prospects. In today's competitive electrical market, the standard of power quality has to be constantly upgraded to ensure business survival.

Controlling Voltage Sags

In order to reduce potential financial losses arising from voltage disturbances, voltage sags in particular, several options could be adopted. The first certainly would be to reduce the causes of disturbances. Since power system faults are the major cause of voltage sags, action can be taken by transmission and distribution companies to reduce the occurrences of faults. This includes, at transmission system level, adjusting transmission tower footing resistance, installing line arresters, regular insulator washing, installing fast switches with instantaneous protection systems and arc-suppression coil earthing with time grading protection. At the distribution level these action include regular tree trimming, installing animal guards, installing arresters, loop schemes, *i.e.*, meshed rather than radial network topology, modified feeder designs and modified protection co-ordination. Even though these preventive measures could significantly reduce the occurrence of faults, and ultimately voltage sags, they could not be completely removed. Therefore, further measures should be taken to deal with remaining faults and to insure that equipment rides through disturbances. At the utility level, FACTS based conditioning devices can be used including Dynamic Voltage Restorer (DVR), Solid State Transfer Switch (SSTS), STATCOM, SSSC, SVC, UPFC. The problem with wide application of these devices, however, is very often their excessive cost. So very careful techno-economic assessment of possible mitigation options has to be carried out prior to proposing a final solution. Bearing in mind the power level involved, *i.e.*, protecting the whole feeder or part of the network, and resulting cost of mitigating device and the fact that not all customers are equally sensitive to voltage disturbances, it is often (if not always) much more economically justifiable to protect only customers plant or selected feeders. At this level, similar mitigation devices can be used as before (*e.g.*, DVR, SSTS, STATCOM, *etc.*); however, due to the much lower rating of protected loads, the cost of solution would be much lower. In addition to these, some other devices and techniques could be used since the powers involved are smaller. These include flywheels, Superconducting Magnetic Energy Storage (SMES), Magnetic Synthesizers, on-line or rotary UPS, Motor Generator (MG) sets, private generation, static and active voltage conditioners, electronic tap changers and even constant voltage transformers (for smaller static/fixed loads). If, however, as often is the case, only particular equipment or devices have to protected and quite often its control system, then smaller ratings and simpler devices can be used. For protecting control systems, devices such as constant voltage (ferro resonant) transformers, UPSs, SMES and MG sets can be used. Individual loads can be protected by improved design for voltage tolerance (including internal control algorithms and protection settings), improved starters, improved DC supplies (larger capacitors to insure adequate level of DC voltage), lower drop-out characteristics or again using similar lower rated devices as for protection of control circuits, *e.g.*, constant voltage (ferro-resonant) transformers. The cost of mitigation typically increase exponentially as the solution is applied to higher voltage level and as it involves more than one device or one control circuit (due to the rating of mitigation device required). It has to be pointed out though that even though modern CUPS (*e.g.*, DVR, STATCONM, UPFC, *etc.*), whose origin is in FACTS technology and which can synthesize required missing voltage and almost fully compensate for the disturbance and as such provide very good if not excellent protection of equipment and processes, may not always be economically the best solution due to their very high costs. In many practical cases a simple modification of control algorithms, improved design of the equipment or use of simple device such as constant voltage transformer can significantly reduce the number of process interruptions and consequently financial losses incurred to end use customers. In cases when this is not possible, however, more sophisticated and more expensive solutions have to be sought and careful techno-economic consideration is required. In such cases, additional benefits (voltage regulation, power factor correction, voltage and angular stability protection provided to other buses/customers in the network, *etc.*) that may arise from the application of sophisticated mitigation device have to taken into account when deciding on its techno-economic merits.

Controlling Harmonic

Harmonic currents typically go through various stages of attenuation before they reach the utility source. One of the common, naturally occurring phenomena is the cancelation of harmonic currents due to diversity in phase angles resulting from different load mix and operating conditions. Others include deliberate supply/connection of loads through transformers having different winding connections (*e.g.*, one VSD supplied through delta/delta transformer and the other through delta/wye transformer) as these introduce additional phase shift which then enhances harmonic cancelation. Addition of small inductors (chokes), typically rated at about 3% of the drive rating, in variable speed drives further contributes to reduction in harmonic currents flowing in the network.

The best, and often the cheapest, way of solving harmonic problems is at the source through equipment specification and design. This includes use of Pulse Width Modulation (PWM) switching techniques (standard practice in modern VSDs and inverter technologies) resulting in much less distorted current and voltage waveforms, choke inductance built in VSDs (which can half the THD with moderate, about 20%, increase in cost), use of 12 or 24 pulse inverters instead of 6 pulse inverters and reducing the overall THD and in particular the content of critical 5th and 7th harmonic, installation of PFCC whose size is less then 20% of the rating of supply transformer, or if larger size is needed, use tuned capacitor banks, *etc.* In case of synchronous machines as harmonic sources, appropriate winding pitch should be used to reduce harmonic content in output voltage and current (*e.g.*, 2/3 winding pitch minimizes 3rd harmonic component).

One of the common harmonic problems in commercial facilities is in-crease in neutral conductor currents. Those currents, which should ideally be zero in balanced, symmetrical and non-distorted systems, could rise to unacceptably high levels due to high 3rd harmonic component in particular (triplen harmonic currents, 3rd, 9th, 15th, *etc.*, coming from different phase conductors do not cancel in a neutral conductor as do fundamental currents). In situations like that the size of the neutral conductor should be increased (at least to the same size as phase conductors or double the size of phase conductor) to prevent overheating or 3rd harmonic filter installed at each load or in neutral conductor. Zig-zag transformers are also often used in cases like this as they provide low impedance path for 3rd harmonic current (and high impedance path to positive sequence current so the load is not affected) and it gets trapped inside transformer.

If, however, the required reduction in THD cannot be achieved in one of the above ways, or through a combination of those, than harmonic filters (passive or active) need to be installed.

Passive filters are an economical method for mitigation of harmonic currents in distribution network [127–129]. In addition to filtering harmonic currents, they can also be used as a reactive power source. Ideally, passive filters should be placed at the point of common coupling by the respective customer, so that harmonic currents penetration to utility system is kept at the minimum level. However, there are also cases where distribution feeders carry a significant amount of harmonic currents, resulting in unacceptable harmonic voltage distortion at network buses, so filters should installed at all strategic network buses.

A passive harmonic filter is cheap as it is built from passive network elements, *i.e.*, resistors, inductors and capacitors. There are two approaches to suppress undesired harmonic currents using passive filters. The series harmonic filter uses series impedance to block harmonic currents while the shunt filter act as harmonic current sink and diverts harmonic currents by providing low impedance shunt path [128]. Series filters are more expensive (and therefore less frequently used) than shunt filters as they must carry full load current. Three types of shunt filters, namely, single-tuned first order, high-pass second-order, and double band pass (double-tuned) filters are most commonly used. The function of passive shunt harmonic filters is usually two-fold to supply the necessary reactive power and to divert harmonic currents by providing a low impedance path at the tuned frequency. Hence, the major criterion of harmonic filter design is to select suitable capacitor size that results in desired power factor at fundamental frequency. The harmonic filter is then tuned by selecting appropriate values of inductances and resistances that result in the filter effectively diverting all or part of the specified harmonic currents. The reactance of the inductance is typically tuned based on the reactance of capacitor corresponding to the reactive power requirement and tuned harmonic order of the filter. The resistance is determined based on the desired quality factor (measure of the sharpness of the tuning frequency) of the filter. Although the common practice is to limit the resistance of the filter to the reactor's resistance, external resistance is often added to modify the sharpness of tuning or change the bandwidth of the impedance *vs* frequency [128]. The single-tuned filter is the most frequently used filter as it is the simplest and the cheapest [128, 129]. It consists of a series combination of a capacitor, inductor and resistor and it is typically tuned to low harmonic frequencies. The main advantage of the high pass second order filter is that it provides low impedance for a wide range of frequencies [129] when tuned to a low Q-factor of between 0.5 and 5. However, one of its disadvantages is that its minimum impedance is never as low as that of the single-tuned filter [130], and therefore it is not as effective in diverting harmonic currents. The double tuned band pass filter consists of a series combination of a main capacitor and reactor, and a tuning device, which is a tuning capacitor connected in parallel with a tuning reactor. The reactances of both, the series and parallel circuit, are tuned to the mean geometric frequency of two specific harmonic frequencies that are to be controlled. It is in general very effective in diverting harmonic currents of two specific frequencies, particularly when tuned to a higher Q-factor (between 5 and 10). Finally, harmonic filters that are tuned to frequencies slightly lower than the resonant frequency, so that the minimum impedance does not occur at the exact order of harmonic, are referred to as detuned filters. They are cheaper as current ratings of their components could be reduced accordingly. In addition, dielectric materials of capacitors typically degrade over time and therefore it is necessary to compensate for this phenomenon by tuning the filter to a lower resonant frequency [129].

The second, more expensive, option of filtering harmonics is to apply active harmonic filters. They are connected in shunt and handle lower currents (typically only the distorted component of total load current) and therefore cannot be overloaded. They are typically of smaller size (up to about 150 kVA) than passive filters and installed close to the harmonic source. Active filters are power electronic based devices, similar to inverters, which first measure/detect harmonic content in load current and then synthesize, through appropriate switching, a signal that negates the harmonic currents injected by the non-linear loads connected to them. When injected back into the network this signal virtually eliminates harmonics from the load current and the combination of the active filter and non-linear load looks like a resistive load to the power system with low distortion and unity power factor. Active harmonic filters can be incorporated as an active front end of VSDs and as such eliminate almost completely harmonic distortion resulting from VSD operation. The cost of VSD with an active front end that acts as a harmonic filter, however, is about double the cost of passive front end VSD.

References

[1] Czarnecki LS, (1997) Powers and compensation in circuits with periodic voltage and currents. Part 2 — Outline of the history of power theory development. (in Polish), vol.III-2:37–46

[2] Steinmetz CP, (1897) Theory of alternating current phenomena. New York

[3] Budeanu CJ, (1927) Puissances reactives et fictives. Institut Romain de l'Energie, Bucharest

[4] Fryze S, (1932) Active, reactive and apparent power in electrical circuits with non-sinusoidal current and voltage. (in Polish), Przegląd Elektrotechniczny, 7:193–203, 8:225–234, 22: 673–676

[5] Firlit A, (2006) Comparison of control algorithms based on selected power theories. (in Polish), AGH-University of Science & Technology

[6] IEEE, (2000) Std. 1459–2000: IEEE Trial-use standard definitions for the measurement of electric power quantities under sinusoidal, non-sinusoidal, balanced or unbalanced conditions

[7] Czarnecki LS, (2000) Energy flow and power phenomena in electrical circuits: illusions and reality. Springer-Verlag, Electrical Engineering, vol.82:119–126
[8] Czarnecki LS, (1984) Considerations on the reactive power in non-sinusoidal situations. IEEE Transactions on Instrumentation and Measurement, IM-34:399–404
[9] Czarnecki LS, (1991) Scattered and reactive current, voltage, and power in circuits with non-sinusoidal waveforms and their compensation. IEEE Transactions on Instrumentation and Measurement, IM-40:563–567
[10] Pasko M, Maciążek M, (2004) Contribution of theoretical electrical engineering to power quality improvement. (in Polish), Wiadomości Elektrotechniczne, no.7-8:37–46
[11] Pasko M, Dębowski K, (2002) Symmetrisation of three-phase and multi-phase systems supplied from sources of periodic non-sinusoidal voltages. (in Polish), Monograph, Gliwice
[12] Shepherd W, Zakikhani P, (1972) Suggested definition of reactive power for non-sinusoidal systems. IEE Proceedings, no.119:1361–1362
[13] Emanuel AE, (1974) Suggested definition of reactive power in non-sinusoidal systems. IEEE Proceedings, vol.121, no.7:705–706
[14] Sharon D, (1973) Reactive power definitions and power factor improvement in nonlinear systems. IEE Proceedings, vol.120, no.6:704–706
[15] Kusters NL, Moore WJM, (1980) On the definition of reactive power under non-sinusoidal conditions. IEEE Transactions on Power Applications, PAS-99:1845–1854
[16] Page C, (1980) Reactive power in non-sinusoidal systems. IEEE Transactions on Instrumentation and Measurement, IM-29:420–423
[17] Czarnecki LS, (1985) Power theories of periodic non-sinusoidal systems. Rozprawy Elektrotechniczne, vol.31, no.3-4:659–685
[18] Czarnecki LS, (1995) Power related phenomena in three-phase unbalanced systems. IEEE Transactions on Power Delivery, vol.10, no.3:1168–1176
[19] Akagi H, Kanazawa Y, Nabae A, (1983) Generalized theory of the instantaneous reactive power in three-phase circuit. Proceedings of the International Power Electronics Conference, Tokyo/Japan:1375–1386
[20] Akagi H, Kanazawa Y, Nabae A, (1984) Instantaneous reactive power compensators comprising switching devices without energy storage components. IEEE Transactions on Industrial Applications, IA-20:625–630
[21] Akagi H, Nabae A, (1993) The p-q theory in three-phase systems under non-sinusoidal conditions. European Transactions on Electric Power, vol.3, no.1:27–31
[22] Akagi H, Watanabe EH, Aredes M, (2007) Instantaneous power theory and applications to power conditioning. Wiley-Intersciencs
[23] Buchholz F, (1922) Die Drehstrom-Scheinleistung bei ungleichmäßiger Belastung der drei Zweige. Licht und Kraft, vol.2
[24] Czarnecki LS, (2005) Powers in electrical circuits with non-sinusoidal voltages and currents. Publishing Office of the Warsaw University of Technology
[25] Kundur P, (1994) Power System stability and control. McGraw-Hill Inc
[26] DeMello FP, Concordia C, (1969) Concepts of synchronous machine stability as affected by excitation control. IEEE Transactions on Power Apparatus and Systems, vol.88:316–324
[27] Milanovic JV, (1996) The influence of loads on power system electromechanical oscillations. Electrical and Computer Engineering, University of Newcastle:191
[28] Klein M, Rogers GJ, Kundur P, (1991) A fundamental study of inter-area oscillations in power systems. IEEE Transactions on Power System, vol.6

[29] Kundur P, (1993) Investigation of low frequency inter-area oscillation problems in large interconnected power systems. Ontario Hydro
[30] Kosterev DN, Taylor CW, Mittelstadt WA, (1999) Model validation for the August 10, 1996 WSCC system outage. IEEE Transactions on Power Systems, vol.14:967–979
[31] Systems Oscillations Working Group, (1994) Inter-area oscillations in power systems. IEEE Power Engineering Society
[32] Larsen EV, Swann DA, (1981) Applying power system stabilizers. Part 1: General concepts. IEEE Transactions on Power Apparatus and Systems, vol.100:3017–3024
[33] Murdoch A, Venkataraman S, Lawson RA, (1999) Integral of accelerating power type PSS. Part 1: Theory, design and tuning methodology. IEEE Transactions on Energy Conversion, vol.14:1658–1663
[34] Murdoch A, Venkataraman S, Lawson RA, (1999) Integral of accelerating power type PSS Part 2: Field testing and performance verification. IEEE Transactions on Energy Conversion, vol.14:1664–1672
[35] Cai LJ, Erlich I, (2000) Fuzzy coordination of FACTS controllers for damping power system oscillations. International Symposium on Modern Electric Power System
[36] Song YH, Johns AT, (1999) Flexible AC transmission systems (FACTS). IEE Press
[37] Cai LJ, Erlich I, (2005) Simultaneous coordinated tuning of PSS and FACTS damping controllers in large power systems. IEEE Transactions on Power Systems, vol. 20:294–300
[38] Canizares A, (2000) Power flow and transient stability models of FACTS controllers for voltage and angle stability studies. IEEE/PES World Meeting Panel on Modeling, Simulations and Application of FACTS Controllers in Angle and Voltage Stability Studies
[39] Chaudhuri B, Pal BC, (2003) Robust damping of inter-area oscillations through controllable phase shifters using global signals. PES General Meeting, Toronto, Canada
[40] Chaudhuri B, Pal BC, Zolotas AC, (2003) Mixed sensitivity approach to H-info control of power system oscillations employing multiple FACTS devices. IEEE Transactions on Power system, vol.18:1149–1156
[41] Choi SS, Jiang F, Shrestha G, (1996) Suppression of transmission system oscillations by thyristor-controller series compensation. IEE Proceedings (Generator, Transmission, Distribution), vol.143:7–12
[42] Fang DZ, Xiaodong Y, Chung TS, (2004) Adaptive fuzzy logic SVC damping controller using strategy of oscillation energy descent. IEEE Transactions on Power System, vol.19:1414–1421
[43] Fang DZ, Xiaodong Wennan YS, Wang HF, (2003) Oscillation transient energy function applied to the design of a TCSC fuzzy logic damping controller to suppress power system inter-area mode oscillations. IEE Proceedings (Generator, Transmission, Distribution), vol.150:233–238
[44] Ghandhari M, Andersson G, Hiskensi IA, (2001) Control Lyapunov function for controllable series devices. IEEE Transactions, vol.16:689–694

[45] Gibbard MJ, Vowles DJ, Pourbeik P, (2000) Interactions between, and effectiveness of, power system stabilizers and FACTS device stabilizers in multimachine systems. IEEE Transactions on Power System, vol.15:748–755

[46] Gu Q, Pandey A, Starrett SK, (2003) Fuzzy logic control schemes for static VAR compensator to control system damping using global signals. Electrical Power & Energy Systems, vol.67:73–152

[47] Hiyama T, Hubbi W, Ortmeyer TH, (1999) Fuzzy logic control scheme with variable gain for Static Var Compensator to enhance power system stability. IEEE Transactions on Power System, vol.14:186–191

[48] IEEE Special Stability Controls Working Group, (1994) Static Var Compensator models for power flow and dynamic performance simulation. IEEE Transactions on Power system, vol.9:229–239

[49] Lo KL, Khan L, (2000) Fuzzy logic based SVC for power system transient stability enhancement. Electric Utility Deregulation & Restructuring & Power Technologies, City University, London

[50] Menniti D, Pinnarelli AB, Sorrentino N, (2003) Synchronizing fuzzy power system stabilizer and fuzzy FACTS device stabilizer to damp electromechanical oscillations in a multi-machine power system

[51] Mishra S, Dash PK, Hota PK, (2002) Genetically optimized neuro fuzzy IPFC for damping modal oscillations of power system. IEEE Transactions on Power System, vol.17:1140–1147

[52] Rao PS, (1998) A QFT-based robust SVC controller for improving the dynamic stability of power systems. Electrical Power & Energy Systems, vol.46:213–219

[53] Sanchez-Gasca JJ, (1998) Coordinated control of two FACTS devices for damping inter-area oscillations. IEEE Transactions on Power System, vol.13:428–434

[54] Wang Y, (2001) Nonlinear coordinated excitation and TCPS controller for multimachine power system transient stability enhancement. IEE Proceedings (Generator, Transmission, Distribution), vol.148:133–141

[55] Gibbard MJ, (1988) Coordinated design of multimachine power system stabilizers based on damping torque concepts. IEE Proceedings (Generator, Transmission, Distribution), vol.135:276–284

[56] Pagola FL, Perez-Arriaga IJ, Verghese GC, (1989) On sensitivities, residues and participations: Applications to oscillatory stability analysis and control. IEEE Transactions, vol.4:279–285

[57] Martins N, Lima LTG, (1990) Eigen-value and frequency domain analysis of small-signal electromechanical stability problems. Eigen-analysis and frequency domain methods for system dynamic performance. IEEE Publication no.90TH0292-3-PWR:17–33

[58] Anaparthi KK, Pal BC, El-Zobaidi H, (2005) Coprime factorization approach in designing multi-input stabilizer for damping electromechanical oscillations in power systems. IEE Proceedings (Generator, Transmission, Distribution), vol.152:301–308

[59] Boukarim GE, Wang S, Chow JH, (2000) A comparison of classical, robust and decentralized control designs for multiple power system stabilizers. IEEE Transactions on Power System, vol.15:1287–1292

[60] Hirano S, Michigami T, Kurita A, (1990) Functional design for a system-wide multivariable damping controller. IEEE Transactions, vol.5:1127–1136
[61] D'Andrea R, (2003) A Linear matrix inequality approach to decentralized control of distributed parameter systems
[62] Pal BC, Coonick AH, Jaimoukha IM, (2000) A linear matrix inequality approach to robust damping control design in power systems with superconducting magnetic energy storage device. IEEE Transactions on Power System, vol.15:356–362
[63] Abdel-Magid YL, Abido MA, (2003) Optimal multi-objective design of robust power system stabilizers using genetic algorithms. IEEE Transactions on Power System, vol.18:1125–1132
[64] Abido MA, (2000) Robust design of multimachine power system stabilizers using tabu search algorithm. IEE Proceedings (Generator, Transmission, Distribution), vol.147:387–394
[65] Maslennikov VA, Milanovic JV, Ustinov SM, (2002) Robust ranking of load by using sensitivity factors and limited number of points from a hyperspace of uncertain parameters. IEEE Transactions on Power System, vol.17:565–570
[66] Larsen EV, Swann DA, (1981) Applying power system stabilizers. Part II: Performance objectives and tuning concepts. IEEE Transactions on Power Apparatus and Systems, vol.100:3025–3033
[67] Larsen EV, Swann DA, (1981) Applying power system stabilizers Part III: Practical considerations. IEEE Transactions on Power Apparatus and Systems, vol.100:3034–3046
[68] Gibbard MJ, (1991) Robust design of fixed parameter power system stabilizers over a wide range of operating conditions. IEEE Transactions on Power System, vol.6:794–800
[69] Fleming RJ, Mohan MA, Parvatisam K, (1981) Selection of parameters of stabilizers in multimachine power systems. IEEE Transactions on Power Apparatus and Systems, vol.100:2329–2333
[70] CIGRE Taskforce 38.02.16, (2000) Impact of interactions among power systems. Paris
[71] Bertsekas DP, (1999) Nonlinear programming. Belmont, Massachusetts: Athena Scientific
[72] Fletcher R, (1981) Practical methods of optimization. Vol. 2: Constrained optimization. Wiley-Interscience
[73] Vournas CD, Maratos N, Papadias BC, (1994) Power system stabilizer coordination using a parameter optimization method. International Conference on Control, Coventry, UK
[74] Urdaneta AJ, Bacalao NJ, Feijoo B, (1991) Tuning of power system stabilizers using optimization techniques. IEEE Transactions on Power Systems, vol.6:127–154
[75] Hong YY, Wu WC, (1999) A new approach using optimization for tuning parameters of power system stabilizers. IEEE Transactions on Energy Conversion, vol.14:780–786
[76] Khaldi MR, Sarkar AK, Lee KY, (1993) The modal performance measure for parameter optimization of power system stabilizers. IEEE Transactions on Energy Conversion, vol.8:660–666

[77] Taranto GN, Chow JH, (1995) A robust frequency domain optimization technique for tuning series compensation damping controllers. IEEE Transactions on Power Systems, vol.10:1219–1225

[78] Mendonca A, Lopes JAP, (2003) Robust tuning of PSS in power systems with different operating conditions. IEEE Bologna Power Tech Conference, Bologna, Italy

[79] Kim JM, Moon SI, Lee J, (2001) A new optimal AVR parameter tuning method using online performance indices of frequency domain. Power Engineering Society Summer Meeting, Vancouver

[80] d. Silva APA, Abrao P, (2002) Application of evolutionary computation in electric power systems. Congress on Evolutionary Computation, Honolulu

[81] Abdel-Magid YL, (1997) Simultaneous stabilization of power systems using genetic algorithms. IEE Proceedings (Generator, Transmission, Distribution), vol.144:39–44

[82] Abido MA, (2000) Robust design of multimachine power system stabilizers using simulated annealing. IEEE Transactions on Energy Conversion, vol.15:297–304

[83] Abido MA, (2002) Optimal design of power-system stabilizers using particle swarn optimization. IEEE Transactions on Energy Conversion, vol.17:406–413

[84] Andreoiu A, Bhattacharya K, (2002) Robust tuning of power system stabilizers using a Lyapunov method based genetic algorithm. IEE Proceedings (Generator, Transmission, Distribution), vol.149:585–592

[85] Chang CS, Yu QZ, Liew AC, (1997) Genetic algorithm tuning of fuzzy SVC for damping power system inter area oscillations. International Conference on Advances in Power System Control, Operation and Management, Hong Kong

[86] Miranda V, Fonseca N, (2002) New evolutionary particle swarm algorithm (EPSO) supplied to voltage/VAR control. Power Systems Computation Conference, Seville, Spain

[87] Kim WG, Hwang GH, Kang HT, (2001) Design of fuzzy logic controller for firing angle of TCSC using real type tabu search. International Symposium on Industrial Electronics, Pusan, Korea

[88] Emanuel A, (2000) Harmonics in the early years of electrical engineering: A brief review of events, people and documents. IEEE Conference on Harmonics and Quality of Power, Orlando, Florida, vol.1:1–7

[89] Guo T, Lin J, Liao C, (2000) Taiwan power's experience in power quality monitoring. Conference on Advances in Power System Control, Operation and Management, Hong Kong

[90] Marchand M, (1997) The seven services which EDF provides to small and medium-sized companies. IEE International Electricity Distribution Conference and Exhibition, publication no.438

[91] Javerzac J, (2000) Contracting the quality of electricity: the French experience. IEEE Conference on Harmonics and Quality of Power, Orlando, Florida, vol.2:431–437

[92] CEIDS/EPRI/PRIMEN, (2001) The cost of power disturbances to industrial & digital economy companies. EPRI's Consortium for Electric Infrastructure for a Digital Society (CEIDS), report no.1006274

[93] Bollen MHJ, (2000) Understanding power quality problems: voltage sags and interruptions. IEEE Press Series on Power Engineering, New York
[94] Pereira F, Souto O, de Oliveira J, (1998) An analysis of cost related to the loss of power quality. IEEE Conference on Harmonics and Quality of Power, Athens, Greece, vol.2:777–782
[95] NREL, (2003) Renewable energy: clear, secure, reliable. National Renewable Energy Laboratory, US Department of Energy, Midwest Research Institute, Battelle, Colorado
[96] Semiconductor Business News, (1998) Interruptions can cost $2 million in revenues per day
[97] Frost & Sullivan, (2000) World UPS market. Report 5804-27
[98] Wagner V, (1990) Power quality and factory automation. IEEE Transactions on Industry Applications, vol.26, no.4:620–626
[99] Bollen MHJ, (2000) Understanding power quality problems: voltage sags and interruptions. IEEE Press Series on Power Engineering, New York
[100] Dugan RC, McGranaghan M, Beaty HW, (1996) Electrical power systems quality. McGraw Hill
[101] IEEE, (1998) Recommended practice for evaluating electric power system compatibility with electronic process equipment, IEEE std.1346
[102] Electromagnetic compatibility (EMC), (2003) Part 4-30: testing and measurement techniques – power quality measurement methods. IEC 61000-4-30
[103] Arrillaga J, Watson NR, Chen S, (2000) Power system quality assessment. John Wiley
[104] IEEE Task Force on Harmonics Modeling and Simulation, (1998) Tutorials on harmonics modeling and simulation. IEEE Power Engineering Society
[105] IEEE, (1993) Recommended practices and requirements for harmonic control in electric power systems. IEEE, New York
[106] ITTCC, (1963) Directives concerning the protection of telecommunication lines against harmful effects from electricity lines. International Communications Union, Geneva
[107] Edison Electric Institute, (1943) Engineering reports of the joint subcommittee on development and research of the Edison electric institute and the Bell telephone system. New York
[108] Shuter TC, Vollkommer HT, Kirkpatrick TL, (1989) Survey of harmonic levels on the American electric power distribution system. IEEE Transactions on Power Delivery, vol.4:2204–2213
[109] Hu CH, Wu CJ, Yen SS, (1997) Survey of harmonic voltage and current at distribution substation in northern Taiwan. IEEE Transactions on Power Delivery, vol.12
[110] Emanuel A, Orr JA, Cyganski D, (1991) A survey of harmonic voltages and currents at distribution substations. IEEE Transactions on Power Delivery, vol.6:1883–1889
[111] Emanuel A, Orr JA, Cyganski D, (1993) A survey of harmonic voltages and currents at the customer's bus. IEEE Transactions on Power Delivery, vol.8
[112] Quaia S, Tosato F, (2003) Interruption costs caused by supply volt-age dips and outages in small industrial plants: a case study and survey results

[113] Aziz MMA, Salam GAA, Kozman SM, (2004) Cost and mitigation of voltage sag for industrial plants. Cairo, Egypt
[114] Lamedica R, Patrizio A, Prudenzi A, (2000) Power quality costs and upgrading solutions: the energy centre
[115] Tosato F, Quaia S, (2003) A method for the computation of the interruption costs caused by supply voltage dips and outages in small industrial plants. IEEE Region 8 EUROCON
[116] Sullivan MJ, Vardell T, Johnson M, (1997) Power interruption costs to industrial and commercial consumers of electricity. IEEE Transactions on Industry Applications, vol.33:1448–1457
[117] LaCommare KH, Eto JH, (2004) Understanding the cost of power interruption to US electricity consumers. Berkeley National Laboratory LBNL-55718
[118] Primen, (2001) The cost of power disturbances to industrial & digital economy companies. EPRI CEIDS
[119] Chowdhury BH, (2001) Power quality. IEEE Potentials: the magazine for up-and-coming engineers:5–11
[120] Lamoree JD, (2004) Cost of outages. ENERNEX Corporation
[121] Yin SA, Lu CN, Liu E, (2001) Assessment of interruption cost to high tech industry in Taiwan. Atlanta
[122] Nam KY, Choi SB, Ryoo HS, (2004) Development of criteria and calculation of Korean industrial customer interruption costs. International Power Quality Conference, Singapore
[123] Linhofer G, Maibach P, Wong F, (2002) Power quality devices for short term and continuous voltage compensation. International Power Quality Conference, Singapore
[124] Verde P, (2000) Cost of harmonic effects as meaning of standard limits, Proceedings. International Conference on Harmonics and Quality of Power, vol.1:257–259
[125] Caramia P, Verde P, (2000) Cost related harmonic limits. IEEE Power Engineering Society Winter Meeting, vol.4:2846–2851
[126] Lee B, Stefopoulos GK, Meliopoulos APS, (2006) Unified reliability and power quality index. International Conference on Harmonics & Quality of Power
[127] Chang GW, Chu SY, Wang HL, (2002) Sensitivity based approach for passive harmonic filter planning in a power system. IEEE Power Engineering Society Winter Meeting
[128] Makram EB, Subramaniam EV, Girgis AA, (1993) Harmonic filter design using actual recorded data. IEEE Transactions on Industry Applications, vol.29:1176–1183
[129] Wakileh GJ, (2001) Power systems harmonics – fundamentals, analysis and filter design. Springer
[130] Acha E, Madrigal M, (2001) Power systems harmonics: computer modeling and Analysis. John Wiley

3

Overview of Power Electronics Converters and Controls

Ryszard Strzelecki[1] and Genady S. Zinoviev[2]

[1]Department of Ship Automation,
Gdynia Maritime University,
81-87 Morska Street, Gdynia, Poland.
Email: Rstrzele@am.gdynia.pl

[2]Department of Industrial Electronics,
Novosibirsk State Technical University,
20 Karla Marksa Prospect, Novosibirsk, Russia.
Email: Genstep@mail.ru

3.1 Power Electronics Background

Power Electronics (PE) is the technology associated with efficient conversion, control and conditioning of electric power by static means from its available input- into the desired electrical output form. Electric energy conversions carried out by

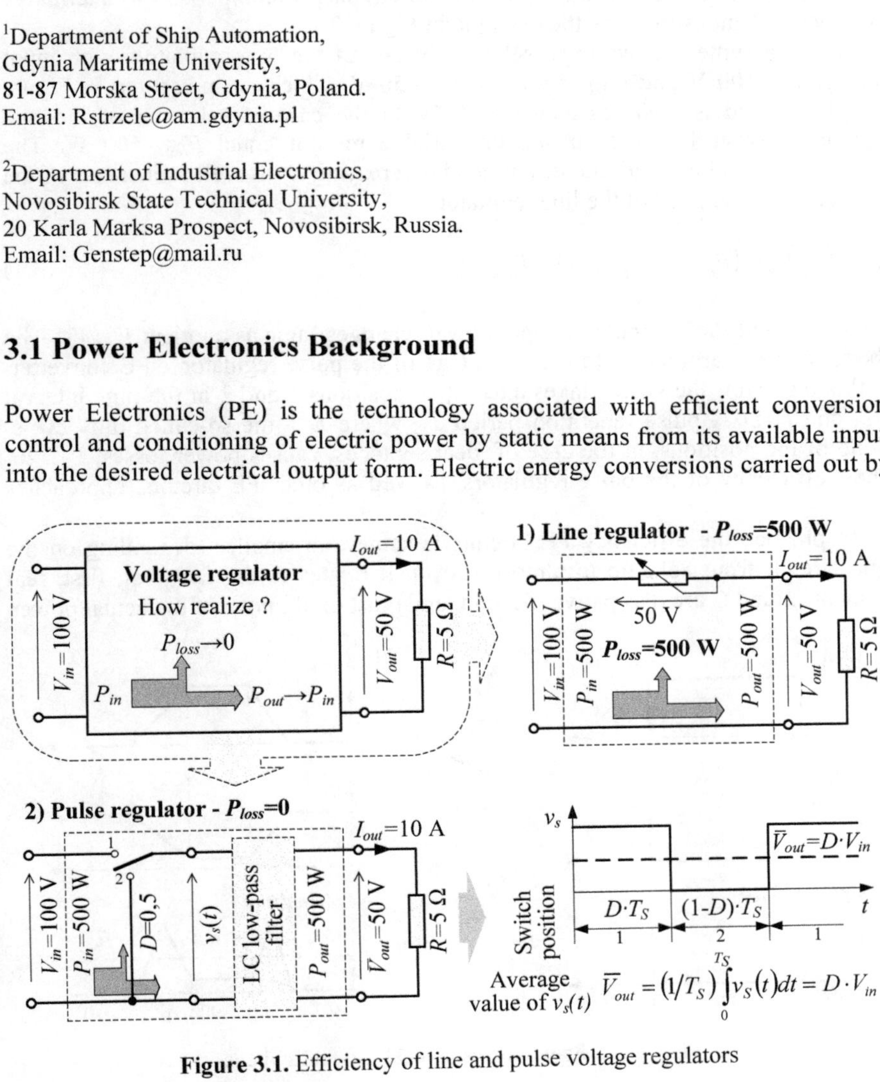

Figure 3.1. Efficiency of line and pulse voltage regulators

PE circuits are diverse and apply to power varying from tens of watts up to hundreds of megawatts. This by particular wide range of power is reflected, among others, in the varying overall dimensions of PE arrangements. Some of the arrangements are hand size, while others require spaces especially designed for their utilization.

Our attention should be drawn to the fact that the definition of PE emphasizes high conversion efficiency. That is connected to the fact that the principle of operation of any PE circuit consists in periodical linking of a power source to the electrical energy consumer – load. In such a case, assuming ideal switches (immediate switching, null resistance in the "*on*" state and infinitesimally high resistance in the "*off*" state) as well as the lack of other dissipation elements, power losses equal zero. High efficiency of the PE circuits, in comparison to alternative solutions is demonstrated by the example in Figure 3.1.

Let us assume that we must select only one of the two regulators with input voltage V_{in}=100 V and output voltage V_{out}=50 V, which are presented in Figure 3.1. Their load is their resistance R=5 Ω. In the case of selection of the line regulator, power losses occurring on variable resistor equal P_{loss}=500 W. The losses increase along with increasing load (decreasing R) as well as increase of the voltage V_{in}. Efficiency of the line regulator

$$\eta = P_{out}/P_{in} = (P_{in} - P_{loss})/P_{in} = 1 - P_{loss}/P_{in} \tag{3.1}$$

where P_{out} and P_{in} – output and input power, changes same as quotient V_{out}/V_{in}. The above provides arguments to the advantage of the pulse regulator – PE converter. In this regulator, the switch takes alternative positions 1 and 2 in the time interval DT_S and $(1\text{-}D)T_S$ pulse repetition period T_S, where D – the so-called duty cycle. None of the positions, in the case of ideal switches, causes power losses, P_{loss}=0. Thus, efficiency of the pulse regulators, as well as other PE circuits, approaches 100%.

In practice the efficiency PE circuits is somewhat smaller (depending on the type circuit, from 85% up to almost 100%). It relates to the fact that, first, real elements L and C are dissipative. Second, and most of all, none of the actual power

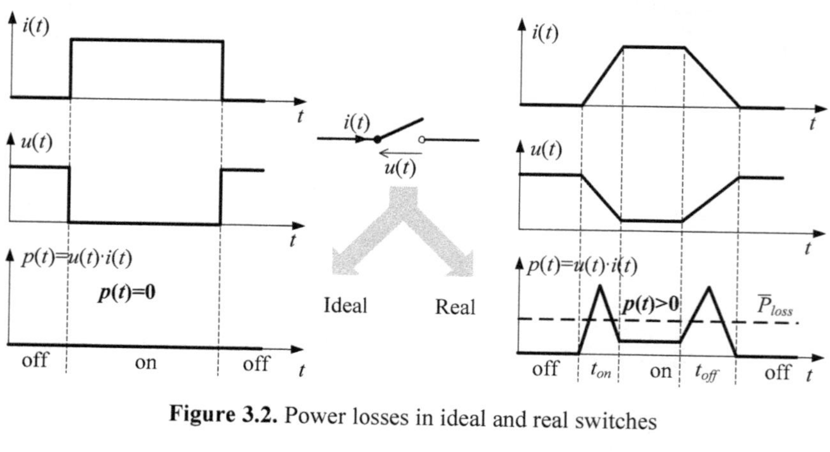

Figure 3.2. Power losses in ideal and real switches

switches switch over immediately, and its resistance in the "*on*" state is positive. That causes extra power losses. Figure 3.2 provides illustrative but simplified demonstration of the process in which the power losses arise.

3.1.1 Historical Perspective

Historical background of PE backs to the end of the nineteenth century, when in 1896 Karol Pollak, honorary doctor of the Warsaw University of Technology, was issued a German patent (DRP 96564) for an electric aluminum rectifier. Description of the patent also included a scheme of one-phase, full-wave rectifier, known today as Graetz bridge. The description was published in the "*Elektronische Zeitung*" no 25 from 1897, with notation from the editor that at that time professor L. Graetz was working on rectifiers of similar principle of operation. However, the solution of Prof. Graetz was published a year and a half after the patent for Dr. K. Pollak had been issued. Other, successive inventions that are of importance to power electronics and therefore we should be aware of are:

1902 – Mercury-arc rectifier (P. Cooper Hewitt);
1903 – Phase angle control (PH. Thomas);
1904 – Vacuum diode (JA. Fleming);
1906 – Triode (L. De Forest);
1908 – Iron vessel rectifier (B. Schäfer);
1912 – Megamp (E. Alexanderson);
1912 – Power rectifier. Sub-synchron cascade one-phase/66 kW, three-phase /300 kW (B. Schäfer);
1922 – Cycloconverter (M. Meyer/LA. Hazeltine);
1923 – Pooled cathode thyratron (I. Langmuir);
1924 – Chopper principle (A. Burnstein);
1925 – Parallel inverter commutation (DC. Prince);
1925 – Field-effect transistor theoretical development (JE. Lilienfeld);
1926 – Hot cathode thyratron (AW. Hull);
1928 – Practical grid-controlled mercury-arc rectifier (I. Langmuir, DC. Prince);
1929 – Thyratron controlled rectifier (A.W Hull);
1931 – Ignitron (J. Slepian);
1931 – Cycloconverter for railways (M. Schenkel, I. von Issendorf);
1932 – Mercury-arc rectifier for wattles power compensation (M. Schenkel 1932);
1932 – First HVDC transmision system (VM. Stör);
1934 – Thyratron motor built and tested (E. Alexanderson);
1935 – HVDC transmission system 287 kV between Mechanicville and Shenectady, NY, USA;
1942 – Frequency changers 20 MW, 25/60 Hz;
1947 – Point contact transistor (J. Bardeen, WH. Brattain, WB. Shockley);
1951 – Junction transistor (WB. Shockley);
1953 – Developed of the germanium power diode 100 A.

The year 1957, which is associated with the development of semiconductor technology, was also adopted as the beginning of modern PE. At that time, Bell Laboratory developed the first p-n-p-n switches – the thyristor or Silicon Controlled Rectifier (SCR). However, the idea of the SCR had been described for the first time by WB. Shockley in 1950. It was referred to as a bipolar transistor with a p-n hook-collector. The operation mechanism of the thyristor was further analyzed in 1952 by JJ. Ebers. In 1956 JL. Moll investigated the switching mechanism of the typical thyristor. Development continued and more was learned about the device such that the first SCR became available in the early 1960s and started gaining a significant level of popularity for power switching. To that day many more innovative and much improved power semiconductor switches were developed [1]. Because of the stimulation of new technical solutions and applications the following arose:

- Triode for Alternating Current (TRIAC) thyristor – developed 1964;
- Bipolar Junction Transistor (BJT) 500 V, 20 A – developed 1970;
- Power Metal-oxide Semiconductor Field-effect Transistor (MOSFET) 100 V, 25 A – developed 1978;
- High power GTO thyristor 2500 V, 1000 A – developed 1981;
- Insulated-gate Bipolar Transistor (IGBT) – developed 1983;
- Intelligent Power Module (IPM) – developed 1990;
- Integrated Gate Commutated Thyristor (IGCT) /emitter turn-off (ETO) thyristor – developed 1997;
- Reverse blocking IGBT (RBIGBT) – developed 2000;
- Matrix converter power module (ECONOMAC) – developed 2001.

However, the present stage of PE development not only results from progress in research on power semiconductors switches [2]. These switches are mostly, and at the same time only, the muscle of PE systems. Also significant are achievements in other related research areas, most of all micro-electronics, control theory and informatics [3]. Without development of these areas we would not be able to equip modern PE arrangements with "*brain and nerves*". All of these areas are interdependent, which is seen in particular on the example of a microprocessor [4]. The application of microprocessors allowed production of practical implementation complex control algorithms, while at the same time stimulating their development. The microprocessor also had significant impact on progress in construction, actual monitoring, diagnostics and remote control of PE systems. Altogether it influenced development of several new technological disciplines [5].

From day to day, changes also occurred in electrical power engineering. The possible place of the PE in the flow of electrical energy from producer to consumer is illustrated in Figure 3.3. Nowadays compensators have become more and more popular on AC transmission and distribution lines as well as feeders, power quality conditioners and power flow controllers [6–11]. It is also unacceptable rationally to apply many renewable energy sources as well as to develop local/distribution generation without PE arrangements [12–15]. The same relates to DC distribution systems and energy storage systems.

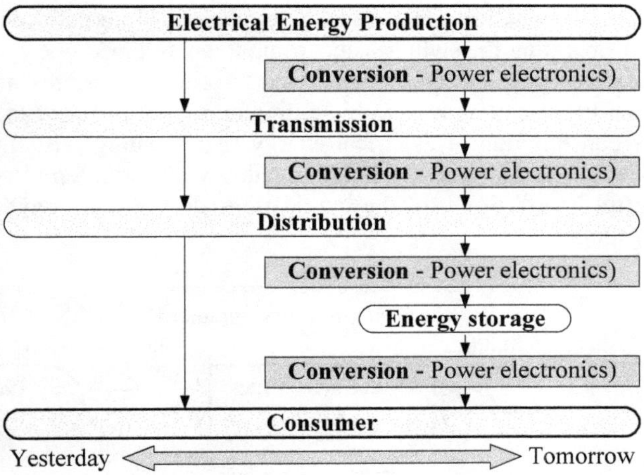

Figure 3.3. Power electronics place in the electrical power engineering

As we see, PE as an area of electrical engineering studies continues to develop intensively. Often, such development is described as a "*quiet revolution*". Utilized in all areas of application of electrical energy, modern PE is a research field of interdisciplinary character (Figure 3.4). It is referred to as industrial electronics, and combines multiple diverse technological disciplines [16, 17].

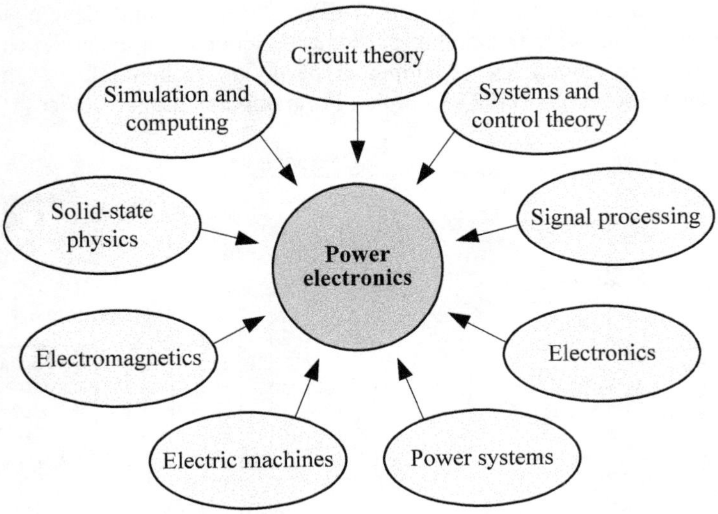

Figure 3.4. Interdisciplinary nature of power electronics

3.1.2 Generic Power Electronics Arrangements

Conventional PE system usually consists of functional modules, delineated as in Figure 3.5. The PE circuit is the central module, and is constructed with application of semiconductor switches. The second module – internal controller – is responsible for operating the switches according to an assumed operation algorithm and on the basis of physical quantities (most by electric currents and voltages), measured in the PE circuit as well as the output and input PE circuits. Supervisory control of the consumption of electrical energy (*e.g.*, heating), usually assured by external controllers, is nowadays realized together with the internal controllers on the same control board. Some of the applications do not even require additional external controllers.

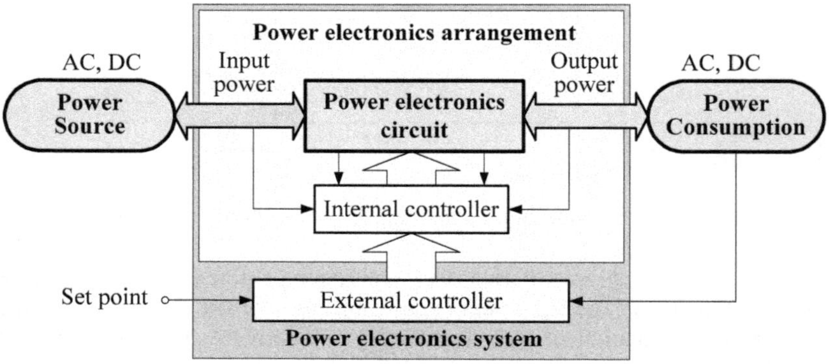

Figure 3.5. Block diagram of a power electronics system

Today PE arrangements (Figure 3.6) differ from solutions developed 10–15 years ago by means of realization of particular functions. For example, changes in control layers resulted from development of digital technologies, in particular Digital Signal Processors (DSP), Complex Programmable Logic Devices (CPLD)

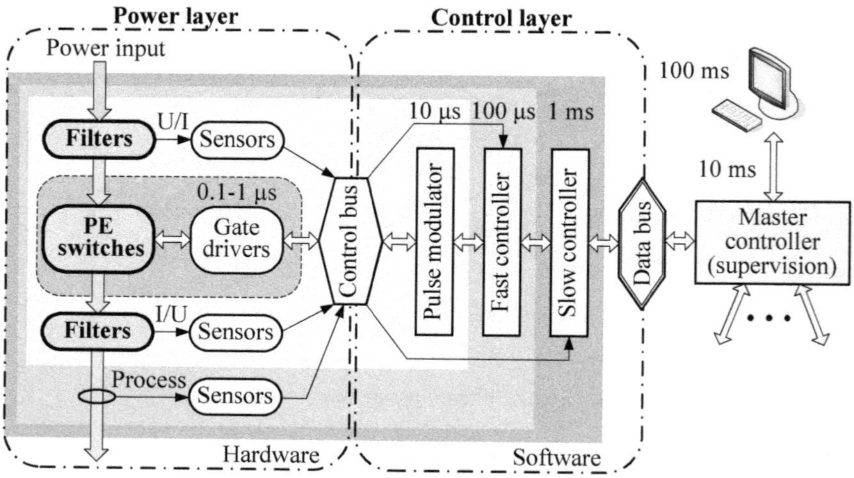

Figure 3.6. Block diagram of a modern power electronics arrangements

and Application Specific Integrated Circuit (ASIC). Here software application is dominating. The majority of the functions (*e.g.*, number of safety devices) that used to be realized by discrete components has been moved to the software level. However, discrete components remain at the power layer, which still requires hardware application. Here, we observe integration or packaging of components, for example, in the form of ready modules [18–21]. Examples of such ready blocks are Power Electronics Building Blocks (PEBB), being constructional closed power layers, as well as block EconoMAC [22–24].

In terms of their functionality, power electronics circuits can be divided into two main groups: contactless switches and converters (Figure 3.7). To the first group belong all modern protection and reconfiguration devices, such as static current limiters, static current breakers, and static transfer switches. Application of contactless switches, when compared to contactors, can be characterized by many advantages and especially by very short operating time, high permitted frequency of switch-overs, long lifetime and lack of electric arc while switching them off. This group is less numerous and is not considered further below.

The second, more numerous group, PE circuits, applies to conversion of the form alternating current into direct current and inversely as well as electrical energy parameters (value of voltage/electric current, frequency, number of phases, reactive power *etc.*). The group can be further divided into basic types of PE converters, which is illustrated in Figure 3.7. AC/AC converters can be realized as single-stage (AC regulators, direct frequency converters) or two-stage through DC link. Similarly, DC/DC converters can be realized. However, rectifiers and inverters are characterized by only a single-stage energy conversion.

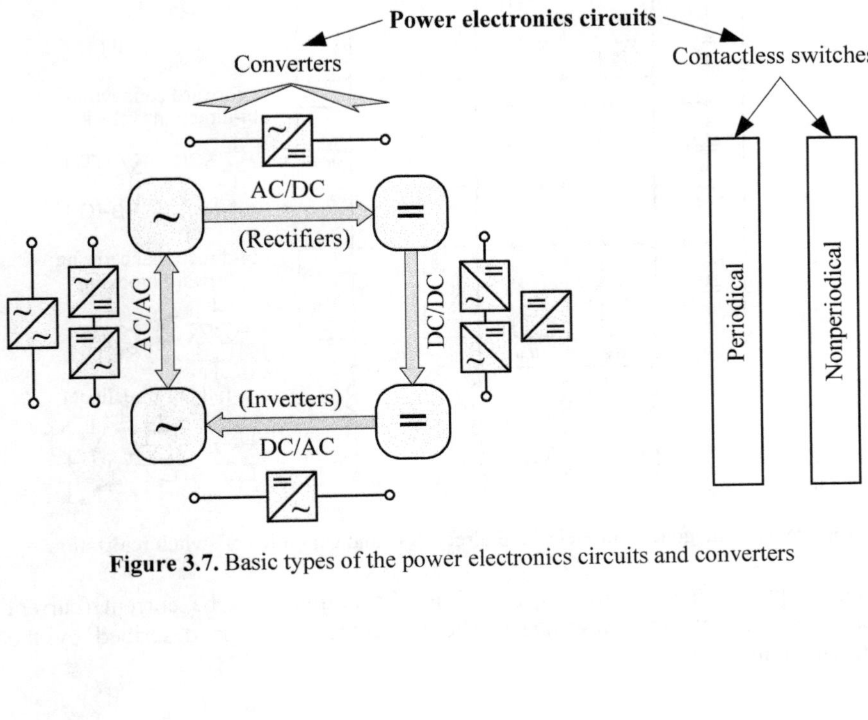

Figure 3.7. Basic types of the power electronics circuits and converters

3.1.3 Switching and Continuous Models of Converters

PE Converters (PEC), independently of their function, construction details and application, can also be divided into two general classes:

- Direct converters – in which main reactive elements are connected only to input or output terminals of the converter and can be considered as part of the source or the load. Rectifiers and voltage inverters with LC filters are an example of such direct converters;
- Indirect converters – including main reactive elements inside their structure. Because they usually have very few elements, indirect converters are mostly analyzed as connections of direct converters with reactive elements among them.

The division, in most cases is consistent with either single-stage or two-stage realization of the PEC.

Figure 3.8 presents the general model of direct PEC in the form of a switch matrix with "*N*" inputs and "*m*" outputs as well as examples of realization of the switches, determining specific characteristics of PEC. Although inputs as well as outputs are changeable, they must remain of different character, *i.e.,* in case of

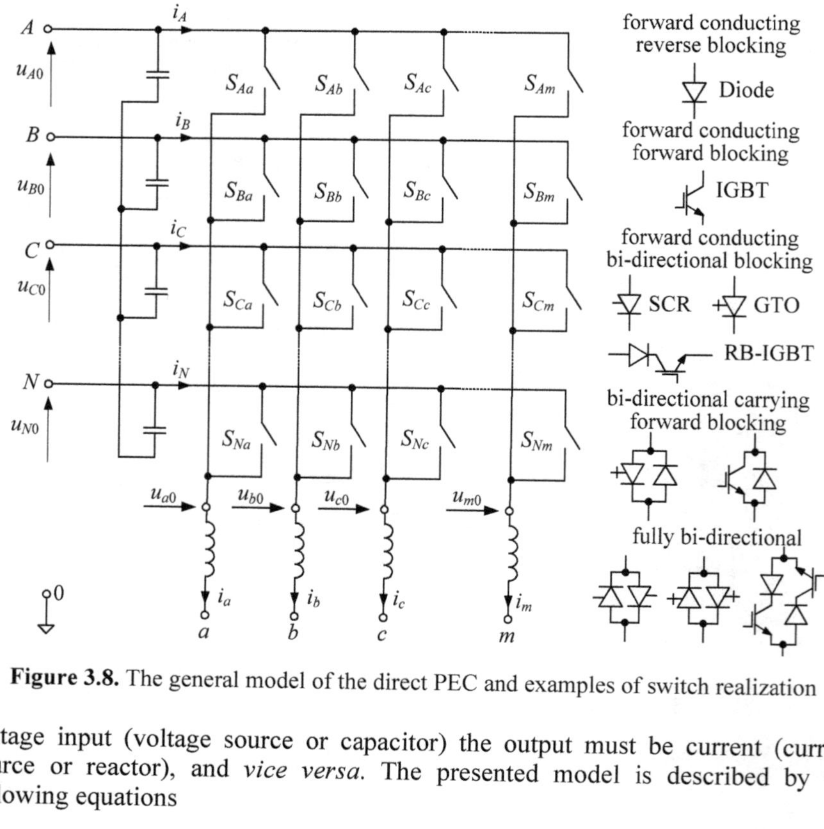

Figure 3.8. The general model of the direct PEC and examples of switch realization

voltage input (voltage source or capacitor) the output must be current (current source or reactor), and *vice versa*. The presented model is described by the following equations

$$\begin{bmatrix} u_{a0} \\ u_{b0} \\ u_{c0} \\ \vdots \\ u_{m0} \end{bmatrix} = \underbrace{\begin{bmatrix} S_{Aa} & S_{Ba} & S_{Ca} & \cdots & S_{Na} \\ S_{Ab} & S_{Bb} & S_{Cb} & \cdots & S_{Nb} \\ S_{Ac} & S_{Bc} & S_{Cc} & \cdots & S_{Nc} \\ \vdots & \vdots & \vdots & \vdots & \vdots \\ S_{Am} & S_{Bm} & S_{Cm} & \cdots & S_{Nm} \end{bmatrix}}_{\|M\|} \times \begin{bmatrix} u_{A0} \\ u_{B0} \\ u_{C0} \\ \vdots \\ u_{N0} \end{bmatrix} \tag{3.2a}$$

$$\begin{bmatrix} i_A \\ i_B \\ i_C \\ \vdots \\ i_N \end{bmatrix} = \underbrace{\begin{bmatrix} S_{Aa} & S_{Ab} & S_{Ac} & \cdots & S_{Am} \\ S_{Ba} & S_{Bb} & S_{Bc} & \cdots & S_{Bm} \\ S_{Ca} & S_{Cb} & S_{Cc} & \cdots & S_{Cm} \\ \vdots & \vdots & \vdots & \vdots & \vdots \\ S_{Na} & S_{Nb} & S_{Nc} & \cdots & S_{Nm} \end{bmatrix}}_{\|M\|^T} \times \begin{bmatrix} i_a \\ i_b \\ i_c \\ \vdots \\ i_m \end{bmatrix} \tag{3.2b}$$

where $|M|$ – connection matrix; $|M|^{\mathrm{T}}$ – transpose of a matrix $|M|$; S_{ij} – state of the switch S_{ij}, where if the switch is "*on*" then $S_{ij}=1$, and if the switch is "*off*" then $S_{ij}=0$, $i=A,B,\ldots, N$ and $j=a,b,\ldots m$.

Switch states S_{ij} can only take values 0 or 1, depending on time. In such a case the natural method to shape output voltage $[u_{a0}, u_{b0}, u_{c0},\ldots, u_{m0}]$ and input currents $[i_A, i_B, i_C,\ldots, i_N]$ in direct converters is pulse modulation. The applied modulation algorithm includes practical limitations [25, 26]. In particular, states of all switches, at any given moment, cannot result in short-circuit or overvoltage. For example, in the presented direct converters model (Figure 3.7) with input voltage and output current, the states S_{ij} of all switches must meet the following requirements

$$\sum_{i=A}^{N} S_{ia} = \sum_{i=A}^{N} S_{ib} = \sum_{i=A}^{N} S_{ic} = \cdots = \sum_{i=A}^{N} S_{im} = 1, \qquad \sum_{i=A}^{N}\sum_{j=a}^{m} S_{ji} = m \tag{3.3}$$

The first condition (Equation 3.3) should be understood as the condition where one switch can be connected to one output only – otherwise, input short-circuit occurs. Meeting the second condition (Equation 3.3) ensures the direction of the output current flow – the number of additional switches must always be equal to the number of outputs. On this basis, the direct converters analysis is made together with synthesis of their algorithms. The analysis of general characteristics of direct converters can also be carried out in a simplified way. It is assumed that the relative time to connect a switch S_{ij} is equal to the instantaneous value of the modulating continuous function $d_{ji}(t)$, such thet $0\leq d_{ji}(t)\leq 1$ and satisfying Equation 3.3, *i.e.*,

$$\sum_{i=A}^{N} d_{ia}(t) = \sum_{i=A}^{N} d_{ib}(t) = \sum_{i=A}^{N} d_{ic}(t) = \cdots = \sum_{i=A}^{N} d_{im}(t) = 1, \qquad \sum_{i=A}^{N}\sum_{j=a}^{m} d_{ji}(t) = m \tag{3.4}$$

In this case, instead of the switch matrix (Figure 3.7) we obtain a continuous model of direct converters, where each switch S_{ij} is exchanged by an ideal transformer with a transformation ratio equaling to the modulating function $d_{ji}(t)$. An example of such a model for 3×3 direct converters with fully bi-directional turn-off switches is presented in Figure 3.9. The same figure shows the fundamental method to determine the state of the matrix switches on the basis of modulating functions. The method is discussed with the example of the simplest Pulse-width Modulation (PWM) and in relation to the output "*a*" of a direct converter.

Additionally, Figure 3.10 presents the basic scheme of a three-phase voltage source inverter supplied by the voltage U_{DC} and the corresponding continuous model. If taking into account this model, the general characteristics of three-phase VSI can be determined from the equations

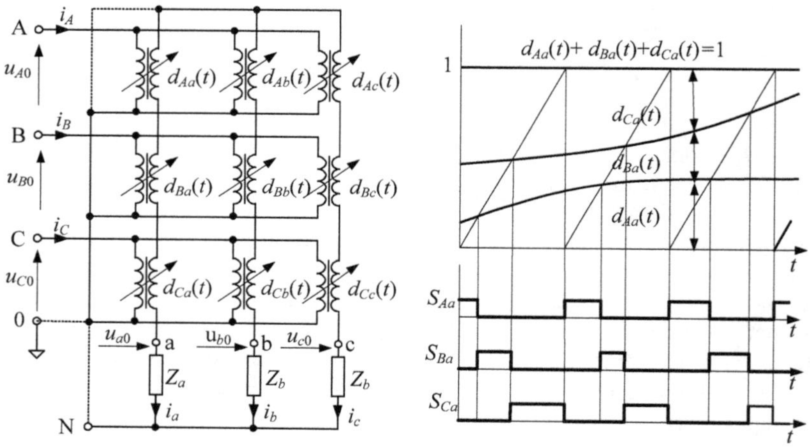

Figure 3.9. The continuous model of the 3×3 direct converters

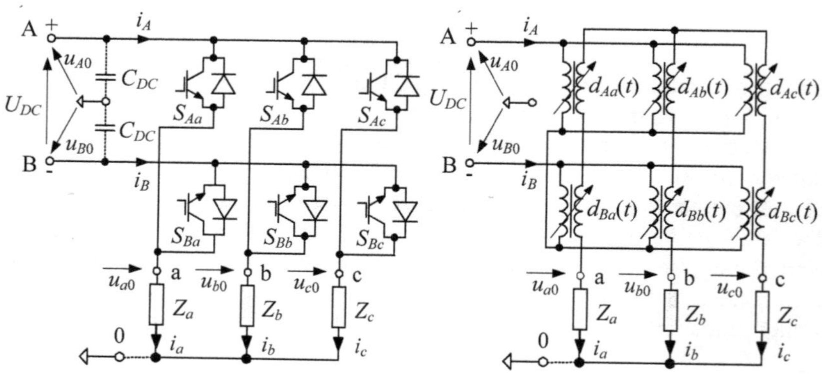

a b

Figure 3.10. a Basic scheme of the three-phase VSI. **b** Its continuous model

$$\begin{bmatrix} u_{a0} \\ u_{b0} \\ u_{c0} \end{bmatrix} = \begin{bmatrix} d_{Aa}(t) & d_{Ba}(t) \\ d_{Ab}(t) & d_{Bb}(t) \\ d_{Ac}(t) & d_{Bc}(t) \end{bmatrix} \times \begin{bmatrix} u_{A0} \\ u_{B0} \end{bmatrix} \tag{3.5}$$

$$\begin{bmatrix} i_A \\ i_B \end{bmatrix} = \begin{bmatrix} d_{Aa}(t) & d_{Ab}(t) & d_{Ac}(t) \\ d_{Ba}(t) & d_{Bb}(t) & d_{Bc}(t) \end{bmatrix} \times \begin{bmatrix} i_a \\ i_b \\ i_c \end{bmatrix} \tag{3.6}$$

where $u_{A0}=-u_{B0}=U_{DC}/2$. Only sinusoidal modulating functions are further considered

$$\begin{bmatrix} d_{Aa}(t) \\ d_{Ab}(t) \\ d_{Ac}(t) \end{bmatrix} = \frac{1}{2}\begin{bmatrix} 1+A\sin(\omega t+0) \\ 1+A\sin(\omega t-2\pi/3) \\ 1+A\sin(\omega t-4\pi/3) \end{bmatrix} \tag{3.7}$$

$$\begin{bmatrix} d_{Ba}(t) \\ d_{Bb}(t) \\ d_{Bc}(t) \end{bmatrix} = \frac{1}{2}\begin{bmatrix} 1-A\sin(\omega t+0) \\ 1-A\sin(\omega t-2\pi/3) \\ 1-A\sin(\omega t-4\pi/3) \end{bmatrix} \tag{3.8}$$

where $\omega=2\pi f$; f – output fundamental frequency; A – modulation factor ($0\le A\le 1$).

From Equations 3.5–3.8 it results that, for such modulating functions, the amplitude of the sinusoidal output voltage cannot exceed $U_{DC}/2$. It should be emphasized that the value is not any boundary value. In the case of vector modulation the amplitude of the sinusoidal voltage can be increased by about 15%. However, if output-voltage overmodulation is allowable (it is even advisable), then taking into account the boundary cases, the amplitude of a component of fundamental frequency may even reach the value $2U_{DC}/\pi$ [26, 27].

On the basis of Equations 3.6–3.8 it is easy to show that the input currents of a three-phase VSI with sinusoidal output voltage are

$$i_A = i_{DC} + (1/2)\cdot i_0\,, \quad i_B = -i_{DC} + (1/2)\cdot i_0\,, \quad i_0 = i_a + i_b + i_c \tag{3.9}$$

where

$$i_{DC} = (A/2)\cdot\left[i_a\cdot\sin(\omega t) + i_b\cdot\sin(\omega t-2\pi/3) + i_c\cdot\sin(\omega t-4\pi/3)\right]$$

The above relations indicate the unique characteristic of a three-phase VSI, which is its ability to generate reactive currents, theoretically without application of any input energy storage such as capacitors C_{DC}. This characteristic, resulting from the lack of reactive power in DC circuits, is used, for example, in D-STATCOM systems [9, 10].

Continuous (average) models, are very supportive when one wants to evaluate usability of PEC in specific application. By avoiding impact of switching process, problems with stiff differential equations are avoided. At the same time, in the case of switching frequency exceeding 5 kHz, simulation error usually does not exceed 5%. Therefore we can successfully focus our attention on functional characteristics of tested application. It is worth noting that application of average models of the PEC usually includes controlled voltage sources and electric current sources, instead of transformers of adjustable ratio of transformation. This results, mainly, from the approach used in the circuits' theory. One example is model three-phase VSI presented in Figure 3.11, corresponding with the model in Figure 3.10b.

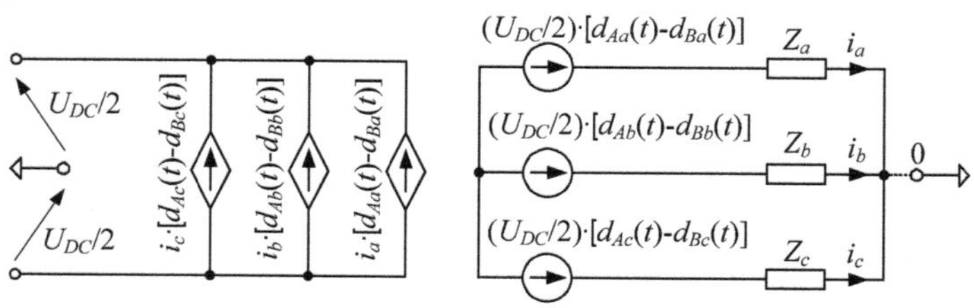

Figure 3.11. Continuous model of the three-phase VSI with of controllable sources

3.2 High Technology of Converters

The conventional circuit elements applied to PE arrangements can be assigned to one of the classes resistive elements, capacitive elements, magnetic devices, semiconductor devices operated in the linear mode, and semiconductor devices operated in the switched mode (Figure 3.12). At the same time, different classes vary in priorities of application. In the case of controllers one usually avoids applying magnetic devices because of relatively large overall dimensions and integration difficulties. Whereas in PEC, with respect to power losses, semiconductor devices operated in the linear mode are not applied. Moreover, application of resistors should also be limited and replaced with other possibilities. Nowadays, resistors remain in use in cases of dissipative snubbers [28–30] as well as in starting systems PEC, *e.g.*, for initial charging of capacitors in DC circuits. As

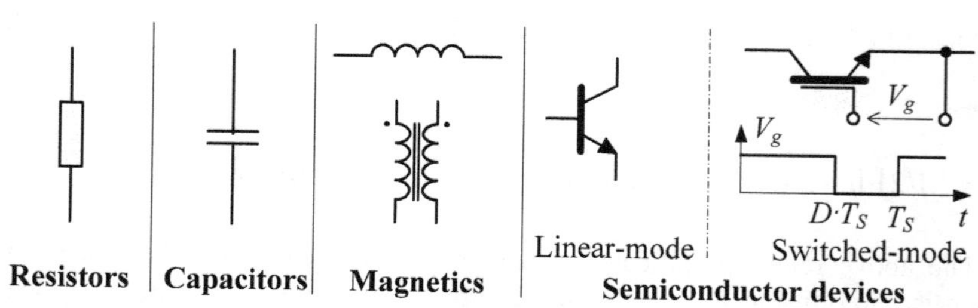

Figure 3.12. Different conventional circuit elements of the PE arrangements

supplementation of the above discussion, Figure 3.13 presents usual quantities, expressed in percentage, of the components in weight and volume of the PE arrangements with medium power. As we can see the critical elements are capacitors, semiconductor switches and magnetic devices, and secondary cooling systems and bus work.

3.2.1 State-of-the-Art of Power Semiconductors Switches

Recent technology advances in PE have been made by improvements in controllable power semiconductor switches. Figure 3.14 presents probably the most important power semiconductors switches on the market today and their power range [31].

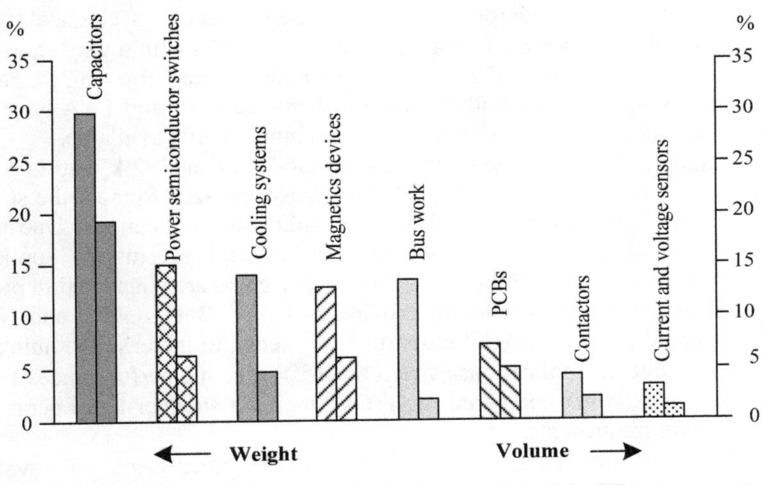

Figure 3.13. Typical components in the construction of the PE arrangements

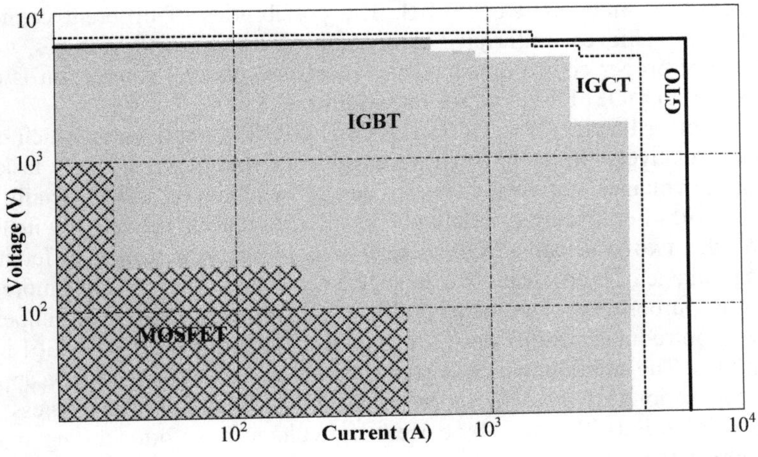

Figure 3.14. Power range of commercially available power semiconductors

MOSFETs and IGBTs have replaced BJT almost completely. A remarkable development in MOSFETs took place during the last few years. Today, available MOSFETs achieve maximum switch power up to about 100 kVA.

Conventional GTOs are available with a maximum device voltage of 6 kV in industrial converters. High state current density, high blocking voltages, as well as the possibility to integrate an inverse diode are considered significant advantages of these devices. However, the requirement of bulky and expensive snubber circuits as well as the complex gate drives leads to replacement of GTOs by IGCTs. Just like GTOs, IGCTs are offered only as press-pack devices. A symmetrical IGCT, for example, is offered by Mitsubishi with maximum device voltage of 6.5 kV. It is technically possible today to increase the blocking voltage of IGCTs as well as the inverse diodes to 10 kV. Due to the thyristor latching structure, the GTO offers lower conduction losses than the IGBT of the same voltage class. In order to improve switching performance of classical GTOs, researchers developed Gate-commutated Thyristors (GCTs) with a very short turn-off delay (about 1.5 μs) [32]. Although new asymmetric GCT devices characterized by up to 10 kV with peak controllable currents and 1 kA have been developed, only devices with 6 kV and 6 kA are commercially available.

Also commercially distributed nowadays are IGBTs from 1.2 kV up to 6.5 kV with DC current ratings up to 3 kA [33]. They are optimized to meet the specific requirements of high-power motor drives for industrial applications. Due to the complex and expensive structure of a pres-pack, IGBT are mainly applied to module packages. In IGBT modules, multiple IGBT chips are connected in parallel and bonded to ceramic substrates to provide isolation. Both IGCTs and IGBTs have the potential to lower overall costs of the systems, to increase the number of economically valuable applications as well as to improve the performance of high-power converters, (when compared to GTOs) due to a snubber-less operation at higher switching frequencies.

In the case of insufficient voltage-current parameters of available semiconductor devices in a given application, it is possible to use their parallel and series connections [30, 34, 35]. In a similar manner it is also possible to connect ready converter modules, *e.g.*, PEBB. The possibilities of different connections depend upon achieved uniformity of division of currents and voltages, with the assistance of proper control mechanism. Therefore, parallel connection is usually applied only to MOSFETs or converter modules.

On the other hand, GTOs, GCTs/IGTCs and IGBTs are devices which connect in series relatively well. In such arrangements the most difficult task is to compensate voltages in dynamic states – during switching on and especially during switching off. For example, relatively small differences between switching-off time of the two transistors IGBT equal Δt_{off}=40 ns can cause differences in dividing voltage of more than 50%. In such a case connecting in series more than *h* devices is purposeless. The same problem can be attenuated in a simper way, mainly by introducing additional Gate Balance Transformers (GTC) [36] into the gate circuits. The solution, which is presented in Figure 3.15, can also be applied to series connections of three or more transistors IGBT. The effectiveness of the proposed solution is illustrated by current and voltage waveforms during switching off (Figure 3.15).

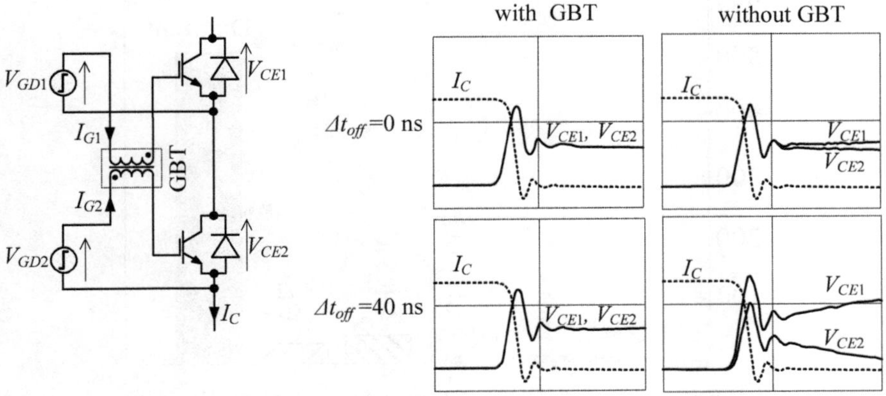

Figure 3.15. IGBTs series connection with gate balance transformer

IGBTs, GTOs, GCTs and IGCTs, as well as Emitter Turn-off (ETO) thyristors, are being improved continuously [37, 38]. Research is mainly focused on increase of permissible voltages, currents and switching frequency, and decrease in conduction and switching losses. Hopes are placed in ETO thyristors, which are distinguished by two gate circuits (Figure 3.16). Because of this characteristic, the turn-off time of the ETO thyristor is shorter than GTOs' or IGCTs'.

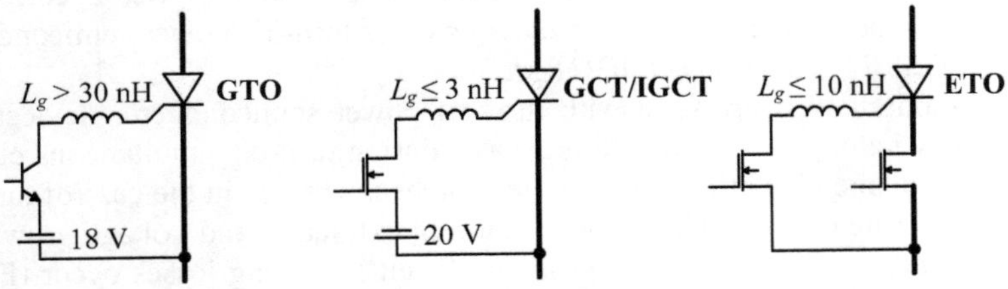

Figure 3.16. Gate drivers of conventional GTOs, GCTs/IGCTs, and ETOs

Research also concentrates on semiconductor devices that would be different to siliceous [39], and which would be characterized by higher voltage breakdowns (Figure 3.17) as well as higher permissible work temperatures. Sooner or later, commercial power-electronic devices based on silicon carbide (SiC), of relatively high power and high mean base voltage [40] should be available on the market. However, today the most important and fundamental semiconductor remains silicon.

3.2.2 Soft-switching *vs* Hard-switching Techniques

At the beginning of modern PE, that is about 10–15 years after the solid state thyristor had been invented, the elementary arrangements were phase-controlled rectifiers, inverters, and cycloconverters that operate on line or load commutation

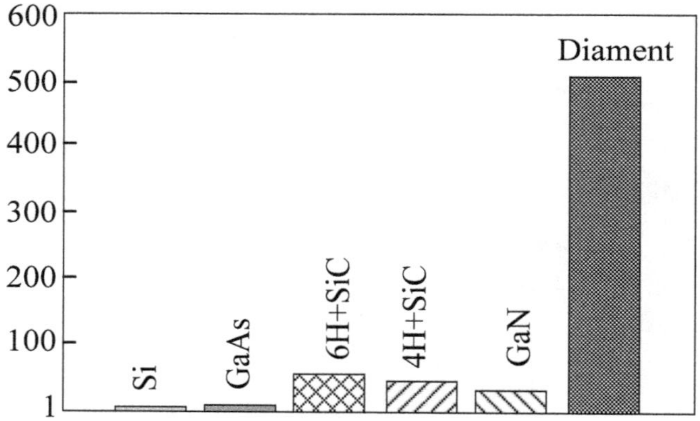

Figure 3.17. Relative breakdown voltage of a p-n junction

principle use soft-switching [41–43]. When an incoming thyristor is turned on, the current is gradually transferred from the outgoing to the incoming device, and then the outgoing device turns off by a segment of reverse voltage. Basically, this is soft switching at zero current for both the incoming and outgoing devices. In fact, the classical force-commutated thyristor inverters could also be defined as soft-switched with the help of auxiliary devices and circuit components [44–46]. Their structures, however, were quite developed, and this fact had negative impact on the overall efficiency of the PE circuits. Force-commutated thyristor converters gradually became obsolete due to the emergence of turn-off power semiconductor devices (MOSFET, IGBT, GTO, IGCT *etc.*).

Many modern PEC are used with turn-off power semiconductor devices that apply hard-switching techniques. In this case, during turn-on simultaneous current growth and voltage extinction occur in the switches, whereas in the case of turn-off the exact opposite occurs – simultaneous current extinction and voltage growth. In both situations, in real power switches, significant switching losses occur (Figure 3.2). For that reason, as well as because of other device stresses and EMI problems [47], the typical PEC switching frequency with application of hard-switching technique is limited to a few tens of kilohertz (depending on the type of power and application PEC).

In order to improve operating conditions of the power devices, in particular in switching processes, the circuits forming the switching trajectory are applied (Figure 3.18). The earliest applied device was dissipative passive snubbers, while later active snubbers with energy recovery were introduced [5, 29, 30]. In addition, supportive LC circuits that realize so-called soft-switching were introduced [48]. This concept consists in utilization of resonant tanks in the converters in order to create oscillatory voltage and/or current waveforms. In such a case, Zero Voltage Switching (ZVS) or Zero Current Switching (ZCS) conditions can be created for the power switches.

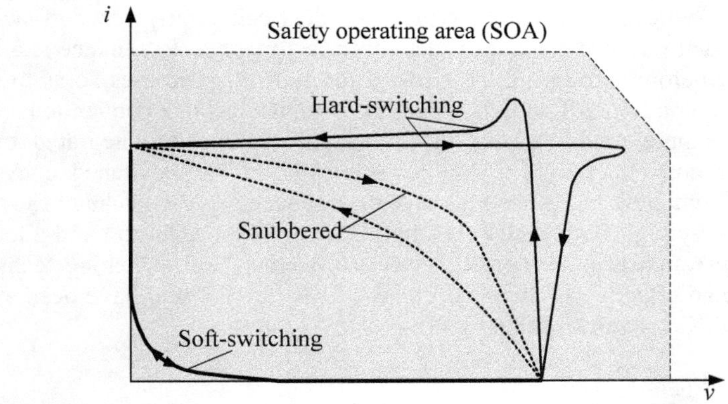

Figure 3.18. Typical switching trajectories of power semiconductor devices

Figure 3.19 shows the typical voltage and current waves at hard turn-on and turn-off of a device in a simple buck converter with and without dissipative snubbers. The turn-on snubber L_1-R_1-D_1 allows decreased maximum value of the transistor current i_T, decreased stress di_T/dt as well as decreased current component i_T caused by reverse current of the diode D_0, limiting turn-on switching losses and transferring them to resistor R_2. However turn-off snubber C_2-R_2-D_2 allows one to

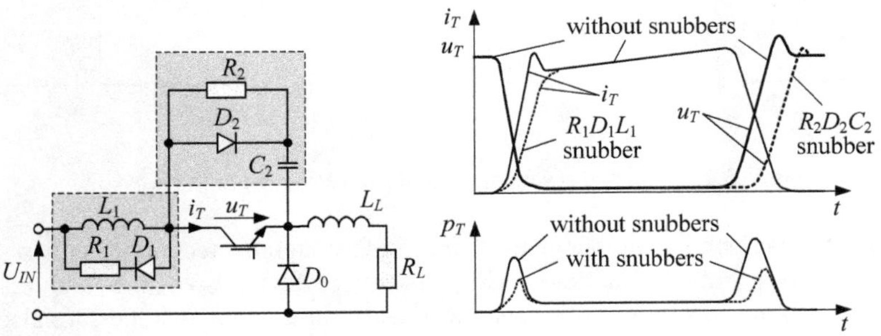

Figure 3.19. Switching waveforms of the converter with and without snubbers

transfer turn-off switching losses from the transistor to the resistor R_2. Therefore it decreases the maximum voltage in the transistor u_T. In such a manner the snubbers produce a more secure switching trajectory of a transistors (Figure 3.18).

Sometimes, in the modern PEC, in particular high power PECs, the number of snubbers is minimized or they are not used at all. This results from the fact that better and better power switches are developed as well as from the pursuit of cost cutting. Obviously semiconductor devices are then more head load and should be over-dimension. Often, however, supportive circuits must be used in order to produce a switching trajectory. In such cases more and more often solutions that allow for soft-switching are utilized [49–53].

Throughout the 1990s, new generations of soft-switched PEC that combine the advantages of conventional hard-switching PWM converters and resonant converters were developed. Unlike typical resonant converters, new soft-switched converters usually utilize resonance in a controlled manner. Resonance is allowed to occur just before and during the turn-on and turn-off processes so as to create ZVS and ZCS conditions. Other than that, they behave just like conventional PWM PEC. With simple modifications, many customized control integrated circuits designed for conventional PEC can be employed for soft-switched converters. Because the switching loss and stress have been reduced, soft-switched PEC can be operated at very high frequencies (reaching even a few megahertz) and allow one to obtain very high packing density (over 10 W/cm^3). Soft-switching techniques also provide an effective solution to suppress EMI [54, 55] and have been applied to different PEC converters [28–30].

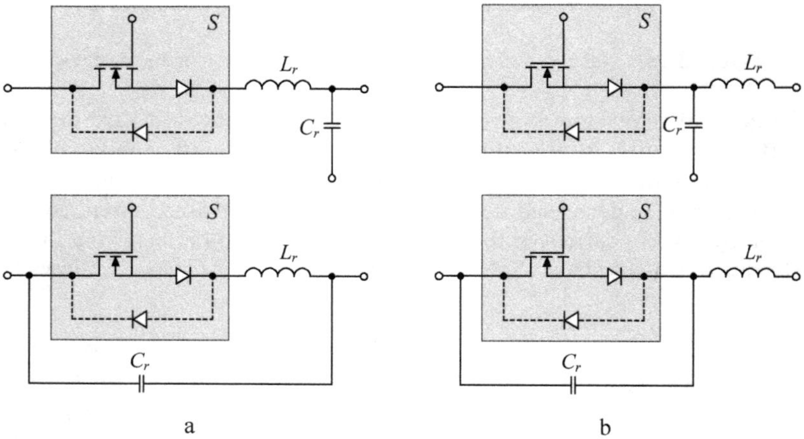

Figure 3.20. Types of resonant switches: **a** zero-current; **b** zero-voltage

The fundamental component used in a soft-switching technique is a resonant switch. It is a sub-circuit comprising a semiconductor switch *S* and resonant elements L_r and C_r. Uni-directional as well as bi-directional switches are also used as switches *S*. In addition, a type of applied switch *S* determines the operation mode of the resonant switch [28, 56]. The basic two types of resonant switches, including Zero-current (ZC) and Zero-voltage (ZV) resonant switches, are shown in Figure 3.20.

In a ZC resonant switch (Figure 3.20a), an inductor L_r is connected in series with a power switch *S* in order to create ZCS conditions. The objective of this type of switch is to shape the switch current waveform during conduction time in order to create a zero-current condition for the switch to turn off. If a uni-directional switch *S* is applied, the switch current is allowed to resonate in the positive half cycle only, that is, to operate in half-wave mode. If a diode is connected in anti-parallel, the switch current can flow in both directions. In this case, the resonant switch can operate in full-wave mode. At turn-on, the switch current will rise slowly from zero. It will then oscillate because of the resonance between L_r and C_r. Finally, the switch can be commutated at the next zero current duration.

In a zero-voltage resonant switch (Figure 3.20b), a capacitor C_r is connected in parallel with the switch S in order to create ZVS conditions. The objective of a ZV switch is to use the resonant circuit to shape the switch voltage waveform during the off time in order to create a zero-voltage condition for the switch to turn on. If the switch S is uni-directional, the voltage of the capacitor C_r can oscillate freely in both positive and negative half-cycle, and the resonant switch can operate in full-wave mode. However, if the switch S is bi-directional (that is when a diode is connected in anti-parallel with the unidirectional switch), the resonant voltage of the capacitor is clamped by the diode to zero during the negative half-cycle. Then the resonant switch will operate in half-wave mode.

This book is not designed to deal with other important aspects of soft-switching in PEC comprehensively. More detailed information about the subject can be found in works referred to in this book.

3.2.3 Construction Arrangement and Cooling Systems

Construction arrangments and dimensions of PECs vary. Distribution of the components and execution of the electric connections significantly influence characteristics of the PECs. Considering reliability, particularly important are electromagnetic screening of the control circuits from power circuits. Therefore, nowadays, connections between the circuits are often realized with fiber optic cable. Reliability of all connections is also very important.

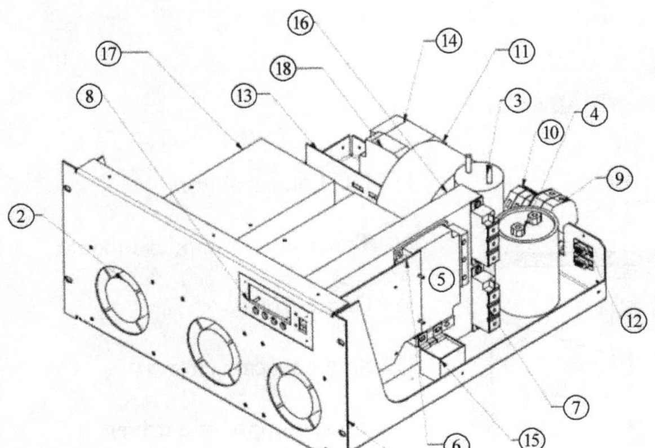

1) Chassis
2) Cooling fan
3) Output filter capacitor
4) DC link capacitor
5) IPM module
6) PCB board
7) Diode bridge
8) LCD display
9) Output terminal
10) Input terminals
11) Output filter choke
12) CANON jack
13) Internal power supply
14) Series contactor
15) Snubbers capacitor
16) IPM heatsink
17) DC/DC converter moduls
18) Time relay

Figure 3.21. Single module 48 V/ 230 V/ 5 kW of a redundant inverters system

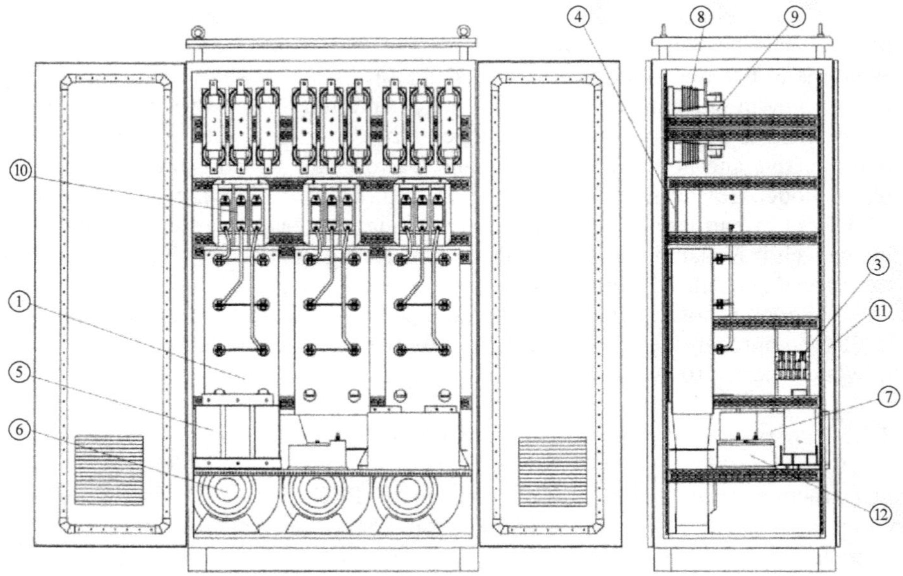

Figure 3.22. Design of an 6 kV / 2 MVA 18-pulse diode rectifier: 1) modules of the three-phase diode rectifier bridge with radiators; 2) power supply of the contactor (invisible); 3) resistors of the auxiliary circuits; 4) starting resistors; 5) transformer of the auxiliary circuits; 6) cooling fans; 7) DC link capacitors; 8) fuse base; 9) fuse; 10) vacuum contactor; 11) cubicle; 12) reactor

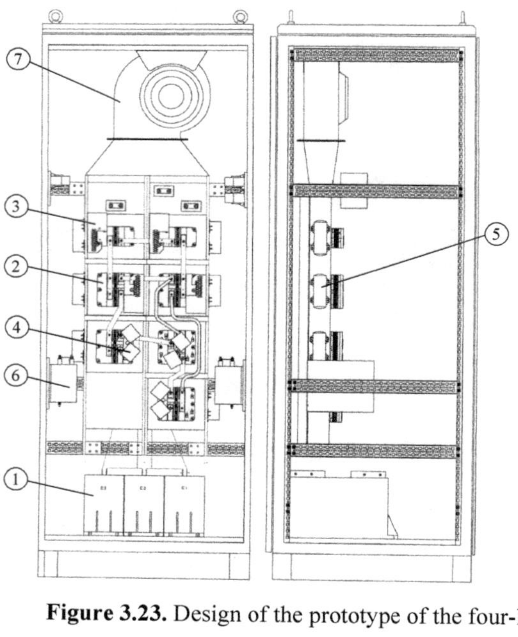

1) DC link capacitors
2) Transistors or diodes module
3) Gates driver
4) Snubbers capacitor
5) Power supply of a driver
6) Auxiliary fuses circuit
7) Cooling fan

Figure 3.23. Design of the prototype of the four-level 6 kV voltage inverter branch

In PECs with power up to about 100 kVA, all components are usually placed in a common box. An example of such a solution could be design of a single modul of a redundant inverters system (Figure 3.21). For high power, particular components or assembly components are placed in a separate cubicle (Figures 3.22 and 3.23).

Relatively often, power semiconductor devices are also offered as complete construction modules – assembly components together with heatsink arrangement. Examples of these modules are shown in Figure 3.24. Similar diode modules were applied, for example, in an 18-pulse rectifier (Figure 3.22). Such a simplifying solution design of the PECs [18, 22, 23, 57] in connection with modern cooling systems [58–60], and modern passive elements [20, 21, 61] can be realized for wide power intervals and for different applications.

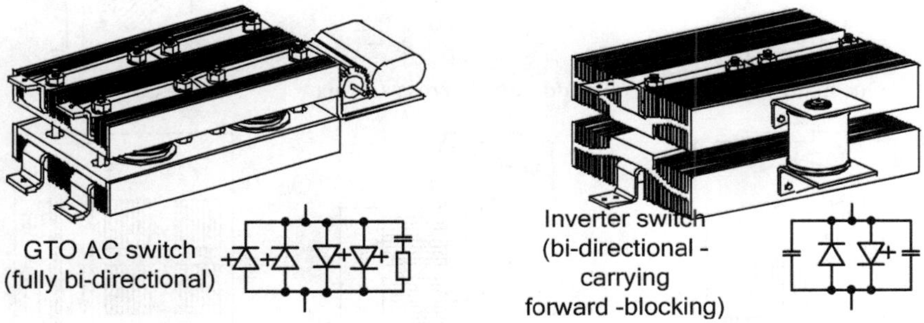

Figure 3.24. Example of a construction of high PE modules

To PE arrangements, air-cooling and liquid-cooling systems – direct and indirect – are applied. During the direct cooling, the medium is in direct contact with package semiconductor devices, heatsink or other electric components. However, with indirect cooling, two cooling media are used, one of which transfers heat from the components to a heat exchanger, second to environment. Coolant can consist of natural circulation or forced circulation (fans, pumps). Sometimes mixed coolant and circuits are also applied.

In low power PECs, air-cooling with natural circulation is almost always used. However, in PECs with power from few to hundreds of kilowatts, it is usual to apply forced air-cooling. The selected solutions of cooling systems of cubicles are presented in Figure 3.25. In all examples, cool air is delivered *via* vents in the bottom of the cubicle or in the lower parts of lateral faces. If PECs are to work under conditions of dustiness, then sealing of the cubicle and application of exchangeable dust filters in the vents are required. Sometimes, in order to achieve the desired goal, indirect cooling with a heat exchanger (HE) is used, for example as presented in Figure 3.25f.

In PECs for very high currents, liquid-cooling systems, for which water is a medium [60], are usually used. At certain, oil is used because of its insulating properties. It should be noted that liquid cooling is also used to carrying away heat from other power components of the PECs. The fundamental problem in liquid-cooling systems is leaking of any connections. Moreover, efficiency of liquid-cooling is much higher if a medium is in a boiling state. Then heat transfer takes place not only by convection but also thanks to the phase transition of liquid into steam. Arrangements which rely on this property are referred to as vapor-cooling systems [58].

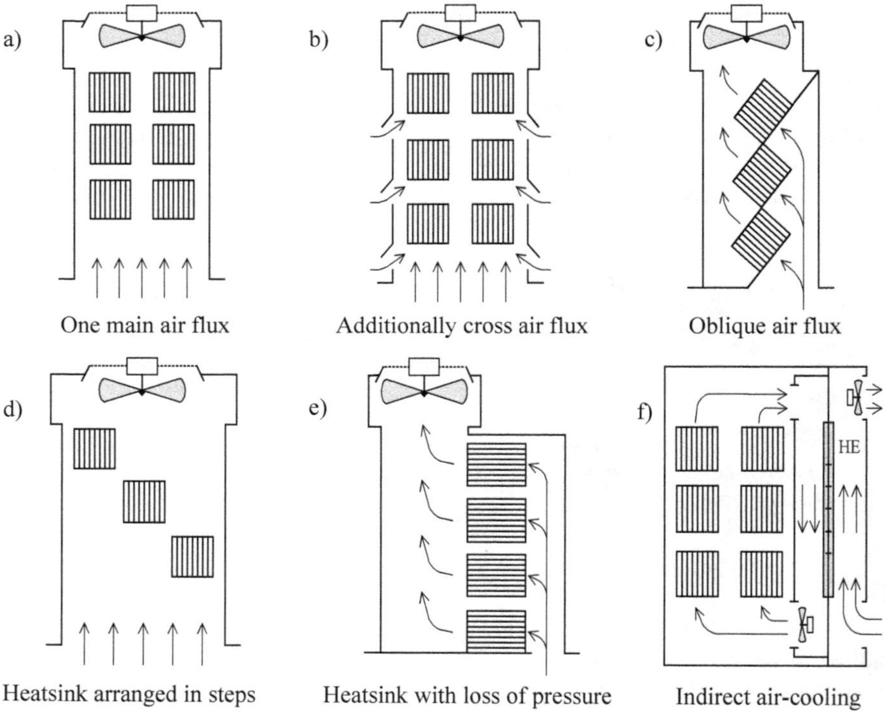

Figure 3.25. Air-cooling systems of cubicles with forced circulation

Variations of vapor-cooling systems are heat pipes (Figure 3.26) [59, 62]. Closed pipe is filled with liquid (water, freon, fluorocarbon *etc.*). On the inner wall are layers of a special material with capillary properties, forming a wick structure.

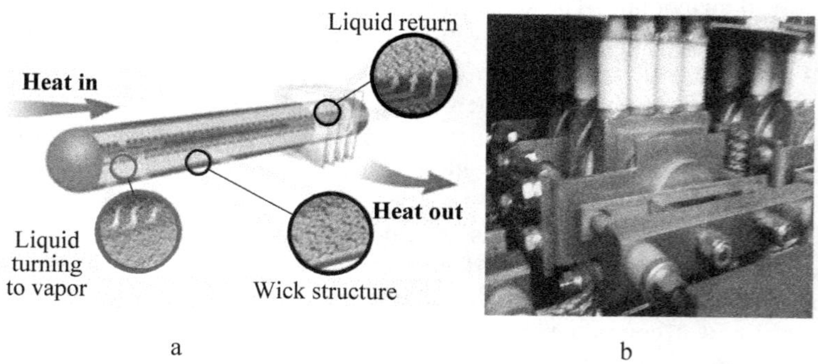

Figure 3.26. **a** Principle of the heat pipe. **b** Example of application in PEC

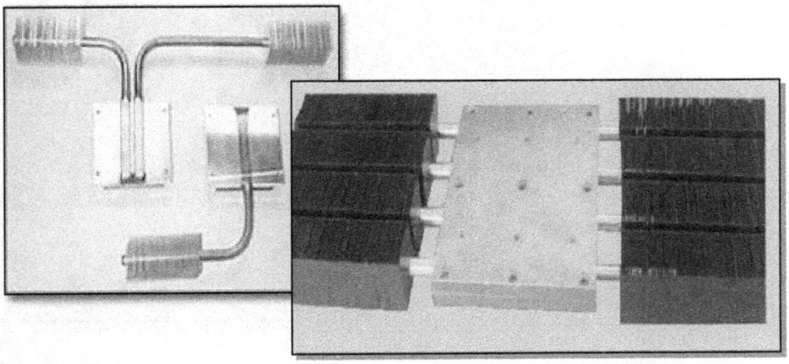

Figure 3.27. Example of the heat pipes assisted heat-sink

The steam formed in the heating sphere is transport to the condensation sphere at low temperature under conditions of pressure difference. In this sphere the steam gives up heat and condenses. Due to capillary forces the liquid comes back through the wick structure to the heating sphere and the cooling cycle is repeated. A significant property of heat pipes is also the option of accommodating almost any arrangement. Thanks to that the constructor can apply a cooling component that enables him to transfer heat to the part from which it is easiest to carry away. One example is the heat pipes presented in Figure 3.27.

3.3 Multi-level Converters

This section briefly discusses selected basic problems of modern multi-level PECs.

3.3.1 Multi-level Converter Concepts

In PECs with PWM of medium/high voltage/power and some specific applications and running conditions, typical solutions (for example three-phase VSI presented in Figure 3.10) are not the most suitable ones. Then too high frequency of the switches in semiconductor devices of high voltage/power (requires a compromise between output-voltage quality and regulation dynamics with application of an output filter), higher voltage stresses, and smaller *dv*/*dt* and EMI problems (without any special countermeasures), and sometimes an insufficient value of the peak voltage in semiconductor devices (for a peak voltage of 6 kV, the recommended voltage is about 3.5 kV) would be the main reasons for a second interest in multi-level PECs in the 1980s–1990s, in particular multi-level VSI [63–69]. Many older solutions would then have limited applications [70–75].

As the main advantages of modern multi-level VSI we can count [76]:

- Increased range of output-voltage amplitude changes;
- Greater accuracy in modeling output voltage and current;
- Ability to decrease transformation ratio and even eliminate the output transformer for medium voltage;

- More easy adaptation to low-voltage energy storage;
- Decreased voltage hazard and current elements (dependent on applied typology);
- Decreased level of common-mode disturbances.

The basic differences regarding conventional two-level VSI and the general principle of wave-forming output multi-level voltage are illustrated in Figures 3.28 and 3.29.

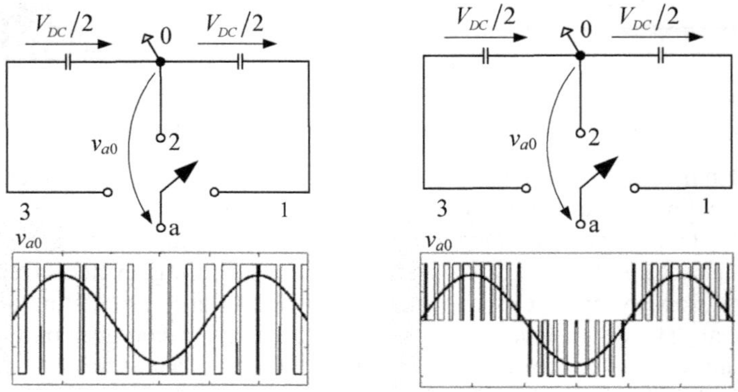

a b

Figure 3.28. Principles of voltage wave-forming in: **a** two-level VSI; **b** three-level VSI

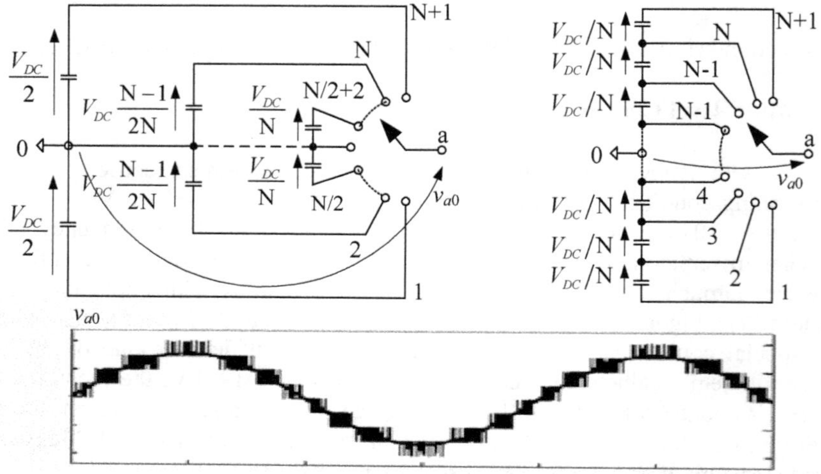

Figure 3.29. General principles of voltage wave-forming in N-level VSI

All known topologies of multi-level VSI in the literature, without isolated (galvanic separated) DC voltage sources, can be synthesized on the basic VSI modules and the multi-layer topology presented for the case of three-layers in Figure 3.30.

The manner of their synthesis is to exclude individual switches properly. However the majority of multi-level VSIs obtained in this manner did not find any applications, either because of the complexity of this typology or because of greater losses. In practice, two special cases of multi-layer topology that are realized are multi-level Diode Clamped Inverters (DCI), with which we also include Neutral Point Clamped (NPC) VSIs, and multi-level Flying Capacitor Inverters (FCI) [76]. Furthermore, also applied are multi-level Cascaded H-bridge Inverters (CHBI), which require, in contrast to the first two, isolated DC voltage sources [69, 76, 77]. The main advantage of CHBIs is the possibility of easy development and independent stabilization of the voltage in the DC circuit for each mode H-bridge. However, none of the three selected topologies of multi-level VSI, presented as one branch in Figure 3.31, has so far gained a leading position.

3.3.1.1 Diode Clamped Multi-level VSI

As a type of VSI configuration, which is important for high-power applications, the diode-clamped inverter provides multiple voltage levels through the connection of the phases to a series bank of capacitors. According to the original invention [64], the concept can be extended to any number of levels by increasing the number of capacitors. Early descriptions of this topology were limited to three levels [70, 71], where two capacitors were connected across the DC bus, resulting in one additional level. The additional level was the neutral point of the DC bus.

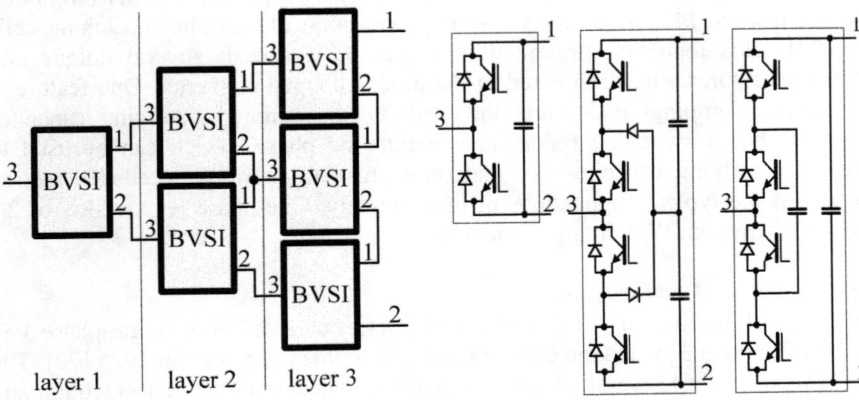

Figure 3.30. a Three-layer VSI. **b** Example of the basic-VSI module

For such multi-level PEC, the terminology neutral point clamp converter was introduced [63].

In case N+1 number of voltage levels of DCI (Figure 3.31), one phase leg consists of 2N active switches (IGBT, IGCT, GTO) and minimum 2(N–1) clamping diodes [78]. The total bus voltage U_{DC} is distributed across the capacitors C_1, ...,C_N. Hence, if voltage pattern on capacitors is uniform, then output voltage of the DCI can take values $U_{DC}\cdot(n-N)/2$ for n=0,1,…,N.

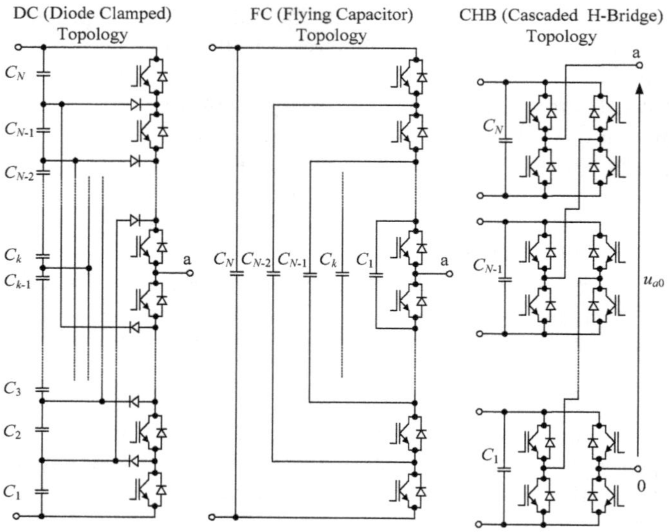

Figure 3.31. Generalized typologies of the most frequently applied multi-level VSI

3.3.1.2 Flying Capacitor Multi-level VSI

Another fundamental multi-level topology, the flying capacitor inverter (and other flying capacitor PECs), involves a series connection of capacitor switching cells [66, 79]. This topology, presented in Figure 3.31, reveals several unique and attractive features when compared to the diode-clamped converter. One feature is that added clamping diodes are not needed. Furthermore, the flying capacitor converter has a switching redundancy within the phase, which can be used to balance the flying capacitors so that only one DC source is needed. Traction converters are typical applications of this topology. One phase leg consists of $2N$ active switches and N–1 flying capacitors.

3.3.1.3 Cascaded H-Bridge VSI

This class of multi-level PECs is based on a series connection of single-phase VSI bridges (Figure 3.31), and the earliest reference to them appeared in 1975 [70]. The CHBI topology has several advantages that have made it attractive to medium and high power drive applications [77]. Since this topology consists of series power conversion cells, the voltage and power level may be easily scaled. The DC link supply for each H-bridge VSI element must be provided separately. The ability to synthesize quality wave-form of the output voltage with excellent harmonic spectrum is one of its main advantages. Additional, a very important advantage of CHBIs is the possibility to utilize low-cost low-voltage power semiconductors, switches and capacitors [80]. However, drawbacks of this topology are the large number of power devices and of voltages required to supply each cell with a complex, bulky and expensive isolated transformer.

3.3.2 Basic Comparison of Multi-level Inverter Topology

Each of the typologies of multi-level inverters presented above differs in the number of semiconductor switches used as well as reactive elements. The cooperative analysis of them could help to decide about appropriate solutions for particular applications.

Based on familiarity with the operation principle we are able to compare topologies of the multi-level voltage inverters according to various criteria. Most often as a criterion we accept a required number of semiconductor and passive components depending on a number N output voltage levels. With these assumptions and one phase of different multi-level VSI, the results obtained are presented in Table 3.1 [81].

Table 3.1. Number of the components for one phase of the N-level VSIs

Topology	DCI	FCI	CHBI
Number of active switches	$2(N-1)$	$2(N-1)$	$2(N-1)$
Number of clamped diodes	$(N-1)(N-2)$	0	0
Number of flying capacitors	0	$(N-1)(N-2)/2$	0
Number of supply capacitors	$(N-1)$	$(N-1)$	$(N-1)/2$

In Tables 3.2 and 3.3, based on [82, 83, 84, 85], are presented results of the comparative analysis and calculation of realization costs of selected topologies of three-phase multi-level VSIs. This analysis and evaluation by the authors of listed publications was conducted for the following comparable topologies:

- FK-L2: conventional VSI with four IGBTs connected in series;
- NPC-L3: NPC VSI with two IGBTs connected in series;
- DCI-L5: five-level DCI with four capacitors;
- FCI-L5: five-level FCI with neutral point;
- CHBI-L9: nine-level IHBI realized as cascaded connection of the four inverter bridges

with the assumption that

- Value of the constant voltage supplying VSI U_{DC}=6.2 kV;
- Inverters should assure line-voltage 4.2 kV;
- IGBTs (3.3 kV, 1200 A) are applied as active switches

with subjective evaluation of degree of complexity of their realization and assuming relative cost per unit for used components: IGBT (generally) – 1 p.u., IGBT for CHBI (lower voltage 1600 V) – 0.5 p.u., power diode with snubbers – 0.5 p.u., diode clamped – 0.3 p.u., capacitor (1.5 kV, 5 mF) – 0.5 p.u., snubber for IGBT – 0.1 p.u.

In typical applications of multi-level inverters in a medium voltage supply network, the number of output voltage levels rarely exceeds four to five. Available commercial turn-off 6.5 kV power switches nowadays allow realization three-phase VSIs of output voltage 6 kV without any problems. Increased number of levels and higher output voltage could be obtained only when cascade VSIs were applied, *e.g.*, CHBIs. Module based construction and isolated power supplies in CHBIs improve their safety in terms of electric shock and ease use. Unfortunately CHBIs have quite large overall dimensions and complicated control and protection.

Table 3.2. Number of components for analyzed multi-level VSI (one-phase)

Topology	FK-L2	NPC-L3	DCI-L5	FCI-L5	CHBI-L9
IGBTs	8	8	8	8	16
Clamped diodes	0	4	12	0	0
Power supply	1	1	1	1	4
Snubbers	8	8	8	0	0
Flying capacitors	0	0	0	6	0
Supply capacitors	4	4	4	4	4

Table 3.3. Comparison of estimated costs of the analyzed multi-level VSI

Topology/costs	FK-L2	NPC-L3	DCI-L5	FCI-L5	CHBI-L9
Semiconductor switches	24	24	24	24	24
Supply capacitors	6	6	6	6	12
Extra (diode/capacitors)	---	3,6	10,8	9	---
Snubbers	2,4	2,4	2,4	---	---

We should note that Table 3.3 does not include the cost of the output filter, which is most expensive in the case of a conventional two-level VSI. The cost of the filter depends to a great extent on harmonic distortion in output voltage. Estimated cost of the filter is equal to the cost of elements of the one-phase VSI, whereas the main component is the cost of the reactor. Also, Table 3.3 does not include realization costs for the controller in this initial charging of capacitors, and the expenses of technological processing of VSI realization and its installation. Therefore the technological-economic evaluation of the presented topologies of multi-level VSIs, without taking into consideration considerable technical problems, cannot be unequivocally final. Each topology (Figure 3.31) and derivative topologies [69, 76–78] have their advantages and disadvantages. Usually the specific application decides which typology should be selected.

3.3.3 Space Vector PWM Algorithm of a Multi-level VSI

For the particular states of the switches of the inverter the appropriate voltage space vector can be selected in stationary coordinates α-β [77, 84, 86]. In the m-level VSIs, the area of the space vector is usually divided into six sectors, in which we can distinguish triangular areas among three nearest locations of the space vector. Single sector and equilateral triangles of a side a is shown in Figure 3.32.

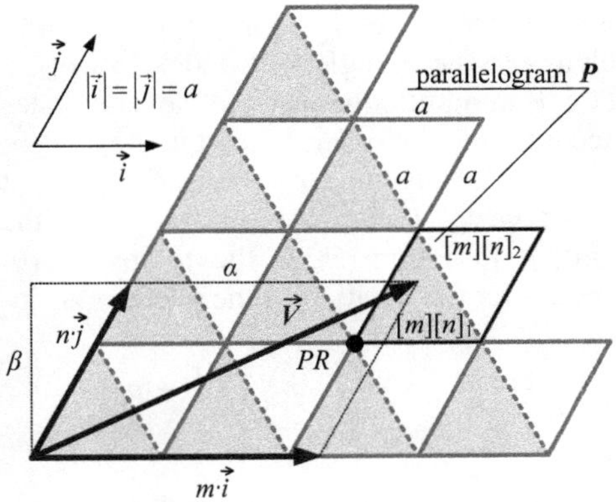

Figure 3.32. Single sector for m-level VSI

Reference voltage vector V can be presented as a linear combination of the vectors i and j. In accordance with Figure 3.34, coordinates $[\alpha, \beta]$ of vector V are as follows

$$\vec{V} = [\alpha, \beta] = m \cdot \vec{i} + n \cdot \vec{j} \tag{3.10}$$

where

$$m = \alpha / a - \beta / \left(a\sqrt{3}\right); \; n = 2 \cdot \beta / \left(a\sqrt{3}\right) \tag{3.11}$$

Integers of the m and n define coordinates of beginning of parallelogram P (point PR on Figure 3.32), where the reference vector occurs at the time. In order to determine the belonging of the reference vector V to one of the two triangles in parallelogram P, the value of the following sum D needs to be found out

$$D = [m - \mathrm{int}(m)] + [n - \mathrm{int}(n)] \tag{3.12}$$

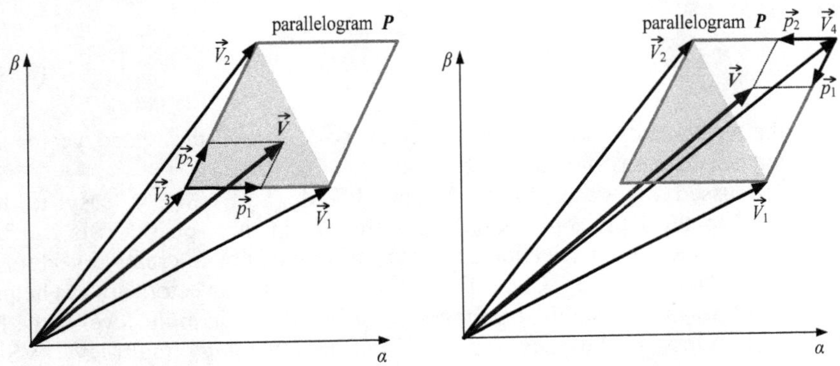

Figure 3.33. Positions of the vector V in parallelogram P

If $D \leq 1$, then the reference vector V occurs in the triangle with index $[m][n]_1$, else the vector V belongs to the triangle with index $[m][n]_2$. Finally, the synthesis of the reference vector V in addition consists of several states of switches, during modulator's work, according to the order defined in the control strategy.

There exist two possible positions of the vector V in the parallelogram P (Figure 3.32). Reference vector V can be projected within the area of each of the two triangles. According to Figure 3.33, the reference voltage vector can be expressed as a vector sum for the position of the vector V as in Figure 3.33 (left)

$$\vec{V} = \underbrace{p_1\left(\vec{V}_1 - \vec{V}_3\right)}_{\vec{P}_1} + \underbrace{p_2\left(\vec{V}_2 - \vec{V}_3\right)}_{\vec{P}_2} + \vec{V}_3 = p_1\vec{V}_1 + p_2\vec{V}_2 + \left(1 - p_1 - p_2\right)\vec{V}_3 \tag{3.13a}$$

or for position of the vector V as in Figure 3.33 (right)

$$\vec{V} = \underbrace{p_1\left(\vec{V}_1 - \vec{V}_4\right)}_{\vec{P}_1} + \underbrace{p_2\left(\vec{V}_2 - \vec{V}_4\right)}_{\vec{P}_2} + \vec{V}_4 = p_1\vec{V}_1 + p_2\vec{V}_2 + \left(1 - p_1 - p_2\right)\vec{V}_4 \tag{3.13b}$$

where p_1, p_2 are relative lengths (durations) of the active vectors V_1 and V_2. Duration of zero vectors V_3 and V_4 result from a difference in carrier period and duration of the active vectors. All space vectors of the four-level VSI are presented in Figure 3.34.

In the selected sector 0 in the Figure 3.34 occur nine numbered regions. Specified position of the space vector is coded as follows – given number defines a point in the linking circuit connected to a load terminal of the particular phase (from left *a*, *b*, *c*). For example, the code "*321*" means that phase *a* was linked to voltage source of value $3\cdot(U_{DC}/3)$, phase *b* was linked to the voltage source of value $2\cdot(U_{DC}/3)$, and phase *c* was linked to the voltage source of value $1\cdot(U_{DC}/3)$. In the case of linear modulation range, value of maximum phase voltage equals $\sqrt{3}\cdot(U_{DC}/3)$, and maximum value of normalized modulation factor m_a in this case equals $3\cdot\sqrt{3}/2$.

Figure 3.35 provides exemplary positions of the space vector of a four-level inverter. When the reference vector V occurs in region 4, relative lengths of the space vectors $V_{331}=V_{220}$ and $V_{321}=V_{210}$ equal

$$p_1 = 2 - n;\ \ p_2 = 1 - m \tag{3.14}$$

Hence, the vector-duty factors for particular positions of the space vector in modulation period are as shown in Table 3.4.

The discussed algorithm Space Vector PWM (SVPWM) is easy to be implemented in the DSP controller. This algorithm in three-phase VSIs can be easily complemented by a selection procedure of one of few alternative vectors – so-called redundancy vectors. Correct selection of one of the vectors always helps but does not always fully stabilize voltages of the capacitors in multi-level VSIs. In particular, it refers to the typology DCI. For this typology multi-level VSIs

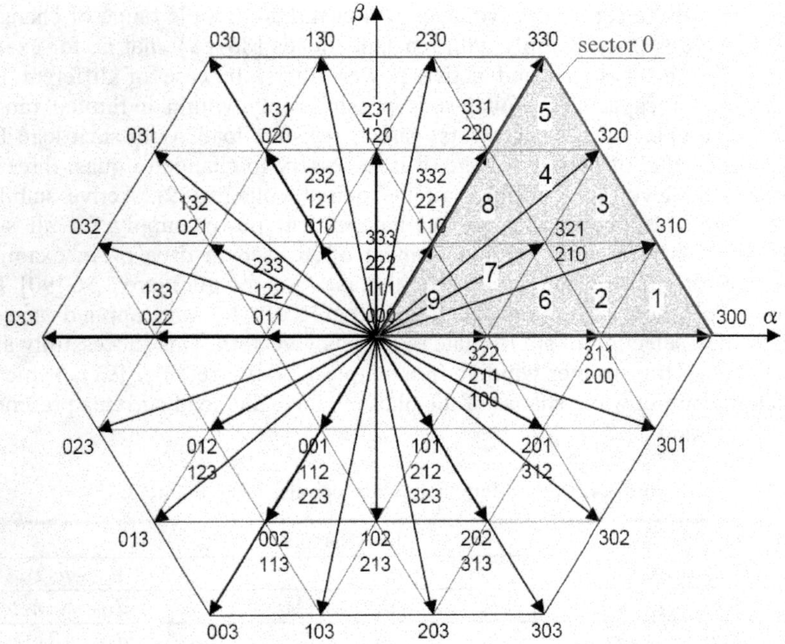

Figure 3.34. Space vectors of the four-level VSI

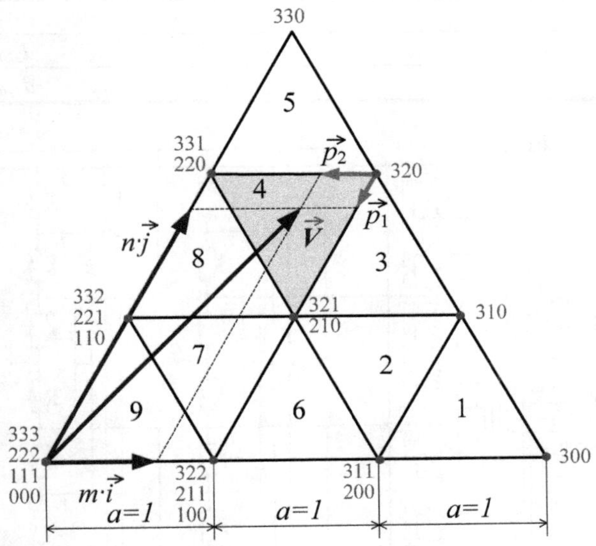

Figure 3.35. Example of the position of the normalized vector V in region 4

stabilization of the capacitors' voltages within full admissible range of changes of output voltage is possible only with reactance loads [86–88], that is, for example, in D-STATCOM systems and active power filters. In case of different loads, stabilization of capacitors' voltages is possible only within a limited range of changes of output voltage. The worst case is resistive load. With such load DCIs, above some value of output voltage, it is necessary to change to quasi-three-level and quasi-two-level wave-forming of the output voltage [89]. Active stabilizing circuits could also be applied, or an independent power supply for all supply capacitor DCIs. The final method is most often used in drives. An example is topology of driving frequency converter that is presented in Figure 3.36 [90]. In the controller of these converters the algorithm SVPVM was applied and then completed by selection of appropriate redundancy vectors. This successfully allows a level load of transformer winding in a simple 12-pulse rectifier [91]. Typical the solution are output wave-shapes of the phase voltage and load currents presented in Figures 3.37 and 3.38.

Table 3.4. Vector-duty factors for regions 1–9 in Figure 3.35

Reg.	Duty factors		
1	$d_{200/311}$=3-m-n ,	d_{310}=n ,	d_{300}=m-2
2	$d_{200/311}$=1-n ,	$d_{210/321}$=2-m ,	d_{310}=m+n-2
3	$d_{210/321}$=3-m –n ,	d_{310}=m-1 ; ,	d_{320}=n-1
4	$d_{210/321}$=2- n ,	$d_{220/331}$=1-m ,	d_{320}= m+n-2
5	$d_{220/331}$=3-m-n ,	d_{330}=n-2 ,	d_{320}=m
6	$d_{100/211/322}$=2-m-n ,	$d_{200/311}$=m -1 ,	$d_{200/321}$=n
7	$d_{110/221/332}$=1-m ,	$d_{100/211/322}$=1-n ,	$d_{210/321}$=m+n-1
8	$d_{110/221/332}$=2-m-n ,	$d_{210/321}$=m ,	$d_{220/331}$=n-1
9	$d_{000/111/222/333}$=1-m-n ,	$d_{100/211/322}$=m ,	$d_{110/221/332}$=n

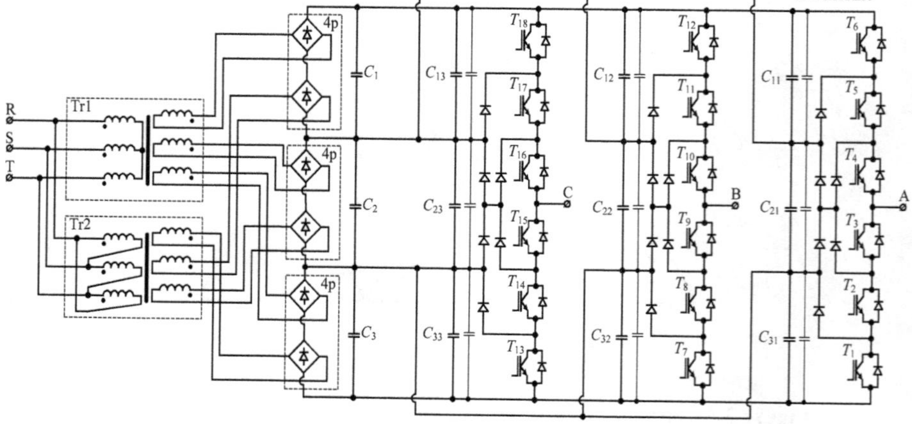

Figure 3.36. Frequency converter with four-level DCI and 12-pulse rectifier

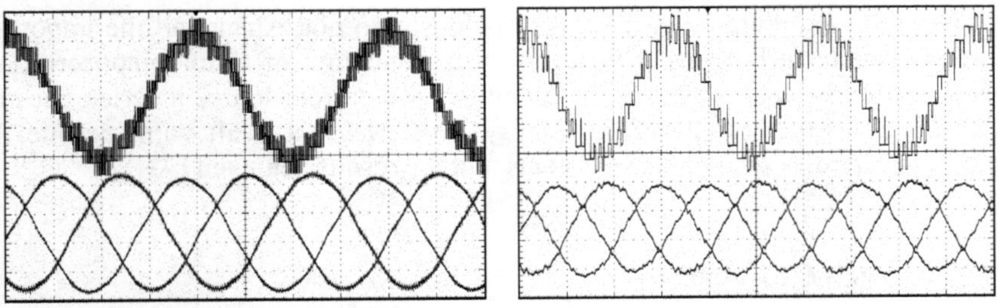

a b

Figure 3.37. Phase voltage and load currents for modulation factor m_a= 2.59 and PWM carriers: **a** f_c= 4 kHz; **b** f_c= 800 Hz

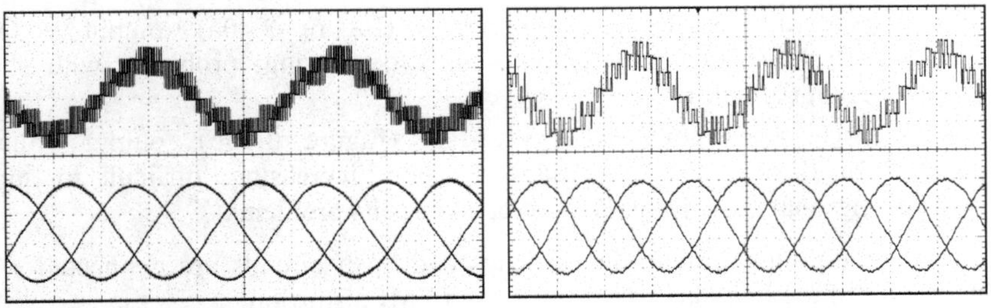

a b

Figure 3.38. Phase voltage and load currents for modulation factor m_a= 1.55 and PWM carriers: **a** f_c= 4 kHz; **b** f_c= 800 Hz

The discussed SVPWM algorithm (slightly modified) is also used in multi-level topology, which is a hybrid connection of typical two-level VSI with additional H-bridge modules in every output [92]. An example of this solution, together with characteristic oscillograms of output line-to-line voltage and phase voltage in individual modes, is presented in Figure 3.39.

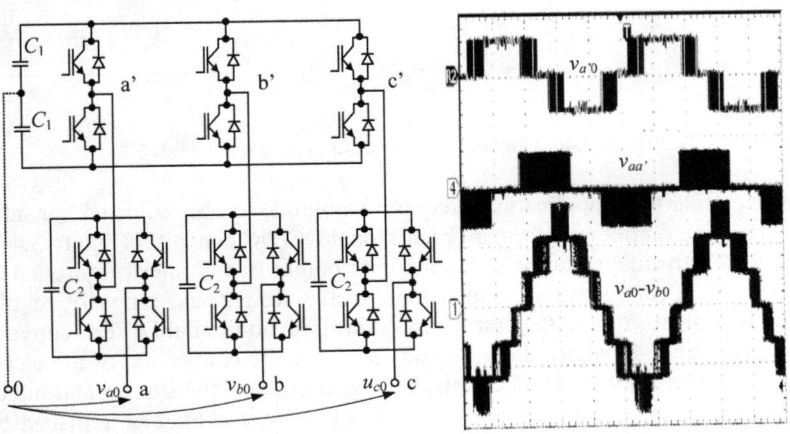

Figure 3.39. Connection of a two-level VSI and three H-bridge module

The discussed multi-level inverters obviously do not exhaust all the important solutions that have been tried in recent years, and first of all they concern only voltage systems – VSI. The issue of the multi-level current source inverter, because of its duality when compared to VSI systems, was not dealt with, despite the increasing interest given to these systems and the results achieved [93].

3.4 Z-source Converters

Many significant problems that occur in the conventional inverters (Figure 3.40) result from their operating principle. These problems are connected to the following disadvantages:

- In case of voltage source inverters (Figure 3.40a): output voltage $V \leq V_{DC}/1.73$; voltage regulation – only decreasing; problems with short circuits in branches;
- In case of current source inverters (Figure 3.40b): output voltage $Um \geq U_{DC}/1.73$; voltage regulation – only increasing; difficult to apply conventional modules IGBT and open circuits problems.

The issues with short circuits in branches and open circuits are connected with vulnerability of the inverters to damages from EMI distortion.

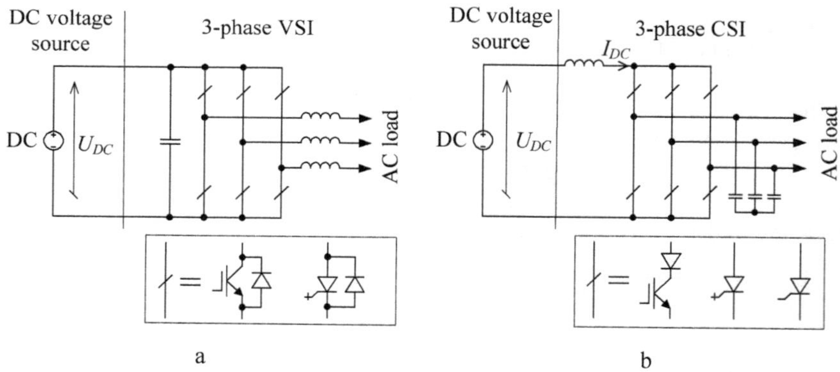

Figure 3.40. Conventional inverter systems: **a** VSI; **b** CSI

If the inverters' applications require amplitude to be adjusted outside the limited region, output transformer or additional DC/DC converter (Figure 3.41) can be used. Disadvantages of the solutions with output transformer (Figure 3.41 left) are most of all large overall dimensions, heavy weight and range of regulation limited by transformer voltage ratio. However, if an additional DC/DC converter is applied (Figure 3.41 right), then it results in two-stage conversion of the electrical energy, and therefore we should consider higher costs of the system and increased losses. Moreover, in such a case, one type of inverter cannot be replaced by an other type (*i.e.*, CSI can not be replaced by VSI and *vice versa*) and short circuits or open circuits and transition processes occur. Therefore, the search continues for new solutions in inverter systems with improved adjustment properties. Especially worth attention seems to be the Z-source inverter patented by F.Z. Peng in 2003 [94].

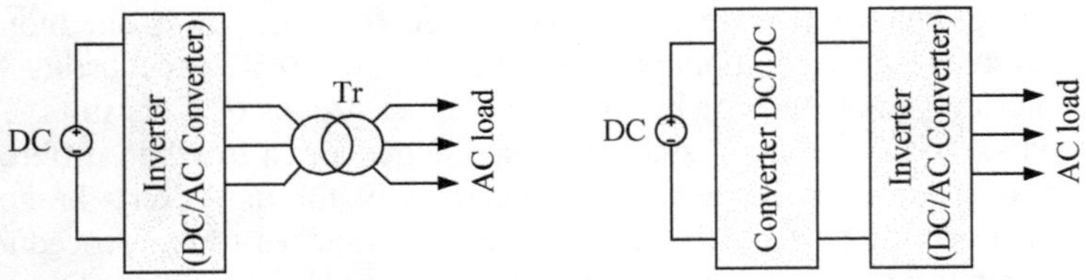

Figure 3.41. The inverter systems with increased range of regulation

Figure 3.42 presents basic schemes of the three-phase Z-inverters: voltage (Figure 3.42a) and current (Figure 3.42b) [94, 95]. In contrast to conventional VSI and CSI inverters, on the DC side of the Z-inverter is a *D* diode and a Z-source of "*X*" shape, composed of two capacitors C_1 and C_2 and two chokes L_1 and L_2. The *D* diode prevents forbidden reversed current flow (for voltage Z-inverter) or reversed voltage (for current Z-inverter). For this reason, application of the basic Z-inverters are possible only if energy return to the input source is unnecessary. Further, this is forbidden in the case of a fuel cell or photo-voltaic cell. It should be noted that the same diode function can be served by other PE systems as well. The main advantages of the Z-converters are:

- Secures the function of increasing and decreasing of voltage in the one-step energy processing (lower costs and decreased loses);
- Resistant to short circuits on branches and to opening of the circuits that improve resistance to failure switching and EMI distortions;
- Relatively simple start-up (lowered current and voltage surges).

We should acknowledge that two-direction energy flow is only possible due to change of a diode of the source on the switch of the inverter.

Because the operation principle of the voltage and current Z-inverter is similar, all the solutions considered below relate only to the voltage Z-inverter.

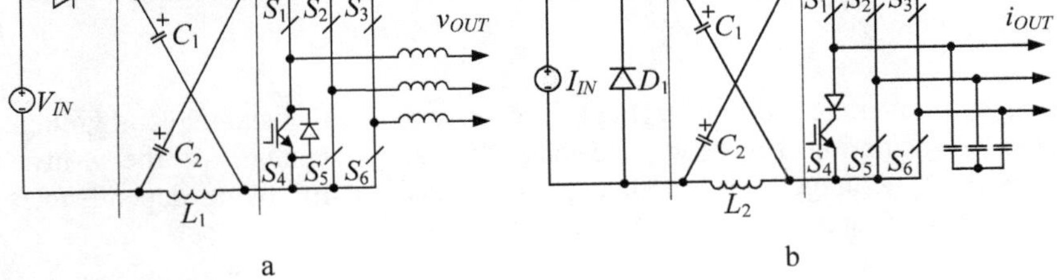

Figure 3.42. Basic schemes of the Z-inverter: **a** voltage; **b** current

3.4.1. Operation Principle of the Voltage Z-inverter

Conventional three-phase VSI system (Figure 3.40a) can assume eight states: six active states (while exchange of instantaneous power between the load and DC circuit) and two null states (when the load is shorted by transistors). Whereas, three-phase Z-inverter (Figure 3.42a) can assume 9 states, that is one more than in the VSI system — the additional nine state is the third 0 state, occurring when the load is shorted simultaneously by lower and upper groups of transistors. This state is defined as "*shoot-through*" state and may be generated in seven different ways, although of equivalent procedures: independently through every branch (three procedures), simultaneously through two of the branches (three procedures), and simultaneously through all of the three branches (one procedure). The main and unique characteristic of the Z-inverter is that the shoot-through state permits one to raise output voltage above the voltage V_{IN}.

Figure 3.43 describes simple equivalent schemes of the Z-inverter examined from the clap site of DC, where a source v_d is modeling inverter S_1–S_6. In the shoot-through states (Figure 3.43a) a D diode is polarized reversely and does not conduct the inverter bridge input voltage v_d=0, and energy stored in capacitors C is transferred to the chokes L. In "*non-shoot-through*" states (Figure 3.43b), where every combination of the switches S_1–S_6 that is allowed in VSI system is also possible, the D diode conducts and the voltage v_d increases stepwise from 0 to its maximum v_d*.

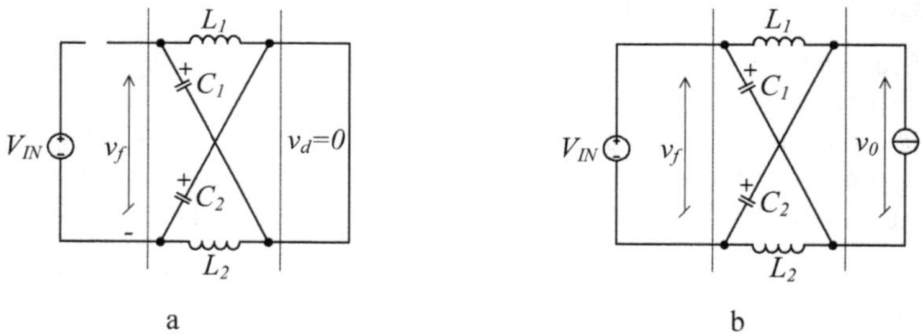

Figure 3.43. Equivalent schemes of the Z-source inverter: **a** "*shoot-through*" states; **b** "*non-shoot-through*" states

Since Z-source are symmetric circuits (Figure 3.43), when C_1=C_2 and L_1=L_2 and low voltage pulsation v_{C1} and v_{C2} during pulse period T,

$$v_{C1} = v_{C2} = V_C \qquad \text{and} \qquad v_{L1} = v_{L2} = v_L \tag{3.15}$$

where V_C is average value of voltage in capacitors, v_L – instantaneous voltage in chokes. Considering Equation 3.15 and equivalent schemes of the Z-inverter (Figure 3.43), voltage v_d is calculated on the basis of following dependencies in "*shoot-through*" states (Figure 3.43a) duration T_Z

$$v_L = V_C, \qquad v_f = 2 \cdot V_C, \qquad v_d = 0 \tag{3.16}$$

in "*non-shoot-through*" states (Figure 3.43b) duration T_N

$$v_L = V_{IN} - V_C, \qquad v_f = V_{IN}, \qquad v_d = V_C - v_L = 2 \cdot V_C - V_{IN} \tag{3.17}$$

where v_f is Z-source input voltage.

Assuming that in a pulse period $T=T_Z+T_N$, in a steady state the average voltage in chokes $V_L=0$, on the basis Equations 3.16 and 3.17, we should conclude

$$V_L = \frac{1}{T}\left(\int_0^{T_Z} v_L dt + \int_{T_Z}^{T} v_L dt\right) = \frac{T_Z \cdot V_C + T_N \cdot (V_{IN} - V_C)}{T} = 0 \tag{3.18}$$

Hence, average input voltage of the inverter bridge input voltage

$$V_C = V_d = V_{IN} \frac{T_N}{T_N - T_Z} = V_{IN} \frac{1-D}{1-2 \cdot D} \tag{3.19}$$

where $D=T_Z/T$ is "*shoot-through*" duty factor, satisfying a requirement $D<0.5$. Similarly on the basis of Equations 3.17–3.19, the value v_d^* of voltage v_d in "*non-shoot-through*" is determined

$$v_d^* = V_C - v_L = 2 \cdot V_C - V_{IN} = V_{IN} \cdot \frac{1}{1-2D} = V_{IN} \cdot B \tag{3.20}$$

where $B=1/(1-2 \cdot D)=T/(T_N-T_Z) \geq 1$ is a peak factor, and the value v_d^* is determined by relative voltage V_{IN}.

Further, the value v_d^* determines output voltage amplitude $V_{OUT(max)}$ of the Z-inverter. When applying sinusoidal PWM the amplitude equals

$$V_{OUT(\max)} = M \cdot \frac{v_d^*}{2} = \frac{M}{1-2 \cdot D} \cdot \frac{V_{IN}}{2} = K \cdot \frac{V_{IN}}{2} \tag{3.21}$$

where M is modulation index, of maximum value limited by inequity $M \leq 1-D$, related to time T_Z of "*shoot-through*" states. As we conclude, based on Equation 3.21, the Z-inverter output voltage amplitude $V_{OUT(max)}$ can be either lower or higher than in typical VSI system with sinusoidal PWM, *e.g.* $V_{OUT(max)}=M \cdot V_{IN}/2$. This possibility is acknowledged when looking at the 3D diagram of the function

$$K = M/(1-2 \cdot D) \tag{3.22}$$

within the acceptable area

$$\Omega \subset \begin{cases} 0 \leq M \leq 1 \\ D < 0{,}5 \cap D \leq 1-M \end{cases} \tag{3.23}$$

This function is presented in Figure 3.44.

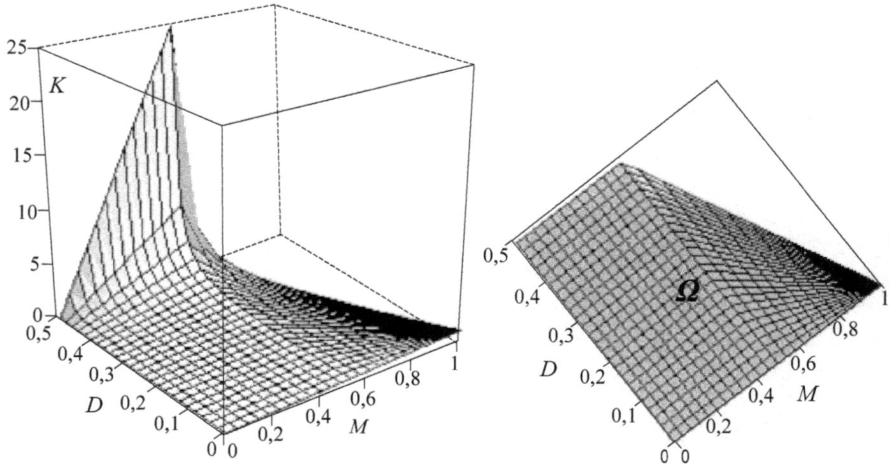

Figure 3.44. 3D diagram of the function K (Equation 3.22) within area Ω

The discussed properties of the voltage Z-inverter are confirmed by research presented in a number of publications from the recent past [96–99]. The properties are also illustrated below in selected results from the authors' research. Basic parameters of the system that were assumed in the research are presented in Table 3.5. In order to control switches V_1–V_6, we applied the algorithm of simple sinusoidal PWM, modified by "shoot-through" states. The essence of this modification is explained in Figure 3.45. Selected results of the research are presented in Figure 3.46.

Table 3.5. Parameters of researched voltage Z-inverter (Figure 3.42a)

Supply DC		U_{IN}	150 V
Z-source	Chokes	L_1, L_2	0.2 mH
	Capacitors	C_1, C_2	0.2 mF
Output filter	Chokes	L_f	100 μH
	Capacitors	C_f	50 μF
Load (resistance)		R_0	60 Ω
PWM frequency carrier		$1/T$	10 kHz

Further research on the Z-inverter confirmed the theoretical analysis. Little error arose from assumed values of loads R_0 and parameters $L=L_1=L_2$ and $C=C_1=C_2$ of Z-source. In addition, a harmonic distortion coefficient in input voltage v_{OUT}, that was calculated every time, has never exceeded 3–4%. The research also showed that transitions in the Z-inverter, resulting from changes of load and factors M and D (in open control system), are relatively fast. Furthermore, they inspired many researchers to elaborate and investigate other Z-source inverters, including NPC inverters and four-wire inverters.

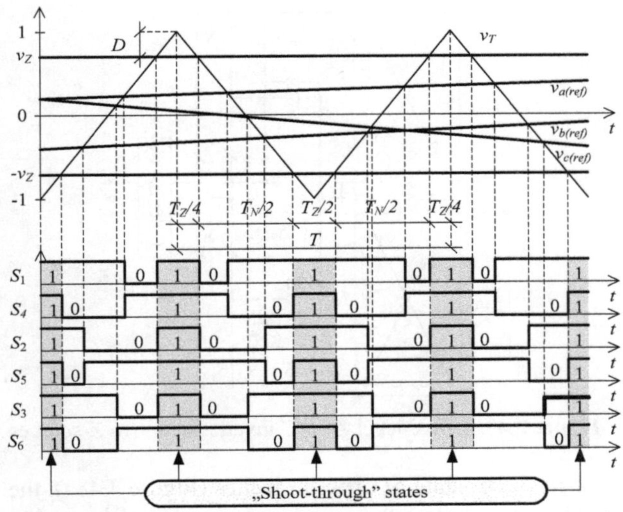

Figure 3.45. Control algorithms implementing of the Z-inverter

3.4.2. Three-level and Four-wire Inverters with Z-source

In the three-level Z-NPC inverter [100, 101], presented in Figure 3.47, two

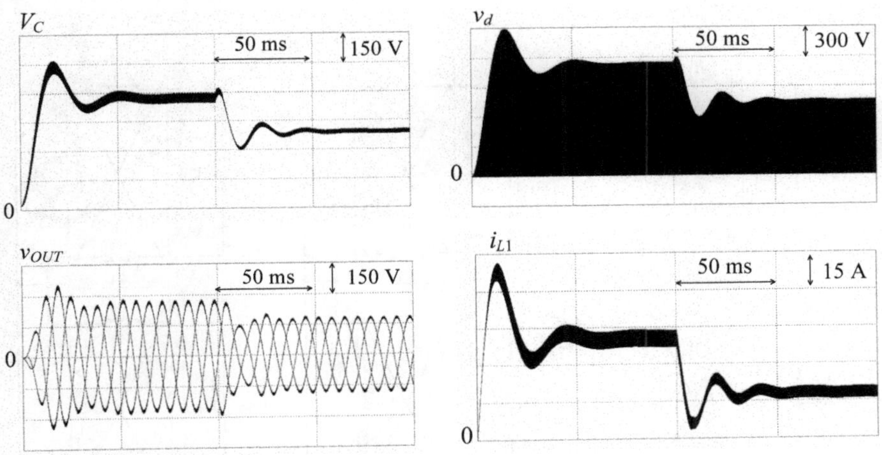

Figure 3.46. Selected currents and voltages waveform before and after change of coefficient $D = 0.43 \rightarrow 0.35$ for the time $t = 10$ ms ($M = 0.48$)

Z-sources with input voltage V_{IN1} and V_{IN2} without common point were applied. That allows joint as well as separated voltage u_{d1} and u_{d2} control. The possibility is explained by equivalent diagrams of the Z-NPC inverter in "*shoot-through*" states, shown in Figure 3.48.

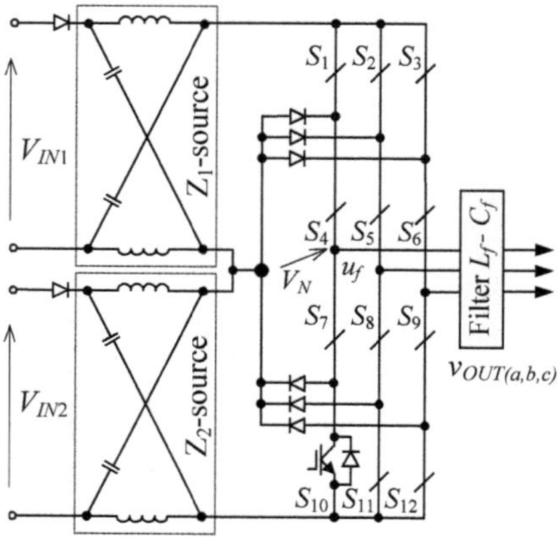

Figure 3.47. Three-level Z-NPC inverter with two Z-sources

In the "*shoot-through*" state of upper branches (Figure 3.48a), the switches S_1–S_6 & S_7–S_9 are turn-on, whereas "*shoot-through*" states of lower branches (Figure 3.48b) switches S_7–S_{12} and S_4–S_6. These two states, together with the duration T_{Z1} and T_{Z2} in pulse period T (Figure 3.49) cause an average voltage increase to V_C and maximal v_d^* up to a value on output of the upper Z_1-source

$$V_{C1} = V_{IN1} \cdot \frac{1 - D_1}{1 - 2 \cdot D_1}, \qquad v_{d1}^* = V_{IN1} \cdot \frac{1}{1 - 2 \cdot D_1} \tag{3.24}$$

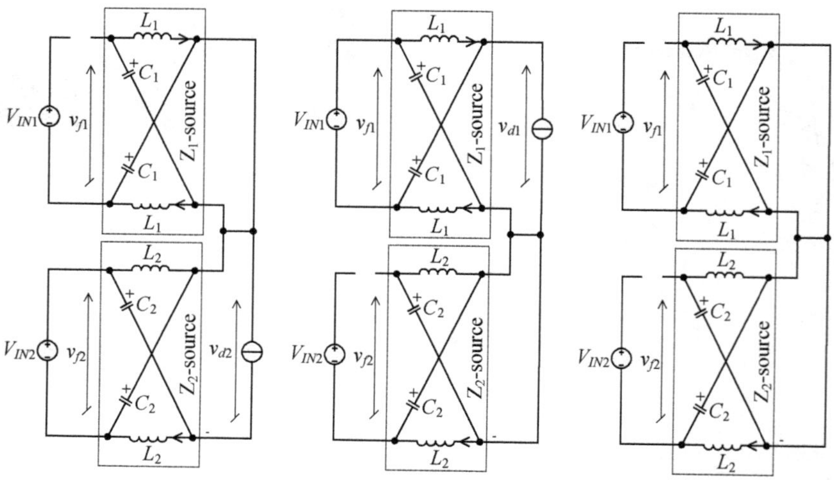

a b c

Figure 3.48. Equivalent schemes of the Z-NPC inverter (Figure 3.47) in states of "*shoot-through*": **a** upper branches; **b** lower branches; **c** full on output o lower Z_2-source

$$V_{C2}=V_{IN2}\cdot\frac{1-D_2}{1-2\cdot D_2}, \qquad v_{d2}^{*}=V_{IN2}\cdot\frac{1}{1-2\cdot D_2} \tag{3.25}$$

where $D_1=T_{Z1}/T$ and $D_2=T_{Z2}/T$ are "*shoot-through*" coefficients of upper and lower branches.

Physical process occurring and following deduced Equations 3.24 and 3.25, are analogous to those in a basic system of Z-inverter (Figure 3.42) and Equations 3.19 and 3.20. It remains unchanged by total "*shoot-through*" state (Figure 3.48c), occurring when short circuits of upper and bottom branched happen simultaneously at the time $T_{Z2}–\Delta T_Z$ (Figure 3.49).

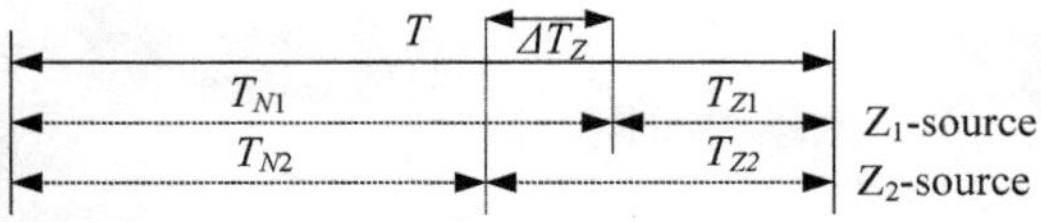

Figure 3.49. Exemplary schedule of branches "*shoot-through*" in T period

When considering the Z-NPC inverter presented in Figure 3.47, which is supplied by a source of different voltage $V_{IN1}\neq V_{IN2}$ and controlled on the basis of sinusoidal PWM, and taking into account Equations 3.24 and 3.25, input voltage peak-to-peak value can be determined on the basis of the following dependence

$$V_{OUT(p/p)}=M_1\cdot\frac{V_{IN1}}{1-2\cdot D_1}+M_2\cdot\frac{V_{IN2}}{1-2\cdot D_2} \tag{3.26}$$

where M_1 and M_2 are modulation indexes for positive and negative halves of the output voltage. Hence, if the following condition is true

$$M_1\cdot\frac{V_{IN1}}{1-2\cdot D_1}=M_2\cdot\frac{V_{IN2}}{1-2\cdot D_2} \tag{3.27}$$

then the voltage v_{com} between reference potential V_N (Figure 3.47) and star point in symmetrical load three-phase load (*e.g.*, DC-offset) is as follows

$$v_{com}=V_N-\left(v_{OUT(a)}+v_{OUT(b)}+v_{OUT(c)}\right)=0 \tag{3.28}$$

If Equation 3.27 is not satisfied, then additional distortion in the output voltage occurs that is mainly related to even harmonics.

Taking into account Equation 3.27, elimination of output voltage distortion and DC-offset, is possible through: (a) selection of different modulation index's M_1 and M_2 for the positive and negative half; (b) selection of different "*soot-through*" coefficient D_1 and D_2 for upper and bottom branches (Figure 3.50). In the second case maximal voltages in switches S_1–S_6 and S_7–S_{12} (Figure 3.47) are equal. Obviously it opens up the possibility to join both procedures.

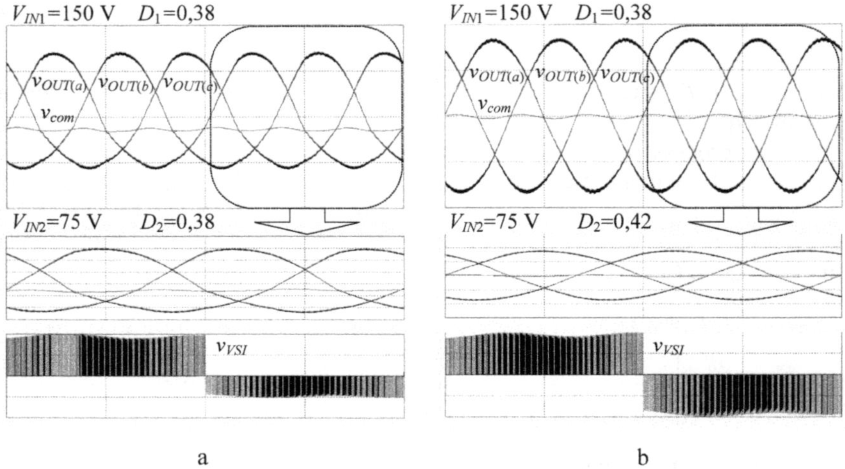

Figure 3.50. The output phase voltages $v_{OUT(a,b,c)}$, DC-offset voltage v_{com} and VSI output voltage v_{VSI} (phase a): **a** in cases of differentiated $V_{IN1} \neq U_{IN2}$ and equal $D_1 = D_2$; **b** after correction of coefficient D_2 on the basis of Equation 3.27

In [102], an alternative for Z-NPC inverter presented in Figure 3.47 topology of the NPC inverter using a single Z-source was proposed. Single Z-source was also tested in application for the DC-link Cascaded Inverter (DCLC). Both solutions that are presented in Figure 3.51 must unfortunately be supplied by two input voltage sources. This disadvantage is not of concern in the Z-NPC inverter, of which the fundamental topology is presented in Figure 3.52 [103]. This inverter can be realized when one of the Z-sources with high frequency transformer of transformation ratio 1:1 is applied. It needs to be supplied from only one source.

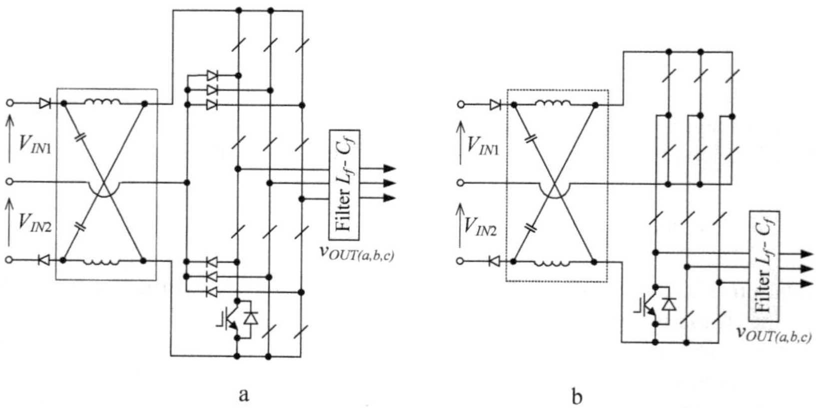

Figure 3.51. Inverters with a single Z-source: **a** NPC; **b** DCLC

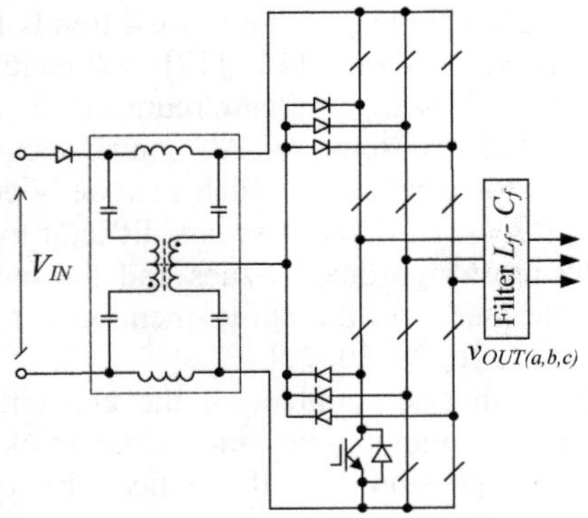

Figure 3.52. Z-NPC inverter supplied only from one source V_{IN}

Presented typologies of Z-inverters for obvious reasons do not consider all detailed solutions, such as other Z-source converters [104–111]. Many are also

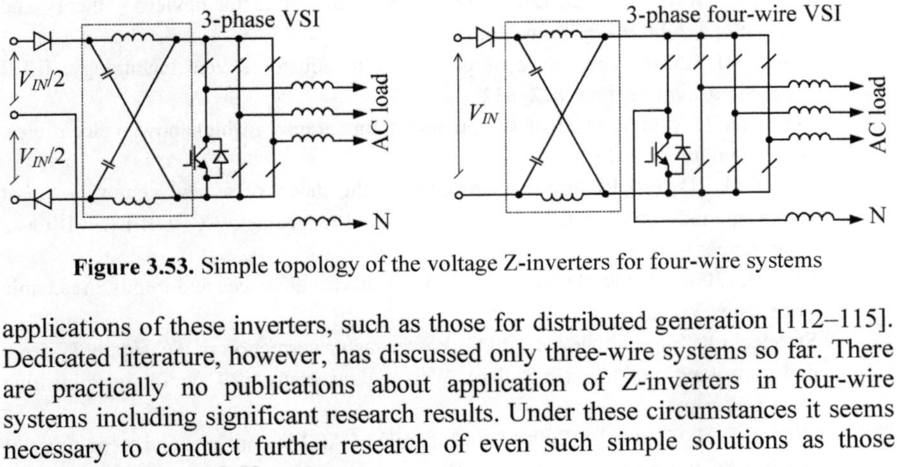

Figure 3.53. Simple topology of the voltage Z-inverters for four-wire systems

applications of these inverters, such as those for distributed generation [112–115]. Dedicated literature, however, has discussed only three-wire systems so far. There are practically no publications about application of Z-inverters in four-wire systems including significant research results. Under these circumstances it seems necessary to conduct further research of even such simple solutions as those presented in Figure 3.53.

3.5 Summary

The following text should only be considered as a comment. Also, in this manner, we would like to explain ourselves for disregarding many significant issues in the chapter. We are aware that relatively important issues discussed in the field of power electronics are not included here.

For example, we did not discuss diode and thyristor converters, including significant and large groups of conventional PECs with different applications, starting with generating systems, power transmission systems and local DC supply networks to improve the power quality for various technological applications. These PECs, however, have already been discussed by the major stream literature for many years, for example [41–43], or the newest trends focusing on issues of improvement of power quality factors [116, 117]. Valuable publications such as [118, 119] continue to be released. Similar circumstances apply with regard to DC/DC converters [120–122], various AC/AC matrix converters [123–129] or even systems of power factor correctors, which include Vienna converters [130–133] and resonant converters [134–137]. It seems difficult to present carefully yet comprehensively all the most significant issues and the most utilized solutions. Well written and valuable guides to the above-mentioned topics are available in books, in particular [1, 5, 11, 26, 28–30, 77].

The authors also hope that the content of the chapter, as well as referred literature, inspires the reader and initiates individual thinking about the issues discussed as well as the possibilities of applications of power electronics converters in smart energy networks.

References

[1] Benda V, Gowar J, Grant DA, (1999) Power semiconductor devices – theory and applications. John Wiley, New York

[2] Baliga BJ, (2001) The future of power semiconductor device technology. IEEE Proceedings, vol.89, no.6:822–832

[3] Stemmler H, (2000) State of the art and future trends in high power electronics. IPEC Conference:4–13

[4] Van Wyk JD, (2000) Power electronics at the dawn of a new century – past achievements and future expectations, Proceedings CES/IEEE IPEMC Conference:9–20

[5] Bose B, (2006) Power electronics and motor drives: advances and trends. Academic Press, New York

[6] Strzelecki R, Supronowicz H, (2000) Power factor correction in AC supply systems and improving methods. (in Polish), Warsaw University of Technology Publishing House, Warsaw

[7] Hingorani N, Gyugyi L, (2000) Understanding FACTS: concepts and technology of flexible AC transmission systems. IEEE Press, New York

[8] Acha E, Agelidis VG, Anaya-Lara O, Miller TJE, (2002) Power electronic control in electrical systems. Newnes, Oxford

[9] Ghosh A, Ledwich G, (2002) Power quality enhancement using custom power devices. Kluwer Academic Publishers, Boston

[10] Moreno-Münoz A, (2007) Power quality: mitigation technologies in a distributed environment. Springer-Verlag, London

[11] Akagi H, Watanabe EH, Aredes M, (2007) Instantaneous power theory and applications to power conditioning. John Wiley-IEEE Press, New York

[12] Simões MG, Farret FA, (2004) Renewable energy systems: design and analysis with induction generators. CRC Press, Boca Raton

[13] Green TC, Arámburo H, (2005) Future technologies for a sustainable electricity system: the role of power electronics in future power systems. Cambridge University Press, Cambridge
[14] Jamasab T, Nuttall WJ, Michael G, Pollitt MG, (2006) Future electricity technologies and systems. Cambridge University Press, Cambridge
[15] Blaabjerg F, Chen Z, (2006) Power electronics for modern wind turbines (synthesis lectures on power electronics). Morgan & Claypool Publishers
[16] Maloney TJ, (2003) Modern industrial electronics. Prentice Hall, Upper Saddle River
[17] Irwin JD, (1997) The industrial electronics handbook. CRC Press LCL, Boca Raton
[18] Lee FC, VanWyk JD, Boroyevich D, Lu GQ, Liang Z, Barbosa P, (2002) Technology trends toward a system-in-a-module in power electronics. IEEE Circuits Syst. Mag., vol.2, no.4:4–22
[19] Van Wyk, JD, Lee FC, Boroyevich D, Zhenxian L, Kaiwei Y, (2003) A future approach to integration in power electronics systems. IECON Conference, vol.1: 1008–1019
[20] Chen R, Canales F, Yang B, Barbosa P, Van Wyk JD, Lee FC, (2003) Integration of electromagnetic passive components in DPS front-end DC/DC converter – a comparative study of different integration steps. APEC Conference, vol.2:1137–1142
[21] Rengang Ch, Canales F, Bo Y, Van Wyk, JD, (2005) Volumetric optimal design of passive integrated power electronics module (IPEM) for distributed power system (DPS) front-end DC/DC converter. IEEE Transactions on Industry Applications, vol.41, no.1:9–17
[22] Ghizoni D, Burgos R; Francis G, Ma X, Guo J, Solero L, Wang F, Boroyevich D, Cartes DA, (2005) Design and evaluation of a 33kW PEBB module for distributed power electronics conversion systems. IEEE PESC Conference:530–536
[23] Wang F, Rosado S, Boroyevich D, (2003) Open modular power electronics building blocks for utility power system controller applications. IEEE PESC Conference, vol.4:1792–1797
[24] Hornkamp M, Loddenkötter M, Münzer M, Simon O, Bruckmann M, (2001) EconoMAC the first all-in-one IGBT module for matrix converter. PCIM Conference:19–21
[25] Holtz J, (1994) Pulse-width modulation for electronic power conversion. IEEE Proceedings, vol.82, no.8:1194–1214
[26] Holmes DG, Lipo TA, (2003) Pulse width modulation for power converters: principles and practice. John Wiley & Sons, Hoboken
[27] Kaźmierkowski MP, Krishnan R, Blaabjerg F, (2002) Control in power converters. Selected problems. Academic Press, San Diego
[28] Skvarina TL, (2002) The power electronics handbook. CRC Press, Boca Raton
[29] Erickson RW, Maksimović D, (2001) Fundamentals of power electronics. Kluwer Academic Publisher, Norwell
[30] Rashid MH, (2006) Power electronics handbook: devices, circuits and applications. Academic Press, San Diego

[31] Tschirley S, (2007) Automatisierte messtechnische charakterisierung von 10kV integrierten gate-kommutierten thyristoren (IGCTs), Technische Universität Berlin
[32] De Doncker RW, (2002) Medium-voltage power electronic technologies for future decentralized power systems, IEEE Power Conversion Conference:927–932
[33] Bernet S, (2006) State of the art and developments of medium voltage converters – an overview. Przeglad Elektrotechniczny (Electrical Review), vol.82, no.5:1–10
[34] Bock B, (2005) Switching IGBTs parallel connection or with enlarged commutation inductance. Ph.D. Thesis, Ruhr-University, Bochum
[35] Thalheim J, (2003) Control strategies for balancing of series and parallel connected IGBT/diode modules. Ph.D. Thesis, Swiss Federal Institute of Technology, Zurich
[36] Abe Y, Matsubara K, Mochida T, Sasagawa K, Matsuse K, (2005) A novel method for loss reduction in high-voltage inverters. IEEE IAS Conference:1849–1854
[37] Wang H, Huang AQ, Chen B, Wang F, (2005) Development of a scalable power semiconductor switch (SPSS). IEEE no.0-7803-8975-1/05:347–353
[38] Zhang B, Huang AQ, Liu Y, Atcitty S, (2002) Performance of the new generation emitter turn-off (ETO) thyristor. IEEE IAS Conference:559–563
[39] Ozpineci B, Tolbert LM, (2003) Comparison of wide-bandgap semiconductors for power electronics applications. Oak Ridge National Laboratory Report for the U.S. Department Energy, Tennessee
[40] Sugawara Y, Takayama D, Asano K, Agarwal A, Ryu S, Palmour J, Ogata S, (2004) 12.7kV ultra high voltage SiC commutated gate turn-off thyristor: SICGT. International Symposium on Power Semiconductor Devices & ICs:365–368
[41] Takeuchi T, (1968) Theory of SCR circuit and application to motor control. Tokyo Electrical Engineering College Press
[42] Shepherd W, (1976) Thyristor control in AC circuits. Bradford University Press, Bradford
[43] Pelly B, (1971) Thyristor phase-controlled converters and cycloconverters. John Wiley & Sons Inc., New York
[44] Larsen AB, Murray JE, (1964) Optimization study of high power static inverters and converters. NASA Report CR-42, Thompson Ramo Wooldridge Inc., Cleveland
[45] Bedford BD, Hoft RG, (1965) Principles of inverter circuits. John Wiley & Sons Inc., New York
[46] Glazenko TA, Goncarenko RE, (1967) Semiconductors frequency converters in electric drives. (in Russian), Energia, Leningrad
[47] Bose BK, (2007) Need a switch? IEEE Industrial Electronics Magazine, vol.1, no.4:30–39
[48] Divan DM, (1986) The resonant DC link converter – a new concept in static power conversion. IEEE IAS Conference:758–766
[49] Azzopardi S, Vinassa JM, Woirgard E, Zardini C, Briat O, (2004) A systematic hard- and soft-switching performances evaluation of 1200V punchthrough IGBT structures, IEEE Transactions on Power Electronics, vol.19, no.1:231–241
[50] Hua G, Lee FC, (1993) An overall view of soft-switching techniques for PWM converters. European Power Electronics Journal, vol.3, no.1:39–50

[51] Cavalcanti MC, da Silva ER, Jacobina CB, Boroyevich D, Dong W, (2003) Comparative evaluation of losses in soft and hard-switched inverters. IEEE IAS Conference, vol.3:1912–1917

[52] Bloh J, De Doncker RW, (1999) Characterizing medium-voltage high-power devices under conventional and soft-switching conditions. IEEE IAS Conference, vol.1:373–378

[53] Citko T, Tunia H, Winiarski B, (2001) Resonant circuits in power electronics. (in Polish), Białystok University of Technology Publishing House, Białystok

[54] Chung HSH, Hui SYR, Tse KK, (1998) Reduction of power converter EMI emission using soft-switching technique. IEEE Transactions on Electromagnetic Compatibility, vol.40, no.3:282–287

[55] Schmidt S, Kempski A, (2004) Analysis of inverters operation with hard and soft switching in context of electromagnetic compatibility (EMC). Electrotechnical Rewiev, vol.80, no.6:620–3

[56] Liu KH, Oruganti R, Lee FC, (1985) Resonant switches – topologies and characteristics. IEEE PESC Conference:62–65

[57] Ericsen T, Hingorani N, Khersonsky Y, (2006) Power electronics and future marine electrical systems. IEEE Transactions on Industrial Applications, vol.42, no.1:155–163

[58] Ponnapan R, Leland J, Beam J, Fronista G, Weimer J, (1995) Effective cooling of MCT and IGBT using Venturi flow. Intersociety Energy Conversion Engineering Conference, paper AP-97

[59] Cao Y, Beam J, Donovan B, (1996) Air-cooling system for metal oxide semiconductor controlled thyristors employing miniature heat pipes. Journal of Thermophysics and Heat Transfer, vol.10:484–489

[60] Lee TY, (2000) Design optimization of an integrated liquid-cooled IGBT power module using CFD technique. IEEE Transactions on Components and Packaging Technologies, vol.23, no.1:50–60

[61] Goldman A, (2001) Magnetic components for power electronics. Kluwer Academic Publisher, Norwell

[62] Peterson GP, (1985) Heat pipe thermal control of electronic components. AIAA Thermophysics Conference:937–944

[63] Nabae A, Takahashi I, Akagi H, (1981) A new neutral point clamped PWM inverter. IEEE Transactions on Industry Applications, vol.17, no.5:518–523

[64] Andrejczak JM, Lescure M, (1987) High voltage converter promising technological developments. EPE Conference:1159–1162

[65] Salutari R, Sciutto G, (1992) A new multilevel PWM method: a theoretical analysis, IEEE Transactions on Power Electronics, vol.7, no.3:497–505

[66] Meynard TA, Foch H, (1992) Multilevel conversion: high voltage choppers and voltage source inverters. IEEE PESC Conference:397–403

[67] Fracchia M, Ghiara T, Marchesoni M, Mazzucchelli M, (1992) Optimized modulation techniques for the generalized N-level converter. IEEE PESC, vol.2:1205–1213

[68] Lai JS, Peng FZ, (1996) Multilevel converters – a new breed of power converters. IEEE Transactions on Industry Applications, vol.32, no.3:509–517

[69] Fazel SS, (2007) Investigation and comparison of multi-level converters for medium voltage applications. Ph.D. Thesis, Berlin Technical University
[70] Baker RH, (1975) Electric power converter. US Patent 3867643
[71] Monin VS, Vojtovic VS, (1977). Transistors inverter. USSR Patent 575751
[72] Kobzev AV, (1979) Multi-zone impulse modulation. (in Russian), Novosibirsk, Russia, Nauka
[73] Monin VS, (1986) Stabilize semiconductor converters. Energoatomizdat, Moscow
[74] Baker RH, (1980) High-voltage converter circuit. US Patent 4203151
[75] Tonkal VE, Mielniciuk LP, Novosielcev AV, Dychenko JI, (1981) Power electronics converters with modulation and high frequency link. (in Russian), Kiev, Ukraine, Naukova Dumka
[76] Rodriguez J, Lai JS, Peng FZ, (2002) Multilevel inverters: a survey of topologies, controls, and applications. IEEE Transactions on Industrial Electronics, vol.49, no.4:724–738
[77] Wu B, (2006) High power converters and AC drives. IEEE Press, John Wiley & Sons Inc., New Jersey
[78] Hartman M, (2006) Multilevel voltage inverters. (in Polish), Gdynia Maritime Publishing House, Gdynia
[79] Meynard TA, Foch H, Forest F, Turpin C, Richardeau F, (2002) Multi-cell converters: derived topologies. IEEE Transactions on Industrial Electronics, vol.49, no.5:978–987
[80] Manjrekar P, Steimer P, Lipo T, (2000) Hybrid multilevel power conversion system: a competitive solution for high-power applications. IEEE Transactions on Industry Application, vol.36, no.3:834–841
[81] Sirisukpraset S, (1999) Optimized harmonic stepped-waveform for multilevel inverter. M.Sc. Thesis, Virginia Polytechnic Institute and State University, Blacksburg
[82] Shakweh Y, Lewis EA, (1999) Assessment of medium voltage PWM VSI topologies for multi-megawatt variable speed drive applications. IEEE PESC Conference:965–970
[83] Shakweh Y, Lewis EA, (1999) The universal medium voltage adjustable speed drive. EPE Conference, CD-ROM:1–8
[84] Walker GR, (1999) Modulation and control of multilevel converters. Ph.D. Thesis, University of Queensland
[85] Shakweh Y, (2001) MV inverter stack topologies. Power Engineering Journal:139–149
[86] Gupta AK, Khambadkone AM, (2006) A space vector PWM scheme for multilevel inverters based on two-level space vector PWM. IEEE Transactions on Industrial Electronics, vol.53, no.5:1631–1639
[87] Pou J, (2002) Modulation and control of three-phase PWM multilevel converters. Ph.D. Thesis, Technical University of Catalonia, Barcelona
[88] Pou J, Pindado R, Boroyewich D, (2005) Voltage-balance limits in four-level diode-clamped converters with passive front ends. IEEE Transactions on Industrial Electronics, vol.52, no.1:190–196

[89] Adam GP, Finney SJ, Williams BW, (2007) Quasi two-level operation of a five-level inverter. Electrotechnical Review, vol.10:120–125

[90] Zymmer K, Zakrzewski Z, Strzelecki R, Szczepankowski P, (2008) 6kV four-level diode clamped inverter. Design and control. Electrotechnical Review, vol.84, no.4:4–9

[91] Lopatkin NN, Usachev EP, Zinoviev GS, Weiss H, (2006) Three-level rectifier fed four-level inverter for electric drivers. EPE-PEMC Conference:775–780

[92] Strzelecki R, Jarnut M, Kot E, Kempski A, Benysek G, (2003) Multilevel voltage source power quality conditioner. IEEE PESC Conference:1043–1048

[93] Xu Z, (2003) Advanced semiconductor device and topology for high power current source converter. Ph.D. Dissertation, Virginia Polytechnic Institute

[94] Peng FZ, (2003) Impedance source power converter. USA Patent no.2003107522, International Publication Number WO 03/107522 A1

[95] Peng FZ, (2003) Z-source inverter. IEEE Transactions on Industry Applications, vol.39, no.2:504–510

[96] Shen M, Wang J, Joseph A, Peng FZ, Tolbert LM, Adam DJ, (2004) Maximum constant boost control of the Z-source inverter. IAS Annual Meeting:142–147

[97] Miaosen S, Peng FZ, (2005) Operation modes and characteristics of the Z-source inverter with small inductance. IAS Annual Meeting, vol.2:1253–1260

[98] Loh PC, Vilathgamuwa DM, Gajanayake CJ, Lim YR, Teo CW, (2007) Transient modeling and analysis of pulse-width modulated Z-Source inverter. IEEE Transactions on Power Electronics, vol.22, no.2:498–507

[99] Rabi BJ, Arumugam R, (2005) Harmonics study and comparison of Z-source inverter with traditional inverters. American Journal of Applied Sciences, vol.2, no.10:1418–1426

[100] Loh PC, Blaabjerg F, Feng SY, Soon KN, (2006) Pulse-width modulated Z-source neutral-point-clamped inverter. Annual IEEE APEC Conference:940–947

[101] Strzelecki R, Wojciechowski D, Adamowicz M, Wilk A, Mosoń I, (2006) Three-level NPC Z-inverter. (in Polish), Electrotechnical Review, no.10:54–60

[102] Loh PC, Lim SW, Gao F, Blaabjerg F, (2007) Three-level Z-Source inverters using a single LC impedance network. IEEE Transactions on Power Electronics, vol.22, no.2:706–711

[103] Strzelecki R, Adamowicz M, Wojciechowski D, (2007) Buck-boost inverters with symmetrical passive four-terminal networks. Compatibility in Power Electronics Conference, CD-ROM:1–9

[104] Loh PC, Gao F, Blaabjerg F, (2006) Topological and modulation design of three-level Z-source inverters, IPEMC Conference, CD-ROM:1–5

[105] Strzelecki R, Wojciechowski D, Adamowicz M, (2006) Multilevel, multiphase inverter supplying by many sources, especially different voltage and non-connection sources. Polish Patent P379977

[106] Strzelecki R, Wojciechowski D, Adamowicz M, (2006) Principle of symmetrization of the output voltage of the multilevel inverter supplying by many different voltage sources, especially with four-terminal impedance networks. Polish Patent P379978

[107] Gao F, Loh PC, Blaabjerg F, Teodorescu R, Vilathgamuwa DM, (2007) Five-level Z-source neutral point-clamped inverter. IEEE PEDS Conference:1054–1061

[108] Xinping D, Zhaoming Q, Yeyuane X, Zhengyu L, (2005) Three phase Z-source rectifier. IEEE PESC Conference:494–500
[109] Xinping D, Zhaoming Q, Yeyuan X, Peng FZ, (2006) A novel ZVS Z-source rectifier. IEEE APEC Conference, CD-ROM:1–6
[110] Fang FX, (2006) Three-phase Z-source AC/AC converter. PEMC Conference:621–624
[111] Chaohua TYZ, Shaojun X, (2007) Single-phase four switches Z-source AC/AC converter. IEEE APEC Conference:621–625
[112] Gajanayake CJ, Vilathgamuwa DM, Loh PC, (2006) Modeling and design of multi-loop closed loop controller for Z-source inverter for distributed generation. IEEE PESC Conference, CD-ROM:1–7
[113] Gajanayake CJ, Vilathgamuwa DM, Loh PC, (2007) Development of a comprehensive model and a multiloop controller for Z-source inverter DGs systems. IEEE Transactions on Industrial Electronics, vol.54, no.4:2352–2359
[114] Gajanayake CJ, Vilathgamuwa DM, Loh PC, Blaabjerg F, (2007) Z-source inverter based power quality compensator with enhanced ride-through capability. IEEE IAS Conference:955–962
[115] Po Xu P, Zhang X, Zhang CW, Cao RX, Chang L, (2006) Study of Z-source inverter for grid-connected PV systems. IEEE PESC Conference, CD-ROM:1–5
[116] Pejović P, (2007) Three-phase diode rectifiers with low harmonics. Current injection methods. Springer Science & Business Media, LLC, New York
[117] Paice DP, (1999) Power electronics converter harmonics: multipulse methods for clean power. IEEE Press, Wiley & Sons Inc., New York
[118] Choi S, Enjeti PN, Pitel IJ, (1996) Polyphase transformer arrangements with reduced kVA capacities for harmonic current reduction in rectifier type utility interface. IEEE Transactions on Power Electronics, vol.11, no.5:680–689
[119] Singh B, Garg V, Bhuvaneswari G, (2007) A novel T-connected autotransformer-based 18-pulse AC/DC converter for harmonic mitigation in adjustable-speed induction-motor drives. IEEE Transactions on Industrial Electronics, vol.54, no.5:2500–2511
[120] Luo FL, Ye H, (2004) Advances DC/DC converters. CRC Press, New York
[121] Mazumder SK, (2001) Nonlinear analysis and control of standalone, parallel DC/DC and parallel multi-phase PWM converters. Ph.D. Thesis, Virginia Polytechnic Institute and State University, Blacksburg
[122] Bo Y, (2003) Topology investigation for front end DC/DC power conversion for distributed power system. Ph.D. Thesis, Virginia Polytechnic Institute and State University, Blacksburg
[123] Gyugyi L, Pelly BR, (1976) Static frequency changers. John Wiley, New York
[124] Wheeler P, Rodriguez J, Clare JC, (2002) Matrix converter: a technology revived. IEEE Transactions on Industrial Electronics, vol.49, no.2:276–288
[125] Fedyczak Z, Strzelecki R, (1997) Power electronics agreement for AC power control. (in Polish), Adam Marszalek Publishing House, Toruń
[126] Ratanapanachote S, (2004) Applications of an electronics transformer in a power distribution system, Ph.D. Thesis, Texas A&M University

[127] Petry CA, Fagundes JC, Barbi I, (2005) Direct AC/AC converters using switching modules. COBEP Conference:94–99

[128] Herrero LC, de Pablo S, Martin F, Ruiz JM, Gonzalez JM, (2007) Comparative analysis of the techniques of current commutation in matrix converters. IEEE ISIE Symposium:521–526

[129] Casadei D, Serra G, Tani A, Zarri L, (2006) A review on matrix converters. Electrotechnical Review, vol.82, no.2:15–25

[130] Bartosa PM, (2002) Three-phase power factor correction circuits for low-cost distributed power systems. Ph.D. Thesis, Virginia Polytechnic Institute and State University, Blacksburg

[131] Cichowlas M, (2004) PWM rectifier with active filtering. Ph.D. Thesis, Warsaw University of Technology

[132] Kwak S, Toliyat HA, (2006) Current-source-rectifier topologies for sinusoidal supply current: theoretical studies and analyses. IEEE Transactions on Industrial Electronics, vol.53, no.3:984–987

[133] Kolar JW, Drofenik U, Zach FC, (1999) VIENNA rectifier II – a novel single-stage high-frequency isolated three-phase PWM rectifier system. IEEE Transactions on Industrial Electronics, vol.46, no.4:674–691

[134] Kazimierczuk MK, Czarkowski D, (1995) Resonant power converters. John Wiley & Sons Inc., New York

[135] Vlatkovic V, Borojevic D, Lee FC, (1994) Soft-transition three-phase PWM conversion technology. IEEE PESC Conference:79–84

[136] Pfisterer HJ, Spath H, (2000) Switching behavior of an auxiliary resonant commutated pole (ARCP) converter. IEEE International Power Electronics Congress CIEP:359–364

[137] Lakshminarasamma N, Swaminathan B, Ramanarayanan V, (2004) A unified model for the ZVS DC/DC converters with active clamp. IEEE PESC, vol.3:2441–2447

4

Quality Problems in Smart Networks

Zbigniew Hanzelka[1], Adam Kempski[2] and Robert Smoleński[2]

[1]Department of Electrical Drive and Industrial Equipment,
AGH University of Science and Technology,
al. Mickiewicza 30, 30-059 Krakow, Poland.
Email: Hanzel@agh.edu.pl

[2]Institute of Electrical Engineering,
University of Zielona Góra,
50 Podgórna Street, 65-246 Zielona Góra, Poland.
Email: A.Kempski@iee.uz.zgora.pl; R.Smolenski@iee.uz.zgora.pl

4.1 Power Quality and EMC

One of the basic notions associated with electrical power utilisation is that of Electromagnetic Compatibility (EMC) – between electric equipment and its environment, or between this equipment and any other equipment; it is, according to the International Electro-technical Commission (IEC) definition "*the ability of an equipment or system to function satisfactorily in its electromagnetic environment without introducing intolerable electromagnetic disturbances to anything in that environment*" [1].

Mutual interaction between electromagnetic environment and a load can be either conducted (associated with solid connection) or radiated (associated with inductive, capacitive or electromagnetic coupling). Currently, unlike in the past, the very concept of electromagnetic compatibility does not exclusively involve issues of information transmission. It also comprises issues of mutual interaction between equipment and systems utilizing electromagnetic phenomena to energy transmission and conversion purposes, as well as impact of these equipment and systems on the natural electro-magnetic environment and biosphere, so it includes all events of any electromagnetic incompatibility.

Electric power is the result of a production process and as a product it should be the subject of evaluation and standardization. It is also a subject of market trade turnover; thus its utility value has to be a subject of evaluation described in an agreement between parties of commercial transaction. In the meaning adopted by the Council of European Energy Regulators (CEER) [2, 3] the quality of electric power supply comprises three main areas; see Figure 4.1:

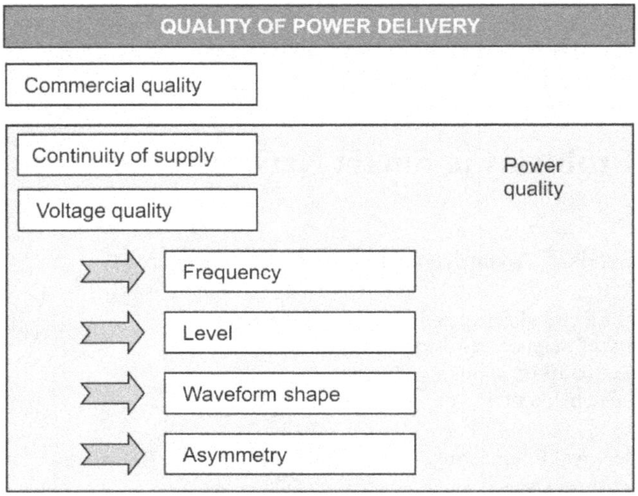

Figure 4.1. Classification of the electric power supply issues

- Commercial quality understood as the quality of commercial relations between the electric power supplier and customer;
- Continuity of supply usually measured in terms of the number of supply interruptions, duration of a single interruption, the cumulative duration of power supply interruption over a specified period of time, *e.g.*, a year, or other continuity indicators defined in this set;
- Voltage quality measured in terms of the difference between the actual waveforms and sinusoidal, three-phase balanced voltages and currents of nominal value.

The voltage quality and continuity of supply are jointly termed the Power Quality (PQ).

Electrical power is often a subject of degradation due to electromagnetic disturbances, *i.e.*, events causing the values of some quantitative indicators – power quality parameters – to differ from their nominal values which refer to the steady states with sinusoidal quantities, occurring in symmetrical polyphase systems. The power quality depends not only on the supply conditions but also on the equipment used (its immunity to disturbances and emission) and installation practices.

For the most part, utilities have adopted the following industry-recognized definition of the power quality problem [4]: "*The quality of electrical power supply is a set of parameters which describe the process of electric power delivery to the user under normal operating conditions, determine the continuity of supply (short and long supply interruptions) and characterize the supply voltage (magnitude, asymmetry, frequency, and waveform shape)*".

It should be emphasized that Electromagnetic Compatibility (EMC) and power quality are two different concepts. According to them, electromagnetic compatibility concerns exclusively the immunity and emission of a device or equipment. In that sense it can be considered to be part of a larger area, namely the quality of electric power. The latter, apart from immunity and emission, also includes the state of a power system. Obviously, the close relationships between them are the reason these notions are often used interchangeably.

In colloquial language electromagnetic compatibility and voltage quality are sometimes considered to be areas of different semantic ranges that require different methods of analysis, different methods for disturbances mitigation, *etc.*, with only a minor overlapping area. In that sense, EMC concerns high-frequency disturbances of radiation nature, whereas the quality of power is associated with low-frequency phenomena propagated through a power system.

With the emission of electromagnetic disturbances and equipment immunity to them are associated emission and immunity levels of equipment, as well as design immunity levels taken with some extra margin, that allow determining the electromagnetic compatibility level of a system (Figure 4.2).

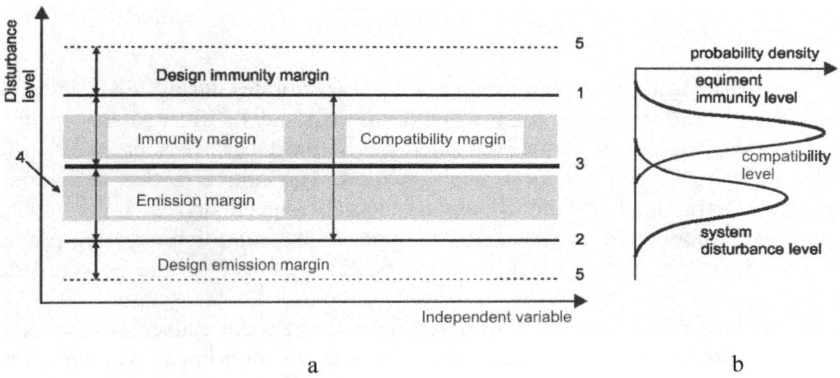

Figure 4.2. a Relations between disturbance levels: 1 – immunity level, 2 – emission level, 3 – compatibility level, 4 – planning level, 5 – design level. **b** Probability density function of equipment performance degradation *vs* disturbance level

Emission level is the value of a given electromagnetic disturbance emitted from a particular load. In the PQ domain its value depends mainly on two factors:

- The impedance of supply network at the point of load connection. It means that the same equipment connected to the network at different locations will exhibit different adverse effects. Hence, the concept of source reference impedance has been adopted in standardisation for the purpose of emission testing;
- Changes in the emission characteristics of the equipment. Emission limit is the value of an electromagnetic disturbance, which must not be exceeded during test measurements. Evidently, it is lower than the compatibility level.

Immunity level is the maximum value of a given electromagnetic disturbance occurring in a supply network that does not cause equipment performance degradation. It is higher than the compatibility level.

Compatibility level (*e.g.*, [5]) is the value of an electromagnetic disturbance which is not exceeded during a given percent of measurements. The compatibility level indicates a high probability of equipment normal operation. The difference between the immunity and disturbance emission level, termed the immunity margin, is a key factor for assessing systems compatibility.

Electric power networks are places where emissions from many, independently operating sources of disturbances add up. In order to develop a method for managing disturbance levels in power systems an additional concept of planning level has been introduced. The planning level is a value adopted by bodies responsible for planning and operation of power networks in a given area, and is used for determining permissible emission levels of large disturbing loads and installations to be connected in this area [6, 7]. The planning level shall not be higher than the compatibility level.

4.2 Power Quality Issues

Since numerical indices of power quality deteriorate during electric power transmission and distribution, mainly due to the customers' influence, the quality of electric power cannot be fully controlled by its producer. This property differs significantly from other products on the market. The source of this deterioration could be electric loads (static converters, variable-speed drives, UPSs, switch-mode power supplies, induction and arc furnaces, air conditioning, high-pressure discharge lamps, saturated magnetic circuits, *etc.*), as well as the power system itself.

Where a group of loads is considered, one of the main causes of poor power quality in power electronic equipment is due to non-linear voltage-current characteristics of semiconductor devices and the increasing switching frequency. A significant factor is also their ever-growing popularity. Power electronic systems are used at every voltage and power level, from *e.g.*, PC power supplies to very large power applications, *e.g.*, hoist machine drives. They can be found in every electromagnetic environment: residential, commercial, services and industrial *etc.* Several classes of electromagnetic environment can be defined, but for simplicity only three are considered and defined as follows [5]:

- Class 1: this class applies to protected supplies and has compatibility levels lower than those on public networks. It relates to the use of equipment very sensitive to disturbances in the power supply, for instance electrical instrumentation in laboratories, some automation and protection equipment, some computers, *etc.*;
- Class 2: this class applies generally to PCCs and to In-plant Point of Coupling (IPC) in the environments of industrial and other non-public power supplies. The compatibility levels of this class are generally identical to those of public networks. Therefore, components designed for supply from public networks may be used in this class of industrial environment;
- Class 3: this class applies only to IPCs in industrial environments. It has higher compatibility levels than those of class 2 for some disturbance

phenomena. For instance, this class should be considered when any of the following conditions are met: a major part of the load is through converters; welding machines are present; large motors are frequently started; loads vary rapidly.

The class applicable for new plants and extensions of existing plants cannot be determined *a priori* and should relate to the type of equipment and process under consideration.

The following quantities are currently regarded to be the most important criteria for the quality of supply voltage: magnitude, waveform, fluctuations, dips and short interruptions. Figure 4.3 depicts the categorization of voltage disturbances in the coordinate system: RMS voltage value – duration of disturbance.

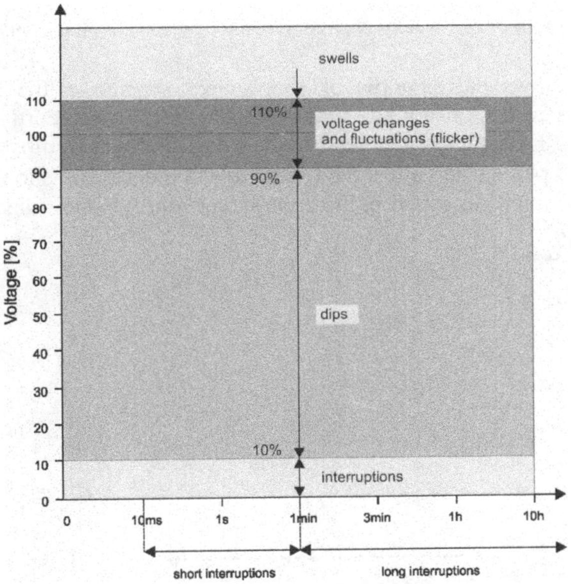

Figure 4.3. Phenomena influencing the RMS voltage value

4.2.1 Magnitude of the Supply Voltage

The main reason for changes in RMS voltage value in a power network is the load variability; see Figure 4.4. Voltage changes in supply network, exceeding tolerable limits (in most cases ± 10% U_N, U_N – nominal voltage), have an adverse effect on loads and may result in their malfunction, and in extreme case, even damage. Voltage reduction causes increased losses in transmission lines, transformers, *etc.*, while the increased voltage magnitude results in increasing magnetizing currents in transformers and motors, reduced service life or damage of equipment insulation, and increases the power dissipated in loads (excessive heating).

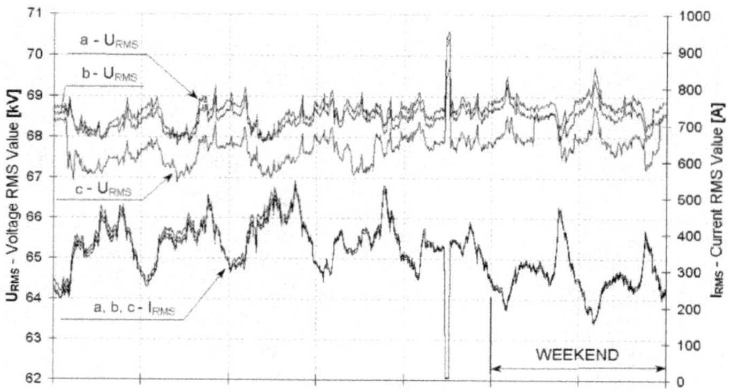

Figure 4.4. Example of one week's change in phase-to-neutral voltages and phase currents

Figure 4.4 shows the example of a one-week change in the RMS phase-to-neutral voltages in a high-voltage network (110 kV). This record is the basis for preparing the CPF characteristic; see Figure 4.5. According to most regulations the CP05 and CP95 percentile values read from this characteristic should be contained within the limit range permitted in the contractual supply conditions.

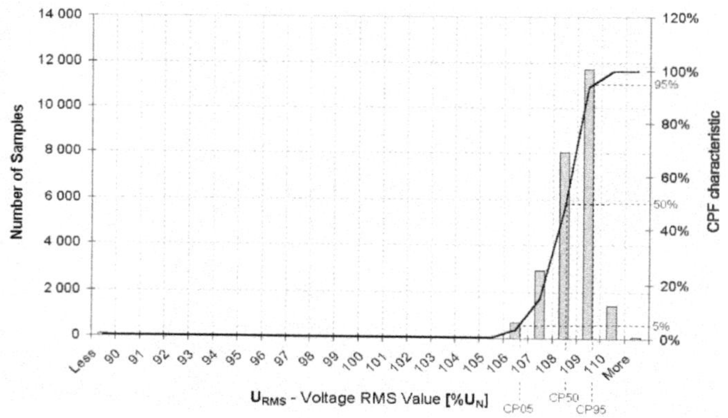

Figure 4.5. CPF characteristic for phase-to-neutral voltage RMS values (phase "*a*", Figure 4.4)

4.2.2 Voltage Fluctuation

Voltage fluctuations are a series of RMS voltage changes or a variation of the voltage envelope; see Figure 4.6. The main source of voltage fluctuations in electric power networks, which decides the phenomenon level, are disturbing industrial high-power loads (electric welders, rolling mill drives, hoisting machines, arc furnaces, *etc.*) and switching operations made in an electric power system in order to change the network configuration, or caused by voltage regulating systems (*e.g.*, by on-load tap changers of transformers).

Voltage fluctuations are a specific type of electromagnetic disturbances as their main effect – the phenomenon of light flicker – appears in the form of a negative direct influence on the human organism. Light flicker is the subjective sensation of variations in luminous flux, whose luminance or spectral distribution fluctuates with time. Physiological effects of voltage fluctuations are dependent on the variation amplitude of the luminous flux (the variation amplitude of a voltage supplying a light source), the sequence of repetitions, a frequency spectrum, and the duration of a disturbance.

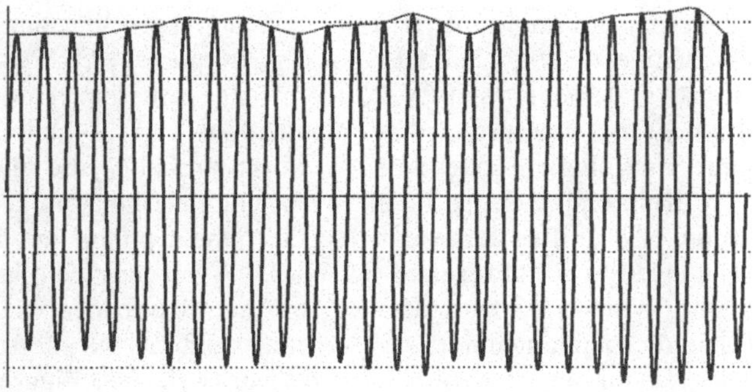

Figure 4.6. Example of voltage fluctuation in the instantaneous value waveform

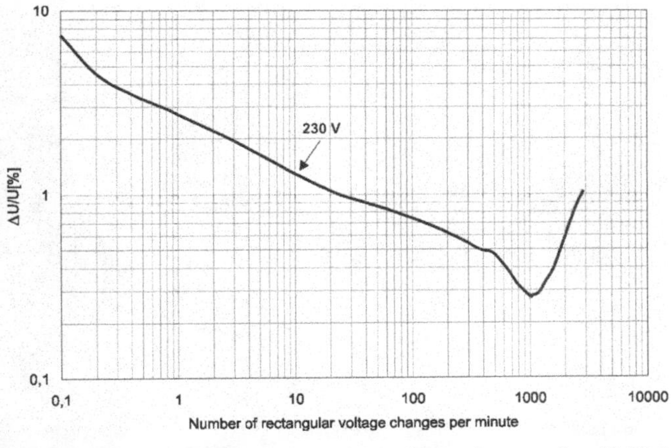

Figure 4.7. Characteristics P_{st}=1 for rectangular voltage changes for 60 W incandescent lamps

Particularly sensitive to voltage fluctuation are incandescent light sources. In their case, even a slight voltage fluctuation may cause a significant difference in the luminous flux Φ, which is proportional to the magnitude of supply voltage according to the relation $\Phi \sim U^{\gamma}$ where γ=3.1-3.7.

The flicker caused by voltage fluctuation adversely influences the general feeling of individuals – it hampers concentration and brings general discomfort and fatigue; this results in reduced work efficiency and quality. Moreover, voltage fluctuations are the cause adverse effects of a technical and economic nature. The effects of voltage fluctuations could be changes in the electromagnetic torque and slip of electric motors, increased mechanical vibrations, reduced life-cycle of electric drives, and increased harmonics emission to the supply network. Quantitative estimation of these effects is difficult, considering other disturbances that exist in an electromagnetic environment at the same time.

Figure 4.7 shows the empirically obtained reference threshold of human perceptibility of bulb light stimuli induced by rectangular voltage fluctuations of specified frequency and amplitude. It can be clearly seen that the human eye is particularly sensitive to voltage fluctuations of frequency 8.8 Hz and amplitude changes of less than 0.3% (for a 60 W, 230 V incandescent lamp).

The instrument that measures voltage fluctuations – the so-called flicker-meter – is a unique metrological phenomenon. It is probably the only measuring instrument, in the broadest sense of electrical engineering, that reconstructs the visual sensation of a unified representative the human population in response to light stimuli induced by a light source (60 W, 230 V incandescent lamp) produced by a fluctuating voltage. Standardization of the disturbance is based on the dimensionless indices obtained at the flicker-meter output: the short term (10-min) flicker severity P_{st} and long term (2-h) flicker severity P_{lt} (see Figure 4.8).

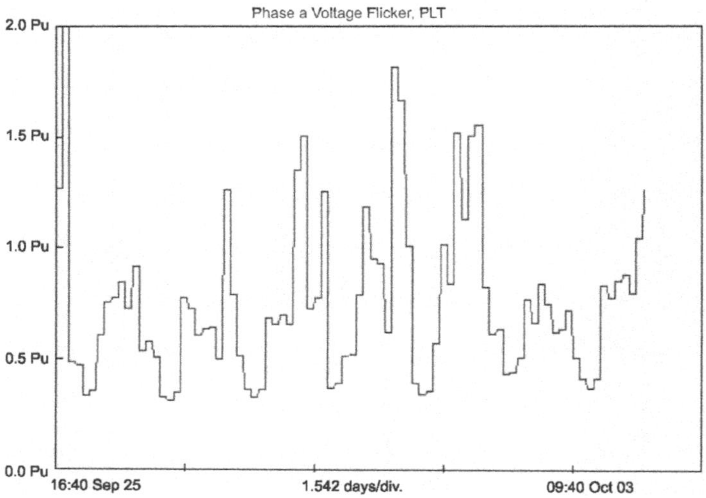

Figure 4.8. Example of one-week recording of long term flicker severity P_{lt}

4.2.3 Voltage Dips and Short Supply Interruptions

Voltage dip is a sudden reduction of voltage at a particular point of an electricity supply system below a specified dip threshold (within a time period not shorter than 10 ms), followed by its recovery after a brief interval. Typically, a dip is associated with the occurrence and termination of a short circuit or other extreme current increase on the system or installations connected to it (*e.g.*, starting of large motors). A voltage dip is mainly treated as a two-dimensional electromagnetic disturbance, the level of which is determined by both voltage (the residual voltage or the amplitude) and time (duration) – Figures 4.9 and 4.10. It can be seen that the disturbance source is at the users side as follows from the nature of the current change during the disturbance.

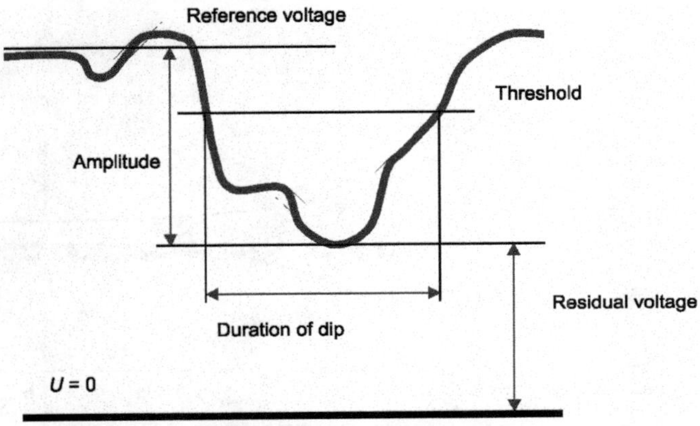

Figure 4.9. Voltage dip and its characteristics

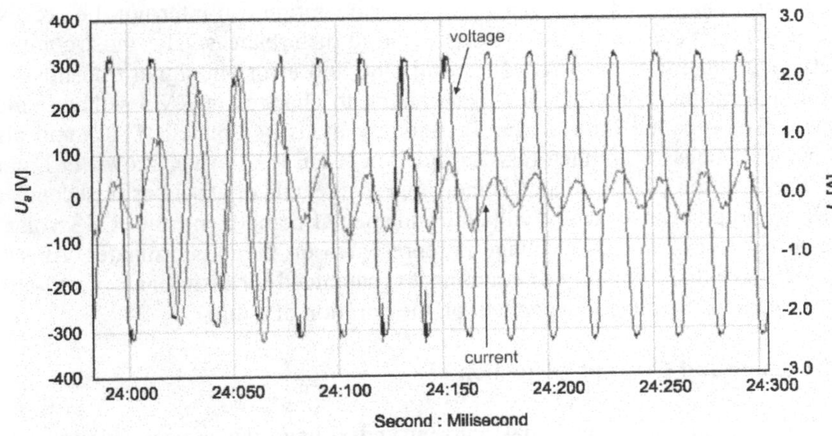

Figure 4.10. The example of a voltage dip recorded at the customer's terminals

Short supply interruption is a sudden reduction of a voltage on all phases at a particular point of an electric supply system below a specified interruption threshold (typically 10% U_N) followed by its restoration after a brief interval. Short supply interruption is a particular case of a voltage dip. Short interruptions are typically associated with switchgear operation related to the occurrence and termination of short circuits on the system or installations connected to it.

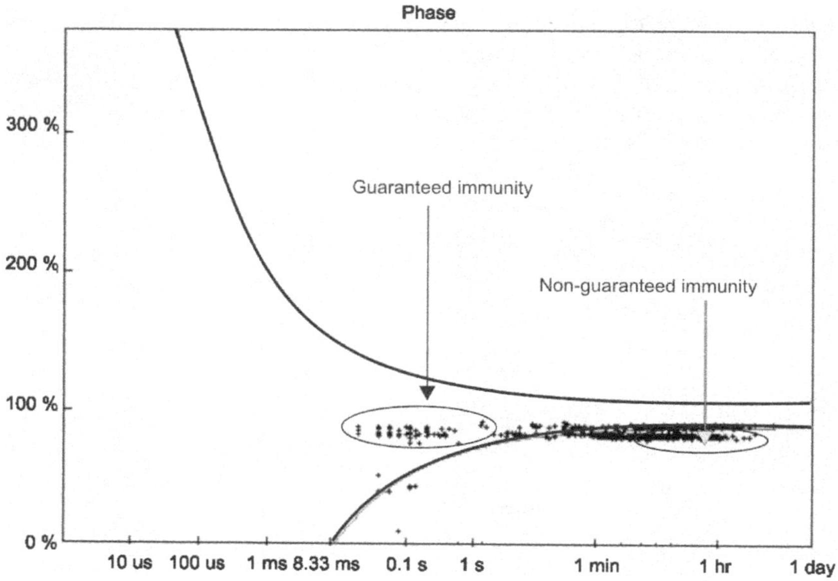

Figure 4.11. CBEMA curve

Microprocessor systems, widely used for process control, are particularly sensitive to voltage dips. Irregularities in their operation can interrupt the process, even if "*power*" equipment is immune to these disturbances. The most common effects of voltage dips are loss of transmission and errors in signal transmission. Information on the immunity of IT equipment to changes in RMS voltage value provides the so-called Information Technology Industry Council (ITIC) and also Computer Business Manufacturers Association (CBEMA) curve, shown in Figure 4.11. The duration of disturbance is indicated on the axis of abscissas in seconds or cycles of the power-frequency voltage fundamental component; the RMS voltage value in percent of nominal voltage is represented on the axis ordinates. As seen from the curve, the immunity of equipment (guaranteed for disturbances within the curve branches) is strongly dependent on the duration of a dip.

4.2.4 Voltage and Current Distortion

The distorted (*i.e.*, non-sinusoidal) current and voltage waveforms (Figure 4.12) became a normal operating condition in today's power system. The commonly adopted measure of distortion involves voltage and current harmonics, *i.e.*, components whose frequencies are integer multiples of the fundamental frequency. The most comprehensive information is obtained from the set which determines the orders, amplitudes (RMS values) and phases of individual harmonics.

Standardization documents adopt various numerical quantities defined on this set. They are mainly the *n*-th harmonic voltage ratio $U_{(n)}/U_{(1)}$ (analogically for the current harmonics) and total harmonic voltage distortion – THD_U (analogically for the current harmonics THD_I) calculated typically up to 50th order harmonic. These factors are the basis for standardization of power quality. Its CP95 values are comparable with the limits provided in standards and regulations.

Among the harmonic sources occurring in a power system, three groups of equipment can be distinguished:

- Equipment having magnetic core, *e.g.*, transformers, motors, generators, *etc.*;
- Arc furnaces and other arc devices, *e.g.*, high-pressure discharge lamps, arc welders, *etc.*;
- Electronic and power electronic equipment.

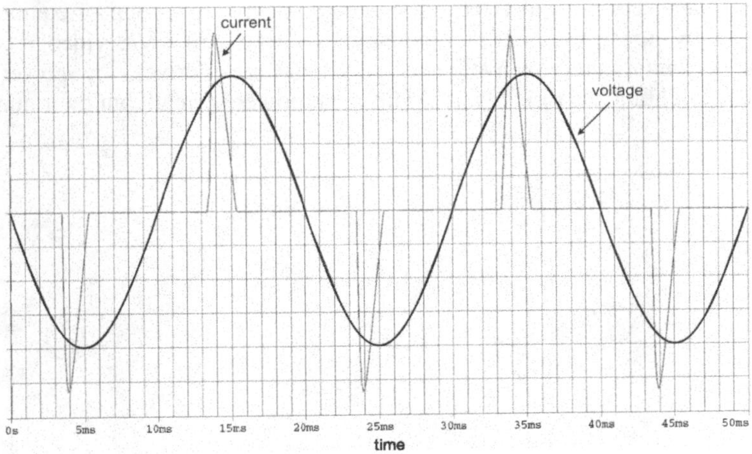

a

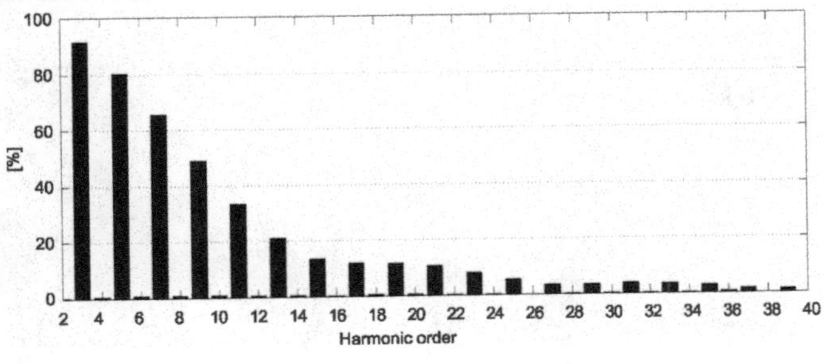

b

Figure 4.12. a The typical waveform (shown against the voltage waveform). **b** Spectrum of the current in a diode bridge with DC side capacitive filter (THD_I=80-130%)

Figure 4.12 shows the waveform and spectrum of a current typical for most of presently used electrical and electronic equipment (the current of a single-phase rectifier with capacitive fitter at DC side). It can be clearly seen that the third, fifth and seventh harmonics values are comparable with the fundamental component value. The massive use of single-phase power converters has increased the problems of power quality in electrical systems.

One of the most important goal of power electronics today is to introduce the AC/DC converters with reduced influence on supply network – active shaping of input currents.

A typical change in the voltage THD during workday and holiday 24 h is shown in Figure 4.13. Changes of this factor are regular in specific days of the week. Over the week the THD achieves its maximum value at the final phase of the weekend (Figure 4.14). Most of the time these values are correlated with the current changes. From this figure it is evident that very often the main sources of harmonics are not industrial loads, but residential loads, particularly power suppliers for electronic and electric home equipment. This group of equipment is used both by households and by industry. It consists of TVs, video recorders, computers, printers, micro-wave ovens, adjustable speed drives (low power), H.F. fluorescent lighting, small UPSs, *etc.* Their unit power is small, but their number is huge.

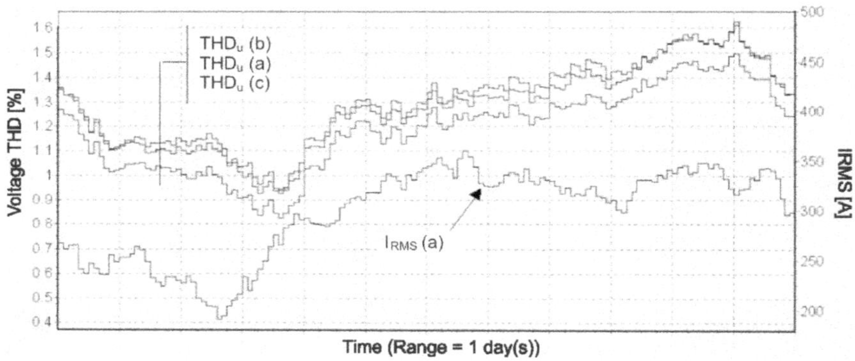

Figure 4.13. Change in the phase-to-neutral voltages distortion factor during the example workday 24 h (110 kV network)

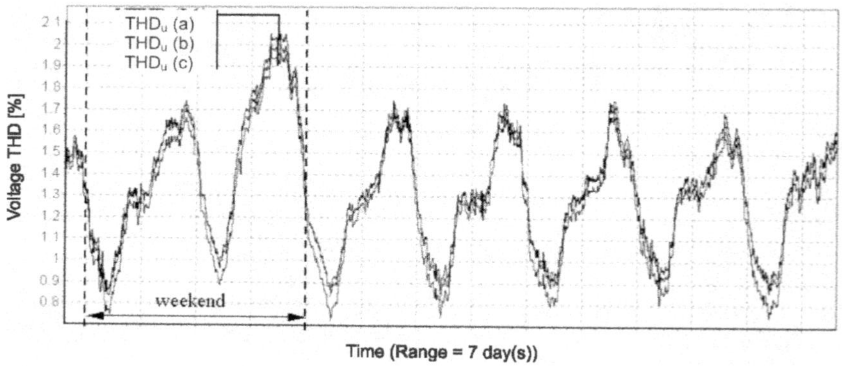

Figure 4.14. Week's variation of the phase-to-neutral voltage THD (110 kV network)

Apart from harmonics, voltage components with frequencies which are not an integer multiple of the fundamental frequency referred to as interharmonics [8]. These components produce a number of adverse effects, analogous to those of the harmonics may also occur in the supply network.

4.2.5 Classification of Electromagnetic Disturbances

Voltage disturbances can be divided into two groups (this differentiation is of particular importance from the point of view of a regulatory authority and responsibilities of the parties) [9]:

- Variations, *i.e.*, small deviations from the nominal or desired value, which occur continuously (slow and fast changes in voltage magnitude, fluctuations, unbalance, harmonics and interharmonics, *etc.*). Their main causes are system load variations or non-linear loads. Voltage variations result from the system operation, and therefore the value of the index describing these disturbances cannot significantly differ from the nominal level. Since the power system is designed to operate optimally under nominal conditions and sinusoidal voltages and currents, voltage variations shall be reduced to a minimum. Systems operators have to take measures to attain this condition which can lead to more efficient system management;
- Events, *i.e.*, sudden and significant deviations from the nominal, desired waveforms. Fast changes, dips, swells and transient overvoltages are the most typical disturbances in a system and, together with supply interruptions, the best-known examples of voltage events. Unlike voltage variations, occurring continuously, the voltage events are incidental. They are identified by continuous monitoring using recording equipment provided with a "*trigger*" function activated when the recorded quantity exceeds a set threshold value. The voltage events represent "*pathology*" in the supply network operation and are of great significance for the end customers' equipment. They can interrupt a production process utilizing electric power, even if a supply interruption does not occur. Since they are of random nature they are dealt with using mainly statistical methods.

Disturbances, measured in terms of their economic effects, may be of different significance so the following ranking can be adopted:

- In an industrial environment: voltage dips; supply interruptions (particularly short interruptions); voltage magnitude; asymmetry; harmonics; voltage fluctuations; frequency variations;
- For small enterprises and services: supply interruptions; voltage magnitude; voltage dips; overvoltages; asymmetry; harmonics; voltage fluctuations; frequency variations;
- In residential environment: voltage magnitude; voltage fluctuations; supply interruptions; voltage dips; overvoltages; harmonics; asymmetry; frequency variations.

4.3 Power Quality Monitoring

4.3.1 Measuring Procedures

Presently there are three basic normalization documents regulating the measurements of power quality parameters [10–12]. For each parameter measured, two classes: A (stands for "*Advanced*") and B (stands for "*Basic*") are defined. For each class, measurement methods and appropriate performance requirements are included. Users shall select the class that they require, based on their application(s).

Class A is used where precise measurements are necessary, for example, for contractual applications that may require resolving disputes, verifying compliance with standards, *etc.* Any measurements of a parameter carried out with two different instruments complying with the requirements of Class A, when measuring the same signals, will produce matching results within the specified uncertainty for that parameter.

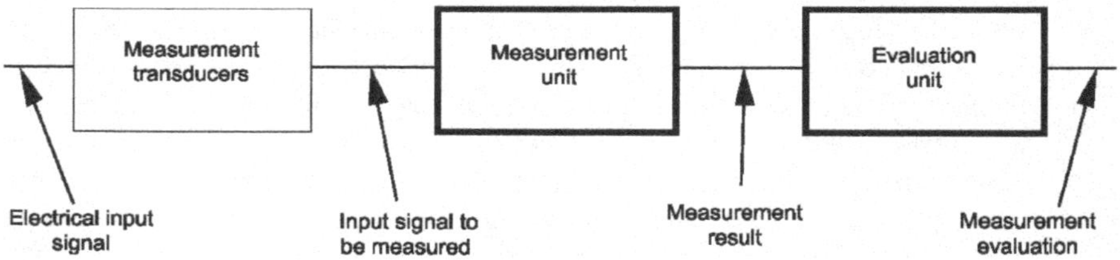

Figure 4.15. Measurement chain

Class B may be used for qualitative surveys, trouble-shooting applications and other applications where low uncertainty is not required.

The electrical quantity to be measured may be either directly accessible, as is generally the case in low-voltage systems, or accessible *via* measurement transducers. The whole measurement chain is shown in Figure 4.15.

4.3.2 Measurement Aggregation Over Time Intervals

For Class A the basic measurement time interval for parameter magnitudes (supply voltage, harmonics, interharmonics and asymmetry) shall be a 10-cycle time interval for a 50 Hz power system. The 10-cycle values are then aggregated over 3 additional intervals:

- 150-cycle interval (for 50 Hz nominal). The data shall be aggregated without gap from 15 10-cycle time intervals. This time interval is not a "time clock" interval. It is based on the frequency characteristic and is free running with the 10-cycle intervals;

- 10-min interval. The data for the 10-min time interval shall be aggregated from 10-cycle time intervals. Each 10-minute interval shall begin on a Universal Time Clock 10-min tick. Any overlapping 10-cycle interval is included in the aggregation of the previous 10-min interval;
- 2-h interval. The data for the 2-h interval shall be aggregated from 12 10-min intervals.

Aggregations shall be performed using the square root of the arithmetic mean of the squared input values.

4.3.3 Flagging Concept

During a dip, swell, or interruption, the measurement algorithm for other parameters might produce an unreliable value. The flagging concept therefore avoids counting a single event more than once in different parameters and indicates that an aggregated value might be unreliable. The flagging concept is applicable for Class A during measurement of power frequency, voltage magnitude, flicker, supply voltage asymmetry, voltage harmonics, voltage interharmonics, mains signalling and measurement under deviation and over deviation parameters.

If during a given time interval any value is flagged, the aggregated value, which includes that value, shall also be flagged. The user may decide how to evaluate flagged data.

4.3.4 Assessment Procedures

The quality of supply is assessed on the basis of a measurement of a specified index over a time no shorter than one week and the index value which was not exceeded during X% of the measurement time, determined from this measurement – the so-called CPX percentile (normally CP95 or CP99).

Figure 4.16a shows the example of one-week THD characteristic (10-min averaging), measurement time 100%. It is the basis for creating the Cumulative Probability Function (CPF) – Figure 4.16b (also Figure 4.5). The percentile, *e.g.*, CP95, a value determined from this characteristic, should not exceed the given limit level.

The quality of power became a trans-border issue. It determines the investment policies of multinational concerns. Hence the need for unification, on an international scale, of the assessment of the quality of power delivery. Normally the indices of supply reliability and voltage quality levels are not monitored individually for a given customer. During the transitional period (until the benchmarking system is developed) the suppliers should create a measurement system, enabling continuous monitoring of the proposed quality factors. The rule "*information is needed, not data*" should be applied. This approach requires processing of measurement results by the instrument or its support software. Power quality monitoring systems have become increasingly sophisticated.

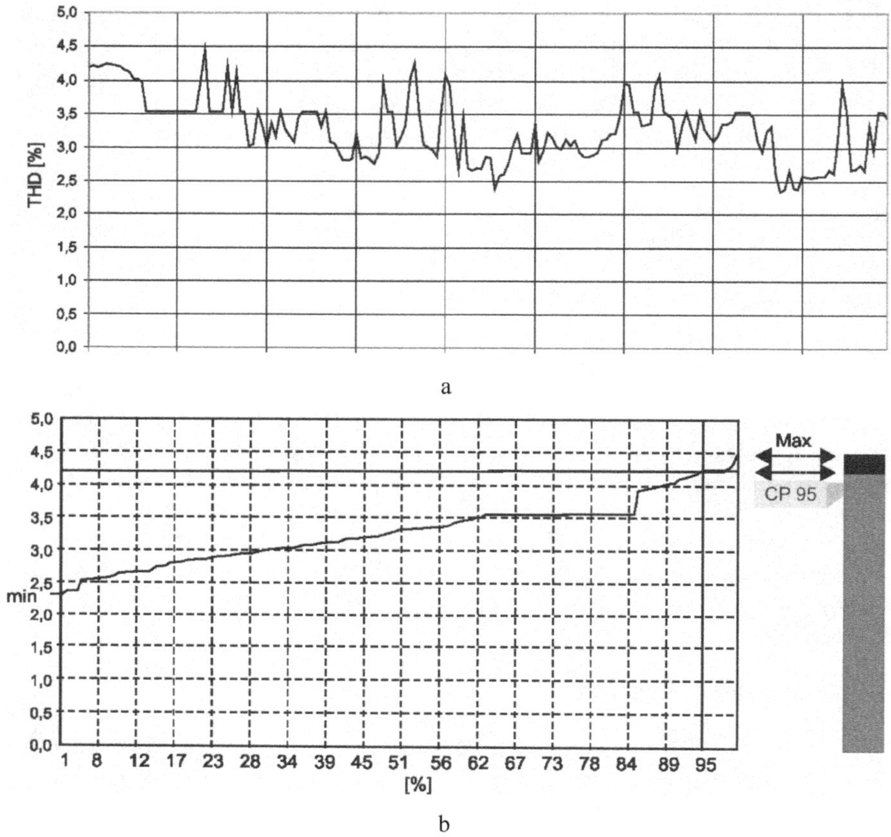

Figure 4.16. Determining the CP95 percentile value for the power quality index chosen as an example – THD: **a** THD characteristic during one week (100%); **b** CPF

Figure 4.17 illustrates an advanced monitoring system configuration that allows analysis of performance in a central database and sharing of power quality information using the intranet and Internet.

Often, in the case of a significant level of disturbance in electrical power system, at the customer's supply terminals, there is a need for locating the source of this disturbance, *e.g.*, harmonics [14–16] or voltage fluctuations [17]. This issue gains particular meaning when formulating contracts for electric power supply and enforcing, by means of tariff rates, extra charges for worsening power quality. In many cases, the quantitative determination of the supplier's and customer's share in the total disturbance level at the PCC is also required. Seeking inexpensive, reliable and unambiguous methods for locating disturbances in power systems, not employing complex instrumentation, is one of the main research areas that require a prompt solution.

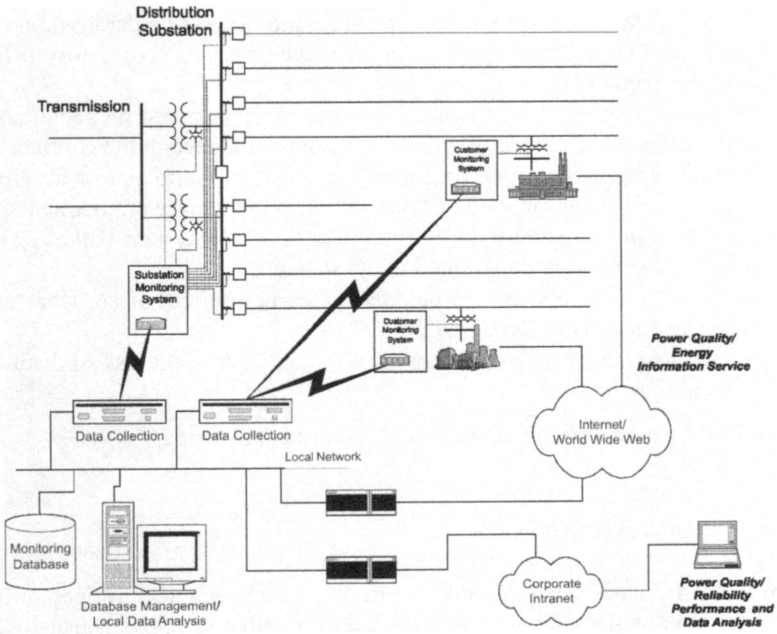

Figure 4.17. Advanced monitoring system configuration with a central database for power quality data management and sharing of information using the Internet [13]

4.4 Legal and Organizational Regulations

The old model in which the problem of power quality involved two partners – the electricity supplier and the customer – is replaced by a new configuration where at least four, mutually dependent parties participate: the customer, supplier of electric power, manufacturer of equipment and electrical installation contractor. The supplier often insists that sources of disturbances are located at the customer's side, whereas the latter complains about causes located in the supply network. It happens that their discussion leads to the conclusion, shared by both parties, that the equipment is not properly installed or adequately designed, to be operated in the given electromagnetic environment.

Is a relationship between these parties of antagonistic nature? Sometimes it is, but it is related to mutual coexistence and cooperation, with no alternative. The solution to this problem requires the precise definition of their own areas of activities and mutual responsibilities *i.e.*, establishing their positions on the electricity market. It is not an easy task. Various countries regulate responsibility for poor quality of electric power in different ways, and answers to the following example questions are differently formulated in the existing regulations:

- What will the financial process of the quality control look like?

- In case of failure to comply with contract provisions/standard requirements, shall the fine be paid on the benefit of the customer, regulatory office or system operator?
- Will the compensation fee paid to the customer be based on actual costs of damages /consequences, or will it be an amount agreed in the contract?
- Will the compensation be paid solely in effect of complaint or by virtue of the law, *i.e.*, always in case of failure to meet the quality requirements?
- Will a "*return fee*" be paid by the operator of the system if the quality of supply is better than determined in the standard?
- Will the annual total of compensation fees be limited? If so, what will be the basis for calculation of the limit value?
- Is the introduction of some form of insurance against the risk of disturbance contemplated?

There are many more questions, most of them still unanswered.

4.5 Mitigation Methods

Where the disturbance level exceeds the limit, various technical mitigation means will be needed to reduce these levels. The set of possible solutions is usually large. The choice of an adequate solution requires cost-benefit analysis of a project. This requires development of methods for accurate cost evaluation of various disturbances which determine the quality of power delivery.

In consequence of the development in large-power semiconductor devices and the microprocessor revolution, enabling more and more sophisticated control algorithms, the power electronics becomes omnipresent. At the same time the power electronics, which has been one of the dominant sources of low-frequency disturbances, now becomes an increasingly successful technical means of their mitigation. Power electronic solutions for power supply improvement, intended for low-voltage and medium-voltage applications, are jointly referred to as the custom power equipment [18, 19] (at the transmission system level such solutions are termed the FACTS [20]). The term "*custom power*" is used to describe advanced technologies that can be employed in the power system to improve the quality of electric power supplied to the end user. This technology includes many various solutions, like SVC, STATCOM, Active Power Filter (APF), series compensator – including DVR, Unified Power Quality Conditioners (UPQC), UPS *etc.* Many of the distribution companies consider this technical solution as a mean of supplying so-called "*premium power*" to the customer at extra charge [18], or rendering additional services to particularly important customers. These solutions may be applied at both the supplier's and the customer's side.

There are three possible ways to mitigate effects of poor quality of power (preferably applied simultaneously):

- Limiting the emission from disturbance sources. It is forced by product standards or by rules of issuing technical conditions of loads connection, and can be achieved by the use of, *e.g.*, power factor corrected converters, passive or active filters, reactive power compensators, front active converters for non-linear loads, proper earthing and shielding techniques, *etc.*;

- Reduction of the coupling level between the supply source and the disturbing load by means of connecting the emitting load at system points with large short-circuit capacity, supplying a disturbing load or a disturbance sensitive load from dedicated lines, elimination of line reactors, the use of parallel lines, ring operation of lines, proper location of compensating capacitors, and the use of autotransformers or load tap changing transformers, *etc.* Other technical means that also improve the quality of a power system, in terms of reducing the disturbances generated in the system and improving its immunity, include a larger share of cable lines with respect to overhead ones, a proper fault clearing practices, overvoltage protection, *etc.*;
- Improving loads' immunity to disturbances by the use of appropriate safety factors in equipment sizing, input filters, appropriate equipment design, shielding, the use of uninterruptible power supply systems and voltage stabilizers, adequate equipment selection to given power supply conditions, appropriate earthing techniques, *etc.*

A more detailed discussion of mitigation of poor power quality effects is possible with respect to specific disturbances, loads, production processes and precisely defined supply conditions. The final result of a case study is a list of possible solutions and evaluation of their implementation. As a general rule, the cost of measures taken for improving power quality is undoubtedly lower than the cost of potential effects of poor quality of power.

Costs of power quality improvement solutions increase with the distance from the source of poor quality of power or from the element most sensitive to it, that is from the final user. Both the severity of effects, and the necessity for the use of more and more advanced means for power quality improvement, depend on the type of electromagnetic environment (the class of electromagnetic environment) that determines the sensitivity of loads in this environment [5].

With regard to effects, it has to be emphasized that there are fields where good quality of power is *sine qua non* condition of their very existence, whereas the cost good quality is an issue of secondary importance. These undoubtedly include life support systems, data banks, emergency services and hospitals, military units, airports, financial institutions, banks, stock exchanges, public utilities like power stations, gasworks, water and sewage utilities, *etc.*

4.6 EMC Related Phenomena in Smart Electrical Power Systems

EMC analyses are increasingly urgent and important in efficient distributed power systems, since greater flexibility of energy conversion and improvement of power quality are achieved by means of the power electronic converters with pulse modulation. In these converters modern fast switching devices are commonly used and nowadays, in small and medium power ranges, thyristors, bipolar transistors and GTOs have been mostly replaced by fast switching IGBTs and MOS-FETs. Reduction of switching losses in the converters increases converter efficiency and makes higher switching frequencies possible. Unfortunately, apart from the positive aspects of the use of fast switching devices, the rapid development and widespread use of power semiconductors and power electronic systems have serious negative side effects.

Intentional power conversion processes in the low frequency band are realized by control processes at a higher frequency range. Due to the wide frequency range of useful electromagnetic processes in power converters, an unwanted EMI spectra spreading over the frequency range from DC to radiated emissions should be expected [21, 22]. However, the main source of EMC problems is the high *dv/dt* or *di/dt* of the nearly square-wave shape of voltage/current waveforms resulting from the switching states of the power converter. The steep slopes of the waveforms excite multi-mode oscillations in parasitic couplings producing, in turn, a wide-band high level HF spectrum of EMI.

Parasitic electromagnetic phenomena in typical power converter circuits have their own peculiarities. There are three key aspects of electromagnetic compatibility in power systems comprising power electronic converters [23]:

- A typical role of power converters and their place in the power system;
- A typical frequency range and high level of generated EMI noises caused by switching of relatively high voltage (in comparison with an EMC study in electronic devices);
- Specific features of the common mode voltage source in three-phase power converter systems.

The increasingly growing number of converters required in efficient distributed power systems mean that special solutions related to the new EMC situation have to be proposed, *e.g.*, concerning EMC standardization, measuring techniques, and EMI mitigation methods.

4.6.1 Origin and Effects of Electromagnetic Disturbances and EMC Terminology

As has been mentioned above, the origin of electromagnetic interference in systems consisting of power converters is the steep slopes of voltage waveforms in the presence of parasitic (mainly capacitive) couplings [24–26]. Because of the short rising and falling times of the voltages and typically long lengths of noise paths, the EMI currents, in most cases, may be treated as a free response of circuits of different resonant frequencies. Thus, electromagnetic emissions produced by power electronic converters are usually broadband in range from operating frequency to MHz band [22]. In EMC terminology such noises are conducted emission.

In order to describe EMI current paths and the appropriate application of emission reduction techniques, the method of splitting EMI current into a Common Mode (CM) and a Differential Mode (DM) is commonly used in EMC terminology [27]. The CM noise is a type of EMI induced on signals with respect to a reference ground. The remaining total conducted EMI is defined as the DM noise. While the CM/DM separation is well defined and understood for the single-phase or the DC system, the same cannot be said of three-phase converter systems and there is no universal CM/DM definition. However, splitting into the CM/DM in three-phase systems is still possible if a symmetrical, linear and time invariant three-phase system is considered [28].

The common mode voltage source in a three-phase system is

$$u_{CM}(t) = \frac{u_a(t) + u_b(t) + u_c(t)}{3} \tag{4.1}$$

where u_a, u_b, u_c – phase voltages.

Figure 4.18 shows the CM voltage at the output of the three-phase two-level inverter and the train of damped sinewave pulses of CM current that flow as a result of each CM voltage step. The CM voltage is a staircase function with the step of 1/3 U_d (U_d – DC link voltage).

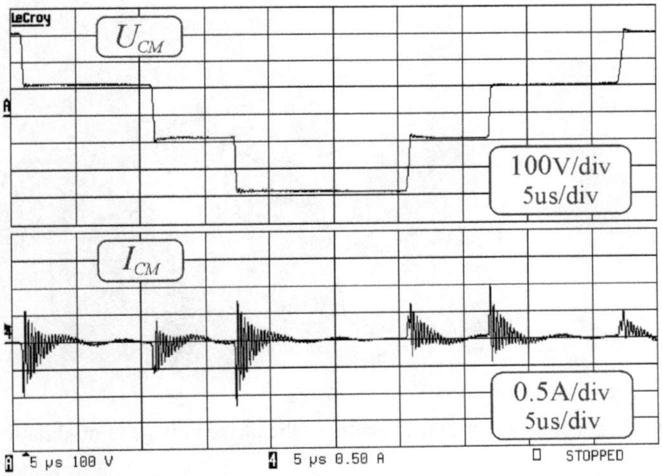

Figure 4.18. Phase voltage at the output of the inverter and the CM current in PE wire

The shape of the single pulse of CM current depends on both *dv/dt* of the CM voltage slopes and resonant frequencies and damping factors of the current path. Usually the oscillatory modes of the CM current are located in the CISPR B frequency band. Control algorithms mainly influence the CM noises spectrum in CISPR A frequency band due to the typical switching frequency of power converters used in efficient power systems.

To determine voltage excitations for both modes the double Fourier integral analysis could be applied. The decomposition has been done by means of the analysis of the sum of phase voltages. The concept has been verified for a three-phase natural sampled (sinusoidal) PWM using double Fourier series analysis.

With this algorithm, phase voltages can be expressed by the well known [29] relation

$$\begin{aligned} u_{iN}(t) &= V_{DC} + V_{DC} M \cos(\omega_0 t + \theta_i) + \\ &+ \frac{4V_{DC}}{\pi} \sum_{m=1}^{\infty} \frac{1}{m} J_0\left(m \frac{\pi}{2} M\right) \sin\left([m+n]\frac{\pi}{2}\right) \cos(m\omega_c t + n[\omega_0 t + \theta_i]) + \\ &+ \frac{4V_{DC}}{\pi} \sum_{m=1}^{\infty} \sum_{\substack{n=-\infty \\ n \neq 0}}^{\infty} \frac{1}{m} J_n\left(m \frac{\pi}{2} M\right) \sin\left([m+n]\frac{\pi}{2}\right) \cos(m\omega_c t + n[\omega_0 t + \theta_i]) \end{aligned} \tag{4.2}$$

where i=a, b, c – phase leg identifiers for three-phase inverter; m, n – harmonic index variables; V_{DC} – half of DC link voltage; $J_n(x)$ – Bessel function of order n and argument x; ω_c – angular frequency of carrier waveform; ω_0 – angular frequency of fundamental component; M – modulation index.

Figure 4.19 shows the theoretical harmonic spectra of the phase voltage for a three-phase natural sampled (sinusoidal) PWM *vs* modulation index.

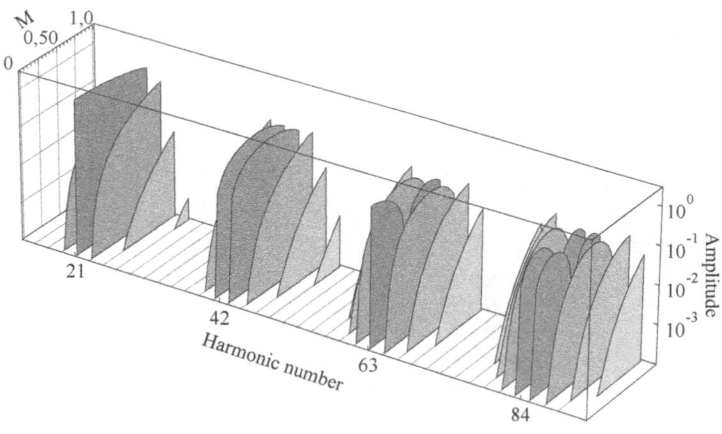

Figure 4.19. Theoretical harmonic spectra of the phase voltage *vs* modulation index

Sideband harmonics with even combination of $m \pm n$ will not appear in the sum because of the term $\sin[(m+n)\pi/2]$. The elimination term $\cos[1+2\cos(n2\pi/3)]$ causes a cancelation of specific harmonics in the CM voltage. Harmonics, which are cancelled in the CM voltage, are represented in the DM voltage because of the term $\sin(n\pi/3)$. Thus, CM and DM components in the phase voltages [30] can be expressed by Equations 4.3 and 4.4

$$\begin{aligned} u_{DMi}(t) &= \frac{8V_{DC}}{\sqrt{3}\pi} \sum_{m=1}^{\infty} \sum_{n=-\infty}^{\infty} \frac{1}{m} J_n\left(m \frac{\pi}{2} M\right) \sin\left([m+n]\frac{\pi}{2}\right) \times \\ &\times \sin n\frac{\pi}{3} \cos(m\omega_c t + n\omega_0 t) \end{aligned} \tag{4.3}$$

$$u_{CMi}(t)=\frac{4V_{DC}}{3\pi}\sum_{m=1}^{\infty}\sum_{\substack{n=-\infty \\ n\neq 0}}^{\infty}\frac{1}{m}J_n\left(m\frac{\pi}{2}M\right)\times \times\sin\left([m+n]\frac{\pi}{2}\right)\left[1+2\cos n\frac{2\pi}{3}\right]\times\cos(m\omega_c t+n\omega_0 t) \tag{4.4}$$

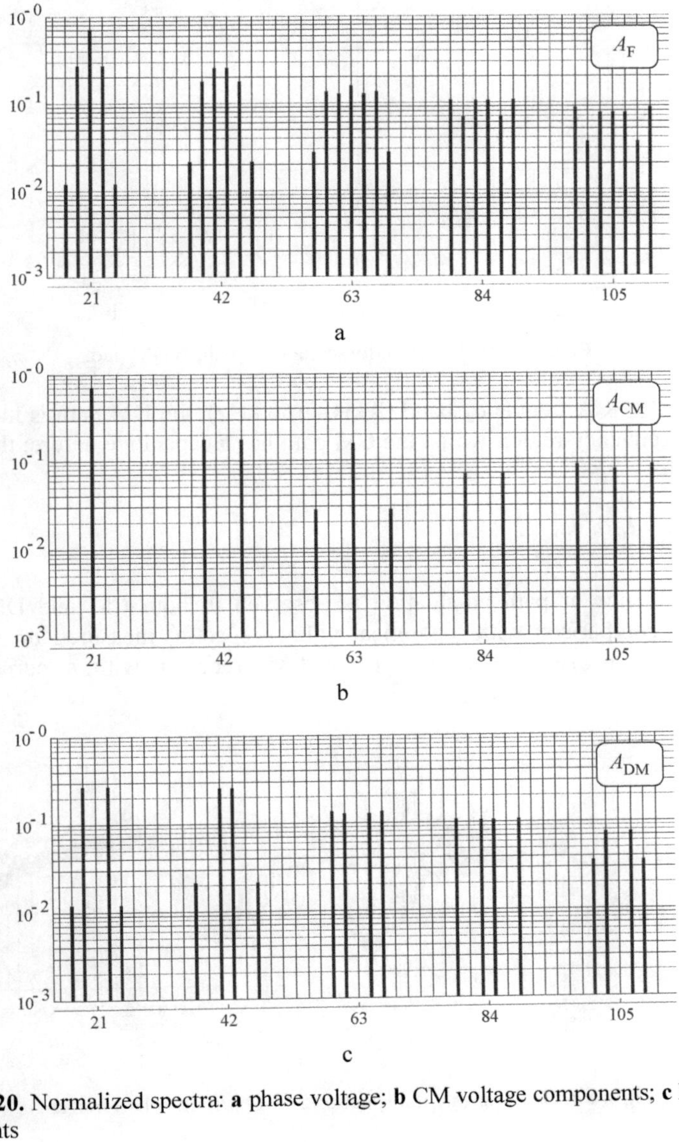

Figure 4.20. Normalized spectra: **a** phase voltage; **b** CM voltage components; **c** DM voltage components

Figure 4.20 shows the result of the analytical decomposition of phase voltages into CM and DM components for arbitrarily selected parameters of the modulation (modulation index M=0.9, f_c/f_o=21).

The shape of CM currents is influenced by parameters of CM current path. Figure 4.21 shows insertion losses of CM and DM paths on the motor side of the drive system.

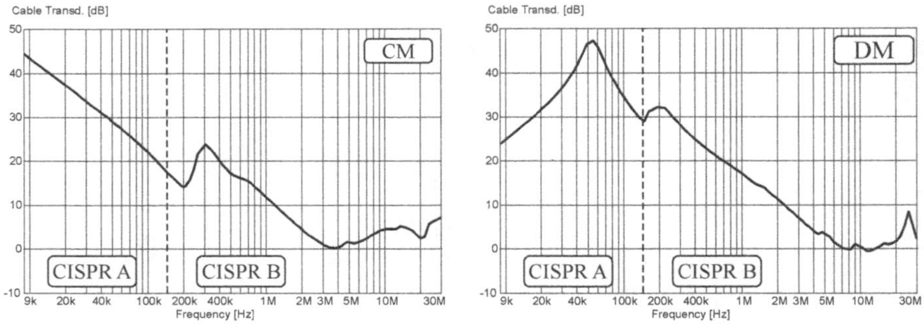

a b

Figure 4.21. Insertion losses: **a** CM path; **b** DM path

In the CISPR A frequency band (9 kHz–150 kHz), the DM path is inductive in the lower frequency region whilst the CM path has a capacitive nature; thus can be expressed as

$$I_{CM}(\omega) = V_{CM}(\omega) \cdot j\omega C \tag{4.5}$$

The influence of path attenuation is observed in both CM and DM current spectra (in comparison with theoretical voltage spectra) measured on the motor side of the converter by means of the EMI receiver with a current probe (Figure 4.22).

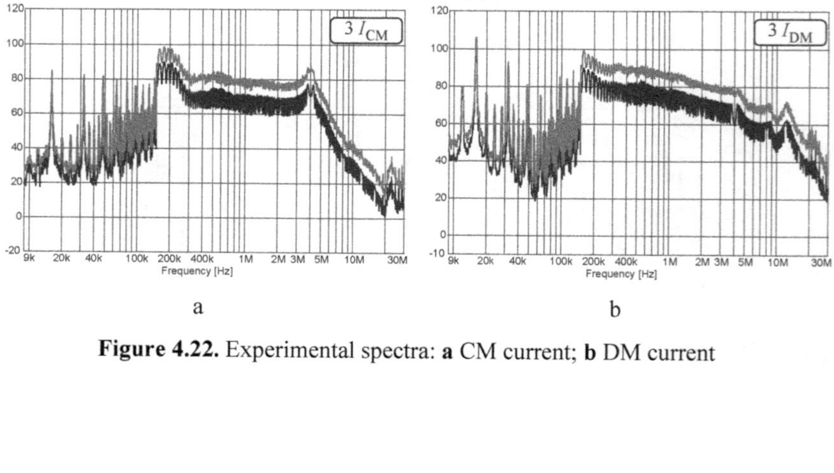

a b

Figure 4.22. Experimental spectra: **a** CM current; **b** DM current

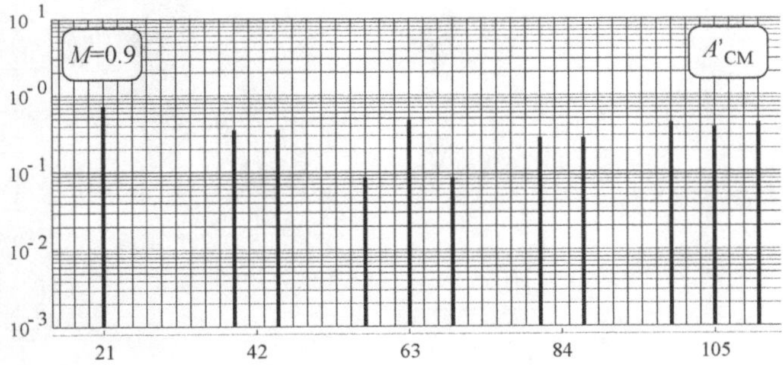

a

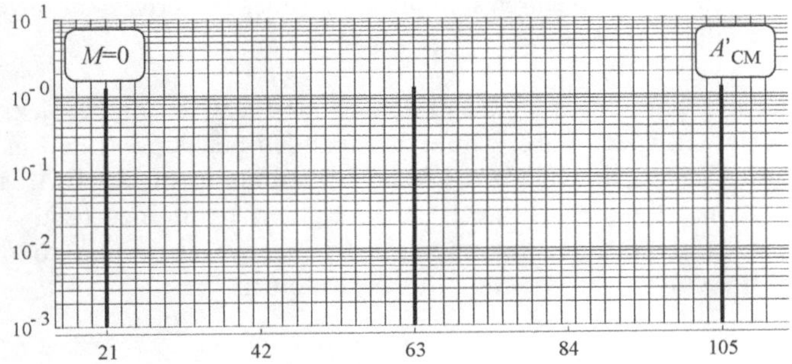

b

Figure 4.23. Normalized weighted harmonic spectra of CM voltage: **a** modulation index M=0.9; **b** modulation index M=0

Harmonics of the CM current spectrum could be approximately expressed as proportional to harmonics of CM voltage weighted by the number of harmonic group (m)

$$A'_{mn}(\omega) = A_{mn}(\omega)m \tag{4.6}$$

where A_{mn} – coefficients of Fourier expansion.

Figure 4.23 shows the spectrum of weighted harmonics of CM voltage (modulation index M=0.9 and M=0). The shape of the analytical spectrum shown in Figure 4.23 (for M=0.9) qualitatively maps the experimentally obtained spectrum of the CM current in CISPR A frequency band, presented in Figure 4.24.

This analytical approach can be useful in the comparative analysis of an influence of inverter control algorithms on the spectra of CM currents in a given system.

Figure 4.24 shows the experimental spectra of CM currents in PWM inverter fed drives measured by an EMI receiver for two different values of inverter output frequency (f_0=50 Hz and f_0=0 Hz) which approximately correspond to theoretical spectra for modulation indexes M=0.9 and M=0 (Figure 4.23).

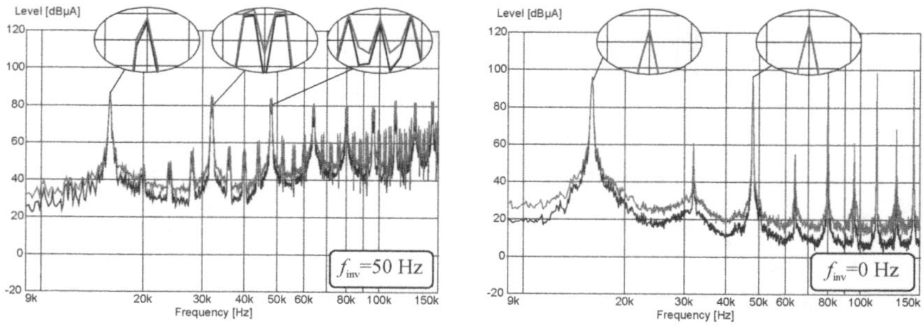

a b

Figure 4.24. Spectra of CM currents in PWM inverter-fed drives for inverter output frequency: **a** f_{inv}=50 Hz; **b** f_{inv}=0 Hz

The sideband harmonics concentrated around carrier harmonics related to theoretical weighted spectra are visible in expanded peaks presented in magnified views (Figure 4.24). The relative levels of common mode EMI noises are consistent with the theoretical spectra of weighted harmonics (and additionally aggregated in bandwidth for intermediate frequency) according to the way in which the EMI receiver "*aggregates*" those harmonics.

4.6.2 EMC Standardisation

Within the IEC, two committees, TC77 (Electromagnetic Compatibility) and Comite International Special des Perturbations Radioelectriques (CISPR), are mainly responsible for EMC standardization; however the others are involved in this subject matter as well. This particular care is required to avoid discrepancies between different standards with different scopes, but dealing with the same technical subject in this situation. The Advisory Committee on Electromagnetic Compatibility (ACEC) was established to ensure coordination between all these special committees and with the outside world. IEC prefers a global approach to EMC and a systematic approach to the classification of electromagnetic phenomena [31] (Table 4.1).

Thus, EMC standardization covers the whole frequency range from 0 Hz up to the GHz frequency region, although traditionally the term EMC was particularly related to high-frequency and/or radio-frequency phenomena, and EMC regulations covered frequency range from 9 kHz to 1 GHz. This frequency range is divided into two major classes of disturbances: conducted emission (9 kHz–30 MHz) and radiated emission (30 MHz–1 GHz). Both conducted and radiated emissions are subdivided into CISPR bands.

In the field of EMC, the IEC as an international organization closely cooperates with other international (ISO, ITU), regional and national organizations, especially with the European Committee for Electrotechnical Standardization (CENELEC). Technical Committee 210 (generally responsible for basic and generic standards) is in charge of European EMC standardization, and in accordance with its agreement with the IEC, in principle, applies the EMC standards of the IEC. However, if the IEC does not accept the new work proposal or is not able to perform it within a required time [32], the CENELEC begins to prepare new standards and develops some provisional documents of its own which are intended to be replaced later by corresponding IEC publications. Moreover, IEC may adopt standards prepared by CENELEC.

Table 4.1. Overview of the principal electromagnetic disturbance phenomena

Conducted low-frequency phenomena
Harmonics, interharmonics
Signalling systems
Voltage fluctuations
Voltage dips and interruptions
Voltage unbalance
Power frequency variation variations
Induced low-frequency voltages
DC in AC networks
Radiated low-frequency field phenomena
Magnetic fields
• Continuous
• Transient
Electric fields
Conducted high-frequency phenomena
Directly coupled or induced voltages or currents
• Continuous wave
• Modulated waves
Unidirectional transients
Oscillatory transients
Radiated high-frequency field phenomena
Magnetic fields
Electric fields
Electromagnetic fields
• Continuous waves
• Modulated waves
• Transients
Electrostatic discharge phenomena (ESD)
High-altitude nuclear electromagnetic pulse (HEMP)

IEC standards are mainly only recommendations representing the state of art and in this sense may serve as reference. They may acquire a legal status, however, if they are introduced into the legislation of a country by law or decree.

The situation is slightly different in the European Union because of the "*New Approach*" Directives brought into effect in order to eliminate the technical barriers hampering trade between the Union Countries [33].

The European Directive 89/336/EEC "*Electromagnetic Compatibility*" is, as is the other "*New Approach*" Directives, limited to defining the essential protection requirements for electric and electronic equipment that must be fulfilled before products may be offered to the EU market. According to the EMC definition Directive requirements are related to both emission and immunity.

The EMC Directive implementation is quite complicated. Standards published by the CENELEC are the basis for "*the harmonization of standardization*" in all EU member countries and must be transferred into the framework of national standards, where they replace all other standards on the same subject. They include particular technical requirements to meet the Directive's demands. "*Harmonized standards*" may assume the status of either a recommendation or a law.

The Directive makes compliance possible by different routes. Although compliance with the standard is not obligatory, the usage of all applicable harmonized standards (self certification to standards) is in most cases the easiest way and is said to give a device the presumption of conformity with the Directive's essential protection requirements.

The "*Harmonized standards*" are grouped in three categories:

- Basic standards: specify the general condition or rules necessary for achieving electromagnetic compatibility and to which the product committees may refer. They are independent of any specific product and are applicable to all products [31].

  Basic standards contain terminology (IEC 60050(161)), descriptions of the electromagnetic phenomena and fundamental EMC definitions and terms (IEC 61000-1-1, IEC 61000-1-2, IEC 61000-2-5), compatibility levels and description of the EM environment in public LV power systems (IEC 61000-2-x), limits for LF conducted disturbances (IEC 61000-3-x), HF conducted and radiated disturbances (CISPR 11/EN 55011, CISPR 14/EN 55014, CISPR 22/EN 55022), emission measurement techniques apparatus (CISPR 16).

  Standard IEC 61000-4-1 gives an overview of immunity tests and immunity testing techniques are included in IEC 61000-4-x standards as well as specifications of compatibility levels, general requirements for the limitation of emission of disturbances, measurement techniques, test techniques and their applicability, installation and mitigation guidelines. The set of basic standards IEC 6100-5-x comprise also the guidelines for EMC compliant installation including earthling, cabling, filtering shielding and surge protection;
- Generic standards: comprise two sets of standards specified to the residential, commercial and light industry environments and for industrial environments. Each of them comprise requirements and tests related to both emission (IEC 61000-6-3, IEC 61000-6-4) and immunity (IEC 61000-6-1, IEC 61000-6-2), respectively. Generic standards are applicable if there is no EMC standard specified to the equipment in a given environment;
- Product standards: provide the requirements and tests specific for certain products or product families.

There are two problems related to EMC standardization, which could emerge as a result of the application of the large number of the power converters in distributed power systems. Broadband and coherent EMI generated by power electronic converters are located in the band from operating frequency up to the range of few megahertz. However, product and product family standards for power electronic equipment (*e.g.*, IEC 61800-3 for PDS, IEC 62040-2 for UPS) usually cover only CISPR B (150 kHz–30 MHz) frequency band, although the lower frequency noises in distributed power systems (50th harmonics of mains frequency – 150kHz) are coming to light more and more; thus they should be more common in EMC standards [22].

Most of the present harmonized standards are related to single items of equipment. For EMC standardization purposes, requirements may be specified in some cases for the whole installation, especially for small systems. It is a challenge for standardisers that they have to describe how a large installation area should be dealt with in the EMC context [31]. Basic standards give only some general installation guidelines. In the EMC Directive an installation is described as a combination of apparatus, components and systems assembled and/or mounted in a given area. In some cases *in situ* measurements should be performed for large system installations.

In other cases, EMC requirements must be specified for each item, in particular for immunity. The supplier of each unit must then indicate the installation conditions for his product (wiring, earthing) that ensure correct functioning of the whole installation [21].

It is evidently difficult and sometimes impossible to carry out emission or immunity tests on a set of units dispersed over a wide area, It is also necessary to take account of the fact that the test may be influenced by environmental conditions and may not be appropriate for certification purposes.

4.6.3 Conducted EMI Spreading Over Distributed Electrical Power Systems

For physical reasons (*e.g.*, long distances between individual elements) and on account of high *dv/dt* of converter voltages the EMI currents in the cable are, in fact, travelling wave phenomena [34]. The understanding of these phenomena is very important for a proper analysis of EMI spectra in dispersed systems [35].

The frequencies of excited noises are formed as results of multiple reflections of the travelling wave and depend on the propagation time between reflections. The parasitic circuits of CM and DM currents can be quite different. However, the velocity of the electromagnetic waves depends only on the physical properties of the propagation environment and in the high frequency region may be expressed by Equation 4.7

$$v = \frac{1}{\sqrt{\varepsilon_0 \varepsilon_r \mu_0 \mu_r}} \tag{4.7}$$

where ε_0 is the dielectric constant; ε_r is relative permittivity; μ_0 is the permeability of the free space; μ_r is relative permeability.

Figures 4.25 and 4.26 show the spectra of CM and DM currents in a shielded and unshielded open-ended cable fed by an inverter. In terms of the travelling wave approach, CM current oscillatory mode in the unshielded cable has higher frequency because electromagnetic wave propagates in the air. In the case of the shielded cable, the frequencies of the main oscillatory modes for both CM and DM are approximately the same, because electromagnetic waves propagate in the material with similar permittivity [36]. The magnitude of CM current spectrum is much smaller because of decreased transverse line-to-ground parasitic capacitances.

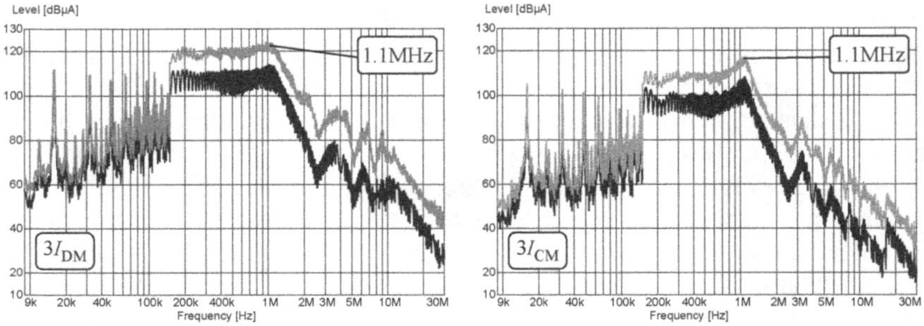

a b

Figure 4.25. EMI currents spectra in a 42 m long shielded cable: **a** DM current; **b** CM current

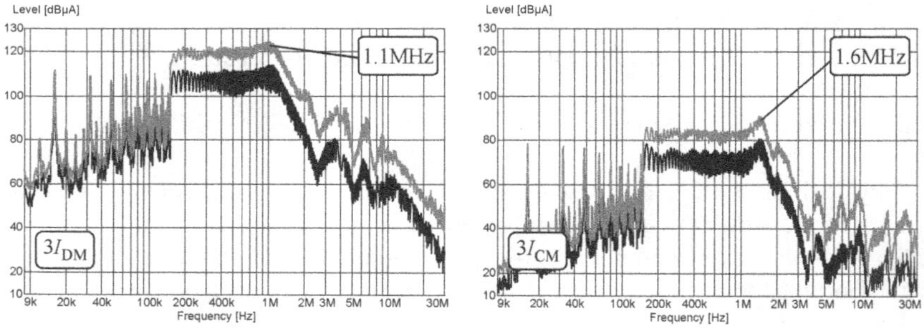

a b

Figure 4.26. EMI currents spectra in a 42m long unshielded cable: **a** DM current; **b** CM current

The observations done on the basis of the spectra presented above have been confirmed by measurements over time. The waveforms of phase voltage, phase current, DM and CM noise currents are shown in Figure 4.27.

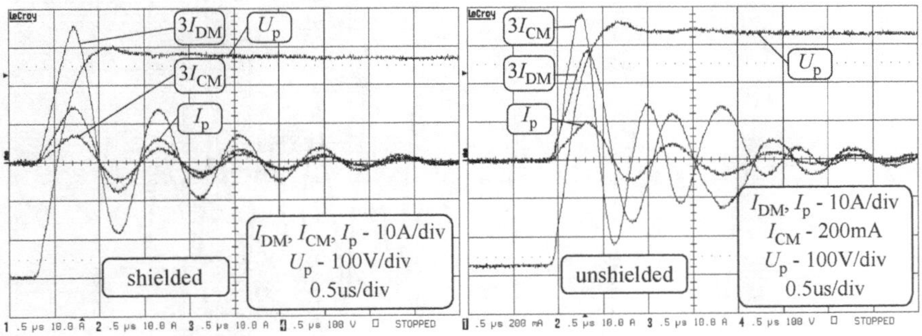

a b

Figure 4.27. Phase voltage, phase current, DM and CM noise currents in: **a** shielded cable; **b** unshielded cable

As can be seen in Figure 4.27 both CM and DM current oscillations are strongly damped. However, cable attenuation strongly depends on the frequency and the key question that appears is how far components of EMI noises forced by almost square wave excitation can penetrate the electric grid.

Figure 4.28 shows frequency characteristics of impedance module and phase for 10 m and 20 m long, typical electric power cable. In fact, high frequency components would be strongly damped. In the lower frequency region where the impedance module is much lower, the cable attenuation would depend on cable length, its parameters and physical properties of the load. The results suggest that transmission line theory could be the right approach. However, in the real situation, the wave impedances of EMI current path are far from matched; thus there is still a lot of research to do in this area.

It is generally believed that because of the cable attenuation, EMI noises cannot penetrate deeply into the electric grid. However our preliminary investigations show that it could be a real problem.
Figure 4.29 shows the spectrum of CM current measured in the PE wire of a feeder cable of a 25-kW, four-quadrant frequency converter (at converter terminals). The level of CM current is relatively high.

In order to demonstrate how deep EMI noises can penetrate a local low voltage power grid, the measurements in a transformer station 200 m away from the converter have been carried out.

Figure 4.30 shows EMI current spectra measured in the PE wire of the power cable on the low voltage side of a power transformer.

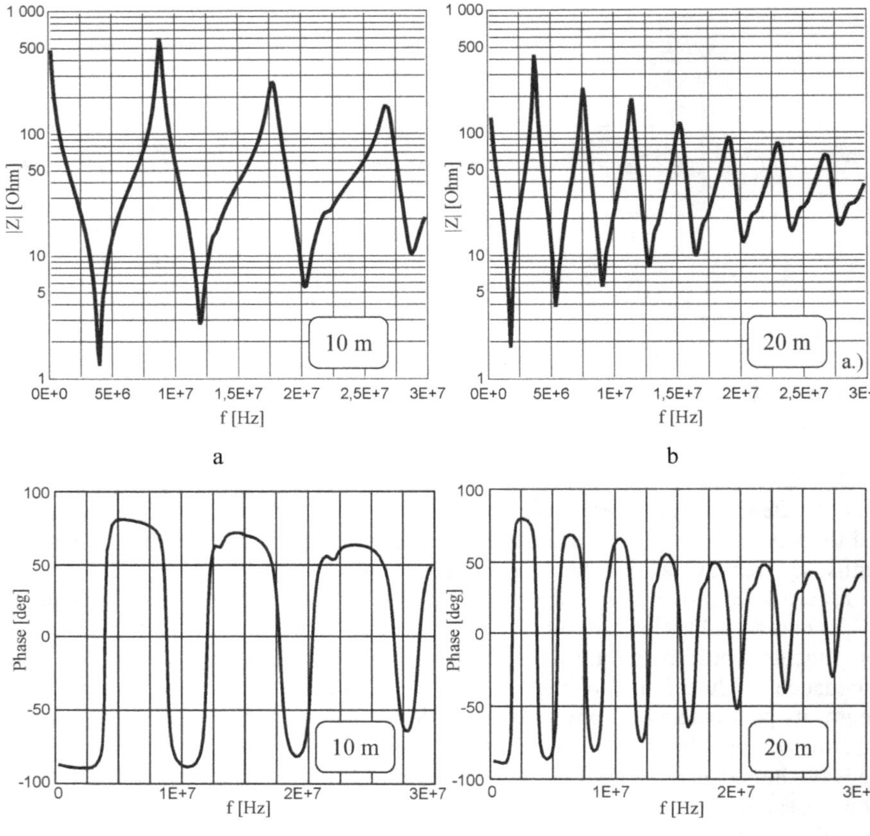

Figure 4.28. Frequency characteristics of impedance: **a** module 10 m long cable; **b** module 20 m long cable; **c** phase10 m long cable; **d** phase20 m long cable

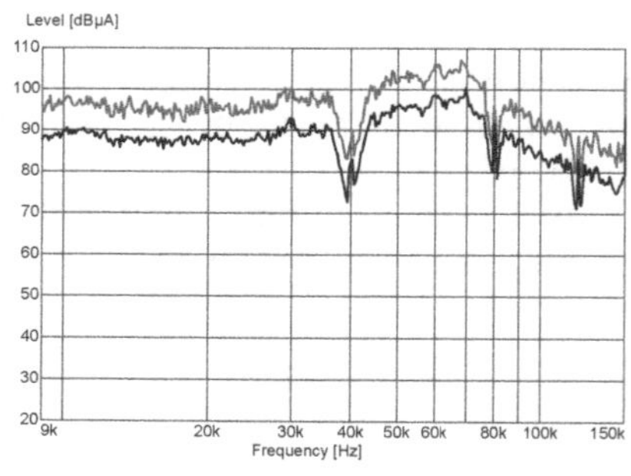

Figure 4.29. Spectrum of CM current measured in the PE wire of four-quadrant frequency converter

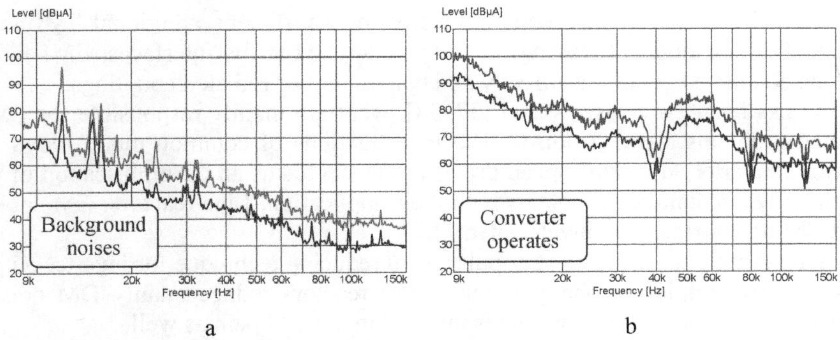

Figure 4.30. EMI current spectra measured in the PE wire of the power cable in transformer station: **a** background CM noises; **b** CM noises when converter operates

The attenuation of characteristic oscillatory mode frequency is only 20dB. This means that CM noises, especially those located in CISPR A frequency range, can spread over a large area in spite of the fact that in a local electric grid there are a lot of alternative CM current paths.

4.6.4 Improving EMC of Distributed Power Systems

An important and useful tool in EMC control for any system is good installation and engineering practice. The descriptions of the methods for EMC compliant installation development have been presented in a valuable book "*EMC for systems and installations*" [21].

Distributed power systems consist of interconnected power modules. The low frequency phenomena are typically DM and may be regarded as functional rather than EMC issues. The power supply connections also act as paths for high frequency disturbances which are much harder to predict and should be filtered at the module supply interface. Because of the complicated interaction between the modules of distributed power systems, only an in-depth EMC analysis may bring the required results on which the EMC control plan should be determined.

The basic methods, according to [21], requires observance of rules of good installation practices, *e.g.*,

- Implementation of protection zones in the installation;
- Bonding of the structural components of rack, cabinets and ducting;
- Layout and connection of apparatus, filters and surge protectors within cabinets;
- Layout, segregation and routing of cables within the system;
- Proper termination of screened cables.

These EMC techniques apply to the majority of terrestrial systems and installations. However, safety requirements are always paramount, and thus safety demands should not be compromised by any EMC driven techniques.

If good installation practice is ineffective or insufficient, additional EMI noise attenuation techniques (filtering) have to be applied at the interfaces. EMI filters reduce conducted couplings directly and help to reduce radiated couplings.

As has been previously stated [37, 38], what are mainly responsible for EMC related problems are CM noises. The investigations of common mode filters for power converter adjustable speed drives are presented as an exemplification of the commonly encountered problems with their application. Line reactors, CM chokes and CM transformers are investigated.

Series reactors are the most popular *dv/dt* reducing technique for inverter output voltages and ripples in phase currents. The reactors reduce mainly DM noises; however their resultant inductance is inserted in the CM path as well.

Another often recommended measure to diminish CM currents is a CM choke. The CM choke has three windings wound on the common toroidal ferrite core and thus acts only for CM noises and does not influence the working current of a DM nature.

A CM transformer is, in fact, a CM choke with an additional tightly coupled secondary winding shorted by a damping resistor [39].

Figure 4.31 shows the waveforms of CM currents (with their RMS values) in the motor PE wire in the drive without filters and with the above described filters.

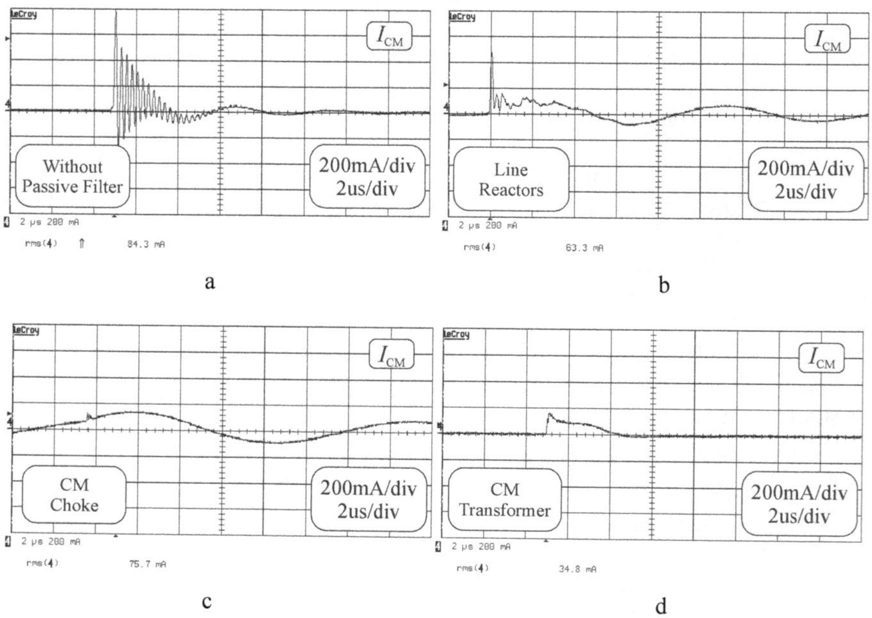

Figure 4.31. Influence of passive filters on CM current shape and its RMS value in system: **a** without filter; **b** with line reactors; **c** with CM choke; **d** with CM transformer

The main part of conducted electromagnetic emissions is formed in resonant (partly parasitic) circuits. The effect of both a CM choke and line reactors on CM current waveforms consists in an insertion of an additional inductance in the CM current path. The differences are caused by various inductances and different values of parasitic capacitances of the devices. Increased values of time constant and characteristic impedance of the resonant circuit permit the reduction of significantly higher frequency components of the CM current on the motor side. However, a weaker damping factor cause that the RMS value is not reduced significantly. High frequency spikes visible especially in the CM current in the drive with line reactors are caused by parasitic turn-to-turn capacitances of inductors, which constitute part of the CM current path.

An aperiodic decay form of the CM current in the drive with the CM transformer assures a reduction of its RMS value.

The comparative results in frequency domain obtained from systems with the various CM current reducing techniques [40] are shown in Figure 4.32 (IF BW=9 kHz for both CISPR A and CISPR B frequency band).

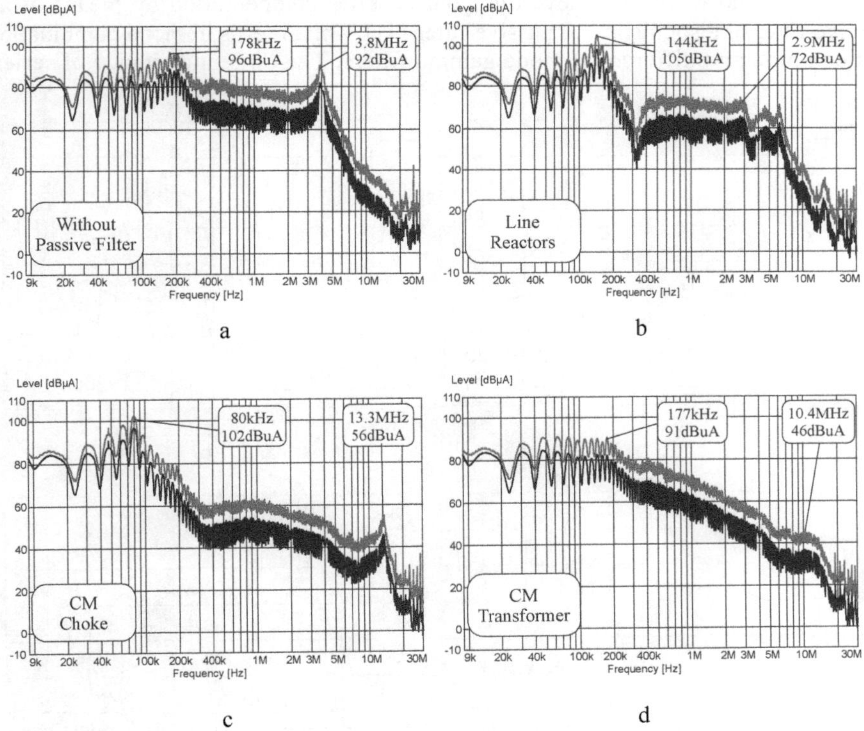

Figure 4.32. Influence of passive filters on spectrum of CM current in system: **a** without filter; **b** with line reactors; **c** with CM choke; **d** with CM transformer

All of the applied filters attenuate the high frequency component (3.8 MHz) in the original spectrum. However, due to a lower damping factor of the CM current path in the system with the line reactors or the CM choke, there has been observed a high level of less damped peaks shifted to lower resonant frequencies (in CISPR

A range). There is no resonant peak in the spectra of CM current in the system with the CM transformer because of the aperiodic waveform of this current.

It would appear that the best solution to meet the requirements of standard EN 61800-3 (CISPR B frequency range) might be a CM choke. However, series resonance between the inductance of a CM choke and a motor-to-ground parasitic capacitance is responsible for the peak at a frequency of 80 kHz. In contrast, the CM transformer acts in the whole conducted emission range (CISPR A and CISPR B).

The flow of the CM current also influences the shape and magnitude of the CM voltage in the neutral point of the load that is shown in Figure 4.33. The increased value of CM voltage in the case of line reactors and CM choke may have adverse effects, especially in drive systems (Chapter 5). It is obvious that, due to the common origin of CM side effects, the most convenient way to eliminate all of them simultaneously is a cancelation of the CM voltage.

One of the ways to achieve this goal is series compensation by means of an active filter or a sophisticated passive filter [41]. This filter is in fact a combination of common and differential mode gamma filters. The large inductance of series

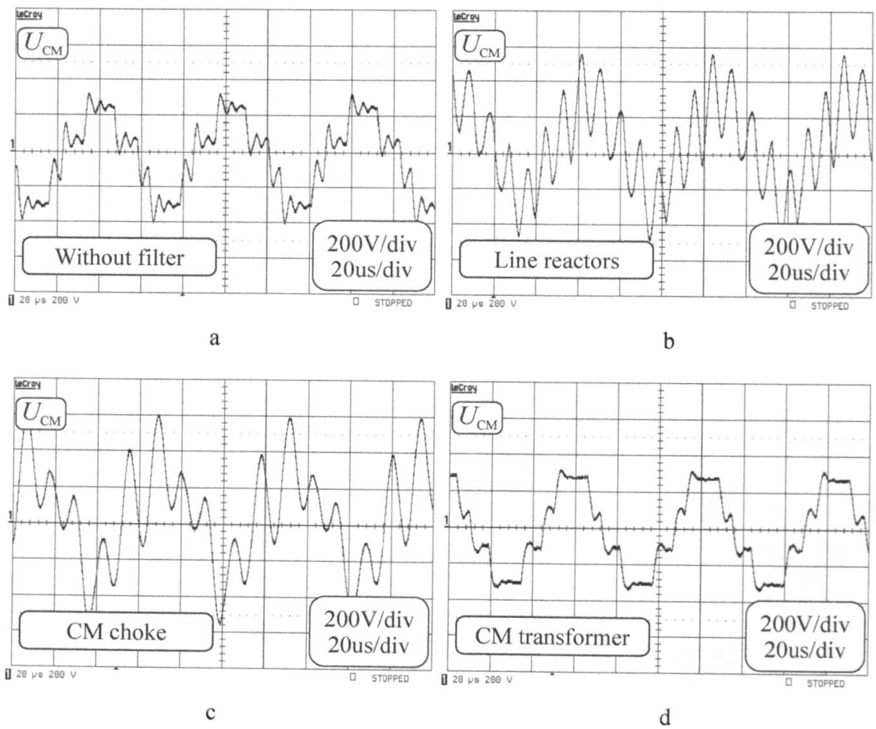

Figure 4.33. CM voltage in system: **a** without filter; **b** with line reactors; **c** with CM choke; **d** with CM transformer

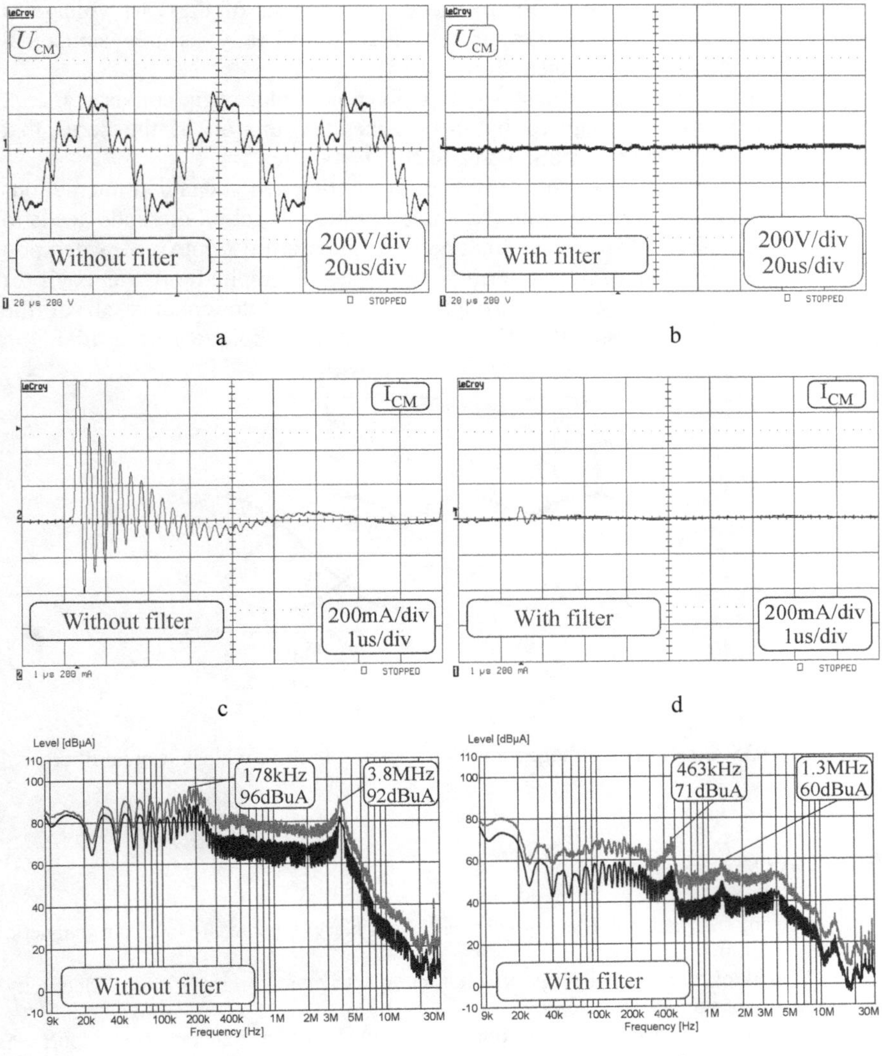

Figure 4.34. CM voltage in drive: **a** without filter; **b** with passive filter; CM currents in drive in time domain: **c** without filter; **d** with passive filter; CM currents in frequency domain: **e** without filter; **f** with passive filter

connected inductors (mainly of a CM choke) in the inverter-motor path causes a CM voltage drop, which is almost equal to the common mode voltage at the output of the inverter [41]. Relevant measurement results of CM voltage at stator winding neutral point, CM current in motor PE wire, both in time and in frequency domain, are shown in Figure 4.34.

The compensation of the high frequency component of the CM voltage is almost perfect, and the DM part of the filter additionally assures sinusoidal waveforms of line-to-line voltages (Figure 4.35).

This filter is not an all-purpose solution for power electronic converters, *e.g.*, there are some problems caused by the magnetic saturation of the cores that significantly reduces the application area [42].

Due to the complexity of parasitic electromagnetic phenomena and interactions between components of the system, the choice of the mitigation methods needs to take various aspects of electromagnetic compatibility into consideration simultaneously. If one of the filtering techniques is applied on the basis of preliminary measurements, it is strongly recommended to conduct all of the measurements again to ensure the effect of a mitigation technique is as desired.

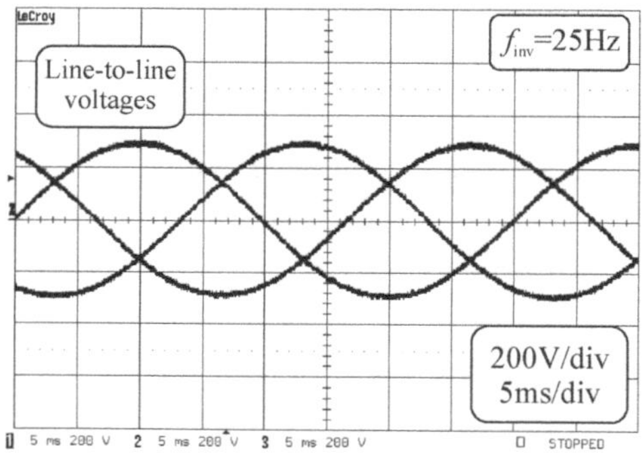

Figure 4.35. Line-to-line voltages at motor terminals in drive with sinusoidal filter

References

[1] IEC 50, International Electro-technical Vocabulary, Chapter 161: electromagnetic compatibility

[2] Council of European Energy Regulators, Electricity Working Group, Electricity Quality of Supply Task Force

[3] Council of European Energy Regulation, (2005) Third benchmarking report on quality of electricity supply

[4] Advisory Committee on Electromagnetic Compatibility, IEC Council

[5] IEC 61000-2-4, Electromagnetic compatibility (EMC). Part 2: environment. Section 4: compatibility levels in industrial plants for low-frequency conducted disturbances

[6] IEC 61000-3-6, Electromagnetic compatibility (EMC). Part 3: limits. Section 6: assessment of emission limits for distorting loads in MV and HV power systems

[7] IEC 61000-3-7, Electromagnetic compatibility (EMC). Part 3: limits. Section 7: assessment of emission limits for fluctuating loads in MV and HV power systems

[8] Hanzelka Z, Bień A, (2004) Interharmonics. Power quality application guide. Leonardo Power Quality Initiative, www.lpqi.org

[9] Bollen M, (1999) Understanding power quality problems: voltage sags and interruptions, Wiley

[10] IEC 61000-4-7, Electromagnetic compatibility (EMC). Part 4: testing and measurement techniques. Section 7: general guide on harmonics and interharmonics measurements and instrumentation, for power supply systems and equipment connected thereto
[11] IEC 61000-4-15, Electromagnetic compatibility (EMC). Part 4: testing and measurement techniques. Section 15: flicker-meter – functional and design specifications
[12] IEC 61000-4-30, Electromagnetic compatibility (EMC). Part 4: testing and measuring techniques. Section 30: measurements of power quality parameters
[13] Dugan RC, Granaghan MF, (2002) Electrical power system quality. McGraw-Hil, New York
[14] Chun L, Wilsun X, Thavatcha T, (2004) A critical impedance based method for identifying harmonic sources. IEEE Transaction on Power Delivery, vol. 19, no.2:671–678
[15] Pyzalski T, Wilkosz K, (2003) New approach to localization of harmonic sources in a power system. Electrical Power Quality and Utilisation Conference
[16] Wilsun X, Xian L, Yilu L, (2003) An investigation on the validity of power-direction method for harmonic source determination. IEEE Transaction on Power Delivery, vol.18, no.1
[17] de Jaeger E, (2000) Measurement and evaluation of the flicker emission level from a particular fluctuating load. Prepared on request of CIGRE/CIRED WG CC02 (Voltage Quality)
[18] IEEE, (2002) Custom power – state of the art. Working Group 14.31
[19] Ghosh A, Ledwich G, (2002) Power quality enhancement using custom power devices. Kluwer
[20] Higorani MG, Gyugyi L, (2000) Understanding FACTS. IEEE
[21] Williams T, Armstron K, (2000) EMC for systems and installations. Newnes
[22] Tichanyi L, (1995) Electromagnetic compatibility in Power Electronics. J.K. Eckert & Company Inc.
[23] Kempski A, ed. by Orlowska-Kowalska T, (2007) Specific aspects of electromagnetic compatibility in power electronics. Power electronics and electrical drives: selected problems. Oficyna Wydawnicza PWr:163–173
[24] Ran L, Casadei D, Clare J, Keith B, Christopoulos C, (1998) Conducted electromagnetic emissions in induction motor drive systems. IEEE Transaction on Power Electronics, vol.13, no.4
[25] Skibinski G, Kerkman R, Schlegel D, (1999) EMI emissions of modern PWM AC drives. IEEE Industry Applications Magazine, vol.5, no.6:47–81
[26] Hu J, van Bloch J, De Doncker W, (2004) Typical impulses in power electronics and their EMI characteristics. Power Electronics Specialists Conference:3021–3027
[27] Grandi G, Casadei D, Reggiani U, (2004) Common- and differential-mode HF current components in *AC* motors supplied by voltage source inverters. IEEE Transaction on Power Electronics, vol.19, no.1
[28] Shen W, Wang F, Boroyevich D, Liu Y, (2004) Definition and acquisition of CM and DM EMI noise for general-purpose adjustable speed motor drives. Power Electronics Specialists Conference:1028–1033
[29] Holmes GD, Lipo TA, (2003) Pulse width modulation for power converters. Principles and practice. IEEE Press
[30] Kempski A, Smolenski R, (2006) Decomposition of EMI noise into common and differential modes in PWM inverter drive system. Electrical Power Quality and Utilization. Journal, vol.12, no.1:53–58
[31] www.iec.ch

[32] Sanhet JL, (1997) Standardization in power electronics. European Power Electronics Conference, vol.1:140–145

[33] Rossetto L, Tenti P, Zuccato A, (1999) Electromagnetic compatibility issues in industrial equipment. IEEE Industry Applications Magazine, vol.5, no.6:34–46

[34] Ogasawara S, Akagi H, (2000) Analysis and reduction of EMI conducted by PWM inverter-fed *AC* motor drive system having long power cables. Power Electronics Specialists Conference, vol.2:928–933

[35] Skibinski G, Kerkman R, Leggate D, Pankau J, Schlegel D, (1998) Reflected wave modeling techniques for PWM *AC* motor drives. IEEE Applied Power Electronics Conference:1021–1029

[36] Magnusson PC, Alexander GC, Tripathi VK, Weisshaar A, (2001) Transmission lines and wave propagation. CRC Press

[37] Weston DA, (1991) Electromagnetic compatibility. Principles and Applications. Marcel Dekker Inc.

[38] Clayton P, (2006) Introduction to electromagnetic compatibility. John Wiley & Sons

[39] Ogasawara S, Akagi H, (1996) Modeling and damping of high-frequency leakage currents in PWM inverter-fed AC motor drive systems. IEEE Transactions on Industry Application, vol.32, no.5:1105–1113

[40] Kempski A, Strzelecki R, Smolenski R, (2004) The influence of passive EMI filters on various aspects of electromagnetic compatibility. Power Electronics Specialists Conference:970–975

[41] Akagi H, Hasegawa H, Doumoto T, (2002) Design and performance of a passive EMI filter for use with voltage source PWM inverter having sinusoidal output voltage and zero common-mode voltage. Power Electronics Specialists Conference:on CD

[42] Kempski A, Strzelecki R, Smolenski R, Benysek G, (2003) Suppression of conducted EMI in four-quadrant AC drive system. Power Electronic Specialists Conference:1121–1126

5

EMC Cases in Distributed Electrical Power System

Adam Kempski and Robert Smoleński

Institute of Electrical Engineering,
University of Zielona Góra,
50 Podgórna Street, 65-246 Zielona Góra, Poland.
Email: A.Kempski@iee.uz.zgora.pl; R.Smolenski@iee.uz.zgora.pl

5.1 Four-quadrant Frequency Converter

The four-quadrant frequency converter is presently commonly applied in novel asynchronous drives, electric power generation using asynchronous variable speed generators and T&D systems [1]. The main circuit of this converter comprises two IGBT three-phase bridges and an intermediate circuit allowing two-way energy flow and four-quadrant operation.

The operation of IGBT devices (high *dv/dt*) causes high level conducted and radiated electromagnetic emissions.

We shall examine an asynchronous drive with the four-quadrant frequency converter in this section as a most interesting EMC case due to different impedance characteristics on the mains and load side. From the point of view of EMC analysis, the motoring and generating quadrants are quite the same. Figure 5.1 shows a four-quadrant asynchronous drive with elements that influence EMC behavior: heatsink-to-DC link capacitance and line reactors [2]. The limits for conducted emissions are contained in EN 61800-3 (EMC product standard for PDS).

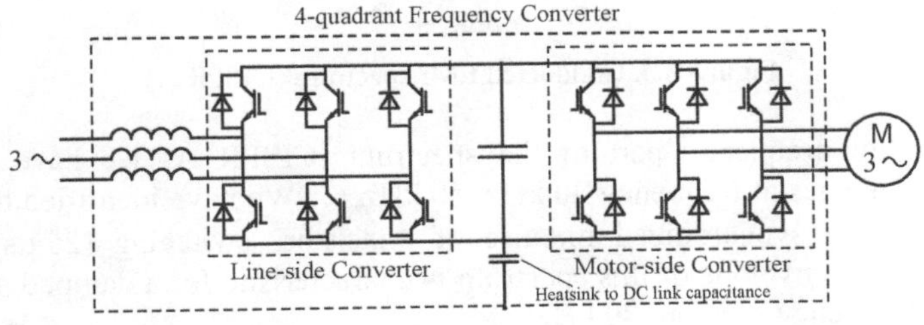

Figure 5.1. Main circuit diagram of the four-quadrant inverter drive system

We have tested a two-pole, 10kW induction motor fed by a typical industrial four-quadrant frequency converter supplied *via* LISN (Line Impedance Stabilization Network). Figure 5.2 shows the results of measurements which have been carried out on the system consisting of LISN and EMI receiver ESCS-30, in the frequency range specified in EN 61800-3. We can observe oscillation modes of EMI current waveforms at frequencies 2.5 MHz, 3.8 MHz and 4.7 MHz. The limits are slightly exceeded at the frequency 2.5 MHz, and significantly at the beginning of the CISPR B band (150 kHz).

The shape of the EMI envelope implies that its source could be located in a lower frequency range.

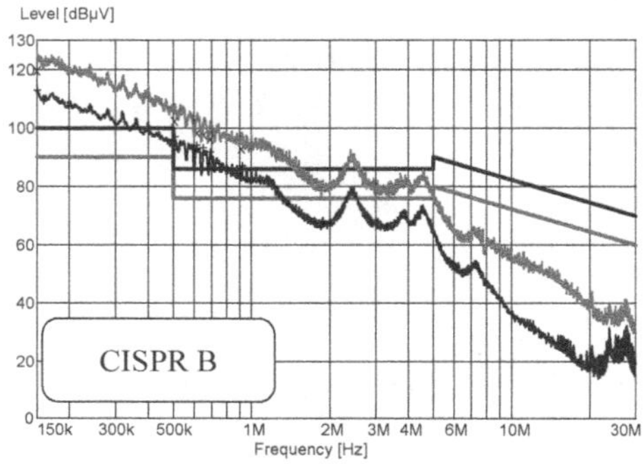

Figure 5.2. Conducted EMI spectrum (drive without filters)

Figure 5.3 shows the results of additional measurements in CISPR A band (not required by the related Standard).

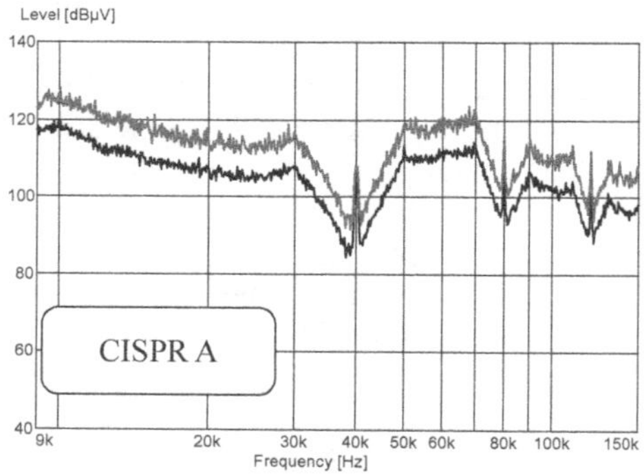

Figure 5.3. Conducted EMI spectrum (CISPR A)

In the low frequency part of the spectrum (CISPR A) we have observed repeatable changes at frequency 40 kHz, 80 kHz, … We have identified them with the time of the synchronized impulse of transistors switching (25 µs) and its harmonics. The envelope of this spectrum is characteristic for a damped sine wave pulse at a frequency of about 70 kHz.

In order to establish the nature of dominant oscillatory modes (CM or DM) measurements using a current probe have been made directly in the PE wires on the line and motor side (Figure 5.4).

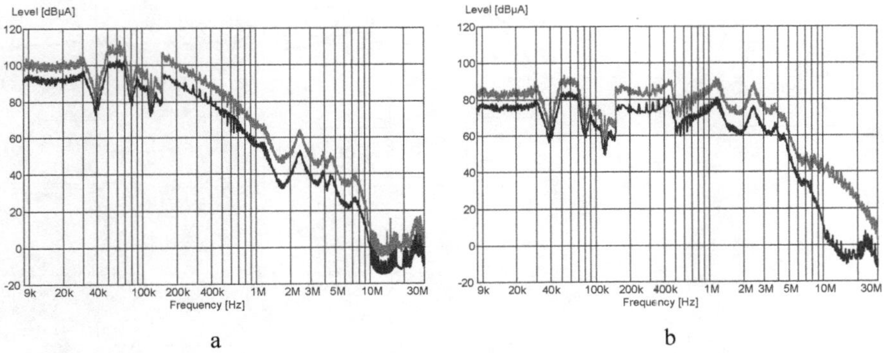

Figure 5.4. Spectrum of CM current on: **a** line side; **b** motor side

In the CM spectrum, it is possible to observe main oscillation modes of the EMI spectrum. On the line side the low frequency component prevails, whereas on the motor side the higher frequency components are more significant.

The measurements in the frequency domain seem to indicate that there are two CM voltage sources in a four-quadrant AC drive system; the first on the line side of the converter and the second on the motor side and the DC link is the interface between both CM parasitic circuits.

The measurements in the time domain have been made in order to confirm this presumption and to localize the sources and paths of CM currents. The additional PE wire, marked as "*inverter PE*", that creates "*virtual grounding point*" makes this localization possible, as shown in Figure 5.5.

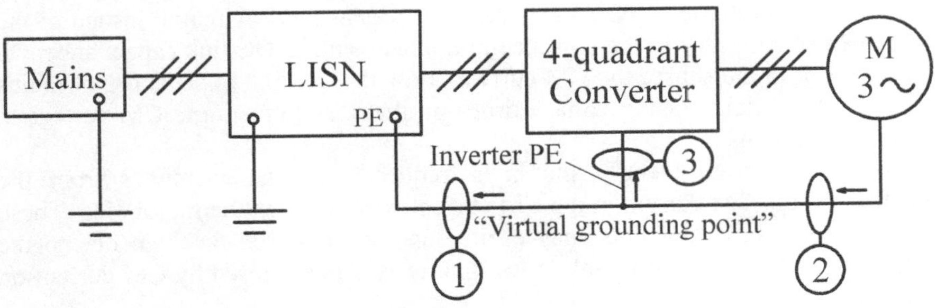

Figure 5.5. Experimental arrangement

Figure 5.6 shows a passage of CM currents through the "*virtual grounding point*". Arrows in Figure 5.5 depict the real CM current paths in the system. All time domain measurements have been made using current probes with a linear frequency range up to 50 MHz.

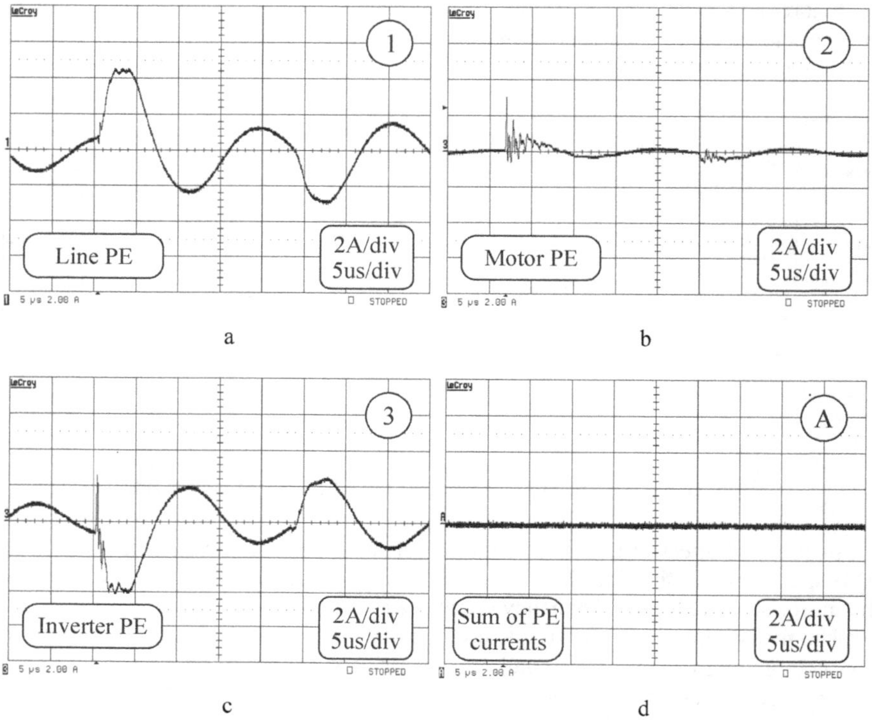

Figure 5.6. Passage of common mode currents through "*virtual grounding point*": **a** line PE; **b** motor PE; **c** inverter PE; **d** sum of PE currents

The CM current on the line side of the converter has a dampened oscillation form with an amplitude up to 10 A, a frequency of about 70 kHz and an RMS value which can exceed 1 A. This current emerges at each switching instant of the controlled rectifier due to the presence of the heat-sink to DC link capacitance. On its way back to the source the CM currents flow through the impedances of mains and the input filter, causing voltage drops on them, and creating a CM voltage at the input terminals.

It is possible to identify the high frequency oscillation modes from the spectrum of the CM current in the CM current waveform on the motor side. These components have a low frequency oscillation superimposed on them due to the voltage drop across the heat-sink to DC link capacitance caused by CM current on the line side.

The source that forces CM current to flow on the motor side is CM voltage at the output of the inverter. The high frequency oscillations are excited by fast changes of the CM voltage in parasitic capacitive couplings, which inherently exist inside the load system (motor and cables). The amplitude of the CM current can reach a value of about 4 A.

The CM currents are closed in two almost independent loops. The heat-sink to DC link capacitance is a common part of these two loops. The shape of the CM current in the inverter PE confirms the essential role of heat-sink to DC link capacitance for both creating and conducting CM noises [2].

The CM current on the line side of the converter is formed in the loop consisting of the heat-sink to DC link capacitance and impedances of series reactors and LISN (or mains). Because of the fact that impedances LISN are much lower than impedances of series reactors there are two ways of suppressing the current, which is measured by LISN [3]. The first is to use a common mode choke. However, the high amplitude of the line side CM current leads to significant technical difficulties in suppressing it to the acceptable level. The second is to use knowledge of the location of the source and paths of the CM current for creating an alternative CM path without appreciable changes of the CM current. In our experimental arrangement we have used star-connected Y capacitors for this purpose, (see Figure 5.7). The value of the capacitances has been estimated on the basis of measurements in the time domain to achieve a relatively small CM voltage drop on them. In such a situation, the small part of CM current, which flows through the impedance of mains can be easily suppressed using a relatively small CM choke. The impedance of converter line reactors, heat-sink to DC link capacitance and CM impedance of mains had to be considered in the filter design. The configuration of this filter may appear to be a typical input equipment CM filter; however, in this case the role of the filter differs from its typical application (the attenuation of input noises from mains). This filter is, in fact, the output filter from an "*equipment point of view*".

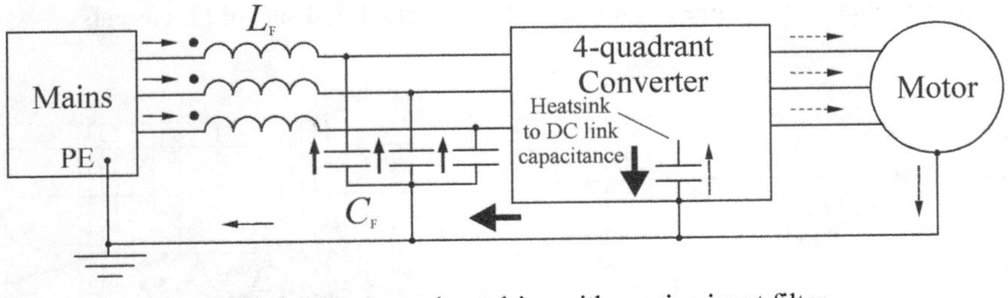

Figure 5.7. Four-quadrant drive with passive input filter

Figure 5.8 shows a passage of CM currents through "*virtual grounding point*" in the drive system with the input filter. The CM current on the line side has been almost entirely suppressed. The motor PE current seems to be wholly closed in the loop on the motor side of the converter. Figure 5.9 shows the EMI spectrum in the drive with the input filter.

It is possible to see a significant attenuation (about 40dB) at the frequency of 70 kHz in comparison with the spectrum of the drive without the CM filter (see Figure 5.3).

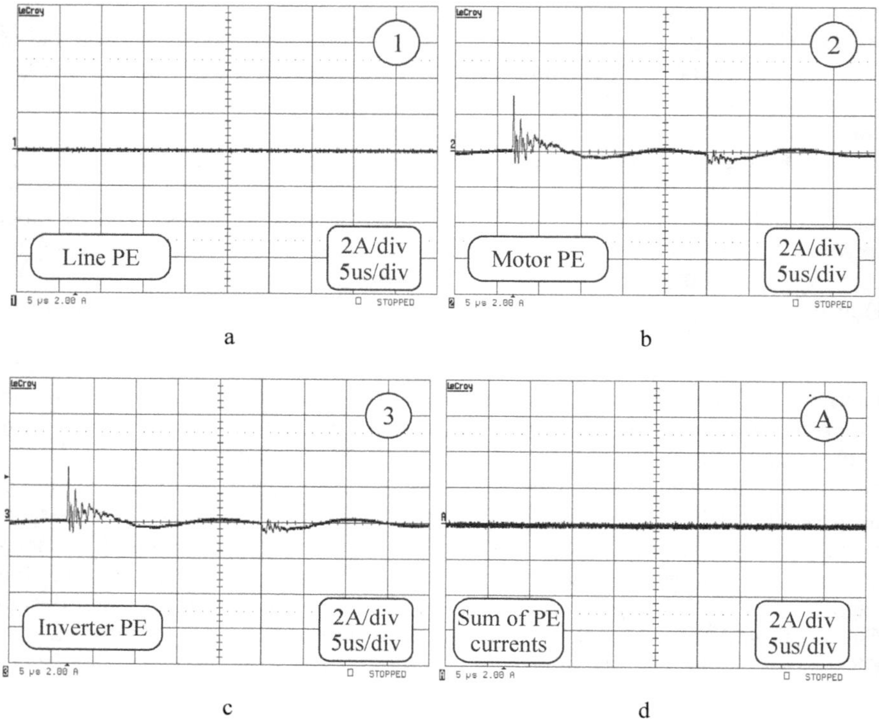

Figure 5.8. Passage of common mode currents through "*virtual grounding point*" in a drive with passive input filter: **a** line PE; **b** motor PE; **c** inverter PE; **d** sum of PE currents

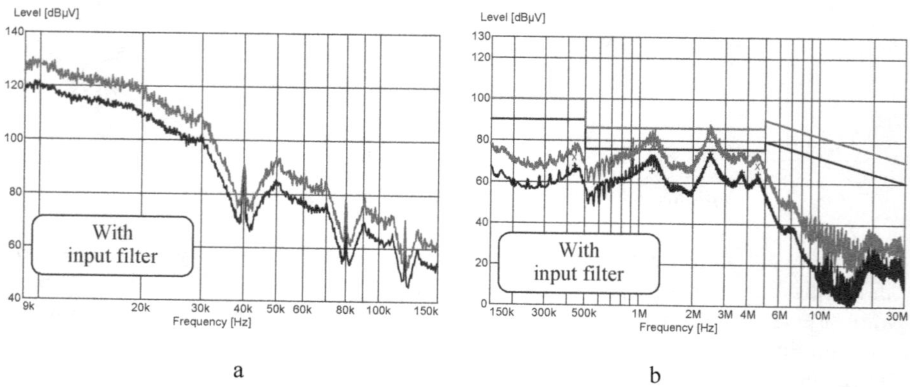

Figure 5.9. Conducted EMI spectrum (drive with input passive filter): **a** CISPR A; **b** CISPR B

The reduction within the CISPR A frequency range has enabled the suppression of the level of conducted emission below the limit lines specified in the standard EN 61800-3 for PDS (150 kHz – 30 MHz). However, in the high frequency range (above 1 MHz) in the EMI spectrum measured on the line side we can still observe an influence of CM current, which flows on the motor side.

Figure 5.10 shows the CM current spectrum, which has been measured using a current probe on the line and the motor side of the converter, respectively.

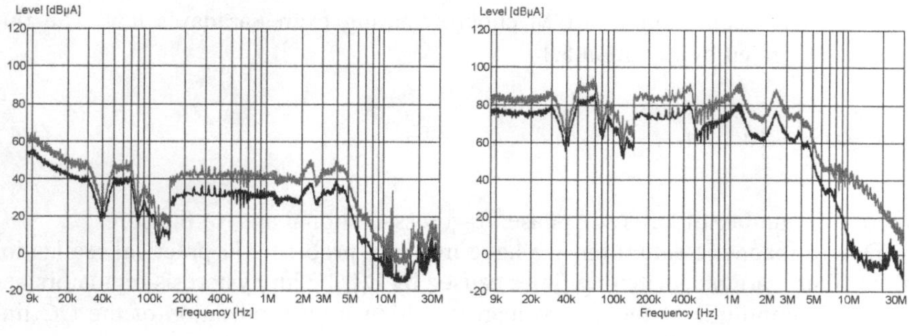

Figure 5.10. Spectrum of CM currents (drive with input passive filter): **a** line side; **b** motor side

Indeed, in the comparison with the spectra presented in Figure 5.4, the input filter has significantly suppressed the CM component in the EMI spectrum on the line side, whilst the CM current spectrum on the motor side has remained practically unchanged.

The best way to mitigate CM currents on the motor side is a cancelation of the CM voltage at the output of the motor side inverter. We have tested the passive CM voltage filter that has been successfully applied to inverter drives with input diode bridge converters and a sinusoidal PWM [4, 5].

Figure 5.11 shows the drive with the sinusoidal passive filter proposed by Akagi *et al.* [4].

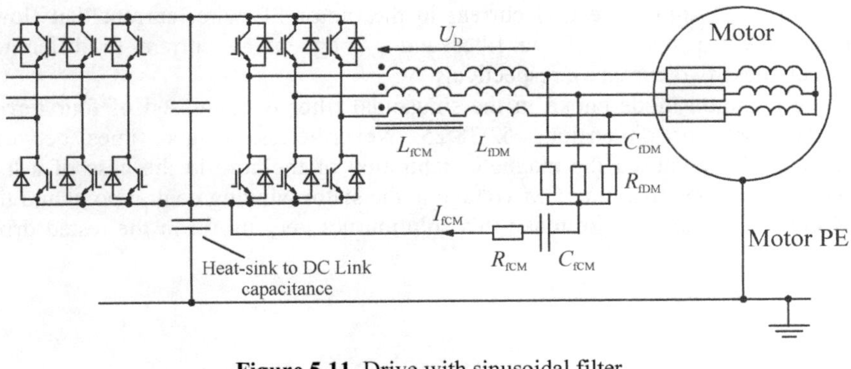

Figure 5.11. Drive with sinusoidal filter

The main goal of this solution is to cause a voltage drop U_D across series connected inductors (mainly CM choke), which should be almost equal to the common mode voltage at the output of the inverter. In such conditions a series compensation of the CM voltage at the output of the filter can be achieved.

Our results obtained in the ASD with an input diode rectifier have been very promising and have encouraged us to transfer the above solution into the four-quadrant frequency converters with DTC [3]. We have taken note of the problems with magnetic saturation of a CM choke resulting from Faraday's law. The flux density B is given by Equation 5.1

$$B = \frac{\phi}{S} = \frac{1}{SN} \int u_{CM}\, dt \tag{5.1}$$

where N – number of turns *per* phase; S – cross-sectional area of the core.

Due to magnetic saturation we have met two problems in practical realization of the filter: variable switching times caused by DTC with hysteresis regulators and very high amplitude of the CM voltage resulting from a pulsation of the DC link voltage with respect to the ground.

Figure 5.12 shows voltages of the DC buses with respect to the ground. The shapes of the waveforms are formed by the implemented control algorithm of the rectifier and an additional oscillation caused by a voltage drop across the DC link-to-heat-sink capacitance. Nevertheless, DC link voltage is clearly constant. We can observe approximately the same oscillation, superimposed on the waveform resulting from DTC algorithm, in the CM voltage in the stator windings neutral point (Figure 5.12).

Figure 5.13 shows the CM voltage in the stator winding neutral point in a drive with sinusoidal passive filter in comparison with a negative DC bus voltage with respect to the ground. This comparison indicates the compensation of the CM voltage with respect to the DC bus midpoint. However, oscillations in the stator winding neutral point with respect to the ground still exist. The voltage drop across the heat-sink-to-DC link capacitance forces the CM current to flow in the loop consisting of distributed parasitic-to-ground capacitances of the motor, motor PE wire and capacitances of the filter.

Figure 5.14 shows the CM current in the motor PE wire, current that flows through the capacitance of the filter and a magnetizing current (a difference between these two currents), respectively.

The common mode choke in the sinusoidal filter is composed of four ferrite toroidal cores of dimensions 63/38/25. Nevertheless, longer times between switchings have caused the magnetic saturation of the core in the case of a low motor speed. The remaining CM voltage in the stator winding neutral point and the CM current in the PE wire make this solution not very useful in the tested drive system.

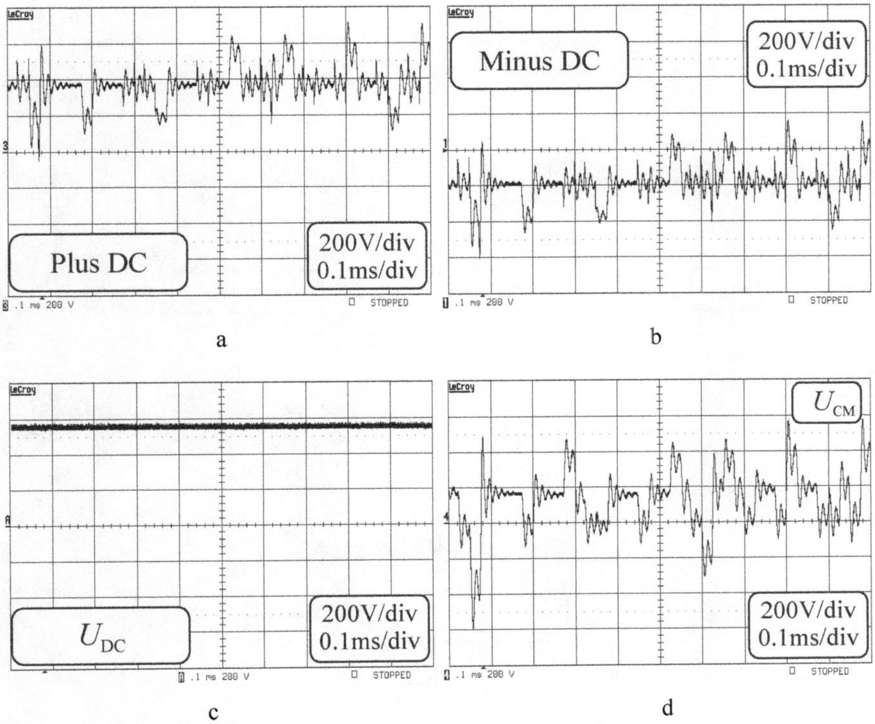

Figure 5.12. Voltages of DC buses (drive without filters): **a** negative DC; **b** positive DC; **c** DC link voltage; **d** CM voltage in stator windings neutral point

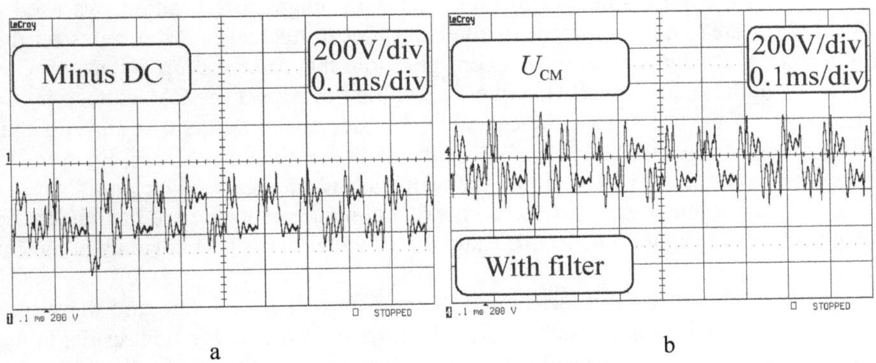

Figure 5.13. Voltages in drive with sinusoidal filter: **a** negative DC bus; **b** CM voltage in stator windings neutral point

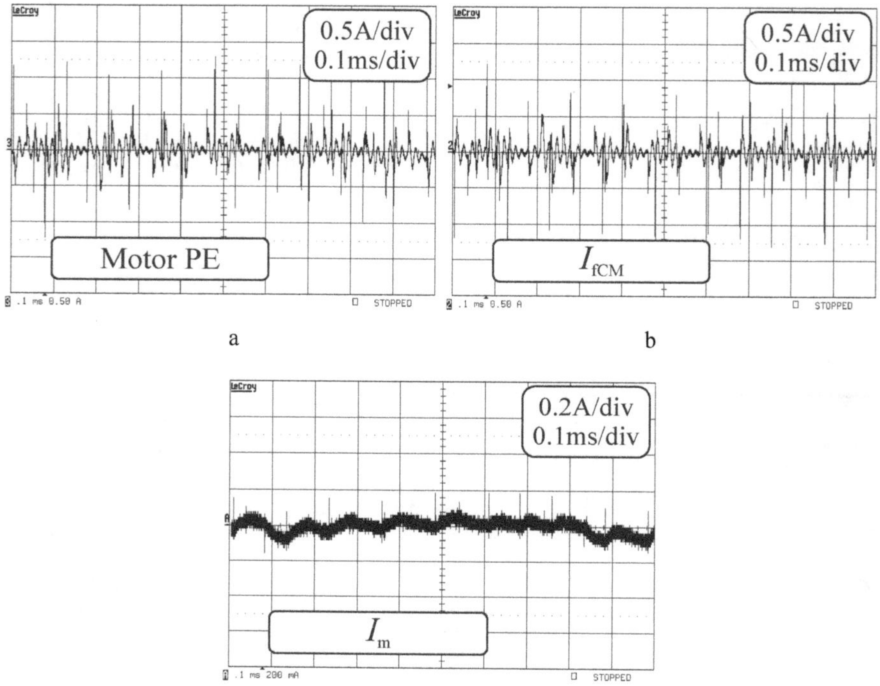

Figure 5.14. **a** CM current in the motor PE wire. **b** I_{fCM} current. **c** Magnetizing current I_m

To avoid the above-mentioned problems with magnetic saturation, we used a CM choke instead of the sinusoidal filter in order to decrease CM currents on the motor side of the converter. In this configuration, the voltage drop across the CM choke would be much lower. However, we could not expect the CM voltage in the stator winding neutral point to be canceled. The decreasing of the CM currents that we have obtained using the same CM choke as in the sinusoidal filter, Figure 5.15 shows the passage of CM currents through the "*virtual grounding point*" in the drive with input and output filters. To make a comparison easier we have used the same scale as in Figure 5.6. In this scale the cancelation of CM currents is almost complete.

Figure 5.16 shows the spectrum of the CM current on the line and the motor side. The significant suppression of the CM current on the motor side results in the lower content of HF components in the CM current spectrum on the line side (see Figure 5.16).

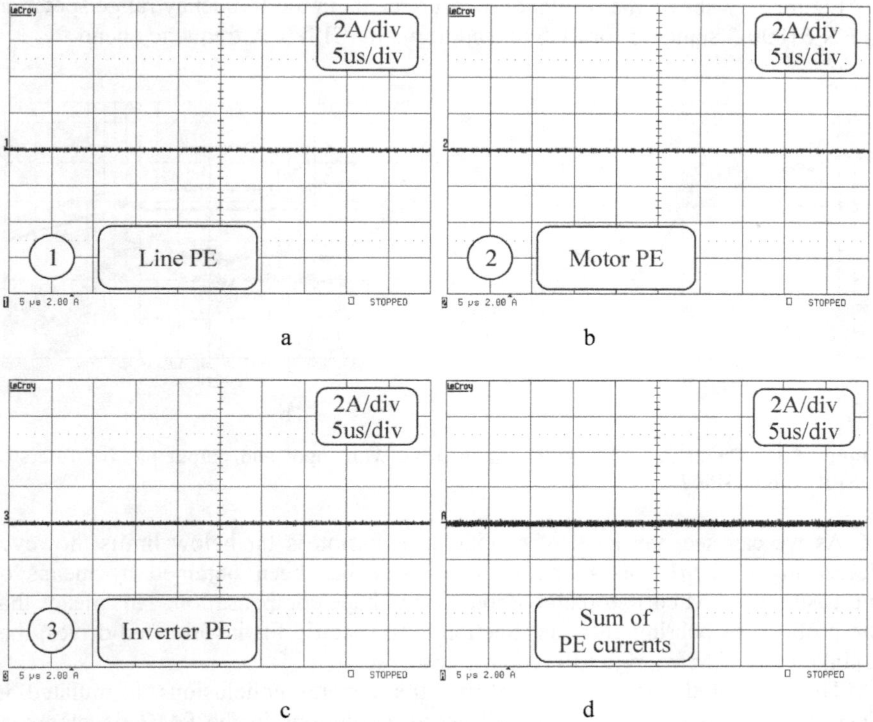

Figure 5.15. Passage of common mode currents through "*virtual grounding point*" in a drive with input and output passive filter: **a** current in PE wire; **b** motor PE current; **c** inverter PE current; **d** sum of PE currents

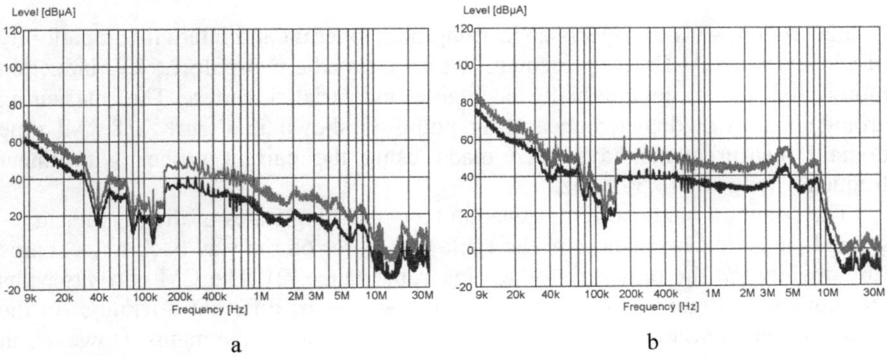

Figure 5.16. Spectrum of CM currents in drive with input and output passive filters: **a** line side; **b** motor side

Figure 5.17 shows the results of measurements in the frequency range specified in EN 61800-3 standard for PDS and additionally CISPR A frequency band.

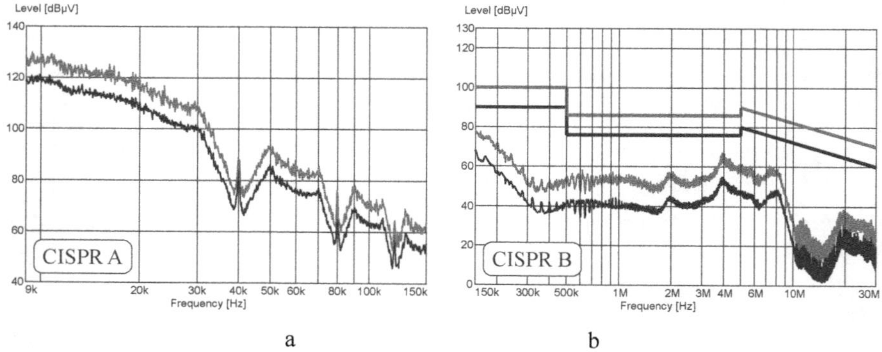

Figure 5.17. Conducted EMI spectrum in drive with input and output passive filters: **a** CISPR A; **b** CISPR B

As we can see, the level of conducted emission is far below limits; however decreasing of EMI noises on the motor side has been obtained by means of increasing the CM current path without CM voltage compensation. This means that the problem of bearing currents (Section 5.2), specific for inverter-fed drives, that is linked with CM voltage still remains unsolved.

The presented case study confirms the general conclusions formulated in Section 4.3. It is difficult to find all-purpose solutions in the EMC development and in-depth case study is always recommended.

5.2 Adjustable Speed Drives

In this section, various aspects concerning both external and internal compatibility of converter-fed ASDs are presented. We have tested a two-pole 1.5 kW induction motor fed by a commercially available industrial inverter. The measuring arrangement with depicted measuring points is shown in Figure 5.18. All time domain measurements have been made using the current probes with linear frequency range up to 50 MHz.

The common mode currents generated by switching states of the inverter in to-ground parasitic capacitances in the cable and motor on the way back to its source flow mainly *via* the heat sink-to-DC link capacitance [2]. The CM currents paths are marked in Figure 5.18 by means of the arrows of different thickness. In the selected scale it seems like almost no CM current flows to the mains. However, in the context of product standard for PDS, the conducted emissions are relatively high, although still below limits. Thus, the investigated drive fulfils requirements for external electromagnetic compatibility in the conducted emission frequency range.

In Figure 5.19 experimental results of CM currents passage through the system are presented.

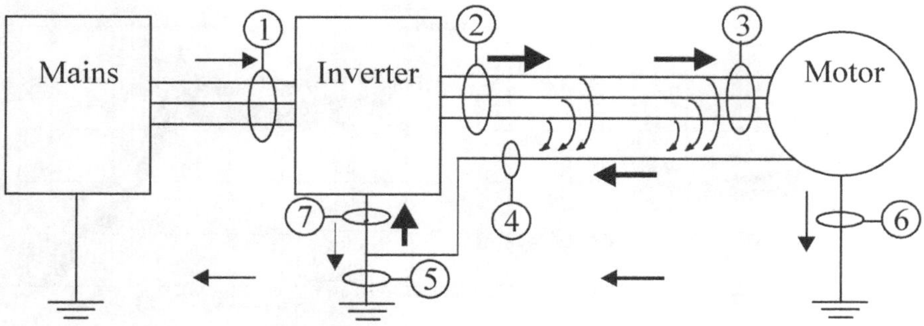

Figure 5.18. Measuring arrangement

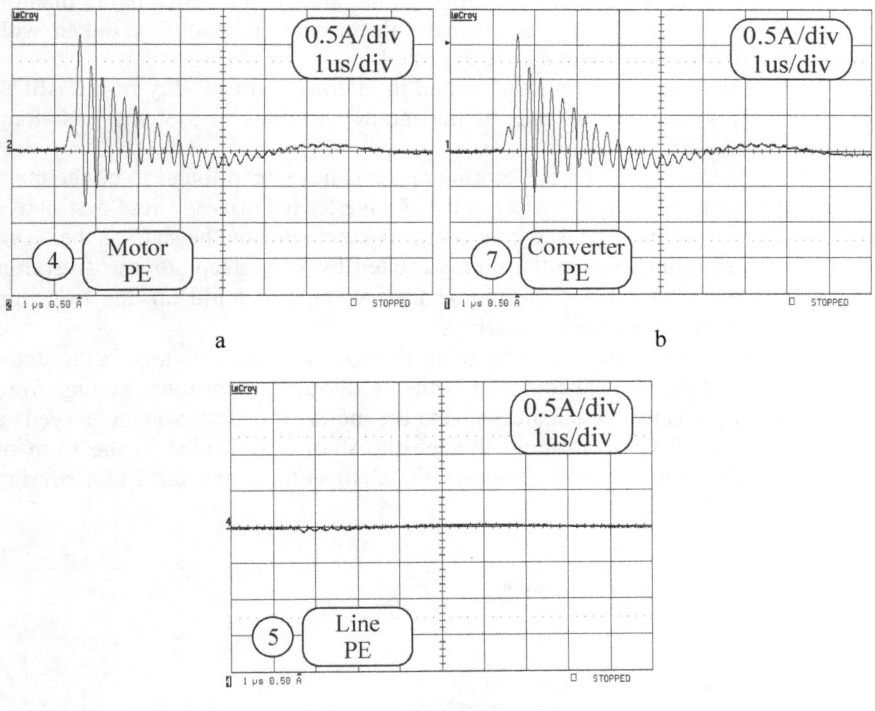

Figure 5.19. Passage of CM currents through the drive system: **a** motor PE current; **b** converter PE current; **c** line PE current

Figure 5.20 shows the spectra of CM currents measured in PE wires by means of the current probes on both sides of the converter (additional measurement not required by the standards).

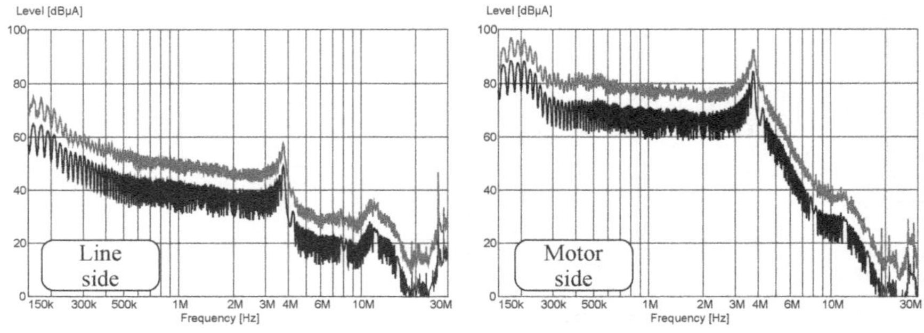

a b

Figure 5.20. CM currents spectra on: **a** line side; **b** motor side

A very high level of CM current on the motor side, about 30dB higher than on the line side, suggests that the problems with internal compatibility linked with functional safety of the system can be expected.

There are two specific problems related to internal compatibility of the ASDs: bearing currents and overvoltages at the motor terminals in systems with long motor feeders [6–9].

Electric Discharge Machining bearing currents have been found to be the main cause of premature bearing damages in PWM inverter fed drives. These destructive currents in spark form are caused by a charge accumulated on the shaft as the result of capacitive couplings inside the motor excited by very steep slopes of inverter output voltages. This enables the motor shaft voltage to build up due to a thin insulating oil film in the rotating bearings.

The shaft voltage almost perfectly maps the common mode voltage in the stator windings neutral point in accordance with the divider proportion resulting from distribution of parasitic capacitances inside the motor. If a shaft voltage exceeds a critical value of oil film threshold voltage the shaft is unloaded in the form of electrical breakdown. Figure 5.21 shows the shaft voltage and the EDM bearing current.

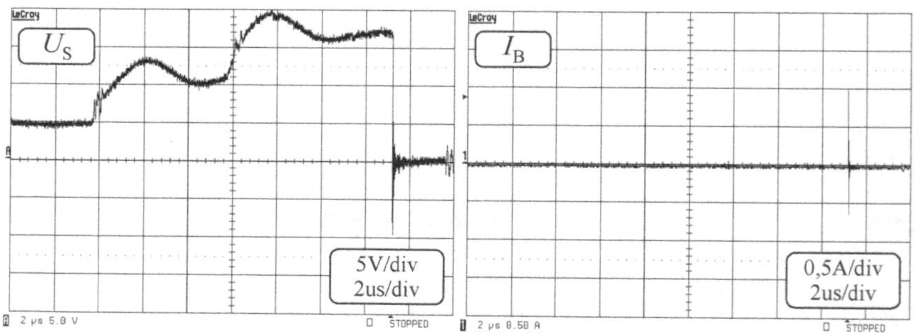

a b

Figure 5.21. EDM current event: **a** shaft voltage; **b** discharging bearing current

In our investigations the amplitudes of EDM currents and moments of breakdown have been treated as random variables because of the delay time occurring in the well-known mechanisms of an electrical breakdown. Due to the high frequency of CM voltage, we observed and analyzed a very large set of discharging events (thousands of events per second). This phenomenon depends on many different factors caused by the modulation strategy of an inverter, such as carrier frequency, inverter output frequency, physical proprieties of the insulating oil film, hydro mechanical bearing behavior, *etc.*

We have examined the influence of the application of the various CM mitigating techniques on EDN current phenomenon, because the CM voltage amplitude and shape (and, in turn, shaft voltage) depend on filter selection (Section 4.6.4). The shaft voltages in systems without filters, with CM choke and CM transformer [10] are shown in Figure 5.22.

Figure 5.23 shows joined distributions in the drive without passive filters, with CM choke and CM transformer [11].

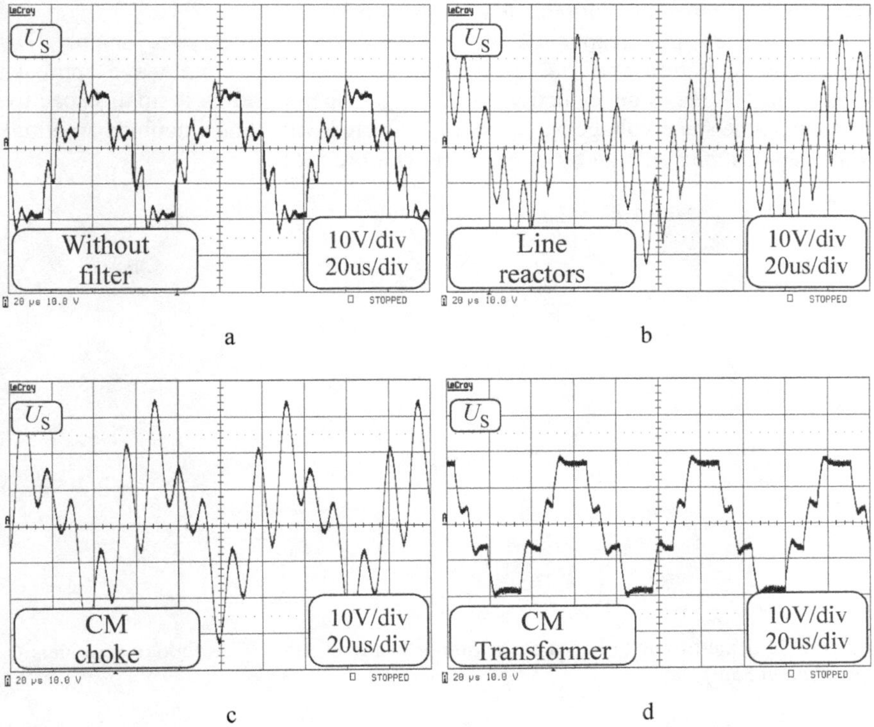

Figure 5.22. Shaft voltages in drives: **a** without filters; **b** with line reactors; **c** with CM choke; **d** with CM transformer

As we would expect from the highest level of the shaft voltage in the drive with the CM choke, we observe the greatest risk of a frequently occurrence of EDM currents of very high amplitudes in this drive arrangement. An attenuation

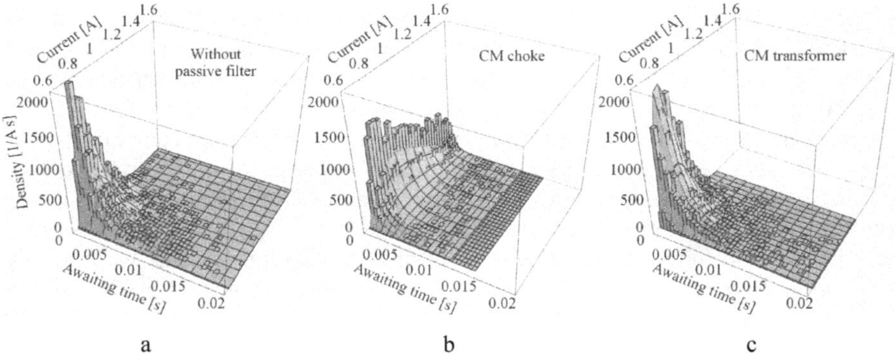

Figure 5.23. Three-dimensional distributions of EDM current amplitudes in awaiting time in drives: **a** without filters; **b** with CM choke; **c** with CM transformer

of resonance oscillations in the drive with the CM transformer can decrease this risk, even in comparison with the drive without passive filters.

Overvoltages in inverter-fed drives with long feeders are mainly DM phenomena in nature and can be discussed using a traveling wave approach. Because in HF range an induction motor can be perceived as a small capacitive impedance, the steep voltage wave front is reflected with almost doubled amplitude (if the cable is long enough for given *dv/dt*) (Figure 5.24).

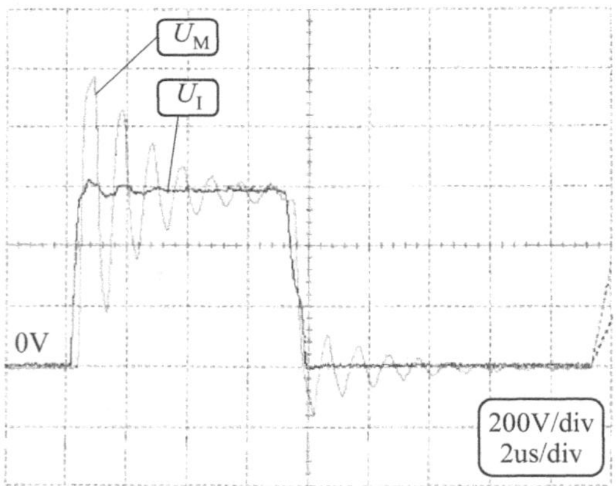

Figure 5.24. Line-to-line voltages at the output of the inverter U_I and motor terminals U_M (cable length 33m)

The wave front that penetrates into windings is unevenly distributed on its first turns. This phenomenon might pose a danger to motor conductor insulation.

The case study presented has once again shown that it is often necessary to consider all internal and external EMC aspects simultaneously.

5.3 Multi-level Inverters

Multi-level inverters, especially cascaded connected, can be used in custom power systems because of the insulated bridges that allow for application of various renewable energy sources of different voltage levels.

It is well known and important in the EMC context that topologies of multi-level inverters offer naturally reduced EMI currents due to decreased values of converter voltage steps (Figure 5.25). Moreover, the CM voltage produced by a multi-level inverter (odd number) can be nearly eliminated by selecting the specific PWM states [12]. However in this case, the reduced number of possible states results in a reduced accessible region for sinusoidal modulation, decreasing in turn the maximum value of phase voltages and increasing their harmonic content [13].

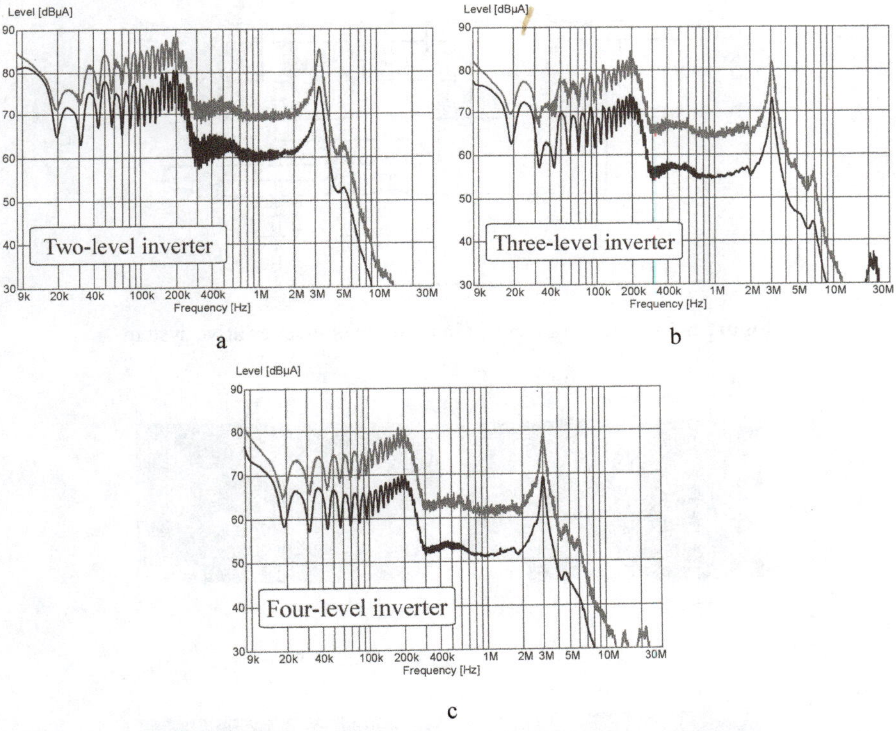

Figure 5.25. CM currents spectra in: **a** two-level; **b** three-level; **c** four-level inverter drive systems

The investigations have been carried out in drive systems with two-, three-, and four-level cascaded multi-level inverters. Figure 5.25 shows spectra of CM currents measured in two-, three- and four-level inverter drives without any filters.

The application of multi-level inverters as part of a power system interface should be preceded by the elimination of unwanted high frequency components of energy conversion. We have tested the possibilities for application of a CM filter with sinusoidal line-to-line voltages (described in Section 4.6.4) in a system with cascaded multi-level inverters [14]. Figure 5.26 shows an example of the filter arrangement in a drive with a four-level inverter. The application of multi-level inverters makes it possible to alleviate problems with magnetic saturation of the CM choke.

Figures 5.27–5.29 show waveforms of phase voltages and CM voltages at the output of two-, three-, and four-level inverters in the case of inverter output frequency 50 Hz with near-unity modulation index and double-edge naturally-sampled PWM [13] with the carrier frequency f_c=12 kHz.

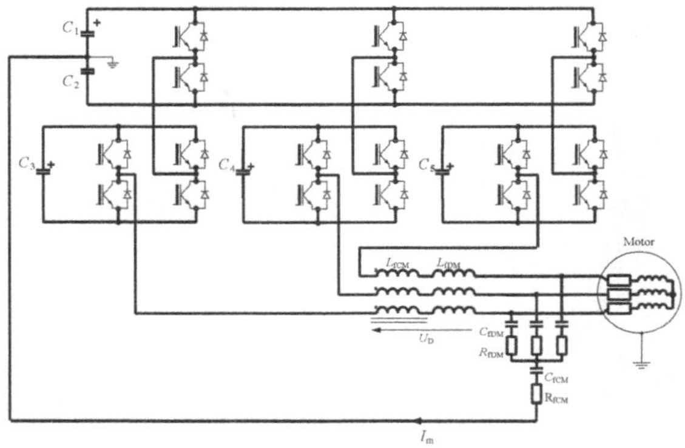

Figure 5.26. Filter arrangement in multi-level inverter drive system

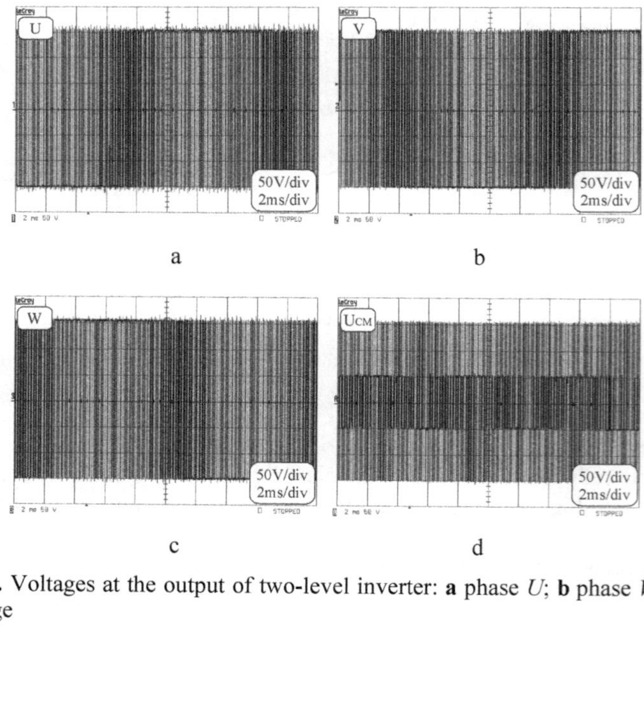

Figure 5.27. Voltages at the output of two-level inverter: **a** phase U; **b** phase V; **c** phase W; **d** CM voltage

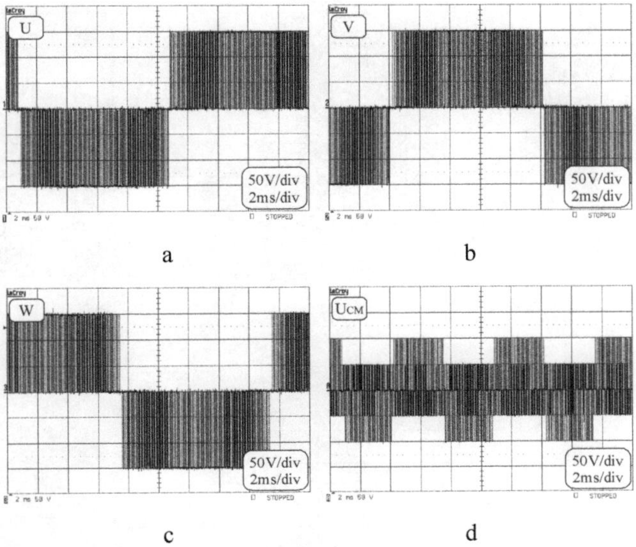

Figure 5.28. Voltages at the output of three-level inverter: **a** phase *U*; **b** phase *V*; **c** phase *W*; **d** CM voltage

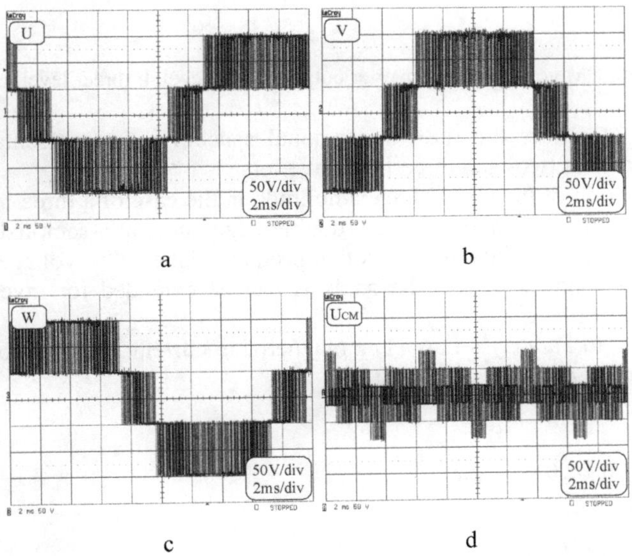

Figure 5.29. Voltages at the output of four-level inverter: **a** phase *U*; **b** phase *V*; **c** phase *W*; **d** CM voltage

In the presented case the CM voltage amplitude decreases inversely proportional to the number of inverter levels. Figure 5.30 shows experimental waveforms of the CM voltages in two-, three- and four-level inverters in an expanded time scale.

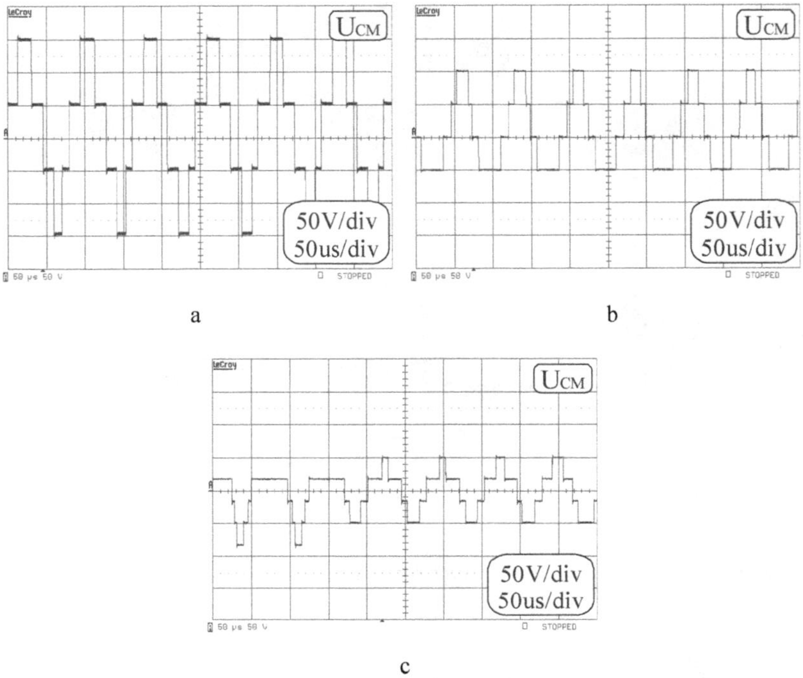

Figure 5.30. CM voltage at the inverter output: **a** two-level; **b** three- level; **c** four-level

Figure 5.31 shows the three-dimensional space vector representation of the switching states of two- and three-level inverters.

The reduction of the CM voltage amplitude in the case of a three-level inverter results from the method of the double-edge naturally-sampled sinusoidal modulation, where the inverter states that produce highest CM voltage levels (111 and -1-1-1) are not selected. The analysis can be extended for inverters with a higher number of levels.

According to Faraday's Law (1) a magnetizing current is proportional to time-integral value of U_{CM}

$$M(t)=\int_0^t U_{CM}(t)dt, 0 \le t \le 2\pi \tag{5.2}$$

In order to estimate the influence of modulation effects onto $M(t)$, Equation 5.3 for $U_{CM}(t)$ at the output of the N -level inverter has been proposed [15]

$$U_{CM}(t)=\frac{2}{3N}\sum_{k=-1}^{1}\sum_{i=1}^{N}\left\{H\left(A\sin\left(\omega_{inv}t+\frac{2}{3}k\pi\right)-C_N^i(f_c t)\right)\right\}-1 \tag{5.3}$$

$$C_N^i(t)=\frac{2}{N}\left\{\left(1-2\left|\langle t\rangle-\frac{1}{2}\right|\right)+\frac{1}{2}N-i\right\}, i=1,...,N \tag{5.4}$$

where $\langle t\rangle$ – gives a fraction part of t ; H - unit step function.

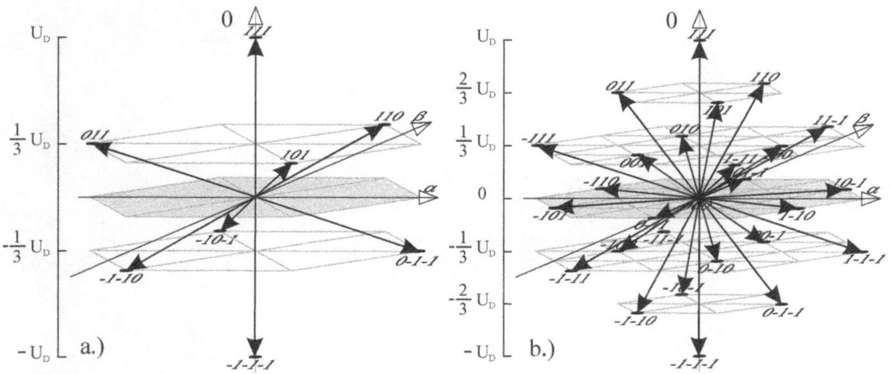

a b

Figure 5.31. Three-dimensional space vector representation of the switching states of: **a** two-level inverter; **b** three-level inverter

This equation concerns a double-edge naturally-sampled PWM of N-number of levels.

Figure 5.32 shows the compensating voltage (equal to CM voltage) and magnetizing current measured in a two level inverter drive with the naturally-sampled PWM and the carrier frequency f_c=16 kHz and arbitrary output frequencies.

The worst case for two-level drives with the scalar control occurs at the output frequency equal to 0 Hz where the beginning of magnetic saturation of the CM inductor has already been observed. In the context of the CM choke saturation and bearing currents the amplitudes and time duration of CM voltage shelves are crucial parameters rather than *dv/dt* of CM voltage steps. In the case presented the CM voltage amplitude decreases inversely proportional to the number of inverter levels. Figure 5.33 shows experimental waveforms of the CM voltages in two-, three- and four-level inverters in an expanded time scale.

Figure 5.34 shows CM voltages at the output of the inverter and magnetizing currents I_m in a two-, three- and four-level inverter drive (f_{inv}=50 Hz, $M\cong1$).

Amplitudes of magnetizing currents decrease as the number of inverter levels increases. Noteworthy are the modulation envelopes that influence magnetizing current waveforms and determine the maximum value reached by the magnetizing current in a given drive. This value is a key parameter when the eventuality of the CM choke saturation is evaluated. To determine "*the worst cases*" resulting from the modulation algorithm a theoretical solution has been developed.

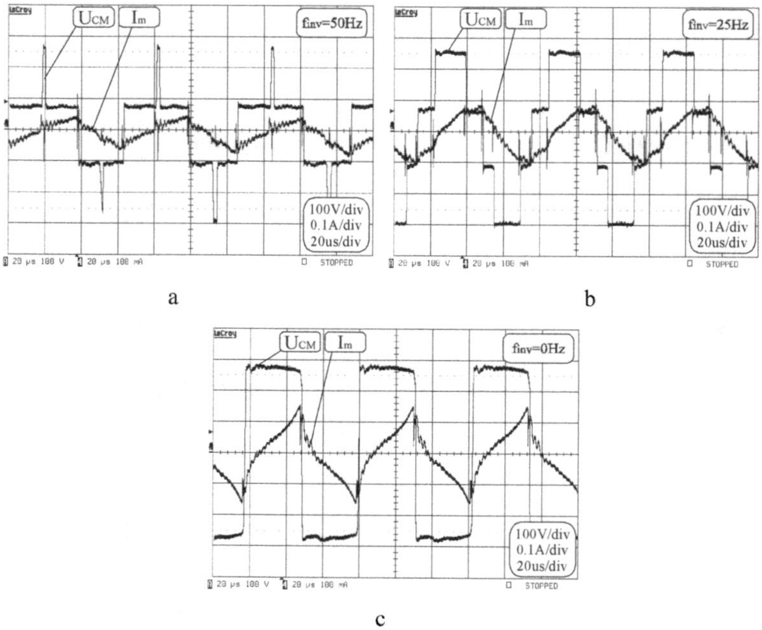

Figure 5.32. CM voltage U_{CM} and magnetizing current I_m for inverter output frequency: **a** 50 Hz; **b** 25 Hz; **c** 0 Hz

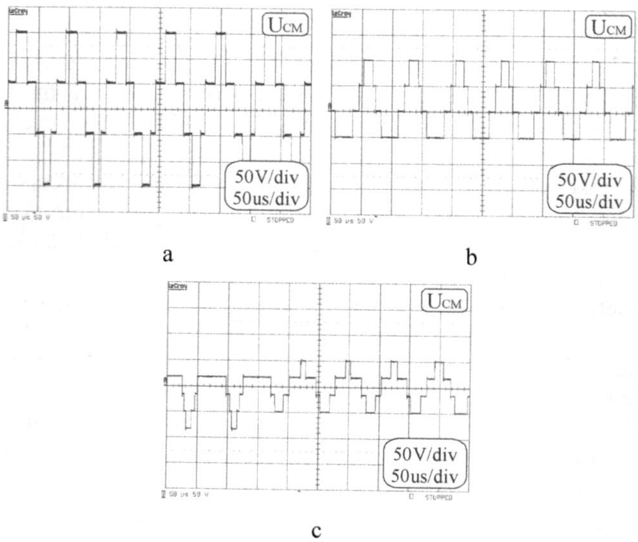

Figure 5.33. CM voltage in: **a** two-; **b** three-; **c** four-level inverters

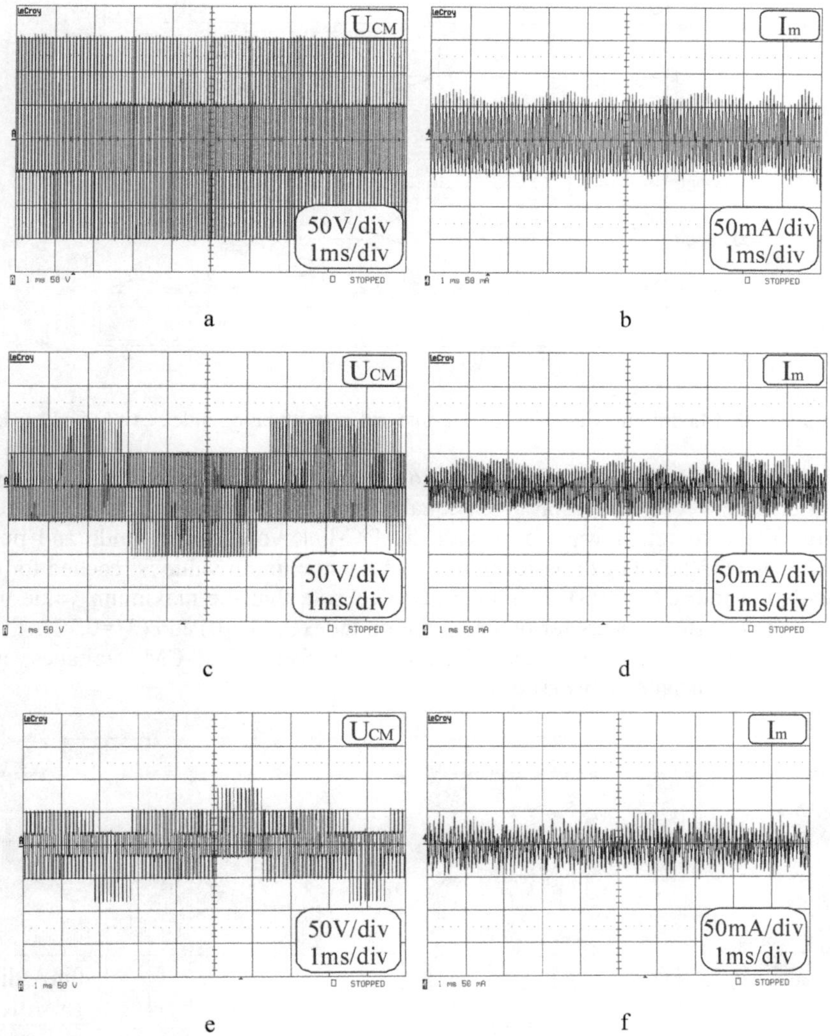

Figure 5.34. **a** CM voltage in two- level inverter. **b** Magnetizing current in two- level inverter. **c** CM voltage in three-level inverter. **d** Magnetizing current in three-level inverter. **e** CM voltage in four-level inverter. **f** Magnetizing current in four-level inverter

The variation of maximum value of U_{CM} time-integral with modulation index for a two-, three- and four-level inverter is shown in Figure 5.35.

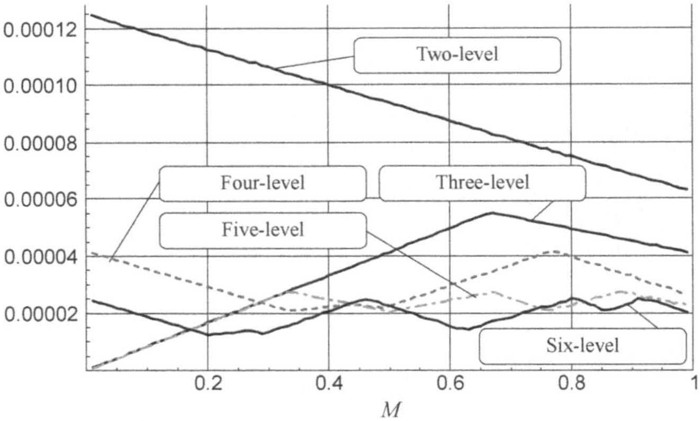

Figure 5.35. Maximum value of U_{CM} time-integral *vs* modulation index (A=1, f_c=12 kHz)

The theoretical solution presented shows that "the worst case" for a two-level inverter occurs for modulation index equal to zero (M=0) where the CM voltage is simply the rectangular waveform with the DC-link voltage amplitude and pulse duty cycle (D=0.5). For three-level inverters the maximum value is reached for the modulation index (M=2/3). It is interesting to note that the maximum value in a four-level inverter appears for two modulation indexes (M=0) and (M=0.77).

Figures 5.36-5.38 show experimental waveforms of CM voltages and magnetizing currents in "*worst cases*".

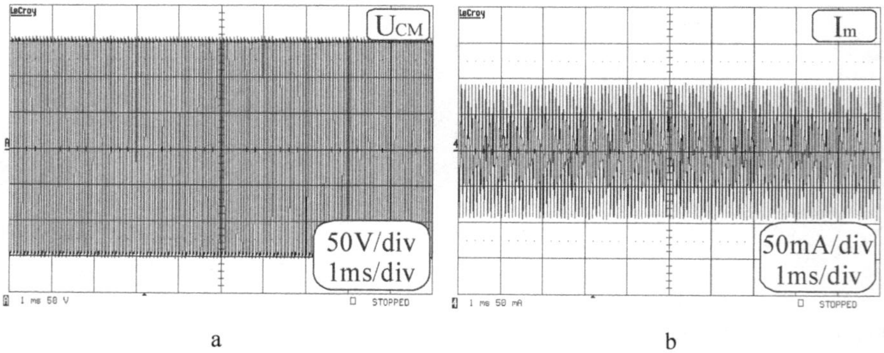

Figure 5.36. a CM voltage. **b** Magnetizing current in a two-level inverter for "*worst case*"

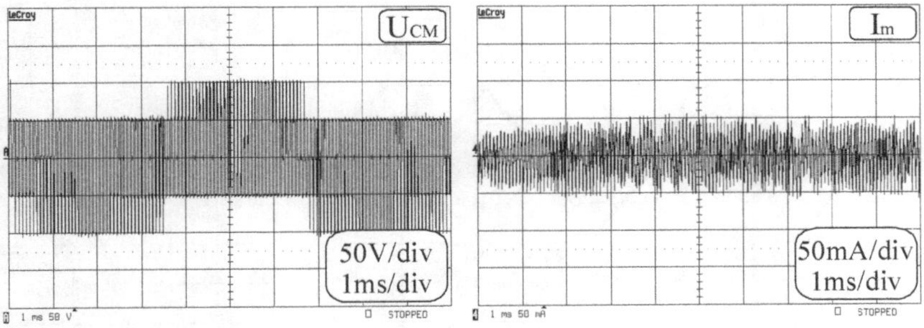

a b

Figure 5.37. a CM voltage. **b** Magnetizing current in a three-level inverter for "*worst case*"

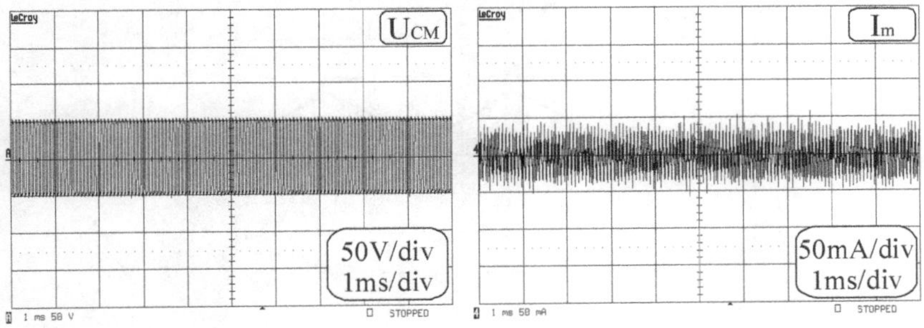

a b

Figure 5.38. a CM voltage. **b** Magnetizing current in a four-level inverter for "*worst case*"

Figure 5.39 show time-expanded scale experimental waveforms of CM voltages and magnetizing currents in two-, three- and four-level converters. In spite of different amplitudes and shapes of the CM voltages in "worst case", amplitudes of magnetizing current in the filter are comparable. These experimental results confirm the results of the theoretical analysis, obtained using proposed Equation 5.3 for double-edge naturally-sampled PWM shown in the graph in Figure 5.35.

The main problem of the passive compensation of the CM voltage in PWM inverter-fed drives is the magnetic saturation of a CM choke. In multi-level inverters the conditions for this compensation become more favorable due to the lower level of the time-integral of CM voltage at the output of the inverter. Consequently, it allows reduction in the size, weight and cost of the inductive components of filters in multi-level drives.

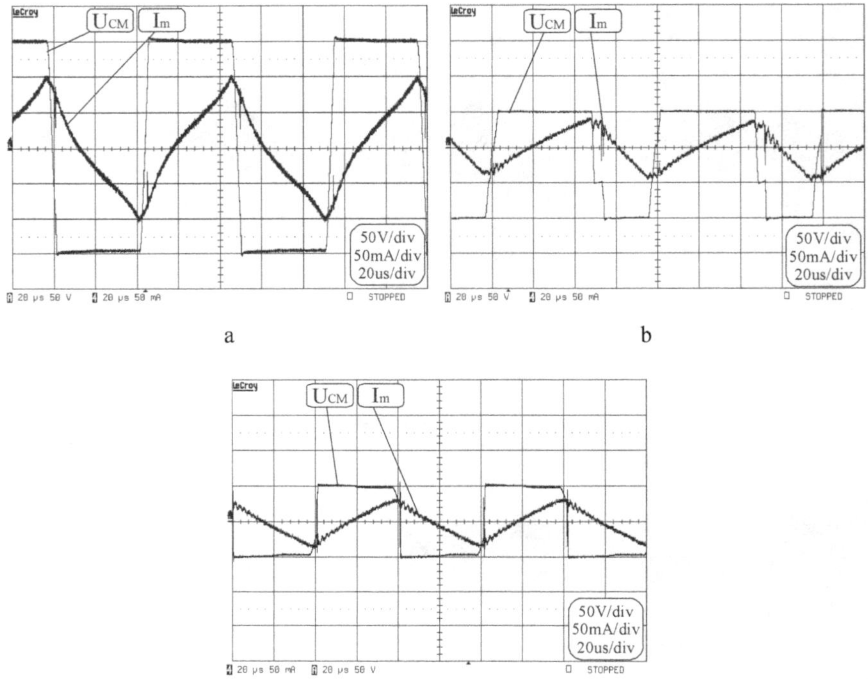

c

Figure 5.39. CM voltage and magnetizing current in inverter for "*worst cases*": **a** two-level; **b** three-level; **c** four-level

When the eventuality of the magnetic saturation of an inductor is analyzed "*the worst case*" that results from modulation algorithms should be taken into account. Substantial work is required (*e.g.*, based on the proposed theoretical formula) to verify unequivocally the influence of different modulation strategies on the eventuality of the saturation of CM inductors.

References

[1] Benysek G, (2007) Improvement in the quality of delivery of electrical energy using power electronics systems. Springer-Verlag

[2] Kempski A, Smolenski R, Strzelecki R, (2002) Common mode current paths and their modeling in PWM inverter-fed drives. Power Electronics Specialists Conference, vol.3:1551–1556

[3] Kempski A, Strzelecki R, Smoleński R, Benysek G, (2003) Suppression of conducted EMI in four-quadrant AC drive system. Power Electronics Specialists Conference:1121–1126

[4] Akagi H, Hasegawa H, Doumoto T, (2004) Design and performance of a passive EMI filter for use with voltage source PWM inverter having sinusoidal output

voltage and zero common-mode voltage. IEEE Transaction on Power Electronics, vol.19:1069–1076

[5] Kempski A, Smolenski R, Kot E, Fedyczak Z, (2004) Active and passive series compensation of common mode voltage in adjustable speed drive system. IAS Conference, on CD

[6] von Jouanne A, Zhang H, Wallace AK, (1998) An evaluation of mitigation techniques for bearing currents, EMI and overvoltages in ASD applications. IEEE Transaction on Industry Applications, vol.34, no.5:1113–1121

[7] Bhattacharya S, Resta L, Divan DM, Novotny DW, (1999) Experimental comparison of motor bearing currents with PWM hard and soft switched voltage source inverter. IEEE Transaction on Power Electronics, vol.14, no.3:552–562

[8] Busse D, Erdman J, Kerkman RJ, Schlegel D, Skibinski G, (1997) Bearing currents and their relationship to PWM drives. IEEE Transaction on Power Electronics, vol.12, no.2:243–252

[9] Kempski A, Strzelecki R, Smolenski R, Fedyczak Z, (2001) Bearing current path and pulse rate in PWM-inverter-fed induction motor. Power Electronics Specialists Conference:2025–2030

[10] Ogasawara S, Akagi H, (1996) Modeling and damping of high-frequency leakage currents in PWM inverter-fed *AC* motor drive systems. IEEE Transaction on Industry Applications, vol.32, no.5:1105–1113

[11] Kempski A, Smolenski R, Bojarski J, (2005) Statistical model of electrostatic discharge hazard in bearings of induction motor fed by inverter. Journal of Electrostatics

[12] Zhang H, von Jouanne A, Dai S, Wallace AK, Wan F, (2000) Multilevel inverter modulation schemes to eliminate common mode voltages. IEEE Transaction on Industry Applications, vol.36, no.6:1645–1653

[13] Holmes GD, Lipo TA, (2003) Pulse width modulation for power converters. Principles and practice. IEEE Press

[14] Chiang L, Holmes DG, Fukuta Y, Lipo TA, (2003) Reduced common-mode modulation strategies for cascaded multilevel inverters. IEEE Transactions on Industry Applications, vol.39, no.5:1386–1395

[15] Kempski A, Smolenski R, Kot E, Strzelecki R, (2005) Series passive compensation of common mode voltage in multilevel inverter drives. Power Electronics Specialists Conference:1833–1838

6

High Frequency AC Power Distribution Platforms

Patrick Chi-Kwong Luk and Andy Seng Yim Ng

Defence College of Management and Technology,
Cranfield University,
Shrivenham Wiltshire SN6 8LA United Kingdom.
Email: P.C.K.Luk@cranfield.ac.uk; S.Ng@cranfield.ac.uk

6.1 Introduction

High Frequency AC (HFAC) power distribution system concerns the delivery of power at multi-kHz frequency *via* electric cables. Early work on HFAC has demonstrated the many potential benefits of HFAC systems in terms of flexibility to meet loads at different voltage levels, ease of electrical isolation using compact high frequency transformers, and the prospect of significant savings in component count and system integration. High frequency operation can also improve the dynamic response of the system, and reduces or eliminates acoustic noise. However, despite the many perceived benefits, the uptake of HFAC has been very slow. To date, HFAC power distribution has been confined to some pockets of applications in communications, computer and information technology systems. The ever increasing demand for more power at higher efficiency by our environmentally conscious economies means that electrical power will be delivered, where viable, at higher voltage, higher current and higher frequency. In view of advances in resonant power converter technologies, advanced soft switching (zero-voltage, zero-current) schemes, compounded by recent developments in high frequency cable and magnetic materials technologies, this chapter aims to give a timely appraisal of this very promising means of power distribution. It gives a detailed account of HFAC work in a diverse range of applications, from space to automotive, and from computer systems to microgrids. It also discusses the future prospects of this new power distribution system, and explores the new concept of embedding power and data within new cable architectures. It will also highlight key challenges such as EMI, and key drivers for HFAC to become a viable power distribution means in the future.

6.2 High Frequency in Space Applications

One of the earliest attempts to use HFAC for power distribution was by NASA Lewis Research Center in 1983, for the Space Station Freedom program [1]. Then the growing electrical power needs for the spacecraft systems were estimated to increase significantly into the megawatt range by the year 2000. The forecast was also motivated by the trend for bigger spacecraft and stations like the space shuttle and the international space station project. Existing DC based power systems were perceived to face limitations in the form of excessive copper cable weight to compensate for the I^2R losses. Figure 6.1 shows an AC power distribution system proposed in the early space program, employing solar arrays and the Brayton generator as power sources, and using a rotary transformer and series resonant inverter to generate high frequency power to an AC link.

However, high frequency operation would mean the additional challenges of higher EMI and higher crosstalk. This led to the design of novel power cables that carry high frequency current without incurring significant skin and proximity effect losses [2]. The cable must also have low inductive reactance to minimize voltage drops and radiated magnetic fields.

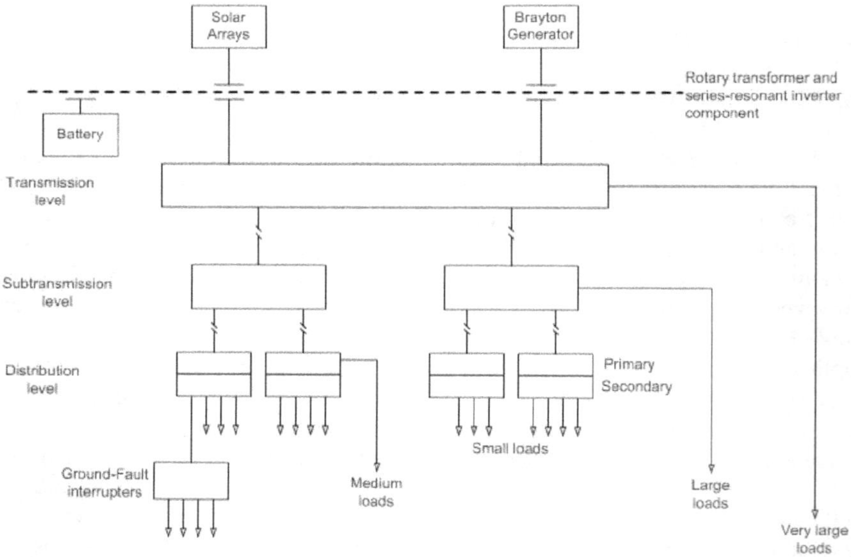

Figure 6.1. Structure of the HFAC power distribution system for the Space Station program [1]

Sood and Lipo [3] proposed another 20 kHz HFAC system for space application, as shown in Figure 6.2. The interconnections are implemented with interface converters, which are half or full bridge resonant converters.

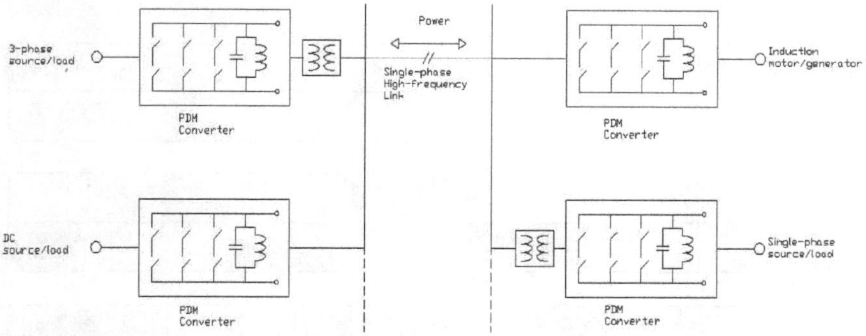

Figure 6.2. High frequency link power conversion system for Space Station [3]

A strategy of Pulse Density Modulation (PDM) called Area-comparison Pulse Density Modulation (AC-PDM) was chosen to control the amplitude of the power output. The area comparison refers to the minimization of volt-time area difference between the reference signal and the synthesized output signals. Figure 6.3a shows that the scheme can be readily realized by a simple logic circuit involving an integrator and a comparator. The circuit aims to produce the correct proportions of positive and negative half cycle pulses to minimize the difference and achieve regulation. Figure 6.3b shows the simulation results of the output for a simple DC reference output, based on the logic circuit.

Using half cycle of 20 kHz sinusoids as the basic building blocks, the PDM can synthesize arbitrary output signal waveforms with sufficient resolution for most applications. A parallel connected LC tank circuit across the AC link is used as the transient energy storage for sudden power demands, a similar function to that of a bulk capacitor across a DC link. Based on the same AC-PDM scheme, Figure 6.4 shows the simulation results of the output waveforms for a typical 400 Hz three-phase power supply system for aircraft applications. It should be noted that the harmonic content of the waveforms is better than that derived from the conventional PWM approach.

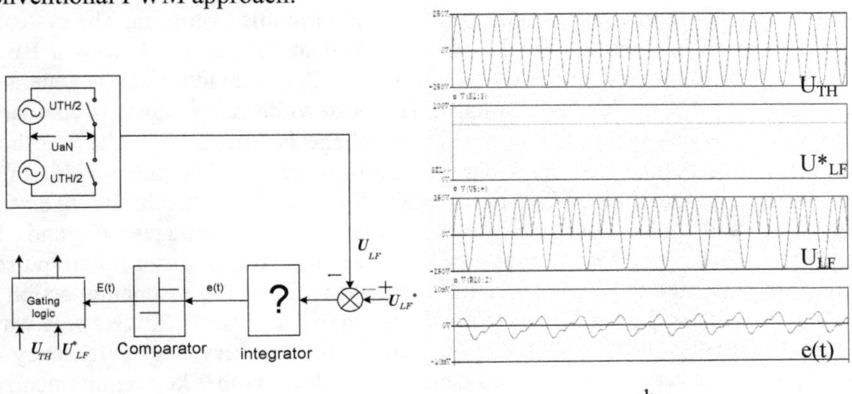

Figure 6.3. Area comparison pulse density modulation (AC-PDM) for a DC reference voltage: **a** implementation circuit; **b** waveforms

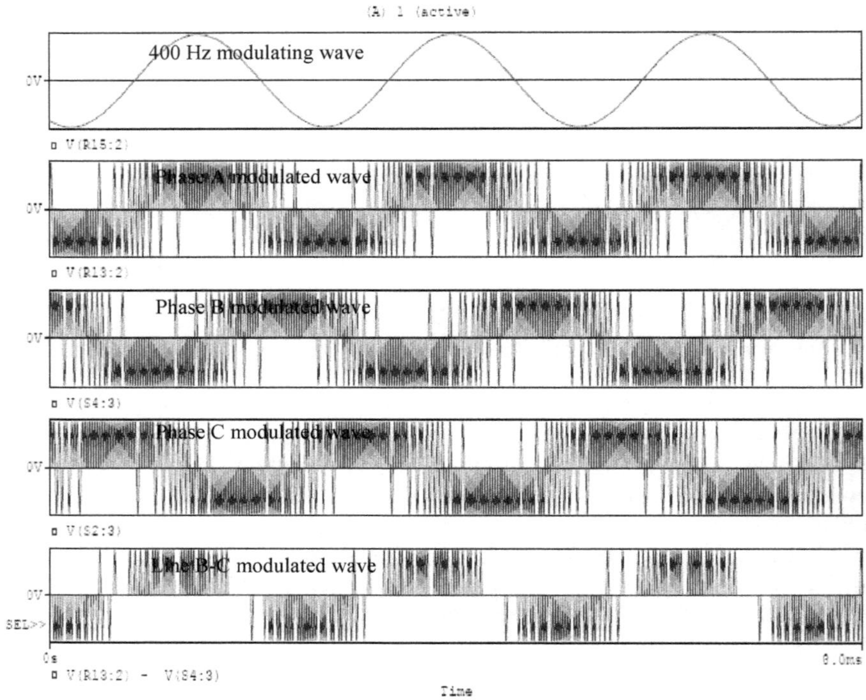

Figure 6.4. Three-phase 400 Hz AC voltages by the AC-PDM method

Another HFAC power system for a low power application using a "*hybrid*", or double tuned resonant inverter, was developed by Jain and Tanju[4]. The hybrid resonant inverter system aims to meet the steady-state operating requirements of a power source for the proposed International Space Station mobile servicing system. The key requirements for the mobile servicing system are high efficiency at varying load, good voltage regulation and low harmonic distortion. The system's circuit diagram and the control block are shown in Figure 6.5. It uses a PWM based method to implement the AC/AC converter. A pulsewidth ramp is generated and compared with a threshold voltage. The pulsewidth ramp signal is compared with the threshold signal V_{th}. If the ramp voltage is greater than the threshold voltage, a zero voltage command signal S_{OFF} is generated. The pulse-width angle can be readily determined by the signals S_{ON} and S_{OFF}. Both the operating frequency and pulsewidth can be controlled by the voltages V_C and V_{th} respectively. The inverter's output consists of bipolar voltage quasi square pulses, which are presented to a series-parallel resonant LC filter. The resonating action of the network filters the harmonics to ensure a "*clean*" and "*stiff*" 20 kHz sine wave output. Another advantage of the system is that a very high efficiency is maintained over varying output-load demand, which is another key requirement. In Figure 6.6, the key waveforms that illustrate the operating principle of the inverter are shown.

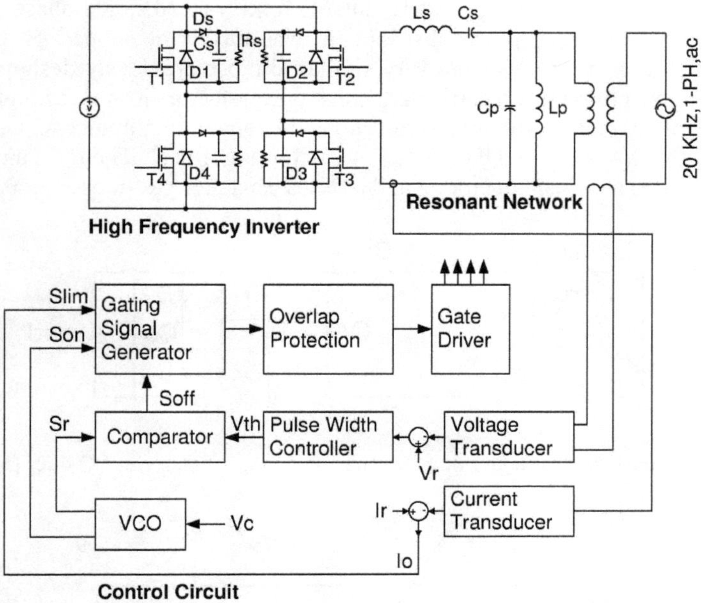

Figure 6.5. Key components of the hybrid resonant inverter [4]

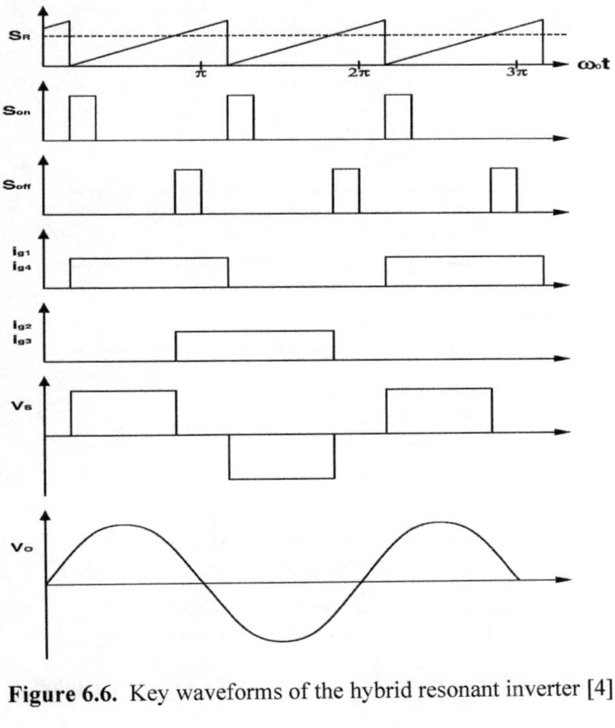

Figure 6.6. Key waveforms of the hybrid resonant inverter [4]

A new class of AC/DC converters suitable for advanced single-phase sine wave voltage, high frequency power distribution, was further developed by Jain *et al.* [5]. The target applications were very similar to those previously designed for the Space Station program. In particular, unity power factor, low mass and volume, and controlled output voltage were among the main requirements. One of the proposed topologies, called Type1A, is shown in Figure 6.7. The resonant network here comprises a series inductor *Ls* and a series capacitor *Cs*. In operation, the high

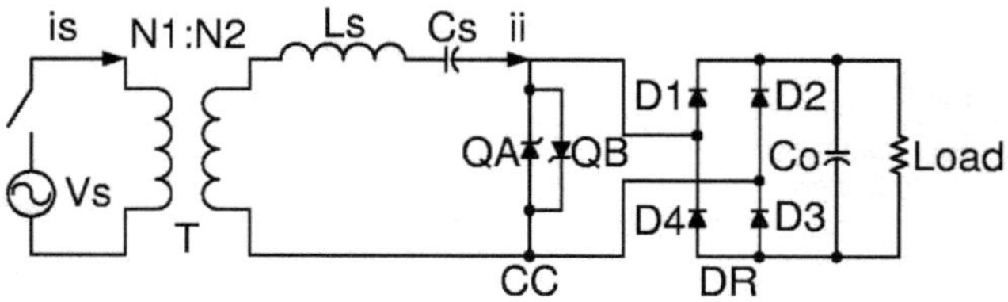

Figure 6.7. Configuration of resonant network with a series LC filter [5]

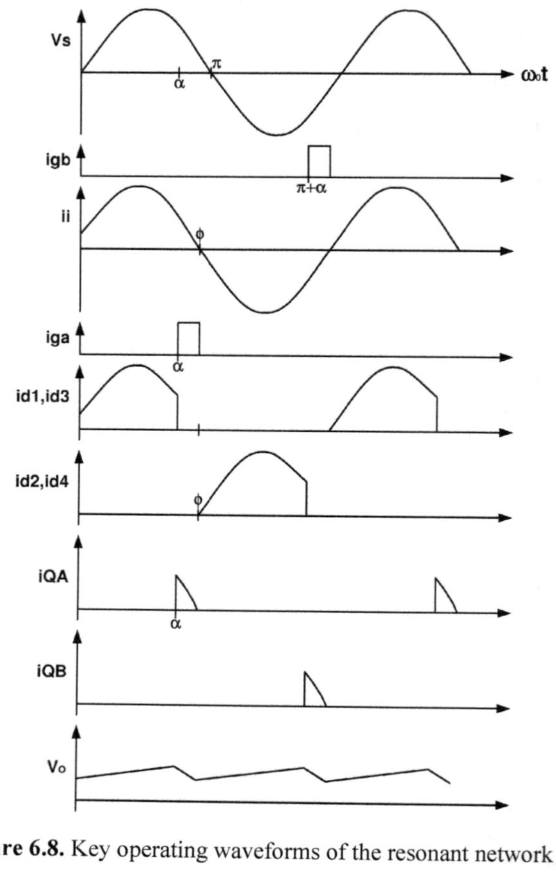

Figure 6.8. Key operating waveforms of the resonant network [5]

frequency sine wave is fed to the network through the input transformer *T*. The resonant input network is tuned such that the fundamental frequency component sees a low input impedance, whereas other harmonics will be blocked with very high input impedance. The operational waveforms for the circuit are shown in Figure 6.8. It can be seen that at the start of the cycle, when $\omega_0 t=0$, the diodes *D1* and *D3* are conducting and current i_i is charging *Co* till $\omega_0 t=\alpha$, when *QA* is turned on. Currents in *D1* and *D3* are transferred to *QA*. When this reaches zero at $\omega_0 t=\pi-\varphi$, *D2* and *D4* will start to conduct and charge *Co*. The output voltage *Vo* can therefore be controlled effectively by the angle α. It is shown that the converter achieved a very high power factor (>0.98), high conversion efficiency (>96%) and low harmonic distortion (<5%).

6.3 High Frequency in Telecommunications

Similar to the space systems, power distribution in telecommunications has essentially been in the DC domain. Although the DC power distribution system has been a proven technology in telecommunication applications, future requirements of ever increasing load levels, higher complexity, higher reliability and compactness, have posed significant challenges for a DC distribution system. This has prompted the exploration of alternative means of power distribution to meet these future challenges.

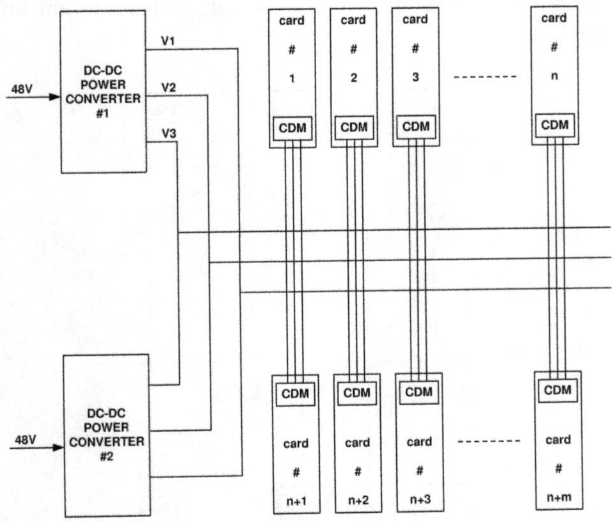

Figure 6.9. Centralized DC/DC power distribution

However, it would be appropriate to study first the existing DC systems, and to appreciate their limitations. There are two types of power distribution that have been used: centralized DC/DC and distributed DC/DC [6]. Figure 6.9 shows the centralized DC/DC power distribution system. The centralized DC/DC method is simple to implement where there is a single DC/DC converter converting the 48 V battery input into different output voltages according to the requirements by various system cards. However, this configuration presents some problems. The current that needs to be distributed along the bus-bar or backplane may be very high. This would result in the need for larger conductors to offset extra power losses. Otherwise excess heat will develop. A related problem is that the output voltage line would experience voltage drop during heavy current loading. This needs to be compensated by remote voltage sensing, which results in additional circuitries.

Due to increasing systems complexity and hence increased risks of failure, it becomes crucial that these risks are mitigated at different levels. Power supply systems are one of the most common causes for system failure. To increase reliability for the DC/DC power supply, it is necessary to provide redundancy by paralleling another similar power converter at the outputs. Since these are invariably implemented by means of OR-ing the diodes with logic gates, there will be further voltage drop and losses due to conduction in the diodes. All these losses contribute to reducing the overall power conversion efficiency, as well as adding extra space and cost for components. Additional heavy conductors are also needed to wire the redundancy circuits.

Thermal management is also an important area. The heat generation for power conversion would come mainly from the DC/DC converter. This may create some localized hot spots that would require a substantial amount of forced air cooling or other similar thermal management measure. For hot insertion and swapping of system cards within the power distribution system, inrush current limiting circuits should to be employed.

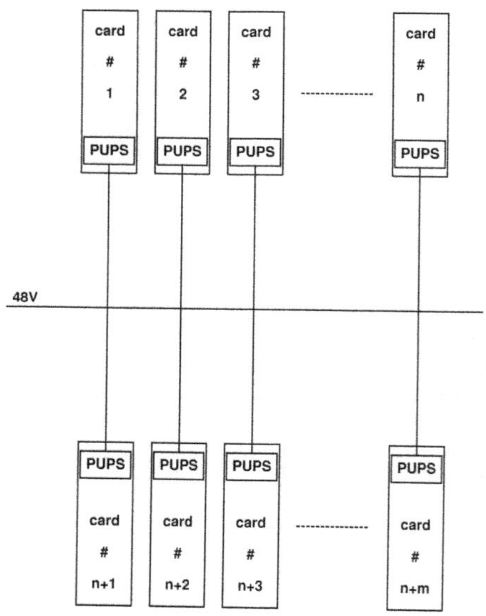

Figure 6.10. Distributed DC/DC power distribution

The other type is a distributed DC/DC power distribution system. This is shown in Figure 6.10, where the battery input of 48 V is distributed without any power conversion stage to each system card in the system. A dedicated DC/DC converter is built on each system card to perform this power conversion to generate locally the output voltages on the system card itself. The DC/DC converter is called Point of Use Power Supply (PUPS) or, alternatively, a power module which is available as a separate PCB subassembly to be mounted on the system card.

This approach has the advantage of distributed heat generation because the heat generation is not concentrated at a main centralized DC/DC converter. It also results in better load regulation without the voltage loss because the output voltage is produced within the system card itself. Heavy load currents are constrained within the system card itself and so the conductors and connectors leading to the system card input can be made smaller in size than in the case of a centralized DC/DC system.

However, the disadvantage is the higher cost of having to implement a dedicated PUPS for each system card. In addition, for hot insertion or swapping, there is still a need to have current limiting circuitry in each of the system cards. This will increase overall complexity and costs.

Thus, it can be seen that the existing DC power distribution systems face significant challenges to cope with future telecommunication systems, where there will be increasing demand for more power, increased reliability and system simplicity.

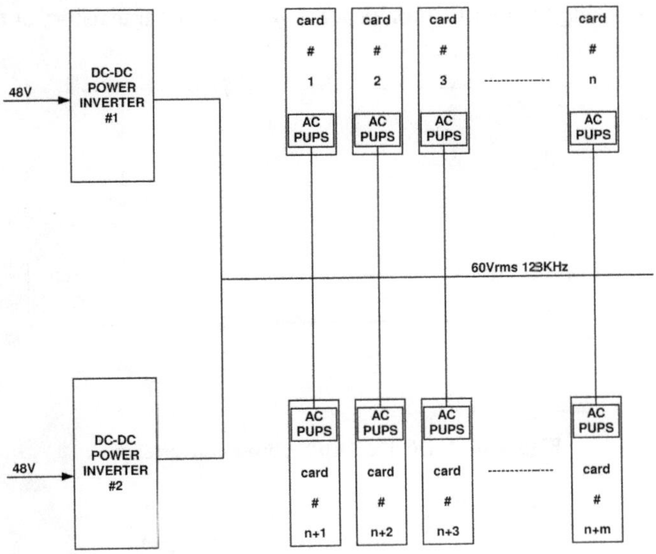

Figure 6.11. A HFAC power distribution for telecommunications [6]

Drobnik [6] proposed a high frequency AC power distribution system that aims to address limitation of the DC supplies. Figure 6.11 shows the configuration of the proposed HFAC power distribution system. There is a centralized DC/AC converter that produces a sine wave output voltage of 60 V at 128 kHz. The high frequency power is distributed along the bus bar or backplane. At the system card, a simple AC/DC PUPS converts the AC power into the required DC voltage output. The AC/DC converter is just a simple transformer passive full-wave rectifier with a series LC input. The resonant circuit provides good load regulation and will limit inrush current implicitly. Since there is no LC low-pass filter, there will be excellent transient response. Thus, the HFAC system would have both the advantages of simplicity and distributed heat generation and solves the inrush current issue.

Because of the resonant circuits used for the main DC/AC inverter and the AC/DC PUPS, the inrush current is inherently limited without need for additional circuits. The lesser component count in the AC/DC converters, compared to both the centralized DC/DC and distributed DC/DC approach, translates to higher reliability calculations and therefore paralleling of outputs for redundancy purposes is not necessary.

Looking at the DC/DC PUPS, even in the simplest implementation there is always an inherent DC/AC/DC to get a required voltage. Figure 6.12 shows an example of a DC/DC PUPS implemented as a forward converter. Each stage of this conversion will generate power loss. For example, in a DC/DC forward topology, the DC is converted into AC pulses in the converter to be pumped across the transformer for step down or step up purposes. Power loss happens during the primary side transistor switching and transformer core and copper loss. At the secondary side, the AC pulses need to be rectified back into DC again. Again there is power loss during the rectification process at the output transistors or diodes.

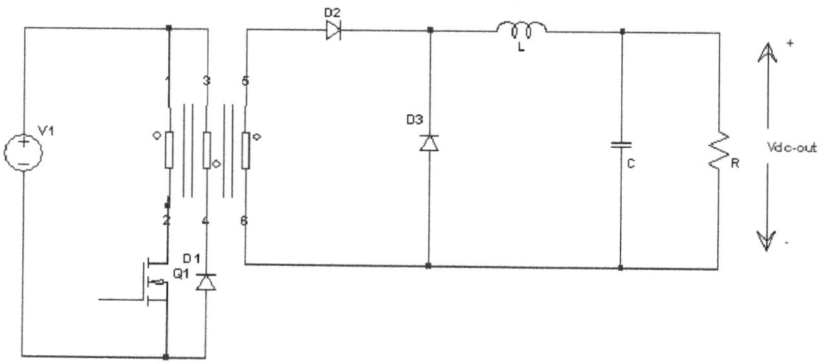

Figure 6.12. DC/DC PUPS power converter [6]

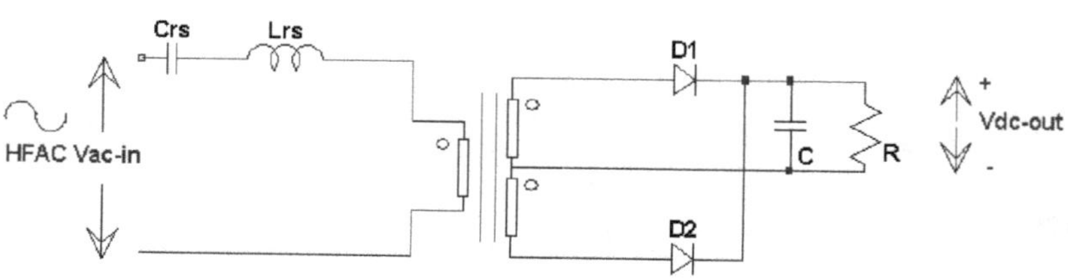

Figure 6.13. AC/DC PUPS power converter [6]

Compared to the case of the simplest AC/DC PUPS, there is just one stage of power conversion which is AC/DC only. Figure 6.13 shows an example of AC/DC PUPS implemented as a full wave rectifier. Therefore there is only the unavoidable transformer core and copper loss, and the rectification power loss in the output diodes. Because of the single power conversion step, losses are less compared to DC/DC and therefore power efficiency is relatively higher.

In HFAC, connector-less power transfer is easily achieved. In a telecom power system, the HFAC link would be fed into primaries of transformers. The secondaries of the transformers would be residing as the input of the system cards. This provides the advantages of electrical isolation in the event of faults, and increased reliability. Figure 6.14 shows the contact-less implementation concept of a telecom load card.

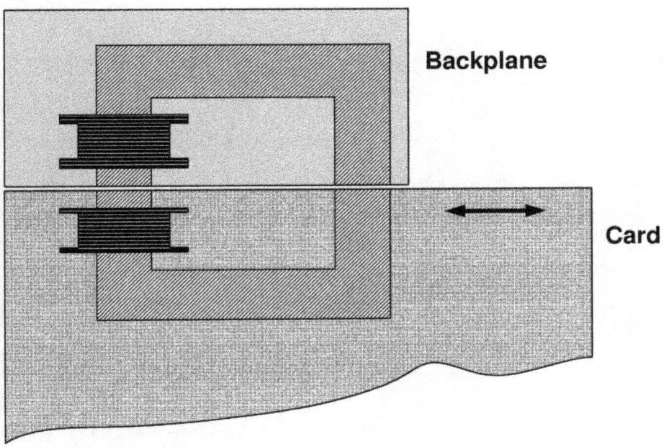

Figure 6.14. Connector-less power transfer used in a telecommunication card [6]

The details about such a telecomm HFAC power system was proposed by Jain and Pinheiro [7]. It is based on the double tuned series-parallel resonant inverter proposed in [6], and a new AC/DC converter was also used. The system attempts to combine the strengths of constant voltage type systems and constant current type systems without their drawbacks. Figure 6.15 shows examples of power distribution, which can be of either constant voltage or constant current type.

Constant voltage systems were reported to be better in terms of EMI and system efficiency for low to full load, but not capable of connector-less power transfer. Constant current systems were found to be conducive to connector-less power transfer. However, performance of EMI is generally poor, and high efficiency can only be attained at high load.

Figure 6.16 shows the circuit diagram details of the proposed new hybrid system. It demonstrates the capabilities for limited power transfer, hence less fire hazard. It has inherent short circuit current limit that effectively provides a fuseless protection. It also has a no-load primary current limit, which results in higher no-load efficiency and higher no-load power factor.

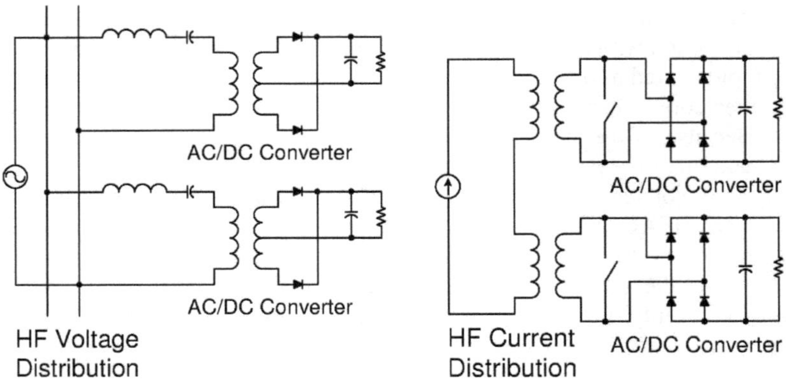

a b

Figure 6.15. Distribution systems used in telecommunication systems [7]: **a** constant voltage; **b** constant current

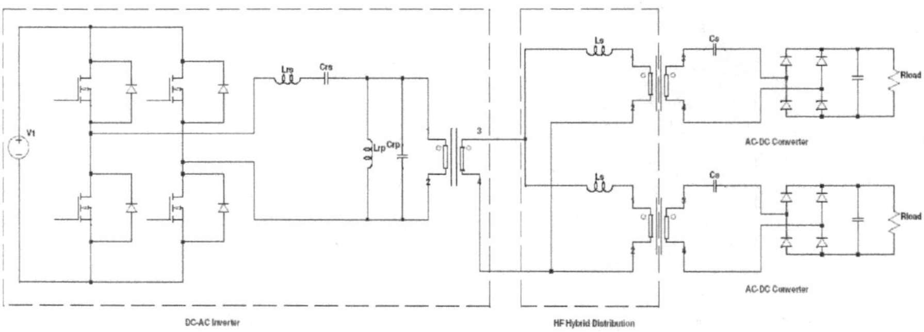

Figure 6.16. Hybrid high frequency AC distribution system [7]

6.4 High Frequency in Computer and Commercial Electronics Systems

The first known case of an HFAC Distributed Power Architecture (DPA) implementation in the commercial sector was found in 1994, by Hewlett Packard in their HP70000 series spectrum analyzers [8]. The system, shown in Figure 6.17, is a modular system consisting of a mainframe and various plug-in measurement modules. The mainframe has a thyristor inverter that accepts 147 VDC rectified input from the AC line, and generates a regulated 27 V 40 kHz sine wave voltage for the plug-in modules [9]. Figure 6.18 shows the block diagram details of this mainframe [10].

The load power converter of the plug-in module is a transformer isolated rectifier with LC output filter. Additional accuracy in output voltage regulation is obtained by a linear post regulator.

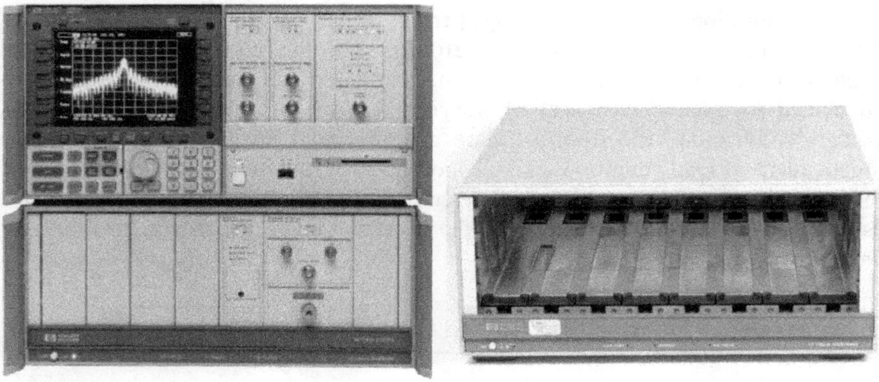

a b

Figure 6.17. HP 70000 series spectrum analyzer: **a** modular system; **b** mainframe power (courtesy of Hewlett Packard)

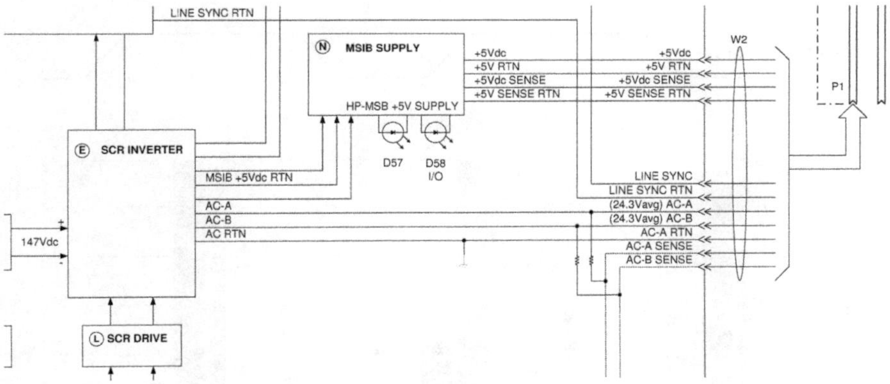

Figure 6.18. A part of the mainframe block diagram showing the 40kHz inverter supplying HFAC power [10]

In 1996, Hewlett Packard sponsored the Virginia Power Electronics Center (VPEC) to investigate further the potential use of HFAC as a commercial DPA [11]. Watson *et al.* proposed a 400 W 300 kHz trapezoidal voltage as the preferred specifications over the use of sinusoidal waveform, claiming a number of benefits of using trapezoidal waveform. Although a sine waveform has the lowest harmonic content, it is reported that the challenge has been the complexity in ensuring the sinusoidal waveform is maintained with low harmonic distortion and good power factor at all load ranges. Using a trapezoidal waveform would simplify the converter design to achieve a satisfactory result. Although the harmonics would not be as good as for a pure sine wave, the harmonics could be limited to an acceptable level by controlling the soft transition times of the trapezoidal waveform to within

10–15% duration of the switching period. Figure 6.19 shows different soft transition times for trapezoidal waveforms, and the corresponding harmonic content power. It can be seen that the harmonic content can be effectively controlled to a desirable level by varying the soft switching times. Also, the voltage conversion ratio from a square or trapezoidal inverter input to output is independent of load conditions. This is not the case for sine wave based waveform inverters. Moreover, the trapezoidal inverter design does not need a resonant tank as in a sine wave inverter. Therefore the inverter circuit design can be simplified and more cost effective.

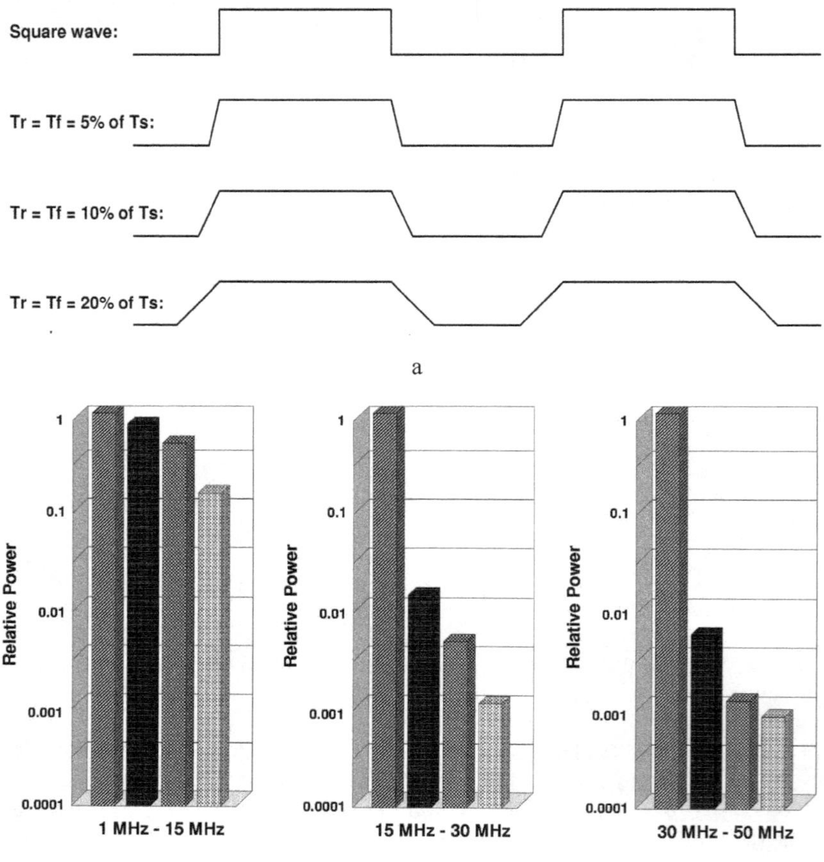

Figure 6.19. Variation of harmonic powers with soft switching times in a trapezoidal waveform [11]: **a** different soft transition times; **b** harmonic power for different soft transition times

The implementation of the trapezoidal HFAC is shown in Figure 6.20. The system consists of a front end PFC combined with half bridge resonant inverter. The inverter makes use of transformer gapping to get the leakage energy to perform Zero Voltage Switching (ZVS) at light loads. An auxiliary circuit provides a holdup time feature to the converter. The back end consists of Series Resonant Rectifiers (SRR) where the resonant inductance *L* is controlled to achieve output voltage regulation.

However, the main disadvantage of the trapezoidal system is the instability of the bus voltage. The transition times of the bus voltage are affected by the load current.

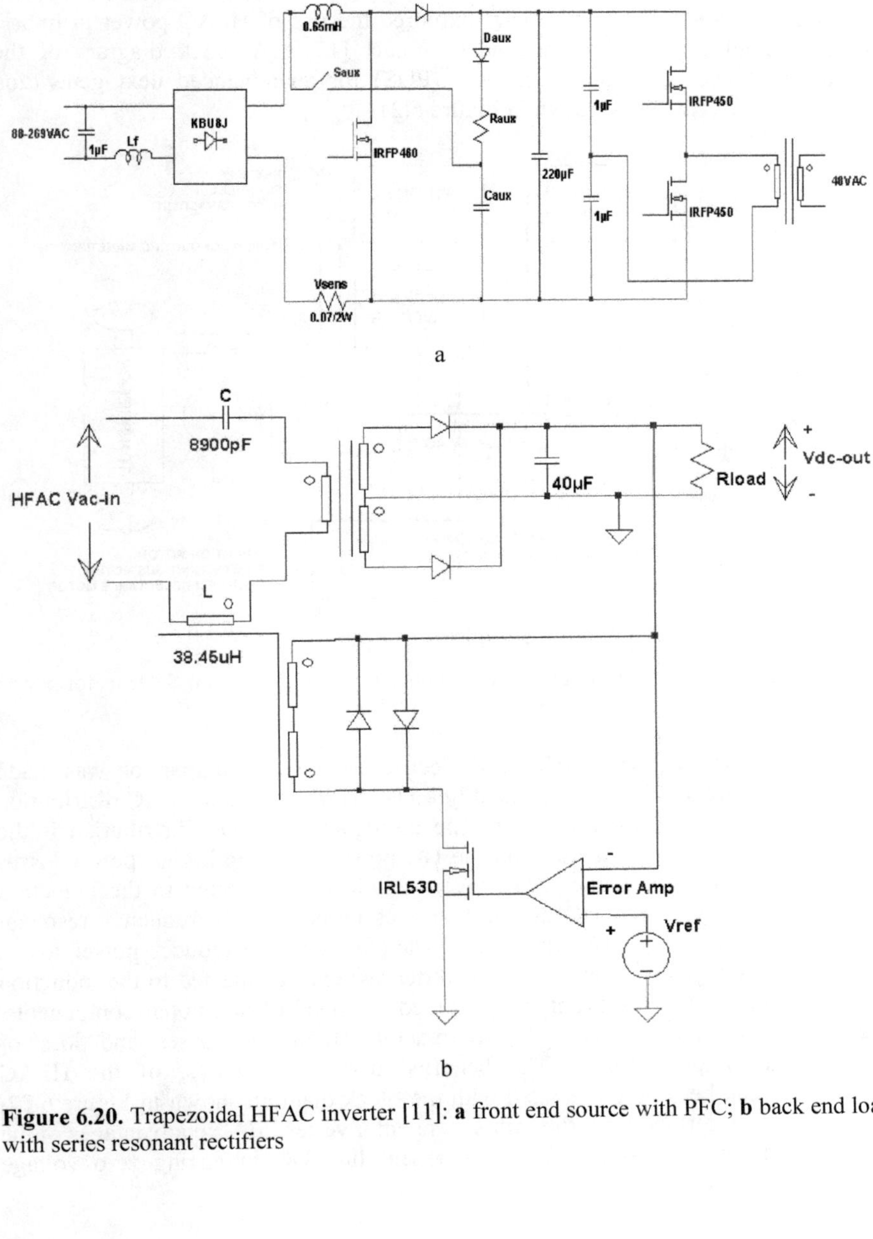

Figure 6.20. Trapezoidal HFAC inverter [11]: **a** front end source with PFC; **b** back end load with series resonant rectifiers

6.5 High Frequency in Automotive and Motor Drives

6.5.1 Automotive

Applications of HFAC in automotives have been closely linked with early work on the space programs. An early advocate for the use of HFAC in automotives was Chan [12], who noted earlier work by Sood and Lipo [3] could be usefully applied to electric vehicles. Bose *et al.* later explored the use of HFAC power in hybrid electric vehicles in an NASA funded project [13]. A block diagram of the propulsion Power Distribution System (PDS) for an advanced next generation hybrid electric vehicles is shown in Figure 6.21.

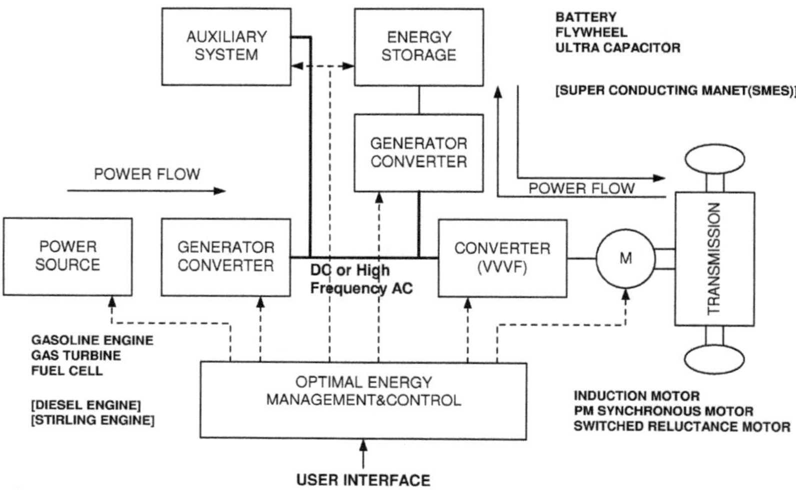

Figure 6.21. Block diagram of advanced propulsion power distribution System for hybrid vehicles [13]

Since the applications emphasized electric propulsion, comparison was made between DC link distribution, resonant link DC distribution and HFAC distribution used in the converter for the motor. The disadvantage of DC distribution is the need for large storage capacitors on the DC power link, and higher power losses due to hard switching in the IGBT inverter switches connected to the induction motor. The resonant link DC distribution uses unipolar high frequency resonant voltage pulses as the power distribution waveform, which reduces power losses due to zero voltage switching for the inverter switches connected to the induction motor. However, the disadvantages included the need of additional components, higher device voltage stress, higher machine harmonic losses and loss of fundamental output voltage. The benefits, and key features, of the HFAC distribution can be best appreciated with the block diagram shown in Figure 6.22, which is in fact derived from the Jain's resonant inverter [5]. The advantages of the HFAC distribution is similar to the resonant link DC in having zero voltage switching that leads to low switching losses. In addition, clamping or snubbing switches are not necessary, leading to further reduction of switching losses. Also, the sinusoidal voltage at the inverter output has lower EMI, and lower *dV/dt* stress on the switches. The use of the high frequency transformer enables the operating voltage of the AC machines to be stepped up and the operating current to be stepped down. This allows for lower copper loss possibility for the machine windings. It also enables auxiliary power supplies to be easily designed by tapping from the transformer. The disadvantages are the need for additional components and complexity compared to the resonant link distribution. The AC switch has to block or allow bidirectional voltage and current, and has to be made with two reverse-connected IGBTs, thus doubling the number of power switches.

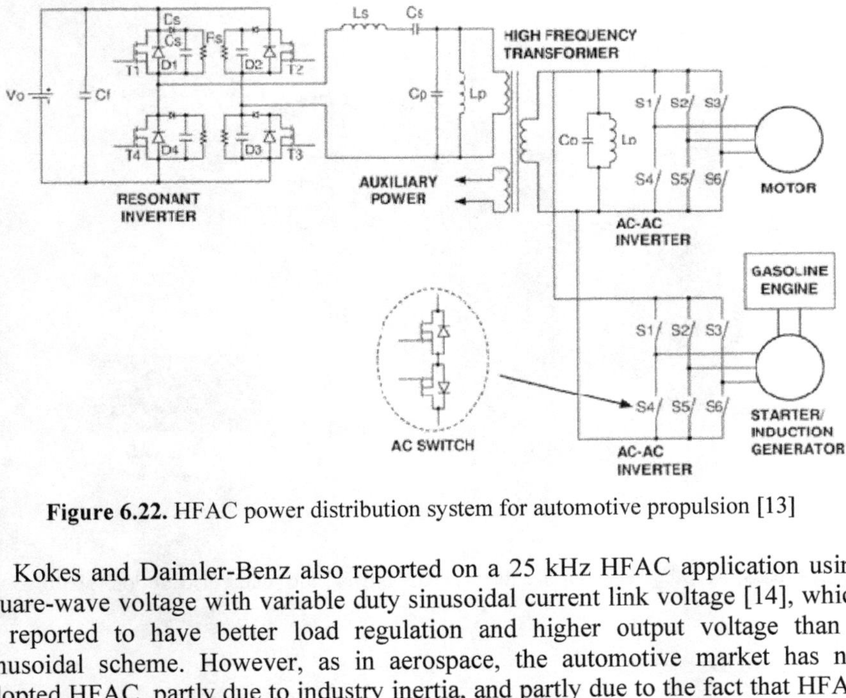

Figure 6.22. HFAC power distribution system for automotive propulsion [13]

Kokes and Daimler-Benz also reported on a 25 kHz HFAC application using square-wave voltage with variable duty sinusoidal current link voltage [14], which is reported to have better load regulation and higher output voltage than a sinusoidal scheme. However, as in aerospace, the automotive market has not adopted HFAC, partly due to industry inertia, and partly due to the fact that HFAC still experiences significant challenges in addressing the issues of EMI and losses for higher power level applications. At a typical power level of 20 kW or above, electric propulsion in a normal hybrid passenger vehicle remains a challenging application for HFAC distribution.

There are, however, other low-power automotive applications that pose less EMI and efficiency challenges, and may present lower market entry barriers. Figure 6.23 shows the concept of an HFAC power and data distribution platform for non-propulsion loads in a vehicle. A proprietary HFAC flat cable structure, called *J1* from Tunewell Group, UK, is used to suppress EMI and facilitate embedded power and data distribution. Three key groups of non-propulsion electrical loads: in-car electronics, actuation and lighting, that can most benefit from HFAC power and data distribution, are examined. In-car electronics and entertainment systems, also called the "*hotel loads*", will grow rapidly for modern vehicles, but impose extra wirings for both signals and power. Actuation loads, that include power steering and active suspension, will be increased to improve driver ability and passenger safety. Optimized actuation can be readily achieved with dynamically adaptive voltage rather than a fixed voltage. The use of LED for automotive headlamps and cabins, for higher luminous efficacy and energy saving, has gained much interest recently. Control and power distribution to all these devices in a cost-effective manner will become increasingly challenging for conventional fixed DC distribution. The use of HFAC power cable that carries control signals and data provides a compact, efficient and highly integrated cable harnessing system for future vehicles.

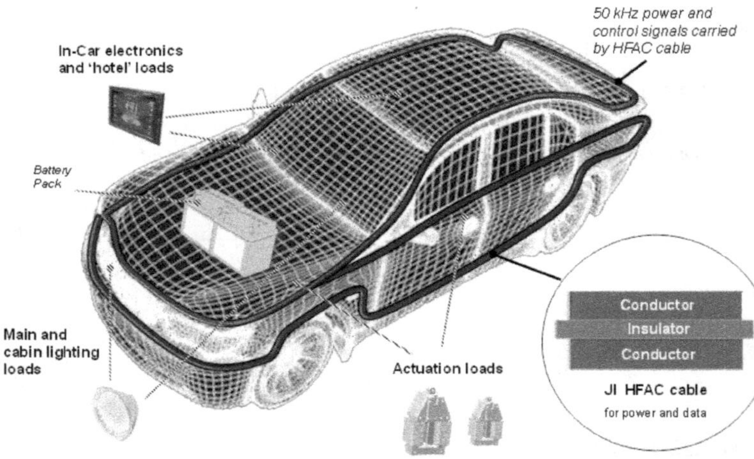

Figure 6.23. A 50 kHz high frequency AC power and data distribution platform for automotives

6.5.2 Motor Drives

HFAC applications in motor drives have historically been closely associated with automotive propulsion, as evident from the previous section. In an HFAC motor drive, Pulse Density Modulation (PDM) is preferred over conventional PWM. Operating voltage and frequency for the AC machines can be readily constructed from high frequency carrier sinusoids. The realization of the modulated waveform is by means of a 12-switch AC link converter as shown in Figure 6.24. This system is examined by Ma *et al.* [15, 16] to drive an induction motor. The density of the link voltage pulses impressed onto the motor terminals is modulated to output a quasi-square wave.

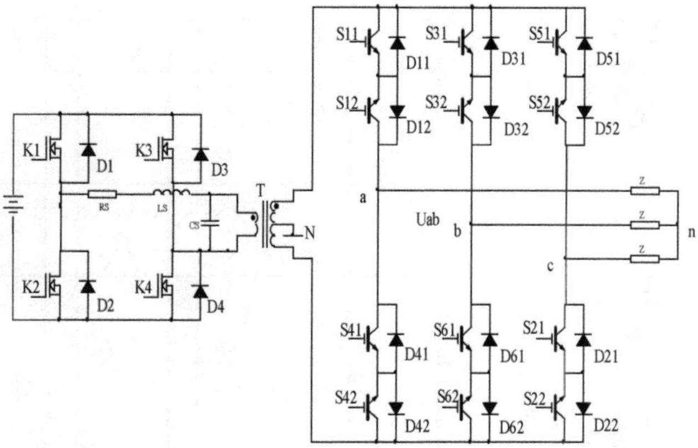

Figure 6.24. The 12-switch HFAC AC-link converter for an induction motor [16]

Switching patterns in Table 6.1 are used to generate the quasi-square voltage waveforms shown in Figure 6.25.

More recently, Bendyk *et al.* [17] explored a new Direct Torque Control (DTC) scheme using PDM modulation on an HFAC power converter. DTC essentially traces the output torque of the motor by means of hysteresis. This makes the control structure significantly much simpler than that of transformation-based Field Oriented Control (FOC). It also achieves dynamic responses comparable to those of the FOC technique. There are however two major problems associated with conventional hysteresis-based DTC drives. First, switching frequency varies with operating conditions, which can cause unpleasant audible noises. Second, the torque ripple is usually very high unless a narrow hysteresis band is used, which results in very high switching frequency in the inverter [18]. More advanced forms of DTC schemes, using constant and high switching frequency, have been proposed. However, these schemes tend to involve more complex control strategies.

The proposed HFAC-DTC scheme [17] demonstrates the potentials of superior performance over conventional DTC schemes, with no extra system complexity or software overhead. Reduction in flux and torque ripples during steady state operation, and lower inrush current during startup for a 35 kW induction motor, are demonstrated.

Table 6.1. Switching table for generating quasi-square waveforms for the induction motor by means of an HFAC converter

Time period	Phase voltage levels	Conducting devices
$0 \sim \omega t_1$	$U_{aN}>0$, $U_{bN}<0$, $U_{cN}>0$	S11, S61, S51, D12, D62, D52
$\omega t_2 \sim \omega t_3$	$U_{aN}>0$, $U_{bN}<0$, $U_{cN}<0$	S11, S61, S21, D12, D62, D22
$\omega t_4 \sim \omega t_5$	$U_{aN}>0$, $U_{bN}>0$, $U_{cN}<0$	S11, S31, S21, D12, D32, D22
$\omega t_6 \sim \omega t_7$	$U_{aN}<0$, $U_{bN}>0$, $U_{cN}<0$	S41, S31, S21, D42, D32, D22
$\omega t_8 \sim \omega t_9$	$U_{aN}<0$, $U_{bN}>0$, $U_{cN}>0$	S41, S31, S51, D42, D32, D52
$\omega t_{10} \sim \omega t_{11}$	$U_{aN}<0$, $U_{bN}<0$, $U_{cN}>0$	S41, S61, S51, D42, D62, D52

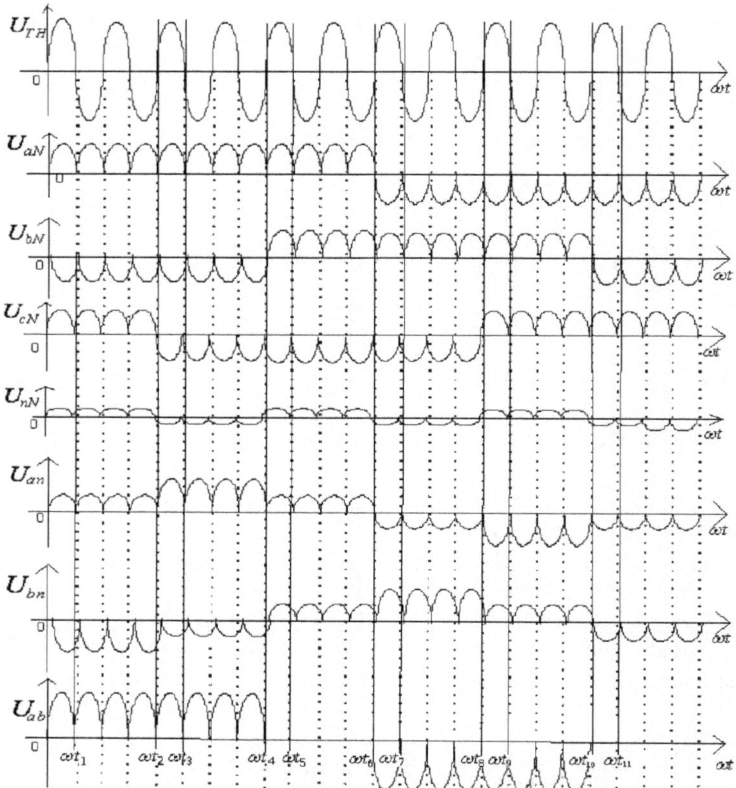

S_{11}	S_{42}	S_{11}	S_{42}	S_{11}	S_{42}	S_{11}	S_{42}	S_{11}	S_{42}	S_{11}	S_{42}	S_{41}	S_{12}	S_{41}	S_{12}	S_{41}	S_{12}	S_{41}	S_{12}	S_{41}	S_{12}	S_{41}	S_{12}
S_{61}	S_{32}	S_{61}	S_{32}	S_{61}	S_{32}	S_{61}	S_{32}	S_{31}	S_{62}	S_{31}	S_{62}	S_{31}	S_{62}	S_{31}	S_{62}	S_{31}	S_{62}	S_{31}	S_{62}	S_{61}	S_{32}	S_{61}	S_{32}
S_{51}	S_{22}	S_{51}	S_{22}	S_{21}	S_{52}	S_{21}	S_{52}	S_{21}	S_{52}	S_{21}	S_{52}	S_{21}	S_{52}	S_{21}	S_{52}	S_{51}	S_{22}	S_{51}	S_{22}	S_{51}	S_{22}	S_{51}	S_{22}

Figure 6.25. Quasi-square waveforms by high frequency half cycle sinusoids [16]

To demonstrate further the advantages of the HFAC-DTC over conventional PWM-DTC schemes, performance comparisons based on a common switching frequency of 20 kHz are made. Figure 6.26 shows that PDM using HFAC is spectrally better spread out, and more EMC friendly, than the conventional PWM based on hard switching.

In Figure 6.27, the instantaneous voltage, current, and flux levels of the two schemes are compared. The high device utilization factor as a result of soft switching means that the HFAC converter is more compact and efficient compared with the PWM converter. It is easy to appreciate from the waveforms that device stresses are much less in the HFAC-DTC scheme, which leads to improved system reliability and extended component life. Efficiency is also significantly improved due to reduced switching loss.

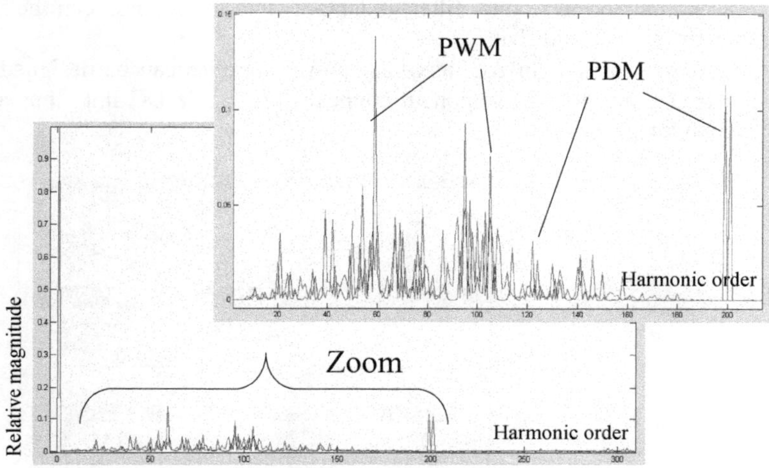

Figure 6.26. Comparison of harmonic contents of PWM and PDM switching schemes [17]

6.6 High Frequency in Microgrids

The microgrid is an emerging form of electrical power distribution in the modern power scenario, where deregulation policies and environment concerns encourage the exploitation of distributed generation of power close to the end-users. A microgrid consists of small power sources called microsources, which are usually derived from renewable power sources such as micro wind turbines, solar panels and fuel cells. A microgrid has a power level usually from sub-kWs up to several MWs, and operates as a single controllable system providing power and heat to a localized area. It can, however, operate both independently or connected to the grid. These conditions provide an ideal application platform for HFAC technology demonstration. Building services loads, such as compressor motors, lighting, and interfacing issues with the grid, can be incorporated in the demonstration. Since 2003, Chakraborty *et al.* have researched on HFAC power distribution for a single phase microgrid [19–21]. Their study shows an HFAC link microgrid offers the following additional advantages over a 50/60 Hz link or DC link:

- Power quality is easier to improve at high operating frequency;
- Acoustic noise can be minimized if the link frequency is above 20 kHz;
- For fluorescent lighting, luminous efficiency will improve, flicker will reduce, and dimming is easier to implement;
- The possibility to use high frequency induction motors;
- Harmonic ripple current in electric machines will decrease, improving efficiency;

- Soft switching can be exploited to reduce power losses, and to allow the use of lower rated power switching electronics;
- Power transformers, and filter inductors and capacitors, can be made smaller in value and size;
- Because the link is in AC, auxiliary power supplies can be designed with lower cost and lower component count than that of a DC link, thus aiding system integration.

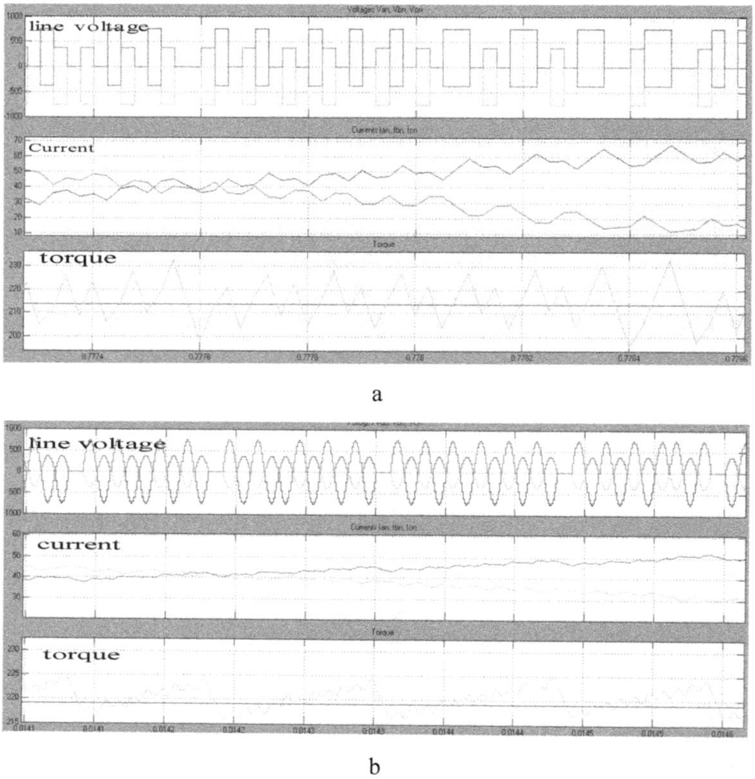

Figure 6.27. Comparison of performance prediction between the two schemes [17]: **a** standard PWM-DTC scheme; **b** proposed PDM-DTC scheme using HFAC

Figure 6.28 shows the schematic of the grid-connected single phase HFAC microgrid system developed by Chakraborty *et al.* [19, 20]. The UPQC is used to compensate for load current harmonics from non-linear loads and voltage distortion from source non-linearity, and provides power factor correction. The UPQC consists of an integrated active shunt and series filter at the AC link between the sources and load of the microgrid. The Universal Active Power Line Conditioner (UPLC) is used for controlling the power flow between the microgrid and the main grid, and also to compensate for the current harmonics due to utility connection. The UPLC also consists of an integrated active shunt and series filter

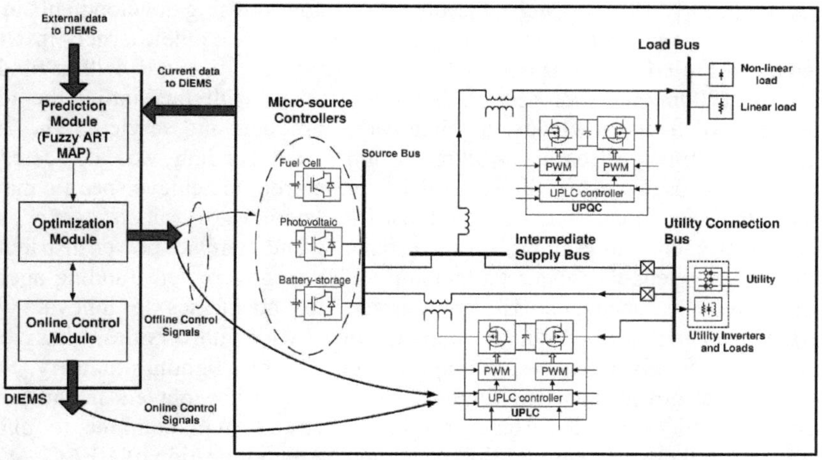

Figure 6.28. Single phase HFAC microgrid showing the HFAC UPQC, UPLC [21]

at the AC link between the source of the microgrid and the source of the utility. Both the UPQC and UPLC controllers use *p-q* theory to reduce the computation time needed for performing the control action. A Distributed Intelligent Energy Management System (DIEMS) is used to optimize the operating costs of the microgrid by controlling power generation of the microgrid sources based on external data patterns from such sources as weather satellites. Neural networks are used to perform such kinds of non-linear tasks which involve data mapping. An energy management approach to neural network architecture known as fuzzy ARTMAP forecasting is proposed, which can selectively learn new patterns when it cannot find a suitable match with existing trained data patterns.

6.7 Future Prospects

6.7.1 Future Drivers and Funding Issues

Whilst most aspects of economic activity rest on the reliable and effective functioning of electrical power, it is increasingly important that future electrical power will be delivered ever more cost-efficiently, embracing not only technologically advanced but also socially and politically acceptable solutions. Environment concerns will continue to be one of the key drivers for the success of any new energy solutions. They will be evaluated holistically by the lifetime environmental impacts of the subsystems, and the system's carbon foot-print, among other things. Increasing energy demands means electrical power will be operated at higher voltage, current and frequency levels. Because of the perceived risks and level of investment, national bodies and government agencies will continue to be key stakeholders in advancing these new electrical power distribution solutions. In the US, NASA was an early applicant of HFAC in motor drives in the early 1980s, which subsequently led to promising developments in the fields of computer and IT systems in the last few decades, albeit largely in North America. In the UK, the business-led Technology Strategy Board, with a funding budget of £200m per year (2007), has been tasked with facilitating the rapid deployment of new knowledge into innovative products and services. The 2006 Technology Strategy Board's Autumn Competition funding was dedicated to power electronics and electrical power delivery, in order to achieve specific market needs in the near future. The Carbon Trust UK, another government agency, also has a similar agenda to promote energy efficiency and compact power distribution solutions. Knowledge Transfer Partnership (KTP), a government funding agency, which provides a managed risk environment for businesses to innovate with emerging technologies through collaboration with universities, has been instrumental in advancing HFAC applications in the lighting industry. It is envisaged that government led funding, together with key players in the power distribution and power electronics products' markets, will continue to play a pivotal role in the successful development and commercialization of HFAC.

6.7.2 Future Trends and Challenges

The technological paths taken by HFAC to the current status can provide useful insights into the future trends it is likely to encounter. The research and development efforts in HFAC applications in the last three decades can be separated into three interesting periods as follows.

High power space applications era. The early applications of HFAC in high power distribution by the space industry in the first decade had encountered significant challenges in distributing high power at high frequency cost-efficiently. In particular, there was no suitable cable technology for large scale, high current, high frequency operations. Research in space and aerospace declined gradually in the mid-1990s. The space station industry has always been extremely conservative due to the immense development cost budget, high reliability and high risks necessary when human lives are involved. There are similar factors in the aerospace industry. The critical issues of EMI impacts on flight control systems and mission critical equipment, together with the relative sufficiency provided by the existing 400 Hz system, have put HFAC out of favor in the industry. However, non-safety-critical in-flight entertainment systems still look promising for HFAC.

Low power IT applications era. Whilst the space programs demonstrated considerable potential advantages of HFAC as a new means of power distribution, it has been in the lower power applications in the telecommunications and computer systems where these advantages have been fully exploited and realized. These consumer-driven market-driven IT industries require more compact and powerful processors to meet consumer's demands for faster and more cost-effective services. These in turn demand power intensive processors and associated high speed electronics. The HFAC power distribution system demonstrates better transient response, reduced power stages, lower current rated power switches, and smaller passives when compared with competing DC/DC power distribution systems. It is noteworthy that the majority of the publications have been from the research team led by Jain of Queen's University, Canada.

Medium power emerging applications era. The success of HFAC in low power IT applications has shown prospects for medium power applications. It is noted that HFAC has attracted interest in some emerging applications such as microgrids and renewable energy. A possible explanation for this growth is based on the increasing demand for more distributed and efficient means of energy delivery, which would be environmentally friendly energy and at a reasonable cost of entry. Most of these sources are of renewable types like solar panels and wind turbines, which have variable voltage, current and frequency outputs. The microgrid solution would enable these power sources to be effectively connected to the main grid and local customers. With the advantages of HFAC such as smaller power transformers, lower current rated power switches and smaller harmonic filters, the HFAC microgrid offers many operational and ownership advantages. Automotives continue to be of interest for HFAC application, although industry inertia means that it is unlikely to be adopted in the near future. However, niche car markets, such as very light electric vehicles, may provide a lower entry barrier for HFAC to demonstrate fully its many potential advantages. After all, the power level and distribution structure of a standalone microgrid can be very similar to that of a light electric car. Although there appears to be no reported HFAC application in building services and lighting, the prospect for these market systems is promising. The advantages of HFAC in fluorescent lighting have been reported in [19]. With lighting accounting for over 20% of electricity usage and the impending regulations on energy efficient lighting, it is potentially an exciting application area for HFAC in the form of fluorescent and LED lamps. Unlike the aerospace and automotive industries, the lighting market provides a very lower risk entry with no hazard to human life or demand for extreme reliability.

From the above observations, it can be seen that, after three decades of development, HFAC has begun to make impacts in some consumer-driven and low risk markets such as telecommunications and IT. In the meantime, the technological advances in high frequency cables and magnetic materials, soft switching and resonant converters have significantly lowered technological risks of applying HFAC. It is envisaged that there will be considerable interest in the application of HFAC in emerging markets such as in microgrids and in renewable energy areas, where there are increasing demands for high power conversion efficiency and system flexibility. Similarly, other new but less risk-averse markets, such as lighting and light mobile vehicles, are expected to make inroads to exploit the many potential benefits that HFAC has to offer.

Acknowledgement

The authors are most grateful to the support by the Knowledge Transfer Partnership, UK. The help from and useful discussions with the following people are gratefully acknowledged: Phil Rimmer of Tunewell Group, UK; Maciej Bendyk, Ken Jinupun, and Steven Lourdes of Cranfield University, UK; Professor Praveen Jain of Queen's University, Canada.

References

[1] Hansen I, (1983) Design considerations for large space electric power systems. NASA Technical Memorandum 83064

[2] Hansen I, (1988) Status of 20kHz space station power distribution technology. NASA Technical Memorandum 100781

[3] Sood PK, Lipo TA, (1986) Power conversion distribution system using a resonant high-frequency AC link. IEEE IAS Annual Meeting, Conference Record:533–541

[4] Jain P, Tanju M, (1989) A 20kHz hybrid resonant power source for the space station. IEEE Transactions on Aerospace and Electronic Systems, vol.25, no.4:491–496

[5] Jain P, Tanju M, Bottril J, (1993) AC/DC converter topologies for the space station. IEEE Transactions on Aerospace and Electronic Systems, vol.29, no.2:425–434

[6] Drobnik J, (1994) High frequency alternating current power distribution. 16th International Conference on Telecommunications Energy:292–296

[7] Jain P, Pinheiro H, (1999) A hybrid high frequency AC power distribution architecture for telecommunication systems. IEEE Transactions on Aerospace and Electronic Systems, vol.35:138–147

[8] Hewlett Packard, (1994) Modular measurement system – hardware design guide. Santa Rosa, CA

[9] Hewlett Packard, (1999) HP70001A mainframe installation and verification manual

[10] Hewlett Packard, (1998) HP70001A mainframe overall block diagram

[11] Watson R, Chen W, Hua G, Lee FC, (1996) Component development for a high frequency AC distributed power system. IEEE Applied Power Electronics Conference and Exposition, vol.2:657–663

[12] Chan CC, (1993) An overview of electric vehicle technology. Proceedings of IEEE, vol.81, no.9:1202–1213

[13] Bose BK, Kin MH, Kankam MD, (1996) High frequency AC *vs.* DC distribution system for next generation hybrid electric vehicle. IEEE Proceedings of the International Conference on Industrial Electronics, Control and Instrumentation, vol.2:706–712

[14] Kokes M, (1997) Resonantes wechselspanungsbordnetz fur kraftfahrzeuge und dessen beschreibung mit zustandszeigern. Rurh-Universitat Bochum

[15] Ma X, Chen Q, (2001) A novel scheme of propulsion system using soft switched high frequency AC/AC converter for electric vehicle. Proceedings of the International Conference on Electrical Machines and Systems, vol.1:496–499

[16] Ma X, (2004) High frequency AC pulse density modulation theory and its application in hybrid electric vehicle drive system. International Power Electronics and Motion Control Conference, vol.2:827–830

[17] Bendyk M, Luk PCK, Jinupun P, (2007) Direct torque control of induction motor drives using high frequency pulse density modulation for reduced torque ripples and switching losses. Proceedings of IEEE Power Electronics Specialists Conference:pp.86–91

[18] Buja GS, Kazmierkowski MP, (2004) Direct torque control of PWM inverter-fed AC motors – a survey. IEEE Transactions on Industrial Electronics, vol.51, no.4:744–757

[19] Correa JM, Chakraborty S, Simoes MG, Farret FA, (2003) A Single phase high frequency ac microgrid with an unified power quality conditioner. IEEE Industry Applications Conference, vol.2:956–962

[20] Chakraborty S, Simoes MG, (2005) Advanced active filtering in a single phase high frequency AC microgrid. IEEE Power Electronics Specialists Conference:191–197

[21] Chakraborty S, Simoes MG, (2005) Fuzzy ARTMAP based forecast of renewable generation for a high frequency AC microgrid. Annual Conference of IEEE Industrial Electronics Society:6

7

Integration of Distributed Generation with Electrical Power System

Khaled Nigim

Institute of Technology and Advanced Learning,
Conetsoga College,
Doon Campus, Doon Valley Drive Kitchener, Ontario, N2G 4M4, Canada.
Email: KNigim@conestogac.on.ca

7.1 Distributed Generation Past and Future

Distributed Generation is a back-up electric power generating unit that is used in many industrial facilities, hospitals, campuses, commercial buildings and department stores. Most of these back-up units are used primarily by customers to provide emergency power during times when grid-connected power is unavailable and they are installed within the consumer premises where the electric demand is needed. The installation of the back-up units close to the demand center avoids the cost of transmitting the power and any associated transmission losses. Back-up generating units are currently defined as distributed generation to differentiate from the traditional centralized power generation model. The centralized power generation model has proven to be economical and a reliable source of energy production. However, with the lack of significant increase in building new generating capacity or even in expanding existing ones to meet the needs of todays mega cities' demand, the whole electrical power industry is facing serious challenges and is looking for a solution.

There are about 12 million DG units installed across the USA, with a total capacity of about 200 GW. The available generating power can be used to power independently the facility demand without external power supply from the local EPS provider – hosting grid, or partially supplying the facility load with a sharing agreement with the local energy provider, or completely supplying electrical demand to avoid extra tariff during peak demand periods or total blackout [1–3].

DG is foreseen to fit well in the new restructured energy market with all stockholders enjoying major benefits:

- To EPS, DG will increase transmission and distribution capacity and, therefore, limit market influence on the energy cost increase;

- For the large or small electricity consumers, DG uses a locally available mix of prime fuel sources, thus decreasing the dependency on importing. It can also be used for emergency backup and, moreover, it can be considered as an income generating vehicle if properly interconnected with a hosting grid.

Today DG come in various power ratings from a few watts to mega watts. DC technologies range from units that only produce electricity such as photovoltaic to those units that produce a combination of heat and electricity. The CHP generation system is a DG that has a higher overall conversion efficiency of 75% compared to the 40% of a mono-converting generating system (electricity-only steam power plant). CHP are designed to utilize fully the process of converting the prime fuel source into thermal and electrical energies in one installation, there by increasing the fuel conversion efficiency.

7.1.1 DG Conversion Energy Systems

A typical DG energy conversion system is simply made of two main energy converting stages. The first stage is the prime fuel converting block in which the prime fuel internal energy is converted into mechanical energy, as in the case of internal combustion engines. The second stage converts the mechanical energy into electrical power using an electromechanical energy conversion device such as synchronous alternator or induction generator, which produces AC power. Another way of converting prime fuel source into electrical energy is through a chemical or photosynthesis conversion process. Fuel cells and photovoltaic solar energy converter are good examples of this category and produce DC power. The interfacing unit is essential to convert the produced DC source into harmonized constant voltage and frequency AC power source. DC to AC power electronic inverter system is incorporated as interfacing unit. One important requirement for the inverter is to produce good quality AC power with frequency supply fluctuation limited to 1.2 Hz and with less than 5% THD of its voltage waveform at the coupling point in accordance with IEEE 519 standard [4].

An additional requirement of the inverter is to have the capability of preventing the DG from islanding (anti-islanding capability) on the hosting grid. Islanding is a condition occurring when a generator or an inverter and a portion of the grid system separates from the remainder of the large distribution system and continues to operate in an energized state. Islanding may pose a safety threat or cause equipment problems; therefore, it cannot be permitted without full coordination and planning between the parties [5–7].

In both conversion types, the output produced must meet the hosting grid electricity voltage and frequency standards. A coupling transformer is needed to interface the DG generator with the grid to match the distribution voltage level at the point of connection. Only when it is safe and exact paralleling conditions (synchronization) exist is the DG interconnected with permission and full coordination with the grid operator.

Another configuration normally adopted for supplying power to sensitive electrical load demand is to use DG armed with a UPS unit. The UPS system normally incorporates an energy storage medium such as batteries to enable power supply continuity as well as improve power quality and reduce the influence of voltage surges, spikes and swells which could cause loss of production. The DG/UPS can be configured in various schemes to meet consumer demand and positively be part of the ancillary support to a hosting grid.

In all the systems described above, the DG output could ultimately be interconnected with the grid. The interconnection takes place at the PCC at the grid side. Once the interconnection is established the hosting utility assumes responsibility of DG operation and contribution and treats it as part of its generation system.

7.1.2 DG Opportunities

Energy investors and utility operators are attracted to the DG role and associated industry for the following foreseen opportunities:

- DG can be fueled by locally available renewable and alternative mix of fuel sources to meet today's energy demand. Renewable sources are wind and solar, while alternative fuels are those produced from waste products or biomass and can be in gas, liquid or solid form. Greater independency from importing petroleum fuel can be achieved by incorporating DG that are powered by various fuel sources;
- DG can support future increase in demand without investment in the expansion of existing distribution network by installing the DG very close to the new load centre;
- Installing DG within the industrial/commercial premises avoids negotiating land use and the need for rights-of-way for electric transmission and distribution, thereby minimizing further capital investment;
- DG can be used in reducing intermittent and peak supply burdens on utilities grid by injecting power as required by the controller;
- DG will have the capability to support the existing energy supply when needed and in a short time (black start) without incurring capital cost;
- DG penetration in the energy market will create overall competitive pricing of energy. The current generation rate ($/KWh) of DG is now competitive with the centralized generation system as more efficient fuel energy conversion units such as fuel cells and micro turbines are continuously improved and diversified;
- DG could contribute to decreasing the vulnerability of the electric distribution system to external threats and hidden undetected faults that may cause wide scale blackout by feeding power to the sensitive infrastructure;
- DG have a flexible feature in the capability to be configured to operate in a stand-by mode, isolated mode, or sharing the load through integration with the existing nearby electric grid.

For the general public, using DG that is fueled by various prime alternative fuel sources will be welcomed. It results in reduced fossil fuel consumption.

Ultimately, DG importance basically depends on its inter-connect-ability with the hosting power provider, its functionality in meeting peak and intermittent load demands, financial feasibility and its environmental impact and alternative fuel processing.

7.1.3 DG Classifications, Location and Sizes

There is no clear cut classification for DG or optimum location and sizing as its role may change from one location to the next. Classification could be according to their rated generation capacity or by ownership, whether it is utility or Independent Power Providers (IPP) ownership. Appropriately, DG could be classified by their functionality in supporting the hosting grid or according to their interconnectivity, power electronics interface or by the used prime fuel source:

- DG equipment owned and operated by the power utilities are strategically located along the distribution grid to support the power demand at specific locations for a specific time. The unit size is chosen to meet the required forecasted ancillary grid support;
- DG owned and operated by IPP identities are scattered along the grid contour and located in either commercial or industrial zones and placed near their load center. They are mainly configured as stand-by or load sharing generating units. The size is only relative to the owner and not to the utility;
- DG rated capacity can span from a few watts to hundreds of kilowatts. Small rated units are normally designed to support residential and small industrial consumers and are likely located in urban surroundings. Many small units come in modular design, enabling series or paralleling cascading of units to increase the available capacity. Large capacity DG units designed for large industrial consumers, are located in industrial zones, and are used for supporting the reliability of supplying power at critical time and cost rather than other distribution grid related issued;
- DG units are energized by various primary fuel sources such as fossil polluting and diminishing and non-fossil clean and renewable.

The main criteria for selecting the type of prime fuel source for a DG are its local availability, conversion system technological advancement, impact on the environment, existing financial incentives and operating cost. Under this sorting, DG are classified into two broad categories depending on the dispatchability nature of the produced energy which is a function of the prime fuel source nature:

- Dispatchable DG are fueled by fossil fuel sources in which the produced electric power can be precisely controlled and dispatched in pre-programmed purchase agreement. Fuel cells, Sterling-engines, and internal combustion engine-generators and many co-generation schemes fall within this classification. Such DG has high capacity rating from a hundred kilowatts to mega kilowatts unit size;
- Non-dispatchable DG fueled by non-fossil fuel sources as the prime fuel sources are a source site specific and moreover the produced electric power is considered to be intermittent in nature. Photovoltaic, micro sized water

turbine and wind turbine grid connected power generating schemes fall within this classification. Such modular DG has low capacity rating from a few watts to kilowatts units' size.

7.2 Interconnection with a Hosting Grid – Parallel Operation

For successful integration of DG with the hosting grid, clear interconnection requirements must be formulated [8–11]. The current engineering practice for DG/distribution network interconnected systems is to revert the distribution network to its original configuration (radial or meshed distribution system) with all interconnected DG units de-energized whenever an unexpected disturbance occurs in the system. Since most distribution systems comprise radial feeders, this practice leads to the discontinuation of the supply for all the downstream customers. In this way DG contribution is restricted to the hosting utility demand and conditions. The following section presents the general issues of interconnecting DG, classified by their prime fuel usage, to a hosting grid.

7.2.1 Interconnection of DG Fueled by Fossil Fuel

AC power is produced by a synchronous alternator; therefore, standard synchronizing units are needed to enable the DG connection to a hosting distribution network. A matching transformer is recommended for connecting the DG to the network at the point of common coupling. The network operator will only guarantee DG interconnection as long as the following challenges are mutually addressed:

- Maintenance worker safety. For a successful integration of load sharing between hosting distribution network and DG, safety protocol and coordination is essential to ensure the safety of working maintenance staff. This entails the existence of maintenance disconnect accessible to the network engineers at the premises of the DG operator;
- Power flow and its direction. Traditionally, electricity distribution allows a unidirectional flow of power with feeders protected with unidirectional protection gear. DG can be integrated with distribution network and own electrical demand in two distinctive modes. The first is the stand-by mode in which DG resources are not fully utilized by the network and only supply its own demand once the supplying source is disconnected. In this case, minimum coordination with the hosting network is required. The second operating mode is to share power supply to its own demand with the hosting network. In case of deficiency it will import from the network and *vice versa* in case of a surplus. This requires a high degree of coordination with both the hosting network and DG allowing bidirectional power flow between them;
- Transmission congestion. First come, first served policy. A late comer to inject power into the transmission lines will be penalized;

- Will islanding be permitted? Intentional islanding is not permitted for various safety and operational issues. The current practice for DG/hosting network interconnection is that all interconnected DG units must be de-energized whenever unexpected disturbance occurs in the system;
- What will be the DG penetration level without compromising the hosting network stability and reliability? Many researchers conclude that 25% penetration is of no harm to the grid and it actually enhances the voltage stability, assuming that the DG is strategically installed, which is not always the case. Moreover, there are insufficient data on the dynamic performance of the hosting grid and the impact on the overall power quality and distribution reliability;
- Where are the interconnection protection and communication infrastructure functionality and reliability? Relays are normally used to provide zoning protection and are independently protecting each zone without any cross-communication between the zones. Moreover, there is no real time communication in the form of wide area network nor intelligent sensing devices protocol standards available that could enhance the interconnection infrastructure.

7.2.2 Interconnection of DG Fueled by Non-fossil Fuel

Renewable energy and alternative fuel sources are used as prime fuel sources for DG [12]. These DG produce either variable DC power sources or variable AC sources. For DC type, a static harmonizing converter/inverter power electronics interface module is needed to convert the DC into stabilized AC source. In the AC type, the output of the generator will fluctuate with the prime source inherited nature; therefore, the converted power needs to be harmonized. In both types a converter/inverter module is needed to keep an eye on the voltage and frequency. In either case, the quality of the AC generated output power source may not be 100% sinusoidal causing harmonics injection to the hosting network. In addition to the discussed points in the above section, DG is powered by alternative energy sources have extra constraints to be added:

- Scheduling. In centralized power management, scheduling is an important factor for the reliability of the energy supply. DG fueled by alternative fuel sources dispatcher cannot send accurate, short term prediction of the available power. For example, wind power variation is faster than demand variations. In this way, the integration of such generating systems needs a special forecasting and dispatching environment;
- Transmission congestion. Again, first come, first served policy. For IPP this is an opportunity for losing income, in particular for generating units that are fueled by renewable sources. Therefore, for effective investment, the DG operator must be at 100% readiness to inject power and be a part of the centralized dispatching center;
- Reactive power flow. Many alternative source generating units do not have sufficient reactive power source (VAR) to meet the load requirements and

once connected in parallel with the hosting network, DG themselves become VAR burden;

- Anti-islanding. Each interfacing module must have anti-islanding capability to prevent devastation to the DG converting units once the distribution network is de-energized.

7.2.3 Interconnection of DG Fueled by Mix of Fossil and Non-fossil Sources

Many of today's DG units are fueled by fossil and non-fossil fuel sources. When a cluster of large and capable generating units exist in a confined locality, they pose a high potential for creating a distributed resources domain. It is quite possible to create a confined distribution network defined as a Micro Grid (MGR) generating and distribution system [13]. The purpose here is to combine all available energy sources and associated DG units in one domain to meet the demand of the well defined energy consumer. Any excess energy could be exchanged with a nearby utility, thereby, filling a slice in the energy market in a very short time.

MGR infrastructure that is based on distributed resources and DG units can have three distinctive grid architectures, namely AC, DC and DC/AC distribution MGR, as depicted in Figure 7.1:

AC architecture. Individually, each DG generates its own high quality AC power. The various DG units are then connected together to form a local AC distribution network as shown in Figure 7.1a. This is a modular structure of connecting DG. More or less DG units could be added to meet the local demand. The demand is continuously supplied through the constructed MG. Such configuration requires ancillary support, particularly VAR requirements. Precise coordination between the DG units is essential for the continuity of the power flow into the grid. Power exchange with existing utility is possible.

DC architecture. Regardless of the prime fuel, all DG units are generating DC power source as shown in Figure 7.1b, creating DC distribution network. The created DC bus can feed directly the individual consumer by standard DC voltage level. Such configuration has high reliability with high redundancy as it requires less sophistication to interconnect DC sources compared with AC sources. The main disadvantage is the risk of having circulating currents between the units. Each consumer may use the DC source or invert into AC source depending on the appliance used. Power exchange with existing utility is also possible by using a grid tie standard inverter unit.

DC/AC architecture. MGR could be structured around individual generating units that produce DC voltage sources connected to one DC bus. Collectively, units are networked to centralized AC interface system for the purpose of integration with the existing grid as depicted in Figure 7.1c. A key weakness in this configuration is the total loss of AC power if the centralized inverter is down.

In all possible architectures, different interconnection or energy management challenges need to be identified and addressed by the MGR operator and by the hosting network. Such a variety of ways DG can be integrated with each other and

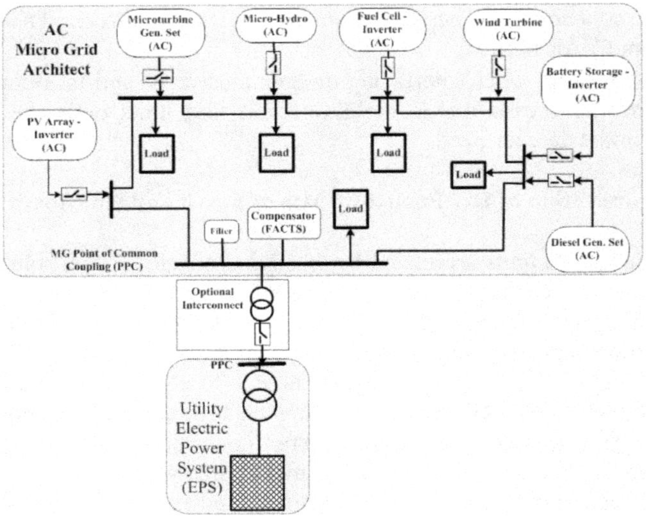

a

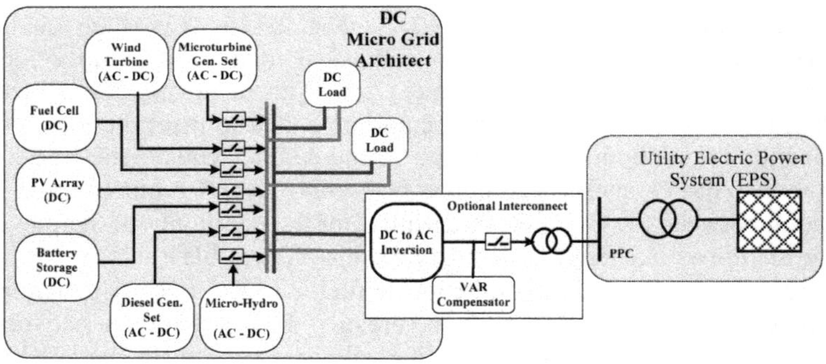

b

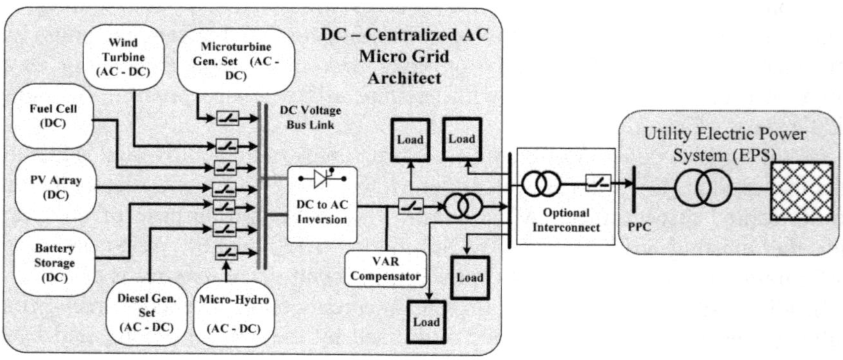

c

Figure 7.1. MGR architectures: **a** AC based architecture; **b** DC current based architecture; **c** DC and centralized AC based architecture with a hosting grid provides several opportunities. It provides support to the energy sector along with new and emerging technologies. It also opens new research areas and assists in creating employment opportunities in the sector.

7.3 Integration and Interconnection Concerns

DG interfacing and regulatory rules may not be established and understood by the local distribution network controller, thereby hampering DG interconnection [14–16]. Obviously, various challenges need to be addressed by all parties collectively within the energy sector regulating body to develop agreed standards and regulations to facilitate successful penetration. P1547 compiled by the IEEE is a good model for required standards. Key elements for the reliability of distributed generation power systems are the performance of the electrical switchgear, interconnection, controls, and communication features. The main components of interconnection according to the protection functions they perform are categorized as follows:

Synchronization. Auto-synchronization is the best choice for DG. Automatic sensing of the voltage and frequency achieves faster interconnection to the hosting grid. Auto synchronizing relays or paralleling switchgear are switching components used to verify the need for synchronizing the DG unit to the hosting network.

Islanding. Islanding protection is considered to be an important feature (a must) that the hosting grid requires from the DG operator. Islanding on part of the hosting network could jeopardize the maintenance crew safety and causes malfunction of nearby coordinated protection units. Relays are normally used to provide protection at both the grid and the DG end of the connection. Many of today's inverters incorporated in DG have built-in features to disconnect from the hosting grid once anti-islanding conditions are violated.

Voltage and frequency tolerance. To ensure high quality power injection, both the voltage and frequency margins should not exceed the recommended grid tolerance. An acceptable voltage limit is ±10% and for the frequency ±0.1%. Both voltage and frequency detection is part of the anti-islanding protection control.

DC injection level. Under abnormal operating conditions, grid tie inverters could inject low level DC current into the hosting grid. Similarly, transformer-less grid-tie inverters may also inject DC current into the grid. The likelihood of DC injection is possible; however, the possibility of cancelation due to opposing polarities when more than one DC source is present is likely. DC injection occurs when the switching devices are not fully operational, causing a large discrepancy between the positive and negative half cycle of the injected current. The generating system should not inject DC current greater than 0.5% of the full rated output current at the PCC in accordance with the IEEE 1547 standard. It is part of the inverter feedback loop to detect the presence of the DC component and adjust the triggering sequence to the switching devices to remedy the situation. A coupling transformer could be used to isolate the DC current from flowing to the AC side. A low cost solution is to incorporate a DC detection device to disconnect the inverter in case of severe DC level injection.

Grounding. Protective grounding is mainly designed to protect the operator. Grounding could also contribute to reducing the magnitude of transient overvoltages and lightning protection. Grounding components must be capable of carrying the maximum available fault current and withstanding a second strike within a few cycles after the first. Grounding cables must be connected directly to the equipment. No impedance, circuit breaker or measuring devices, *etc.*, are permitted to be between the grounding cable and the equipment.

Table 7.1. Interconnection components anticipated protection function

IEEE - ANSI (Device Number)	**Protection function**
Voltage regulation (27 and 27G)	To maintain constant voltage (± 5%).
Synchronization (25)	It is important that the DG produces the same frequency and delivering voltage within the tolerance of the utility grid's voltage. In addition, the phase sequence and phase balance of the DG unit voltage must be the same as those of the grid at the point of interconnection
Over current protection (25, 50, 51)	To ensure fast fault clearing under various possible over current and short circuit conditions
Re-closing (79)	To avoid the continuity of a fault condition, the DG unit should be prevented from re-energizing a de-energized distribution system circuit
Islanding (32)	DG islanding can expose utility workers to hazards by circuits that otherwise be de-energized and may pose threat to the public as well
Voltage Disturbance (59 and 59G or 59N)	When any of the measured effective (RMS) voltages is over or under a utility pre-specified range, the DG should cease to energize the grid within certain clearing time
Frequency disturbance (81 O/U, 81R)	Under and over frequency protective functions are important means of preventing the establishment of a DG island
Power Quality (27TN, 87T)	The DG unit should not impose any power quality problems on the existing grid. Therefore, the DG facility should be equipped with means of ensuring the quality of their exported power
Grounding (64F, 87GD)	Additional grounding current paths are forbidden to avoid malfunctions of the protective relays. Therefore, the DG grounding should be integrated directly with the existing grid grounding scheme. In addition, all DG equipment shall be grounded in accordance with applicable local and national codes

Metering and monitoring. Monitored parameters often include current, voltage, real and reactive power, oil temperature, vibration, *etc.* Metered parameters also include power output, which may be used for billing that requires utility-grade metering accuracy.

Dispatch, communication, and control. These integration and communication components interface the DG units with the utility. Among the functions of these components are:

- Regional load management, work order management, and billing services;
- Distribution automation;
- Feeder switching;
- Short circuit analysis;
- Voltage profile calculations and trouble calls management.

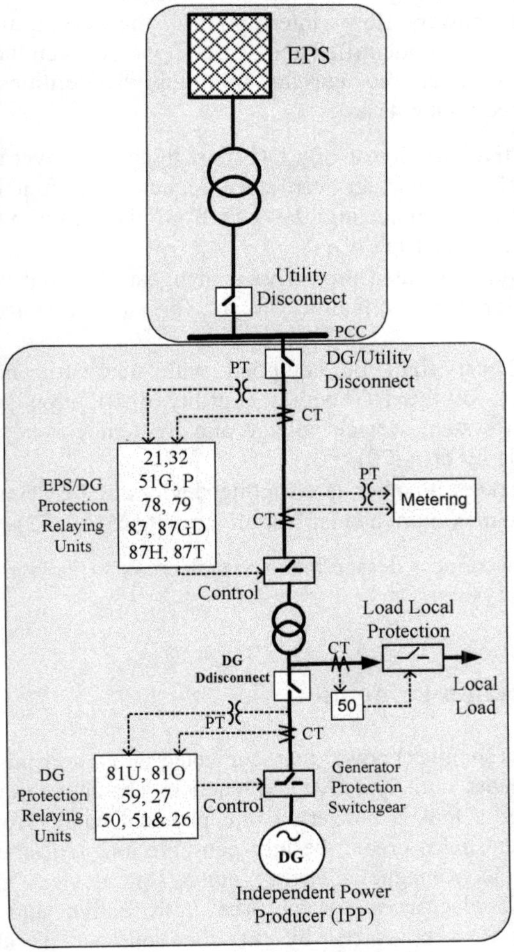

Figure 7.2. DG/utility interconnection protection requirement

The main protection devices that are necessary to the system operation are presented in Table 7.1. In the table, the ANSI protection function number is also indicated.

A typical interconnection line schematic with the protection elements between DG and a hosting grid is shown in Figure 7.2 [17]. Unfortunately, interconnection and protection requirements are established by each individual utility which add to the industry constraints in meeting the required standards. The components used in interconnection are engineered and assembled based on DG technology, size and ratings. Safe and reliable integration requires secure and reliable interconnection between the DG and the hosting grid. For economical operation and reliability, the system control must be coordinated with proper communication protocols using microprocessor based intelligent interfacing switchgear.

To minimize power flow interruptions, the essential interconnection requirements must first be identified and agreed upon between the grid and the DG operators. For example, the current minimum DG/utilities interconnection protections relay requirements are:

- The protective switchgear of a DG must include an over/under voltage trip function (27 and 27G), an over/under frequency trip function (81 O/U), and a means for disconnecting the DG from the utility when a protective function initiates a trip (79);
- The DG and associated protective switchgear shall not contribute, (unless permitted for higher reliability) to the formation of an unintended island (32);
- DG switchgear shall be equipped with automatic means to prevent reconnection of the DG with the utility distribution system unless the distribution system service voltage and frequency is of specified settings and is stable for 60 s (79);
- Circuit breakers or other interrupting devices at PCC must be capable of interrupting maximum available fault current (25, 50, 51).

The manual disconnect device has a visual break to isolate the DG from the utility distribution system.

7.4 Power Injection Principle

The total apparent (complex) power that is injected into a transmission line is made up of two components, namely active and reactive. The active power P component is the part of energy that is converted into physical energy form. The reactive power Q component helps create the indispensable magnetic medium needed for most of today's electromagnetic energy conversion devices and systems. For example, the AC electric motor absorbs both active and reactive power components once it is energized by the AC source. The absorbed reactive component creates the needed magnetic field to allow the energy conversion process to take place inside the motor. The active power component is absorbed and converted into mechanical power that moves the coupled mechanical load such as a mechanical conveyor. The electric motor will store the reactive power as fluctuating magnetic energy in its windings as long as the conversion process continues. The majority of industrial and commercial appliances require both active and reactive power components for operation.

Both *P* and *Q* are needed instantly and in different quantities to meet the requirement of the electrical energy converting device connected to the AC source. Reactive power can be absorbed or supplied depending on the energy medium associated with the electric device. Energy absorbing or supplying components are reactors and capacitors respectively. Reactors absorb reactive power $+Q$ and draw what is defined as lagging current. The consumed energy is stored as a magnetic energy in the reactor turns. Meanwhile, capacitors supply reactive power $-Q$ and draw leading current, storing it as electric charge within its dielectric medium and associated charge plates.

To understand *P* and *Q* flow in a transmission line, consider a simple line that is made of sending and receiving buses with a transmission cables in between as shown in Figure 7.3 [18].

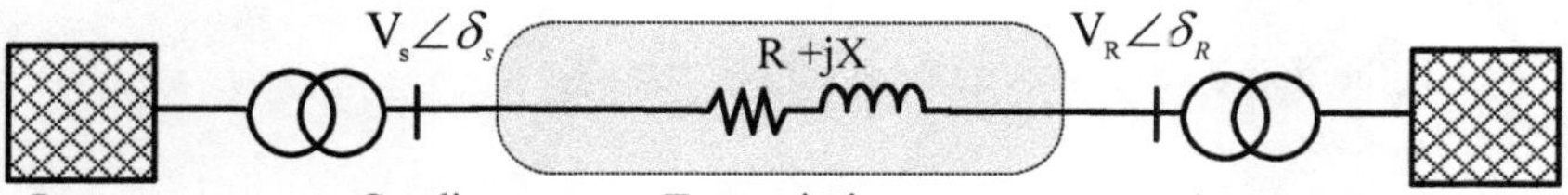

Figure 7.3. Simple line presentation of generating distribution network

Thus for small line resistances, $R<<X$, the active and reactive power components can be approximated to

$$P_S = \frac{V_S V_R}{X}\sin(\delta_S - \delta_R) \tag{7.1}$$

$$Q_S = \frac{V_S^{\ 2} - V_S V_R \cos(\delta_S - \delta_R)}{X} \tag{7.2}$$

It can be seen from the above approximated power components that power flow is dependent on four controlling variables V_S, V_R, X and $\delta_s - \delta_R$.

Employing shunt compensation at midpoint in the transmission line increases both the active and reactive components of the injected power. For lossless compensator and transmission lines $V_S = V_R = V$, the injected power at midpoint is now given by [19]

$$P_{Sh} = \frac{2V^2}{X}\sin((\delta_S - \delta_R)/2)) \tag{7.3}$$

and

$$Q_{Sh} = \frac{4V^2}{X}(1 - \cos((\delta_S - \delta_R)/2)) \tag{7.4}$$

Meanwhile, employing series compensation at midpoint with voltage V_C in quadrature with respect to the line current allows the compensating elements to assist only in the reactive power control. The result in the injected power is given by

$$P_{Ser} = \frac{V^2}{(1-r)X}\sin(\delta_S - \delta_R) \tag{7.5}$$

and

$$Q_{Ser} = \frac{2V^2}{X}\frac{r}{(1-r)^2}(1 - \cos(\delta_S - \delta_R) \tag{7.6}$$

where r is the degree of series compensation ($0 \le r \le 1$).

7.5 Power Injection Using Static Compensators

The majority of static compensators are designed to provide reactive power control. Series and shunt type compensators have been used and proved to be a practical solution for VAR compensation. Power electronic switching devices are used to control the connection of either fixed units or switchable units of reactors and capacitors to the transmission lines [20–23].

7.5.1 Fixed VAR Compensation

Fixed VAR compensation is achieved by shunt capacitors connected to the lines injection point. Shunt capacitors can be switched to meet varying levels of VAR requirements to maintain good levels of voltage regulation and power factors using binary switched capacitors, as shown in Figure 7.4. For constant active power flow and supply voltage V, the required capacitive VAR is the difference between the pre-compensation VAR and the required compensated VAR thus the amount of the capacitive susceptance is then given by

$$B_C = \frac{VAR_{Required} - VAR_{Uncompensated}}{V^2} \tag{7.7}$$

from which the required capacitance value is given by

$$C = \frac{B_C}{\omega} \tag{7.8}$$

Series reactors are normally tuned with shunt capacitors to eliminate undesired current harmonics and to act as an inrush current limiter during switching of the capacitors. Such a configuration is low cost and only provides leading power. The disadvantages are the occurrence of switching transients as different values of capacitors are switched ON and OFF.

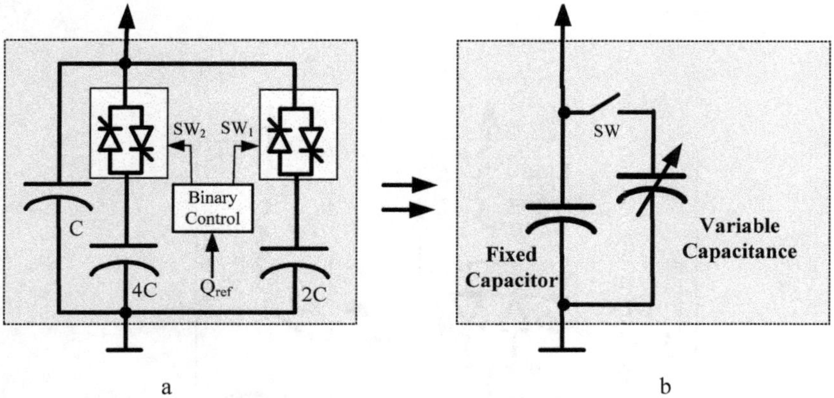

Figure 7.4. **a** Binary switched capacitive VAR controller. **b** Equivalent model

7.5.2 Controllable Dynamic VAR Compensators

Dynamic VAR compensators are generated by using SVC that allow the compensator current to be varied from leading to lagging values. This can be achieved by changing the terminal reactance by the use of controlled capacitive and inductive elements connected in parallel at the point of power feed, *i.e.*, PCC. The delay of current with the respect to the voltage is carried out by using devices such as thyristor and power transistors.

7.5.2.1 Thyristor Switched Capacitor

Thyristor Switched Capacitor (TSC) configuration is shown in Figure 7.5. The switching device in TSC is used only to switch ON or OFF the capacitor banks, and no phase angle control is used. TSC by itself does not produce harmonics but may very well produce switching transients. The switching of the capacitor units is used when load demands capacitive support. This capacitive power support is usually divided into 3 or 4 steps.

In general, series reactor is connected in series with capacitor bank to limit the inrush while switching capacitor and to limit the rate of rise of current of switching device to a safe valve.

7.5.2.2 Thyristor Controlled Reactor

In the Thyristor Controlled Reactor (TCR) configuration, the static switch is connected in series with the reactor as shown in Figure 7.6, TCR acts like a variable susceptance and is dependent on thyristor switching control. By phase angle control of the switch from 90° to 180°, the flow of current through the reactor is varied. The fundamental inductor current is, therefore, a function of the branch susceptance B_{TCR} that is dependent on the firing delay angle α and is estimated by the following dependency [20]

$$I_L = V\frac{1}{\omega L}\left(1-\frac{2\alpha}{\pi}-\frac{\sin(2\alpha)}{\pi}\right) \tag{7.9}$$

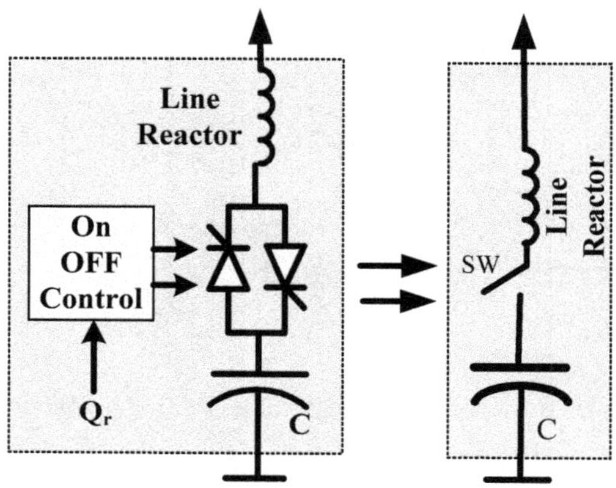

a b

Figure 7.5. **a** TSC. **b** Equivalent model

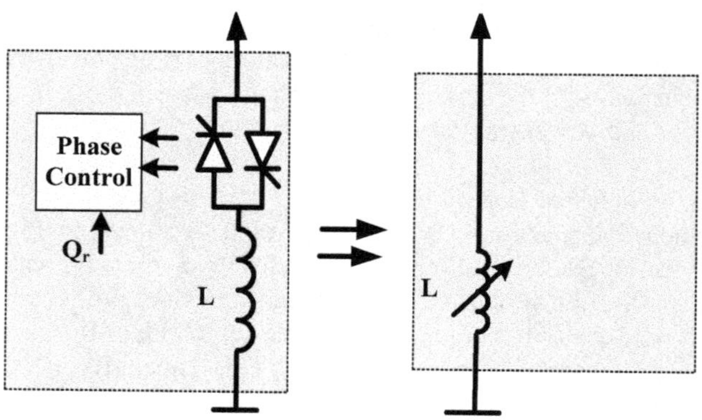

a b

Figure 7.6. **a** TCR. **b** Equivalent model

In three-phase arrangement, TCR is normally connected in delta to minimize harmonics. The harmonic components in reactor current flowing into the network are of the 3rd, 5th, 7th, 9th, 11th and 13th with maximum amplitude of 13.8%, 5%, 2.5%, 1.6%, 1% and 0.7% respectively of reactor fundamental current. If the network is balanced, all the multiples of the 3rd harmonic will be blocked in the delta connected TCR and will not flow in the network. It is also to be noted that in the case of unbalanced load, multiples of 3rd harmonic currents (3rd, 9th harmonics) also flow into the network in addition to other odd harmonics. Hence filtering elements capacity has to be increased while filtering the undesired harmonics created by TCR during its operation.

7.5.2.3 Fixed Capacitor and Thyristor-controlled Reactor

It must be noted that TSC by itself cannot provide smooth variable reactive power. Control is possible only in steps. Step-less reactive power control is therefore achieved by operating the TCR bank in conjunction with TSC with additional filter banks required to filter out harmonic currents. In this configuration a fixed capacitor is connected across a combination of series connected switching device and a reactor as shown in Figure 7.7.

The current in the reactor is varied by the previously discussed method of firing delay angle control. The fixed capacitor in practice is usually substituted, fully or partially, by a filter network that has the necessary capacitive impedance at the fundamental frequency to generate the VAR required. However, this provides low impedance at selected frequencies to shunt the dominant harmonics produced by the TCR. The fixed leading VAR supplied by the capacitor is opposed by the variable lagging VAR consumed by the reactor to yield the required net VAR. The total susceptance seen by the system is made of two parts (fixed and variable) and is given by the dependency

$$B_{C-TCR} = B_C + B_{TCR}(\alpha) \tag{7.10}$$

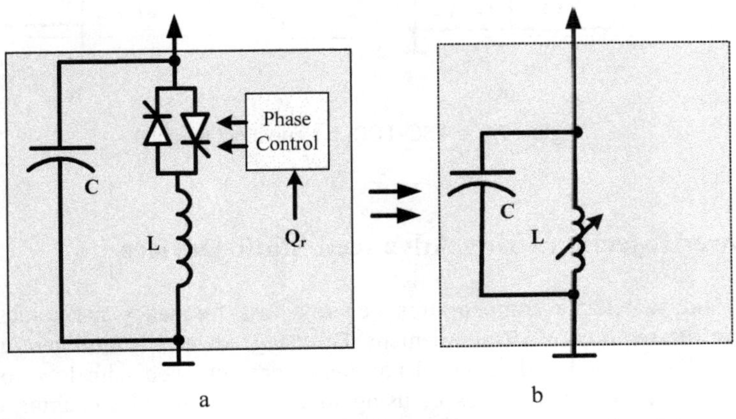

Figure 7.7. a Fixed capacitor and thyristor-controlled reactor. **b** Equivalent model

At the maximum capacitive output, the thyristor-controlled reactor is OFF. To decrease the capacitive output, the current in the reactor is increased by decreasing delay angle. At zero VAR output, the capacitive and inductive currents are made equal. With a further decrease of the delay angle (assuming that the rating of the reactor is greater than that of the capacitor), the inductive current becomes larger than the capacitive current, resulting in a net inductive VAR output.

7.5.2.4 Thyristor-switched Capacitor and Thyristor-controlled Reactor

Thyristor-switched Capacitor and Thyristor-controlled Reactor (TSC-TCR) is similar to the operation of the Fixed Capacitor and Thyristor-controlled Reactor (FC-TCR) but with stepped values of capacitors. A single-phase TSC-TCR is shown in Figure 7.8. For a given capacitive output range, it consists of multiple numbers of TSC branches and one TCR. The number of branches is determined by practical considerations that include the operating voltage level, maximum VAR output, current rating of the switching devices, *etc.* The inductive range can be expanded to any maximum rating by employing additional TCR branches. The total capacitive output range is divided into *n* intervals. In the first interval, the output of the VAR generator is controllable in the zero to *VAR/n* range, where VAR is the total rating provided by all TSC branches. In this interval, one capacitor bank is switched in and, simultaneously, the current in the TCR is set by the appropriate firing delay angle so that the sum of the VAR output of the TSC (negative) and that of the TCR (positive) equals the capacitive output required.

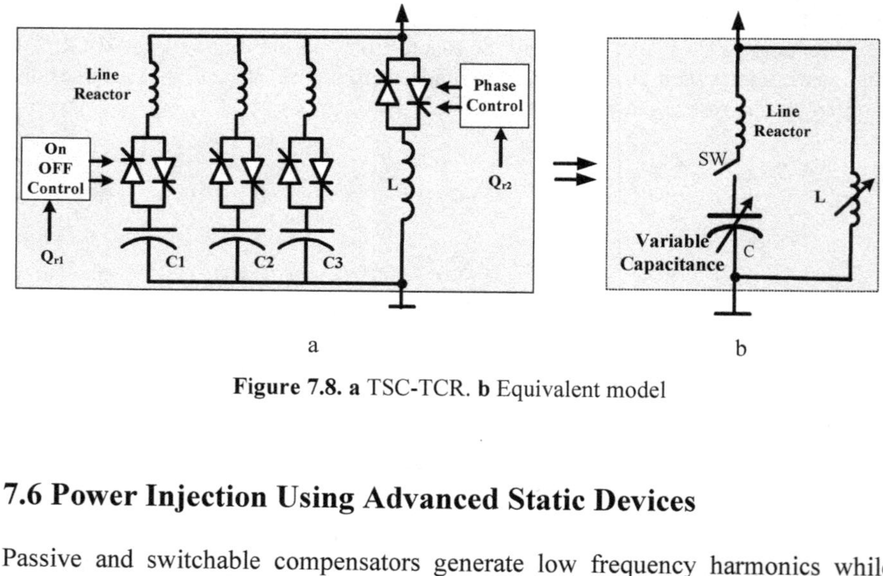

Figure 7.8. a TSC-TCR. **b** Equivalent model

7.6 Power Injection Using Advanced Static Devices

Passive and switchable compensators generate low frequency harmonics while providing the required VAR supplement. By using advanced intelligent and fast switching devices, reduced reflected harmonics are achieved with low cost high pass filter units. The advantages of using forced commutated switching devices that control compensation capacitors and reactors give the compensation system the ability to control both active and reactive power. Among the practical advanced static compensation devices are the STATCOM that is often known as the Dynamic Static Synchronous Compensator (D-STATCOM) and the UPFC [24, 25].

7.6.1 Static Synchronous Compensator

STATCOM is a highly recommended compensation unit for flicker mitigating as well as injecting reactive power components to the distribution system [26]. It is made of six power electronics forced commutating switching devices in a three-phase full bridge configuration. The bridge is controlled to supply current to a large DC capacitor. The power rating of these devices is relatively low. The switching devices are controlled by a PWM switching logic driver circuit to turn them on and off as required. The switching control strategy is to switch the devices rapidly on and off. This in turn results in a pulse modulated line voltage with variable magnitude and phase presented by a large fundamental component and a handful of high frequency harmonics. The high frequency harmonic components can be easily filtered using high frequency tuned passive elements. As the fundamental component of the line voltage varies so the current is drawn by the converter. STATCOM is connected in shunt to the distribution network through a limiting reactor or coupling transformer as shown in Figure 7.9. Therefore, STATCOM can be simply represented by a variable controlled voltage source injecting leading or lagging current to the feeder.

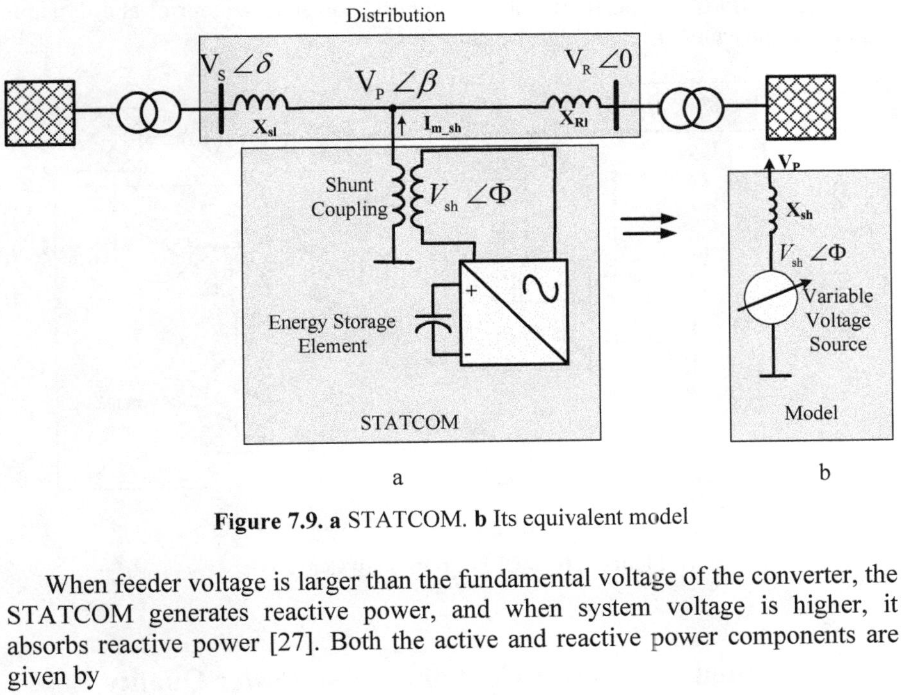

Figure 7.9. a STATCOM. **b** Its equivalent model

When feeder voltage is larger than the fundamental voltage of the converter, the STATCOM generates reactive power, and when system voltage is higher, it absorbs reactive power [27]. Both the active and reactive power components are given by

$$P_{Sh} = \frac{V_{Sh} - V_P}{X_{Sh}} \sin(\Phi - \beta) \qquad (7.11)$$

$$Q_{Sh} = \frac{V_{Sh}^{\ 2} - V_{Sh}V_P\cos(\Phi - \beta)}{X_{Sh}} \tag{7.12}$$

STATCOM allows the device to absorb or generate controllable reactive power with an apparent capacitive or inductive current absorbed by the controlling bridge and is independent of the AC line voltage level with respect to the converter voltage.

7.6.2 Unified Power Flow Controller

UPFC basically consists of two fully controlled converters sharing a common DC link, as shown in Figure 7.10. The converters are controlled to compensate for the voltage as well as the power of the coupled transmission lines. Converter 2 performs the main function of the UPFC by injecting an AC voltage with controllable magnitude and phase angle in series with the transmission line [28]. Converter 1, on the other hand, supplies or absorbs the active power demanded by Converter 2. Both converters can be controlled to generate or absorb controllable reactive power or provide independent shunt reactive compensation for the line. In principle, a UPFC can perform voltage support, power flow control and dynamic stability improvement in one and the same device.

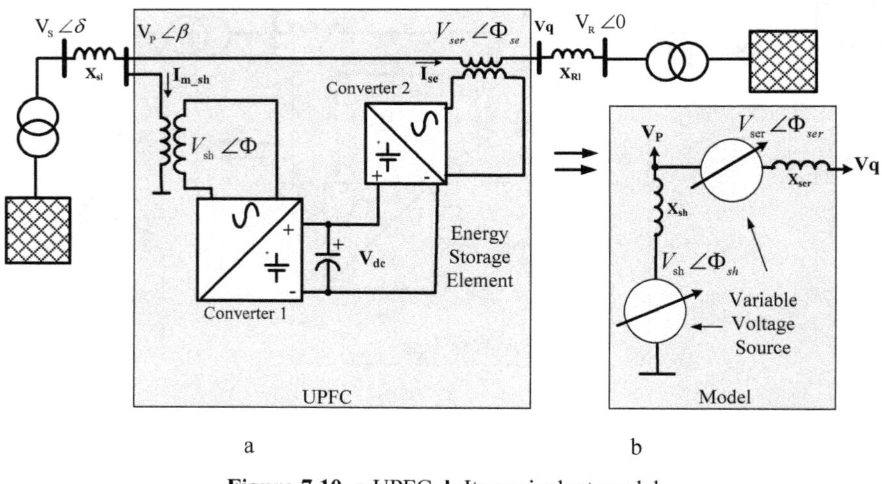

Figure 7.10. a UPFC. **b** Its equivalent model

7.7 Distributed Generation Contribution to Power Quality Problems

Potentially, DG can support the distribution grid in mitigating power surges, voltage dip, brownout and even blackout if properly incorporated as part of the mitigation resources. However, in practical terms this needs a new well coordinated two ways traffic power control management system. The system operator may accommodate the coordination of a couple of distributed DG along the distribution grid, although the task will be extremely challenging if the DG number increases drastically. Importantly, DG are incorporated mainly to supply power at the vicinity of the demand with their control setup programmed to deal with short distribution lines and well defined voltage, frequency tolerance, active and reactive power limits. To interconnect this category of DGs into a larger distribution system entails the re-programming of the DG controlling set values. DG may introduce vulnerability to the hosting grid as follows:

- Unnecessary trip of protection devices. Some of the feeder's protection devices are traditionally coordinated on the basis of protection schemes for faults and grounding caused by defined parameters known to the system planner;
- Frequent connecting and disconnecting of the DG to the feeder may cause unnecessary flickers and voltage variation depending on the DG location and the status of the line;
- Utility could be forced to shed some loads to limit the increase in feeder capacity as a result of active power injection from DG;
- DG may not come on line as scheduled or requested by the hosting utility controller therefore, risking high financial losses.

Certainly, DG can be used to relieve some of the feeder stressful situations if they are located at some distance from the substation [29]. For example, DG can be incorporated to support voltage compensation along a particular feeder through a PCC connection. Typically, voltage regulators incorporated at the distribution grid are located near the designated substation. The best location for voltage regulators is halfway along the distribution line, assuming that the load is evenly distributed. Meanwhile, shunt capacitors used for power factor correction are installed near the load. Both are used to improve the voltage profile.

DG can support the voltage by injecting the required amount of reactive and active power components or by injecting series voltage. Both power components can be supplied using DG units with synchronous alternators as the interfacing part with the grid or those with inverter units as the interfacing part with the grid. For example, a synchronous alternator can be used to inject or absorb reactive power by the control of the excitation source for the alternator. However, such an arrangement is limited to alternator excitation capacity and needs to be coordinated with the hosting utility. DG with inverter output stage can be used as static compensator to mitigate flickers and inject a series voltage to balance the voltage dip as long as it does not cause the grid voltage at PCC to exceed the permitted voltage fluctuation range. Excessive voltage injection could endanger the connected load to the grid.

Yet, if a nearby load exists, it is most likely that the DG will supply power to this load rather than contribute to the voltage profile improvement. Increasing the DG power injection will ultimately increase the voltage profile at some distance along the line if existing grid voltage regulators are also in action at the same time. Without proper coordination between the system operator and the DG operator, the voltage level may increase beyond the limits, triggering overvoltage protection relays incorporated in the grid. Therefore, DG contribution to voltage depends on the injected active and reactive power, its location relative to other compensators, degree of coordination with the system operator and the grid load [29].

Finally, many power quality problems are attributed to commercial and industrial facilities equipment quality and ailing operating conditions. Some are due to fault occurrence and imperfect grounding. Power surges, voltage spikes, switching transients, frequency variations, electrical lines noise, DC current injection, harmonics, power sags, brownout and even blackout are all common power quality problems. Many of the listed power quality problems can be handled within the facility premises using low cost passive components. Surge suppressors, passive and active filters, UPS systems and load controlled transformer tappings are all used to reduce the stress initiated by voltage spikes, switching transients, electrical lines noise and harmonics that may be generated by AC and DC drives.

7.8 DG Current Challenges

Utilities and distribution companies are cautious in integrating dispersed generating units to their system for various technical and non-technical reasons [30–34]:

- Technical challenges: issues such as safety, islanding, power flow, capacity congestion restoration from scheduled and unscheduled shut downs, protection coordination, capacity and reserve management, reliability and power quality liability and development cost in the needed interconnection technologies;
- Non-technical challenges: issues such as pricing, incentives, decision priority, risk responsibility and insurance for new technologies adaptation, interconnection standards and regulatory control and addressing barriers. Despite the fact that their existence sounds practical and provides a feasible solution to increasing energy demand, many regulatory issues are not introduced nor cautiously implemented.

This section will only look at a few issues of the technical challenges. The major obstacle facing DG penetration is how the interconnection mechanism is safely implemented.

Electricity generation has traditionally been under the central control of the centralized system operator, who schedules the generation plant, and issues dispatch instructions to regulate output in the light of changing conditions. They also arrange ancillary services, such as frequency response and voltage control, to keep the grid stable in the light of unexpected changes.

Now, how will the interconnected DG be controlled? Who will be the operator? Is it the system operator or the owner? The issue will not be a complex one if the DG is owned and operated by the utility, as it is most likely that the size, location, purpose and dispatch commands are selected by the system operator. If not, then will the owner allow the system operator to control the DG unit? Will the system operator utilize the full potential of the DG?

When it comes to interconnecting scattered DG units, various issues are hampering the integration process. Mainly utilities are uncomfortable with using dispersed DG along its feeders, as it is not under their control. It is risky, expensive and difficult for the system operator to issue dispatch instructions to scattered DG units along the distribution grid. Also, it is difficult to make decisions as to which DG is the most cost effective among the scattered assortment.

A special kind of set-back for rapid DG implementation is evident when it comes to DG fueled by non-fossil alternative energy resources such as solar and wind energy conversion systems. To illustrate some of the issues, consider a typical wind farm energy conversion and interfacing system incorporated as DG. Various immediate constraints will be facing the system operator:

- The primary fuel (the wind) is intermittent in nature resulting in intermittent, somehow difficult to predict, energy supply;
- The wind farm location may be few hundred miles away from the designated feeder, defeating one of the main objectives of DG implementation as a source of energy available near the demand;
- Nearby transmission lines congestion and power flow constraints may exist, prohibiting DG interconnection to the feeder;
- Many wind farm generators need continuously controlled reactive power components as demand varies; if this is not incorporated effectively within the wind farm, a large amount of reactive power will be absorbed from the hosting network and this cannot be endured.

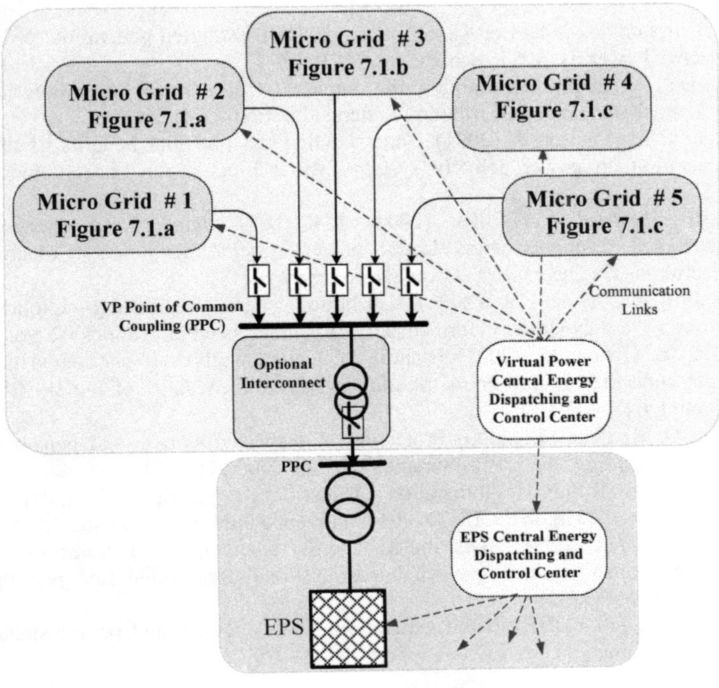

Figure 7.11. Virtual power architecture

Opportunely, many of the alternative and renewable energy fueled DG are incorporated as part of DER systems where various mixes of energy sources are controlled under a central energy management unit. In this way, many of the above constraints can be tolerated and diminished.

DER includes electricity and heat generation, thus increasing reliability and diversity of fuel mix as well as DG operational efficiency. DER can also be incorporated in a wider infrastructure, creating a micro grid generating and distribution domain. Further, various reliable DG units forming a local distribution grid can be interconnected, creating a virtual power pool that utilizes local resources nationally [34]. Virtual power concept was introduced in late 1996 to help utilities to supply additional power from nearby utilities upon the first sign of deficiency in demand. Figure 7.11 shows an architectural layout of the virtual power concept in which nationally existing MGR are coordinated to serve existing energy infrastructure.

The implementation of such a scheme requires a high degree of coordination and knowledge of each MGR attributes and the capability of energy exchange among various MGR domains in real time.

With advances in utility communication and the Internet, such coordination is feasible using web based agents. Further research and vulnerability studies are needed to check the reliability and security of supply.

References

[1] Ackermann T, Andersson G, Soder L, (2001) Distributed generation: a definition. Electric Power System Research, vol.57:195–204

[2] Barker P, (2000) Determining the impact of distributed generation on power systems. Part 1: radial distribution systems. IEEE:1645–1656

[3] Daly PA, Morrison J, (2001) Understanding the potential benefits of distributed generation on power delivery systems. Rural Electric Power Conference:A211–A213

[4] IEEE Standard 1547-2004, (2004) 1547 IEEE standard for interconnecting distributed resources with electric power systems. Institute of Electrical and Electronics Engineers, Piscataway, New Jersey

[5] Walling RA, Miller NW, (2002) Distribution generation islanding – implications on power system dynamic performance. IEEE Conference Publications:92–96

[6] Kim JE, Hwang JS, (2001) Islanding detection method of distributed generation units connected to power distribution system. Proceedings of the IEEE Summer Meeting:643–647

[7] Usta O, Redfem M, (2000) Protection of dispersed storage and generation units against islanding. IEEE PES Summer Meeting 2000:976–979

[8] Willoughby R, (2001) Integration of distributed generation in a typical USA distribution system. IEE CIRED2001 Conference Publication, no.482:18–21

[9] Friedman NR, (2002) Distributed energy resources interconnection systems: technology review and research needs. National Renewable Energy Laboratory, report SR-560-32459

[10] Marks M, (2000) Distributed generation: CEQA review and permit streamlining. California Energy Commission, report 700-00-019

[11] Rifaat RM, (1995) Critical considerations for utility/cogeneration inter-tie protection scheme configuration. IEEE Transactions on Industry Applications, vol.31, no.5:973–977
[12] Simons G, Sethi P, Davis R, (2001) The role of renewable distributed generation in California's electricity system. IEEE Power Engineering Society Summer Meeting, vol.1:546–547
[13] Lasseter R, (2002) Microgrids. Proceedings of the Power Engineering Society Winter Meeting, vol.1:27–31
[14] Dugan RC, Price SK, (2002) Issues for distributed generation in the US. Power Engineering Society Winter Meeting, vol.1:121–126
[15] Brahma S, (2000) Effect of distributed generation on protective device coordination in distribution system. IEEE PES Summer Meeting:115–119
[16] Puttgen HB, MacGregor PR, Lambert FC, (2003) Distributed generation: semantic hype or the dawn of a new era?. IEEE Power and Energy Magazine, vol.1, no.1:22–29
[17] Nigim KA, Hegazy YG, (2003) Intention islanding of distributed generation for reliability enhancement. IEEE PES General Meeting, Toronto:1–6
[18] Rashid MH, (2003) Flexible AC transmission. Chapter13: power electronics circuit devices and applications. Pearson Prentice Hall, third edition
[19] Hingorani NG, Gyugyi L, (2000) Understanding FACTS: concepts and technology for flexible AC transmission systems. Piscataway, IEEE Press
[20] Mathur RM, Varma KR, (2002) Thyristor – based FACTS controllers for electrical transmission systems. IEEE Press, New York
[21] Gyugyi L, (1989) Solid-state control of electronic power in AC transmission systems. International Symposium on Electric Energy Converters in Power Systems, T-IP.4, Capri, Italy
[22] Gyugyi L, Hingorani NG, (1990) Advance static VAR compensator using gate-tran-off thyristor for utility applications. CIGRE, no.23–203
[23] Salles MBC, Freitas W, Morelato A, (2004) Comparative analysis between, SVC and DSTATCOM devices for improvement of induction generator stability. IEEE MELECON, Dubrovnik, Croatia
[24] Gyugyi L, (1992) Unified Power Flow Control concept for flexible AC transmission systems. IEE Proceedings, part-C, vol.139, no.4
[25] Sannino A, Svensson J, Larsson T, (2003) Power electronic solution to power quality problems. Electric Power Systems Research, vol.66:71–82
[26] Al-Mawsawi SA, (2003) Comparing and evaluating the voltage regulation of a UPFC and STATCOM. Electric Power and Energy System, vol.257:735–740
[27] Wang HF, Jazaeri M, Cao YJ, (2005) Operating M and control interaction analysis of unified power flow controllers. IEE Proceedings (Generation, Transmission, Distribution), vol.152, no.2:264–270
[28] Mcdermott T, Dugan RC, (2003) PQ, reliability, and DG. IEEE Industrial Application Magazine:17–23
[29] Kojovic L, (2002) Impact DG on voltage regulation. IEEE Power Engineering Society Summer Meeting, vol.1:97–102
[30] Bhowmik A, Maitra A, Schatz SM, (2003) Determination of allowable penetration levels of distributed generation resources based on harmonic limit considerations. IEEE Transactions on Power Delivery, vol.18, no.2:619–624
[31] CIGRE Study Committee, (2003) Impact of increasing contribution of dispersed generation on the power system. CIGRE Study Committee no.37, Final Report
[32] Quezada VHM, Abbad JR, Roman TGS, (2006) Assessment of energy distribution losses for increasing penetration of distributed generation. IEEE Transactions on Power Systems, vol.21, no.2:533–540

[33] Grijalva S, Visnesky AM, (2005) Assessment of distributed generation programs based on transmission security benefits. IEEE Power Engineering Society General Meeting:1441–1446

[34] Castelaz SA, (2000) Plugging into hidden capacity & networking distributed generation with the Virtual Power Plant™. Standard & Poor's Utilities & Perspectives Newsletter Special Technology Issue:1–5

8

Active Power Quality Controllers

Ryszard Strzelecki[1] and Grzegorz Benysek[2]

[1]Department of Electrical Engineering,
Gdynia Maritime University,
81-87 Morska Street, 81-225 Gdynia, Poland.
Email: Rstrzele@am.gdynia.pl

[2]Institute of Electrical Engineering,
University of Zielona Góra,
50 Podgórna Street, 65-246 Zielona Góra, Poland.
Email: G.Benysek@iee.uz.zgora.pl

8.1 Dynamic Static Synchronous Compensator

Dynamic Static Synchronous Compensator is the most important controller for distribution networks and probably in SEEN. It has been widely used since the 1990s to regulate system voltage precisely, improve voltage profile, reduce voltage harmonics, reduce transient voltage disturbances and load compensation. Rather than using conventional capacitors and inductors combined with fast switches, the D-STATCOM uses a power-electronics converter to synthesise the reactive power output. A D-STATCOM converter is controlled using PWM or other voltage/current-shaping techniques [1–4]. D-STATCOMs are used more often than STATCOM controllers. Compared to STATCOM, D-STATCOMs have considerably lower rated power and, in consequence, faster power-electronics switches; thus the PWM carrier frequency used in a distribution controller can be much higher than in a FACTS controller. It has a substantial positive impact on the dynamics of the D-STACOM.

8.1.1 Topology

D-STATCOM controllers can be constructed based on both VSI topology and Current Source Inverter (CSI) topology (Figure 8.1). Regardless of topology, a controller is a compound of an array of semiconductor devices with turn off capability (*i.e.*, IGBT, GTO, IGCT *etc.*) connected to the feeder *via* a relative small reactive filter. The VSI converter is connected to the feeder *via* reactor L_F and has a voltage source (capacitor C_D) on the DC side. On the other side, the CSI converter is connected on the AC side *via* capacitor C_F and has a current source (inductor L_D) on the DC side. In practice, CSI topology is not used for D-STATCOM. The reason for this is related to the higher losses on the DC reactor of CSI compared to the DC capacitor of VSI. Moreover, a CSI converter requires reverse-blocking semiconductor switches, which have higher losses than reverse-conducting switches of VSI. And, finally, the VSI-based topology has the advantage because an inductance of a coupling transformer *Tr* (if present) can constitute, partially or completely, the inductance of an AC filter. The following text will describe the properties of VSI-topology based D-STATCOM only, but in many respects they are the same as for CSI-based controllers.

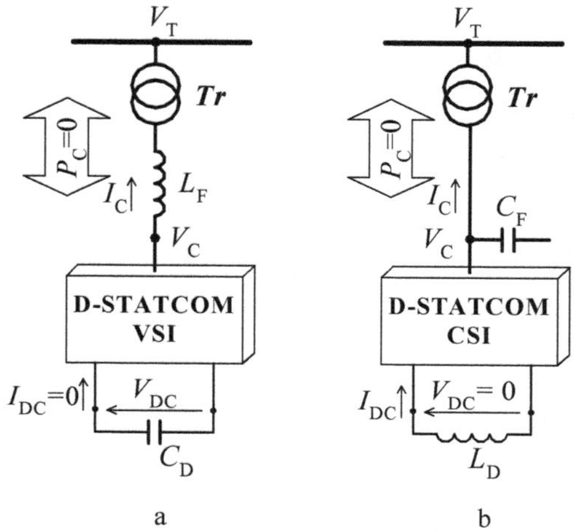

Figure 8.1. General topology of VSI-based and CSI-based D-STATCOM

The VSI converters for D-STATCOM are constructed based on multi-level topologies, with or without use of a transformer. These solutions provide support for operation with a high level of terminal voltage. Additionally, D-STATCOM controllers can be a compound of several converters configured to various topologies, to achieve higher rated power or lower PWM-related current ripples. The exemplary topologies are presented in Figure 8.2. In the parallel configuration (Figure 8.2a) converters are controlled to share the generated power equally, or at a given ratio, for example proportional to the rated power of the particular converter. In this solution it is necessary to provide inter-converter communication at the control level to distribute information about set controller power or currents. The cascade multiconverter topology (Figure 8.2b) is similar to the parallel configuration, but in this case the constituent converters do not share power equally, but successively, depending on the requirement. In this case, no communication between constituent converters is required, but on the other hand it is also not possible to use common PWM strategy. The converters in this case are exactly the same as for standalone operation. In Figure 8.2c,d are presented series and parallel master–slave topologies, respectively. The master–slave topologies require a high degree of integration between constituent converts, including a control system, and are treated and realised as a single, multiconverter controller. The master converter (called a "*slow converter*") has substantially higher rated power and, in consequence, considerably lower PWM carrier frequency than the slave converter (called a "*fast converter*"). The task of the master converter is to cover the requirements for power, while the slave has to compensate AC current/voltage ripples using serial superposition of voltages (Figure 8.2c) or parallel superposition of currents (Figure 8.2d).

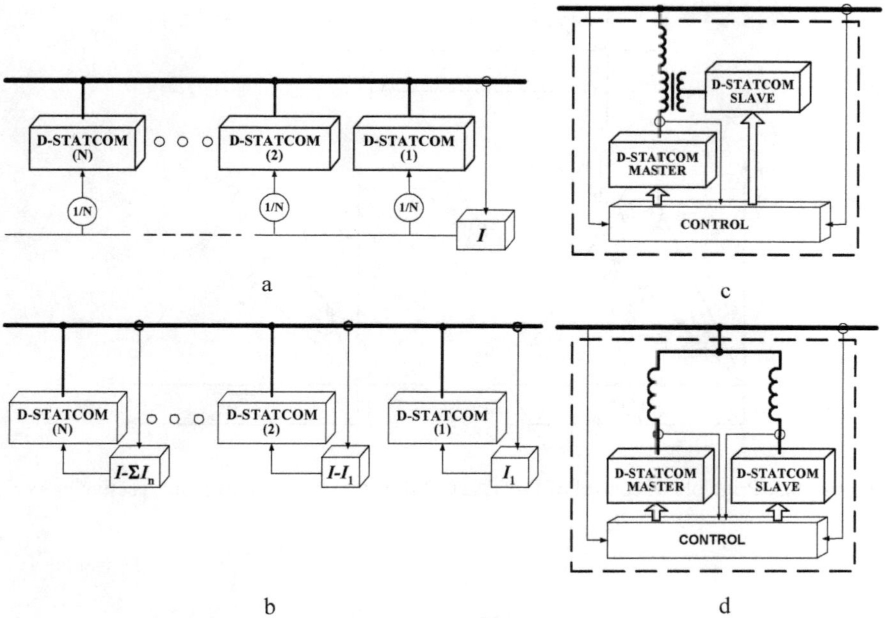

Figure 8.2. Multiconverter topologies of D-STATCOM controller: **a** parallel; **b** cascade; **c** master – slave series; **d** master – slave parallel

8.1.2 Principle of Operation

For the operation analysis of the D-STATCOM converter, it is possible to represent its PWM-controlled VSI with an instantaneous (averaged for PWM period) voltage source. The principle of generating instantaneous active and reactive powers by D-STATCOM is shown in Figure 8.3. In this figure, voltages and currents are represented with instantaneous space vectors obtained using a power-invariant Clarke transform. In Figure 8.3 are presented three cases: the general one, for reactive power equal to zero and for active power equal to zero. From this figure it is clear that, by generating an appropriate AC voltage, it is possible to generate arbitrary instantaneous vectors of both active and reactive power. The real component of currents is related to the equivalent series resistance modeling losses on the AC side. The possible active and reactive powers that can be generated or absorbed by D-STATCOM are limited. This limitation is related to circuit parameters and maximum ratings of VSI components. In Figure 8.3 is presented an exemplary limit for AC voltage, which depends on VSI DC voltage

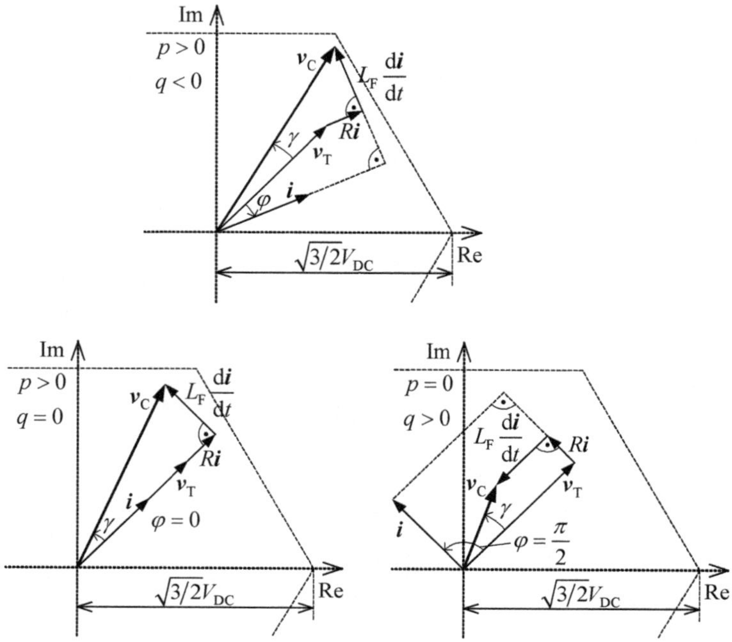

Figure 8.3. Principle of control of D-STATCOM instantaneous active and reactive power

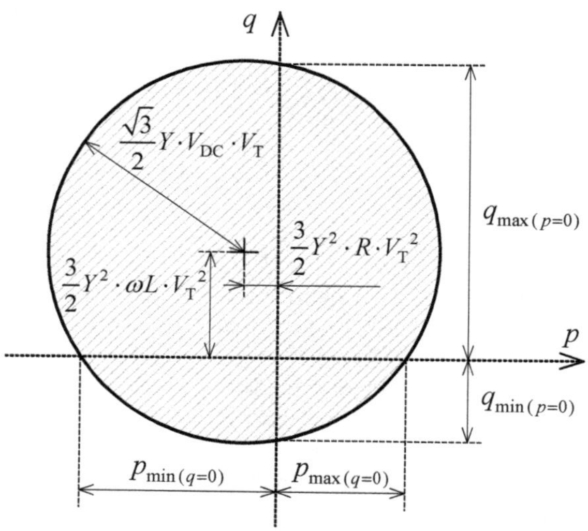

Figure 8.4. Operating region of two-level VSI-based D-STATCOM

V_{DC}. This limit, together with filter inductance L_F and terminal voltage V_T, defines the operating region of a D-STATCOM controller. The operating region of a two-level VSI-based controller is presented in Figure 8.4. In this figure, Y denotes the modulus of admittance on the AC side of VSI. In practice, the operating region does not limit the maximum ratings of VSI semiconductors, so the static *V–I* characteristic of D-STATCOM reactive power is symmetrical (Figure 8.5).

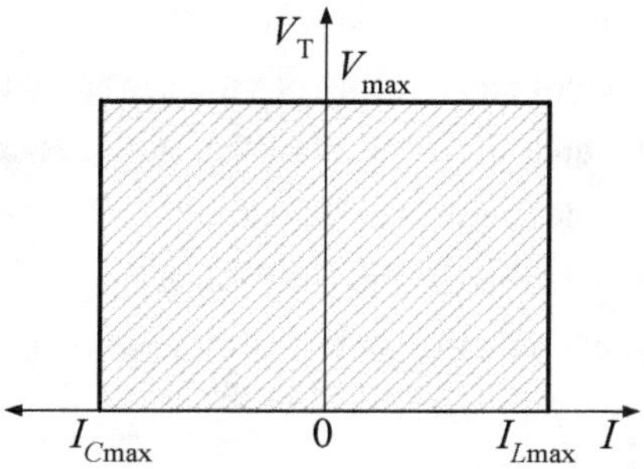

Figure 8.5. The *V–I* characteristic of D-STATCOM

The active power is consumed by the D-STATCOM only to cover internal losses. Assuming lossless operation, the averaged (but not instantaneous) active power has to be zero. There are no similar limitations for reactive power, because it is only exchanged between phases, and is not converted between the AC and DC sides of D-STATCOM VSI.

There are two modes of D-STATCOM operation: load compensation in current-control mode and voltage regulation in voltage-control mode.

8.1.3 Load Compensation

In the load-compensation mode, D-STATCOM is controlled in current mode. The control system of D-STATCOM has to generate reference currents, compensating harmonic, unbalance and fundamental reactive components of non-linear load supply currents. The required rated power of load-compensating D-STATCOM depends only on reactive power, harmonic distortion and power of the compensated load. In general, D-STATCOM is capable of compensating current disturbances from harmonics to long-duration effects, including active power transients. The possibility and effectiveness of compensation of a particular voltage-quality problem depends on the topology and rated power of the controller as well as on the capacity of the energy-storage system connected at the D-STATCOM DC side.

Consider the block diagram of the system that is depicted in Figure 8.6. The instantaneous apparent power noted on that figure is defined as

$$s_{3f} = e^{*}i = p_{3f} + jq_{3f} \tag{8.1}$$

where $p_{3f} = \mathrm{Re}\left(e^{*}i\right)$; $q_{3f} = \mathrm{Im}\left(e^{*}i\right)$.

Based on Equation 8.1, the instantaneous apparent power of a non-linear load can be expressed as follows

$$s_{3f} = \overline{s}_{3f} + \widetilde{s}_{3f} = \overline{s}_{e_1 i_1} + \sum_{n=-\infty}^{\infty} \overline{s}_{e_n i_n} + \sum_{n=-\infty}^{\infty} \widetilde{s}_{e_1 i_n} + \sum_{n=-\infty}^{\infty} \widetilde{s}_{e_n i_1} + \sum_{m=-\infty}^{\infty} \sum_{n=-\infty}^{\infty} \widetilde{s}_{e_m i_n} \quad (8.2)$$

where particular components are related with:

$\overline{s}_{e_1 i_1}$ – fundamental components of voltages and currents
$\sum \overline{s}_{e_n i_n}$ – higher, equal-order harmonics of voltages and currents
$\sum \widetilde{s}_{e_1 i_n}$ – higher-order harmonics of currents
$\sum \widetilde{s}_{e_n i_1}$ – higher-order harmonics of voltages
$\sum \widetilde{s}_{e_m i_n}$ – higher, not equal-order harmonics of voltages and currents

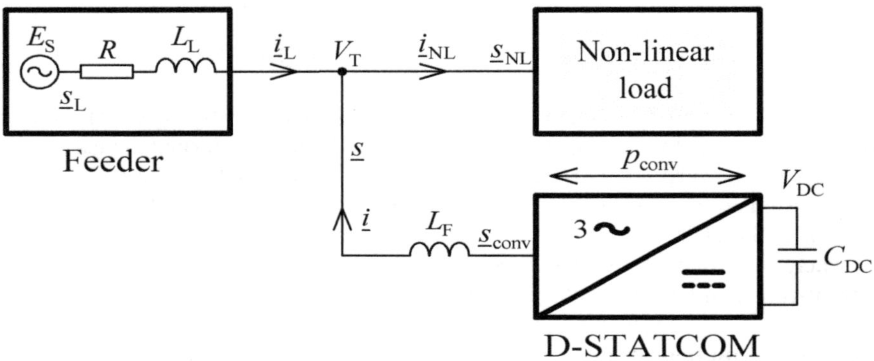

Figure 8.6. Configuration of system with D-STATCOM

There are three possible compensation methods that ensure that feeder currents are:

- Proportional to the fundamental, positive component of terminal voltage (the power components 1 and 4 are present);
- Proportional to the terminal voltage V_T, which is equivalent to resistance (the power components 1, 2, 3 and 4 are present);
- Corresponding to the constant instantaneous active power, and zero instantaneous reactive power (the power components 1 and 2 are present).

Among the possible strategies, the first is usually considered as the best. This is because it provides feeder currents that do not distort voltage at any point of the feeder. From the above considerations it is clear that load compensation also provides a reduction of the voltage distortion related to the feeder voltage drop. The level of distortion reduction depends on the configuration of the distribution network as well as the ratio between the power of the compensated non-linear load and the feeder short-circuit power.

From the control viewpoint, the effectiveness of compensation realised by D-STATCOM operating in load-compensation mode depends on the reference-current calculation method and current controller (Figure 8.7). Among the most important properties of the reference current calculation algorithm are dynamics, exactness, possibility for selective compensation and computational effort. There are two major types of such algorithms: time-based and frequency-based. The exemplary reference calculation algorithms are presented in Figure 8.8. In the time-based algorithms the reference currents are derived directly from the transients of currents and voltages, thus, they provide an "instantaneous" result. The *p–q*, three-wire theory is based on the power definitions given in Equation 8.3. The reference currents are determined using the following equation

$$i = \frac{s}{e^*} = \frac{1}{|e|^2}(p\,\mathrm{Re}\,e - q\,\mathrm{Im}\,e) + j\frac{1}{|e|^2}(p\,\mathrm{Im}\,e + q\,\mathrm{Re}\,e) \quad (8.3)$$

By selecting the appropriate instantaneous power components it is possible to calculate the desired reference currents according to the particular compensation method, with or without reactive power filtering. In a four-wire system it is necessary to define independent zero sequence active power.

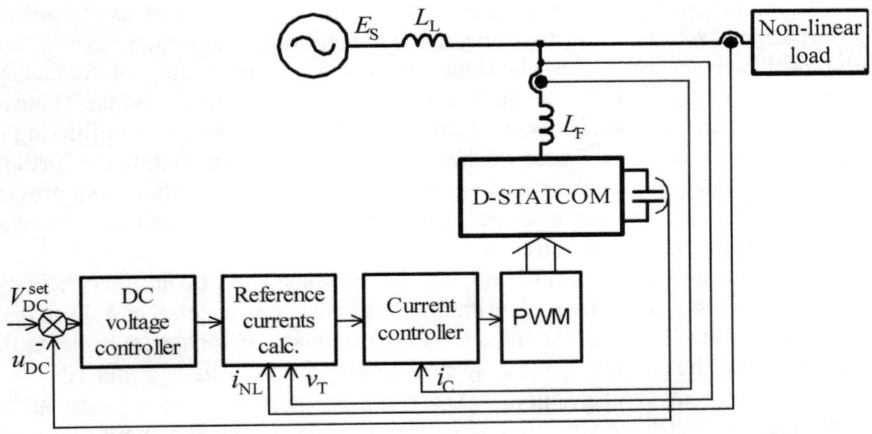

Figure 8.7. Block diagram of a control system for load compensating D-STATCOM

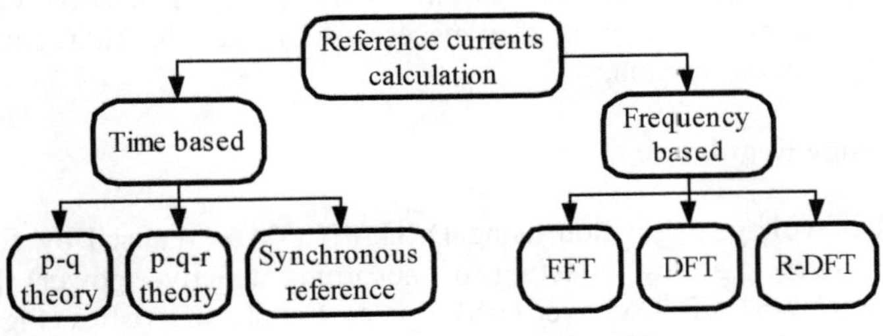

Figure 8.8. Selected reference current calculation algorithms

In the *p–q–r* four-wire method there are calculated three linear independent current components *p*, *q* and *r*, according to the equation

$$\begin{bmatrix} i_p \\ i_q \\ i_r \end{bmatrix} = \begin{bmatrix} \cos\theta_2 & 0 & \sin\theta_2 \\ 0 & 1 & 0 \\ -\sin\theta_2 & 0 & \cos\theta_2 \end{bmatrix} \begin{bmatrix} i'_\alpha \\ i'_\beta \\ i_0 \end{bmatrix}, \; \theta_2 = \tan^{-1}\left(\frac{e_0}{\sqrt{e_\alpha + e_\beta}} \right) \tag{8.4}$$

where $$\begin{bmatrix} i'_\alpha \\ i'_\beta \\ i_0 \end{bmatrix} = \begin{bmatrix} \cos\theta_1 & \sin\theta_1 & 0 \\ -\sin\theta_1 & \cos\theta_1 & 0 \\ 0 & 0 & 1 \end{bmatrix} \begin{bmatrix} i_\alpha \\ i_\beta \\ i_0 \end{bmatrix}, \; \theta_1 = \tan^{-1}\left(\frac{e_\beta}{e_\alpha} \right) \tag{8.5}$$

The synchronous reference method is the family of algorithms that is based on representation of the currents in the rotating, voltage synchronized reference frame. In that frame, currents are filtered using a highpass filter to derive the desired compensating components. It is also possible to calculate particular, higher-order current harmonics using the multiple rotating reference frame method [5].

The frequency-based methods are the family of algorithms that use frequency representation of currents to separate compensating harmonics. The main difference between particular algorithms results from the method of frequency-domain calculation. Methods like fast Fourier transform, discrete Fourier transform, recursive discrete Fourier transform and others, like Kalman filtering or neural networks are used. These methods require higher computational effort and usually have lower response time compared to the time-based methods, but provide the possibility to calculate reference currents compensating particular, selected harmonics of distorted load currents.

Among current-control techniques, several methods, like PI controllers, sliding-mode controllers, predictive algorithms, dead-beat controllers, and hysteresis methods are used [3, 6, 7]. The most important property of such a controller is the dynamics; thus the predictive and dead-beat algorithms are usually preferred.

Despite current compensation, a D-STACOM controller can be used at the same time for AC/DC power conversion, for example providing a supply for a DC feeder or micro-DC distribution system, especially in distributed generation systems. In Figure 8.9 are presented transients of a shunt compensator providing AC/DC conversion. This solution can allow overall costs of D-STATCOM to be reduced in cases where it is also necessary to provide bidirectional AC/DC electrical power conversion.

8.1.4 Voltage Regulation

The idea of voltage regulation using D-STATCOM is realised by compensating reactive power (*i.e.*, by injecting or absorbing reactive power) [8, 9]. The advantage of D-STATCOM over D-SVC is its *V–I* characteristics (Figure 8.5) and dynamics, but this controller is more expensive. It is more important in the case of voltage-regulation mode because it requires, in general, higher compensating power than for load compensation.

The block diagram of a voltage-regulating D-STATCOM is presented in Figure 8.10. The reference AC voltage of VSI is calculated based on the output of a DC-voltage controller and terminal voltage controller. The DC-voltage controller defines the component of voltage across a filter inductor L_F that is perpendicular to the terminal voltage V_T, while the terminal-voltage controller determines the component of voltage at L_F that is parallel to V_T. Thus, the active power of D-STATCOM depends on the DC voltage error and the reactive power of D-STATCOM depends on the error of the terminal voltage V_T (see Figure 8.9). The dynamics of voltage regulation depends on terminal-voltage controller. State-feedback or output-feedback controllers can be applied for this purpose.

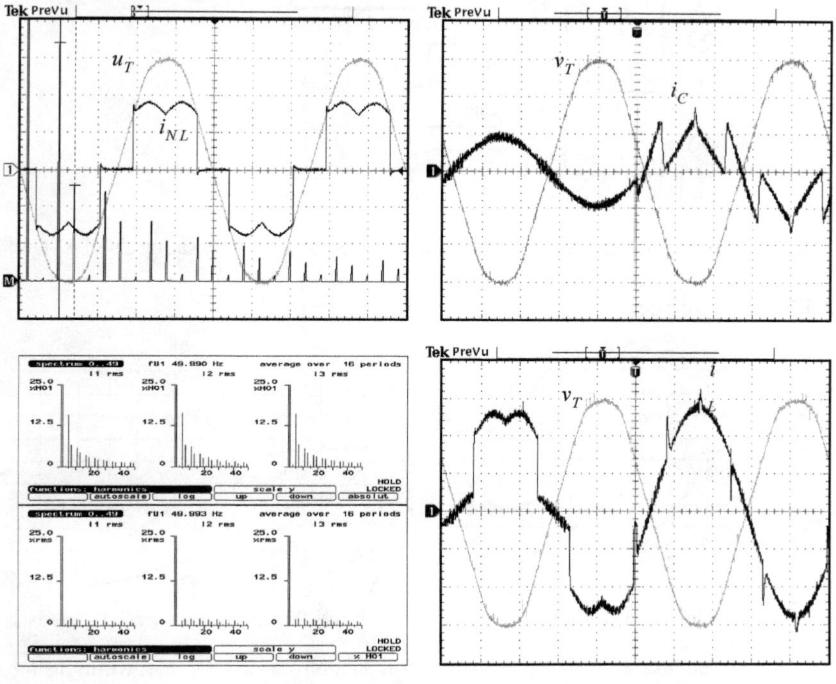

Figure 8.9. Transients and feeder-current harmonics without and with compensation for D-STATCOM compensating load and converting electrical energy from the AC to DC side

In fact, D-STATCOM in voltage-regulation mode compensates disturbances of voltage V_T that are the result of all non-linear or unbalanced loads connected to a distribution system, which have an influence on this voltage. Although shunt compensators are used more often than series compensators, the latter are better suited for voltage regulation from the viewpoint of the required rated power of the converter. The reason for using shunt controllers more often than series controllers is their immunity to feeder short circuit.

D-STATCOM controllers are especially useful for compensating voltage disturbances like flicker or voltage dips, caused by rapidly changing loads, like arc furnaces or high-power electrical machines during direct start up. In Figure 8.11 are presented the exemplary results of voltage-flicker compensation using a D-STATCOM controller.

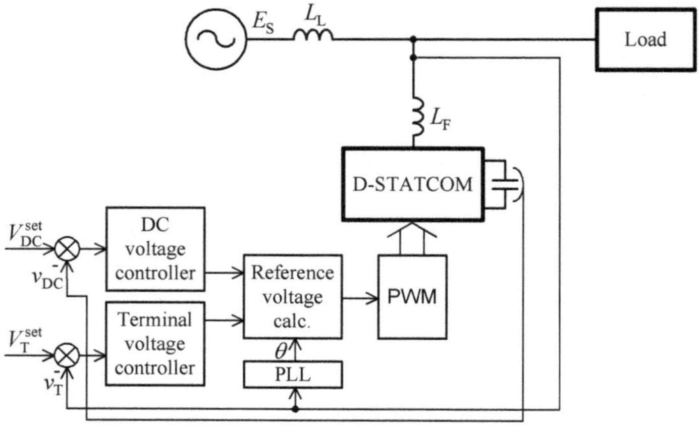

Figure 8.10. Block diagram of a control system for voltage-compensating D-STATCOM

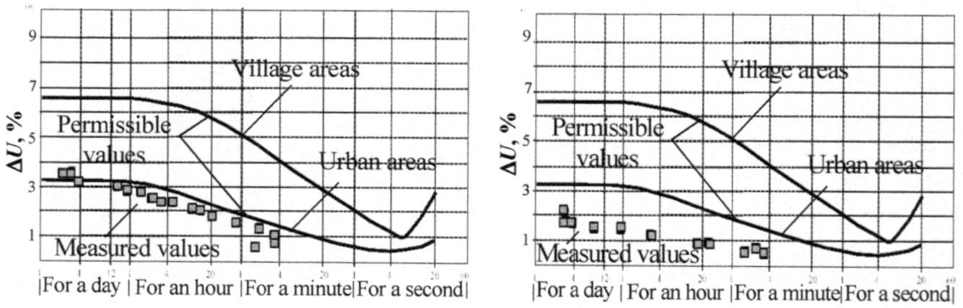

Figure 8.11. The example of voltage-flicker reduction caused by a sawmill using D-STATCOM installed near the industrial facilities. Flicker level without and with compensation

8.2 Other Shunt Controllers Based on D-STATCOM

The compensation capability of D-STATCOM controllers can be extended using combined topologies. In general, there are two major topologies extending the D-STATCOM compensation capability: hybrid arrangements and energy-storage system applications [10, 11].

8.2.1 Hybrid Arrangements

The D-STATCOM-based hybrid arrangements can be used for both voltage-regulating and load-compensating controllers. D-STATCOMs are usually combined with D-SVCs or passive harmonics filters. The former are most often used for voltage regulation, while the latter are utilised for load compensation. Both provide improvement of compensation capabilities. Additionally, in hybrid topologies the rated power of D-STATCOM constitutes a part of a hybrid controller's rated power thus they allow the installation costs to be reduced.

In Figures 8.12–8.14 are presented general topologies and *V–I* characteristics of D-STATCOM D-SVC hybrid arrangements. The D-SVC part extends the operating region of D-STATCOM. The combination of D-STATCOM and D-TSC (Figure 8.12) extends the operating region towards the generation of reactive power (capacitive region). This property is important in practice, because it is often necessary in distribution systems to compensate inductive-type loads to provide terminal-voltage regulation. Extension of *V–I* characteristics of D-STATCOM towards absorption of reactive power (inductive region) is possible by combining it with D-TSR (Figure 8.13). The symmetrical extension of *V–I* characteristics is provided by the hybrid arrangement presented in Figure 8.14. Despite improvements of the *V–I* characteristics, a hybrid arrangement can be used to optimize losses, cost and performance for a particular application.

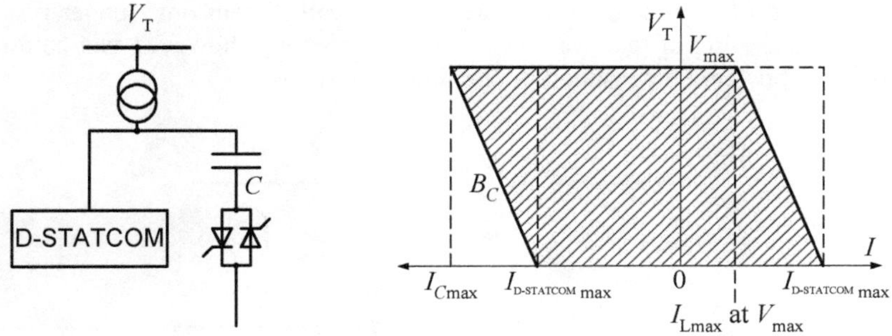

a b

Figure 8.12. a Hybrid D-STATCOM D-TSC controller. **b** Its *V–I* characteristics

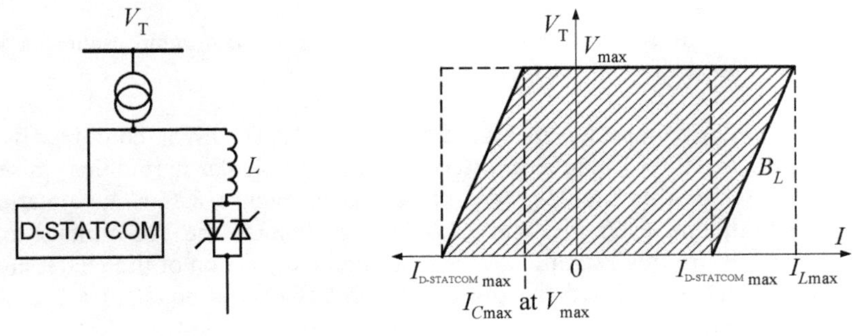

a b

Figure 8.13. a Hybrid D-STATCOM D-TSR controller. **b** Its *V–I* characteristics

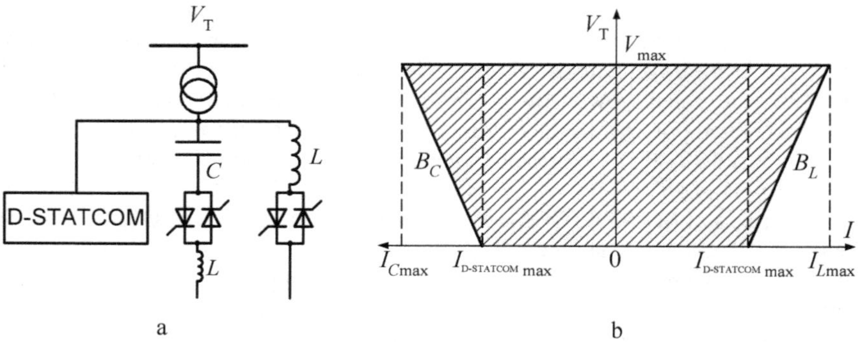

Figure 8.14. a Hybrid D-STATCOM D-TSC D-TSR controller. **b** Its $V-I$ characteristics

In Figure 8.15 are presented hybrid-D-STATCOM based topologies of load-compensating controllers. Among the advantages of hybrid arrangements in this case are better compensation performance and considerably lower rated power of D-STATCOM (lower voltages or/and currents). On the other hand, the disadvantages of hybrid topologies are their overall dimensions, longer response time and generating of reactive power. The last property, however, can be utilised for reactive power compensation in distribution systems.

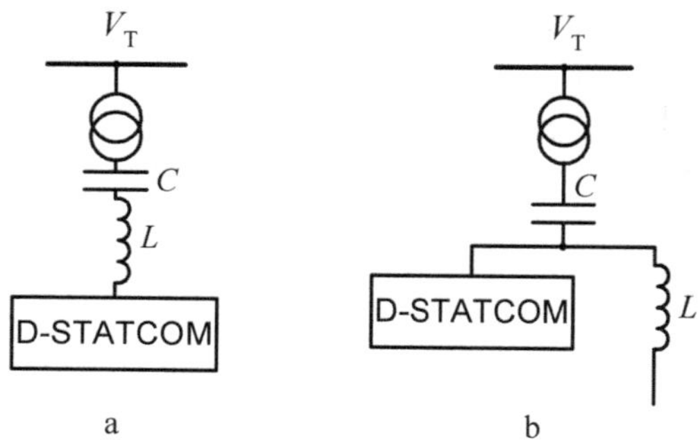

Figure 8.15. Hybrid D-STATCOM topologies intended for load compensation: **a** series combination; **b** parallel combination

In the topology presented in Figure 8.15a, D-STATCOM is combined with a passive harmonic filter (or several filters tuned for particular harmonics) in series. The transients for this topology are presented in Figure 8.16. The notation is consistent with that used in Figure 8.6, and u denotes the AC voltage of D-STATCOM VSI. In this example the total harmonic distortion of the feeder current is equal to THD=1.87% and the power of D-SATCOM is equal to 8.2% of the compensated load power.

The hybrid topology presented in Figure 8.15b combines a passive harmonic filter with D-STATCOM connected in parallel with a passive filter inductor. The transients are presented in Figure 8.17. The feeder current THD=3.44% and D-STATCOM power is equal to 1.2% of the compensated load power.

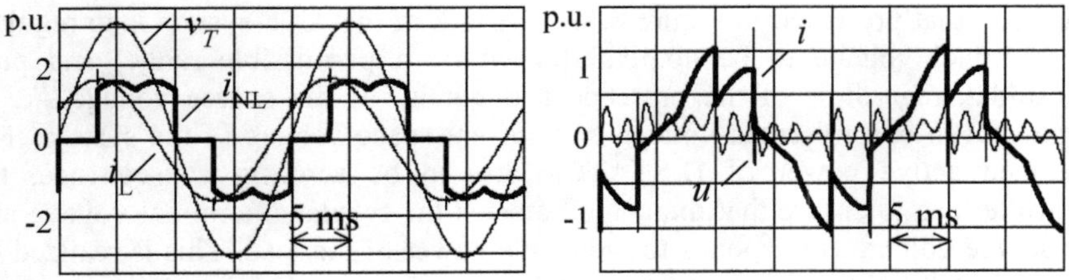

Figure 8.16. Exemplary transients of currents and voltages in steady state for hybrid topology presented in Figure 8.15a

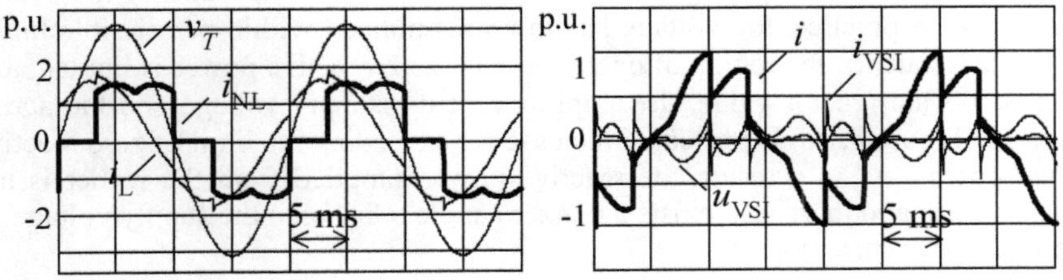

Figure 8.17. Exemplary transients of currents and voltages in steady state for hybrid topology presented in Figure 8.15b

8.2.2 Controllers with Energy-storage Systems

A D-STATCOM controller can be supplemented with an ESS connected to the DC side of the converter to extend the compensation capabilities. Depending on its type, ESS are connected by employing static converters or electromechanical converters in machine/generator systems. For solutions with ESS the averaged real power of D-STATCOM still needs to be zero, but the period of averaging can be considerably longer. This means that D-STATOM with ESS can compensate (level) considerably slower fluctuations of feeder active power. In fact, a combined D-STATCOM/ES controller can compensate disturbances of active power from harmonics to fluctuations up to several minutes long (for example feeder overloads).

There are many types of ESS used. Among the most important are supercapacitors energy storage, superconducting magnetic-energy storage, flywheel-energy storage and battery-energy storage. The description of these system is presented in [12–14]. Figure 8.18 summarises the power and energy capabilities of the selected ES systems.

The combined D-STATCOM ES systems can be controlled as a UPS in the line interactive configuration [6, 15] to provide high quality terminal voltage V_T for critical loads, regardless of supply interrupts and the quality of the feeder voltage. The block diagram of such a system is depicted in Figure 8.19a. The reactor L_T has to be connected in series to lower the short-circuit power of the feeder. This makes it possible to lower considerably the rated power of VSI and to reduce harmonic distortion of supply currents. In this application, the operation principle of D-STATCOM is the same as in voltage-regulation mode, and the control system is similar to that presented in Figure 8.10. The task of this UPS system is to regulate the terminal voltage to be nominal (or within a permissible range) and pure sinusoidal, regardless of the presence and quality of the source voltage V_S. In Figure 8.19b are presented voltage and current space vectors of the system. The averaged active power of D-STATCM has to be zero. In consequence, the controller has to ensure that the angle between the regulated terminal voltage and the source voltage corresponds to the active power of the load. This is realized by the D-STATCOM's DC voltage-control loop. To provide zero reactive power consumed from the feeder, it is necessary that the voltage across the reactor is perpendicular to the source voltage V_S. This means that the amplitude of the regulated terminal voltage V_T has to be greater than the amplitude of the source voltage V_S. In practice, the voltage has to be maintained within permissible limits and in consequence the region of operation with zero reactive power is limited, and depends on the reactor value, the amplitude of the supply voltage and the active power. In Figure 8.19b is presented the case for exceeding the limit of zero reactive power operation. As a result, the reactive power supplied from the feeder is not equal to zero, and there also exists a nonzero angle α between the supply voltage V_S and the supply current i_S.

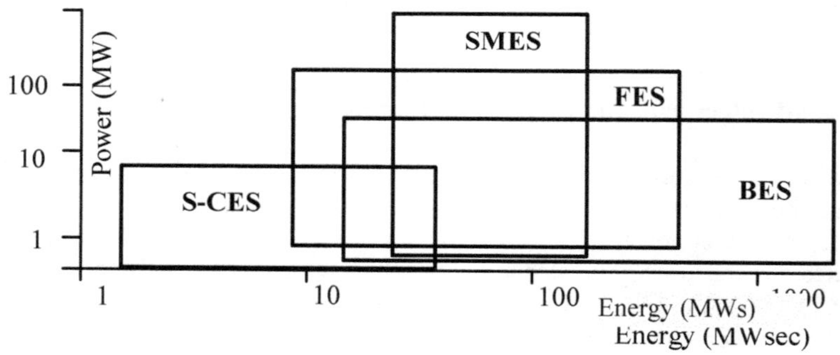

Figure 8.18. Power and energy ranges of the selected ESS

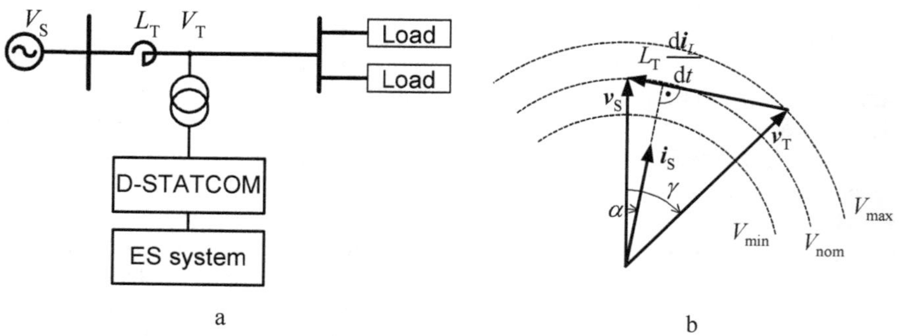

Figure 8.19. The line interactive UPS with D-STATCOM controller: **a** block diagram; **b** space vectors

From the principle of operation of a line-interactive UPS it is obvious that a distorted source voltage causes a distorted voltage across the reactor (for a sinusoidal, regulated terminal voltage) and, in consequence, a distorted supply current. Thus, it is not possible to regulate the terminal voltage and supply current at the same time. Figure 8.20 demonstrates the case for distorted supply voltage, which determines distorted voltage across reactor v_L. This simultaneous control of terminal voltage and supply current can be realised by a line-interactive delta UPS, but it is a combined series-shunt controller.

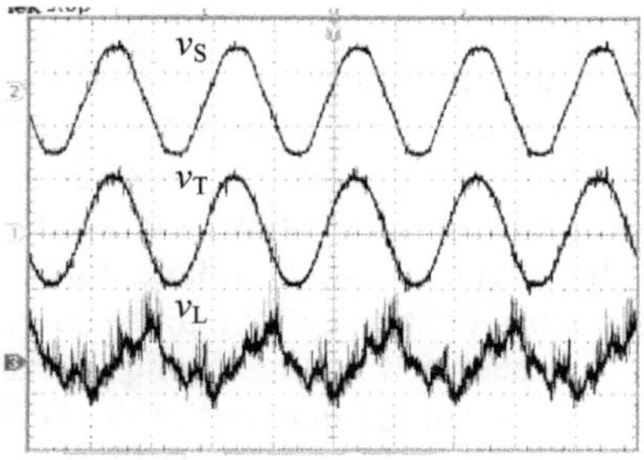

Figure 8.20. The source, terminal and reactor voltages of a line-interactive UPS controller

8.3 Dynamic Static Synchronous Series Compensators

Figure 8.21 presents examples of fundamental structures of Dynamic Static Synchronous Series Compensators (D-SSSC) realized by transistorized three-phase VSI topologies. We decided to pass over overvoltage protections connected across the secondary winding of the transformer. From the viewpoint of the device, the structure D-SSSC is similar to SSSC [16]. The difference between the two results from a place of installation (distribution network or transmission system) and rated power and control method. The SSSCs are flexible as transmission-system devices and are used to enhance controllability and increase power-transfer capability in transmission systems the latter are custom power-compensating series controllers used in distribution systems.

The D-SSSC, which is presented in Figure 8.21a is intended only for three-wire supply systems; however, the other two D-SSSCs (Figure 8.21b,c) may be applied to four as well as to three-wire systems. Among them, interesting characteristics are present in the structure D-SSSC, realized on the basis of three-phase VSI topology with an additional fourth branch of switches (variant II – Figure 8.21b). The additional branch of switches allows a significantly decreased capacity of capacitors when compared to the application of sectional capacitors (variant I – Figure 8.21a). However, the structure D-SSSC shown in Figure 8.21b allows application of unipolar PWM to the waveform of the required boosting voltage u_C. To this, the unipolar PWM allows application of a smaller filter L_f–C_f, preventing switching frequency harmonics from entering the power system, and thereby improve the transient and accuracy in output-voltage control.

The D-SSSC structures, which are presented in Figure 8.21, obviously do not present all the possible arrangements. Sometimes, instead of the L_f– C_f filter, we place capacitor C_f across the secondary winding of the transformer. The function of reactor L_f is then taken over by the leakage inductance of the transformer. This results not only in simplification of the arrangement, but also in significant increase of losses in the transformer, which is a very significant further disadvantage of the solution. In the feeders MV we can also apply D-SSSC with multi-level VSI. Basically, we could also apply D-SSSC realized on the basis of CSI topology, although industrial applications of these solutions are not known to the authors.

In the following part of this chapter we will consider D-SSSC as a controlled-voltage source with or without additional energy storage. We are allowed to do so because of the permissibility of continuous models of PE converters with PWM in the area of their evaluation from a functional perspective and the possibility of very careful wave-forming of the required output voltage, for example by application of output-voltage feedback control [17]. We will mainly concentrate on issues related to the identification of compensated voltage components and determination of reference control signals in selected applications of the D-SSSC, and issues related to the energy balance between the power source, D-SSSC and load.

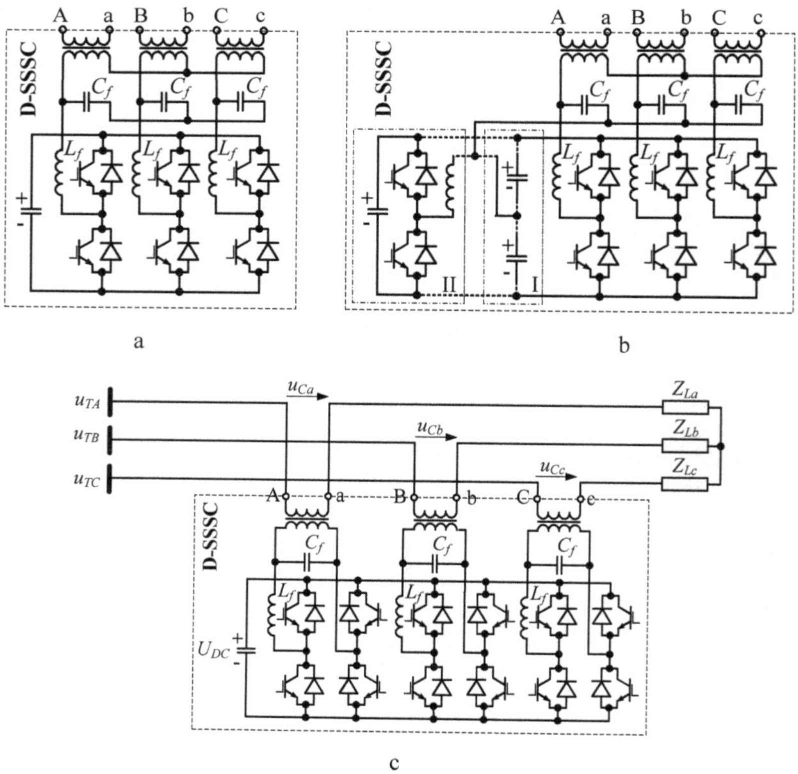

Figure 8.11. Examples of the D-SSSC structures realised by three-phase VSI topologies

8.3.1 Identification of Separate Components of the Supply-terminal Voltage

Among various methods of identification of individual components of unbalanced and distorted supply voltage, some of the most important methods are those based on *d–q* transformation [18, 19]. In many aspects they are convergent to similar methods used for control of D-STATCOMs in shunt active and hybrid power-filter arrangements [20, 21].

In Figure 8.22 we present a block diagram of exemplary identification of separated voltage and current components in three-wire systems, conducted on the basis of a *d–q* transformation. Initially we used (not marked in Figure 8.22) space-vector transformation [22–24], investigating three-phase voltages and currents to an orthogonal system α–β on the basis of the equation

$$\begin{bmatrix} u_\alpha + ju_\beta \\ i_\alpha - ji_\beta \end{bmatrix} = \sqrt{\frac{2}{3}} \begin{bmatrix} u_a & u_b & u_c \\ i_a & i_c & i_b \end{bmatrix} \cdot \begin{bmatrix} 1 \\ a \\ a^2 \end{bmatrix} \tag{8.6}$$

where $a=exp(j2\pi/3)$, j – imaginary unit.

After transformation of the vectors $u_\alpha + j \cdot u_\beta$ and $i_\alpha - j \cdot i_\beta$ to coordinates $d - q$ by using operators $\exp(-j\omega_S t)$ and $\exp(j\omega_S t)$, where ω_S – fundamental angular frequency, and then filtration of the obtained output functions by using simple low-pass filters (LPF), and after application of inverse operators $\exp(j\omega_S t)$ and $\exp(-j\omega_S t)$, we obtained the current and voltage components of positive (+) and negative (–) sequence and fundamental frequency (^) within coordinates α–β. From here we can easily select higher harmonics (~).

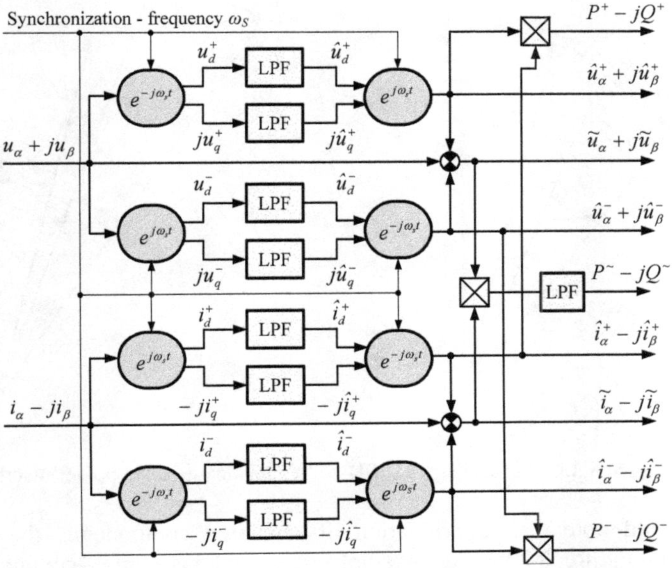

Figure 8.22. Block diagram of exemplary identification of separated components

Based on the conducted identification, as presented in Figure 8.22, we can determine three average active power components $P = P^{+} + P^{-} + P^{\sim}$ and three average reactive power components $Q = Q^{+} + Q^{-} + Q^{\sim}$. These power components are directly related to the voltage and current components of positive (+) and negative (–) sequence and fundamental frequency (^), and also higher harmonics (~). As results from the energy balance [25], D-SSSC arrangement, which compensates unbalanced and/or distorted supply voltage, must insert active power $P_C = P^{-} + P^{\sim}$ into the distribution system. The average active power P_C obviously depends upon unbalanced and/or non-linear load. If, after compensation, the load-side voltage U_L of positive sequence (+) is contained within allowable limits of the nominal voltage U_{nom} changes, then a relatively small active power P_C can almost always be generated without an additional energy source, supplying D-SSSC. The voltage u_C, inserted by D-SSSC, must then include a positive sequence component $U^{+} = U^{P} + jU^{Q}$ of the low real part U^{P}, which is illustrated in Figure 8.23a. The opportunity presented above is, however, very limited in situations where the load-side voltage U_L of the positive sequence is much lower than the voltage U_{nom} (Figure 8.23b) and when restoration of voltage U_L to value U'_L located within the allowable limits of deviation from the rated voltage is required. In this case, the real part U^{P} of a component U^{+} is most often high enough that D-SSSC must be supplied by an additional energy source. We also recommend readers see references [26–29].

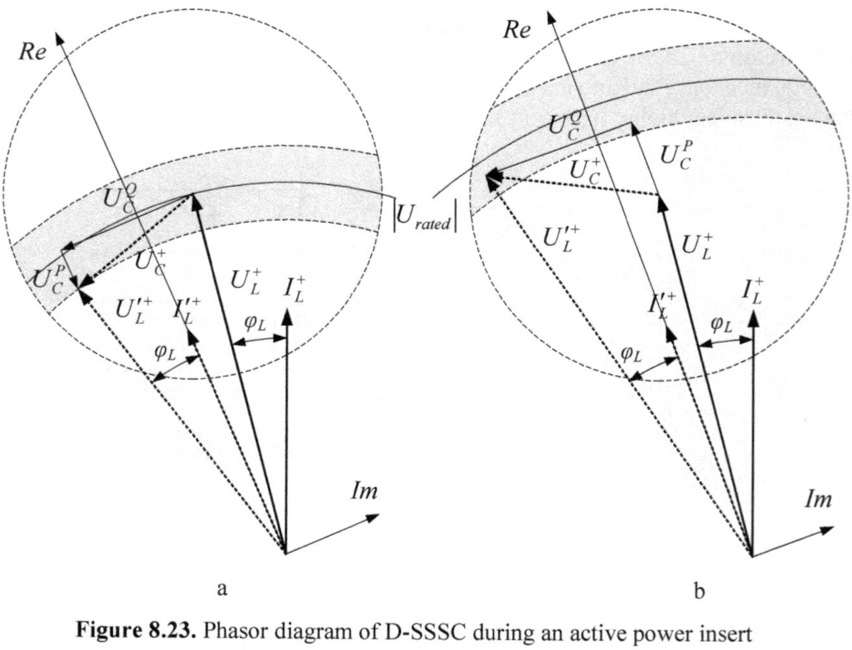

Figure 8.23. Phasor diagram of D-SSSC during an active power insert

We should note that identification of separated components, the way it is presented in Figure 8.22 as an example, does not take into account the zero-sequence voltage and zero-sequence current, and thereby is not sufficient in four-wire three-phase systems. However, in such systems, identification is not much harder and the process not more complicated. We could easily prove this, for example, on the basis of the instantaneous PQ theory [30] or other similar theories [31, 32].

8.3.2 Harmonic Filtration and Balancing of the Voltage in Three-wire Systems

To three-phase three-wire distribution systems we can apply simpler D-SSSC structures than those presented in Figure 8.21. In these systems the D-SSSC agreement can be, for example, realized on the basis of two independent one-phase H-Bridge VSI topologies [33]. Another example of a simpler structure D-SSSC is shown in Figure 8.24.

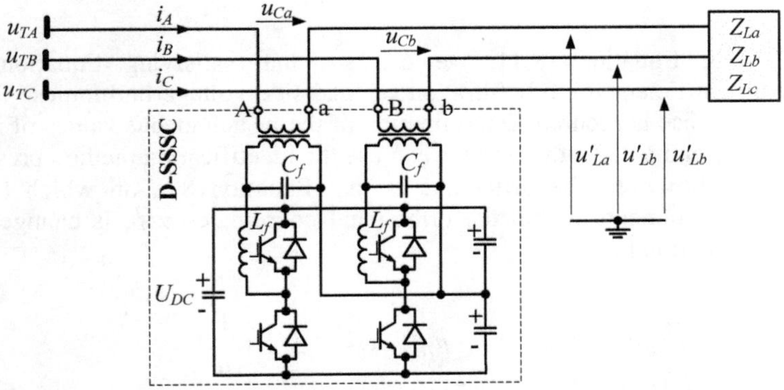

Figure 8.24. Simple structure D-SSSC for three-phase three-wire system

8.3.2.1 Harmonic Voltage-supply Filtration

The D-SSSC structure presented in Figure 8.24 can be used for filtration of voltage harmonics in a three-wire system, since the sum of line-to-line voltage is always equal to zero, that is

$$u_{TA-TB} - u_{TB-TC} - u_{TC-TA} = (u_{TA} - u_{TB}) + (u_{TB} - u_{TC}) + (u_{TC} - u_{TA}) = 0 \tag{8.7}$$

When required, instantaneous values of inserted voltages u_{Ca} and u_{Cb}, compensating components $\tilde{u}_{TA}$, $\tilde{u}_{TB}$ and $\tilde{u}_{TC}$ of higher harmonics of phase voltages u_{TA} , u_{TB} and u_{TC}, must satisfy equations

$$u_{Ca} = \tilde{u}_{TC} - \tilde{u}_{TA} \text{ and } u_{Cb} = -\tilde{u}_{TB} + \tilde{u}_{TC} \tag{8.8}$$

to which

$$u_{TA} = \hat{u}_{TA} + \tilde{u}_{TA} ,\ u_{TB} = \hat{u}_{TB} + \tilde{u}_{TB} ,\ u_{TB} = \hat{u}_{TB} + \tilde{u}_{TB} \tag{8.9}$$

where $\hat{u}_{TA}$, $\hat{u}_{TB}$, $\hat{u}_{TC}$ – components of the fundamental frequency phase voltage.

Taking into account Equation 8.5, phase voltages on the load terminals after compensation are described by the equations

$$u'_{La} = \hat{u}_{TA} + \tilde{u}_{TA} + u_{Ca}\,,\ u'_{Lb} = \hat{u}_{TB} + \tilde{u}_{TB} + u_{Cb}\,,\ u'_{Lc} = u_{TC} = \hat{u}_{TC} + \tilde{u}_{TC} \tag{8.10}$$

Therefore, the line-to-line voltages on the load terminals are

$$\begin{aligned} u'_{La-Lb} &= u'_{La} - u'_{Lb} = (\hat{u}_{TA} + \tilde{u}_{TA} + u_{Ca}) - (\hat{u}_{TB} + \tilde{u}_{TB} + u_{Cb}) \\ u'_{Lb-Lc} &= u'_{Lb} - u'_{Lc} = (\hat{u}_{TB} + \tilde{u}_{TB} + u_{Cb}) - (\hat{u}_{TC} + \tilde{u}_{TC}) \\ u'_{Lc-La} &= u'_{Lc} - u'_{La} = (\hat{u}_{TC} + \tilde{u}_{TC}) - (\hat{u}_{TA} + \tilde{u}_{TA} + u_{Ca}) \end{aligned} \tag{8.11}$$

Following Equations 8.11, we can see that satisfying Equation 8.8 is completely sufficient for full filtration of load-side voltage harmonics. The most fundamental case is accurate determination of the instantaneous values of inserting voltages) u_{Ca} and u_{Cb}. To this end we can use the identification method presented in Figure 8.22 however, the difference is that Equation 8.6, on which basis we determine components within the orthogonal coordinates α–β, is changed to the following equation [34]

$$\begin{bmatrix} u_\alpha + ju_\beta \\ i_\alpha - ji_\beta \end{bmatrix} = \sqrt{\frac{2}{3}} \begin{bmatrix} u_{TA} - u_{TB} & u_{TB} - u_{TC} & u_{TC} - u_{TA} \\ i_A - i_B & i_C - i_A & i_B - i_C \end{bmatrix} \cdot \begin{bmatrix} 1 \\ a \\ a^2 \end{bmatrix} \tag{8.12}$$

Based on identified $\tilde{u}_\alpha$, $\tilde{u}_\beta$, components and in accordance with the equation

$$\begin{bmatrix} u_{Cb} - u_{Ca} \\ -u_{Cb} \\ u_{Ca} \end{bmatrix} = \begin{bmatrix} \tilde{u}_{TA} - \tilde{u}_{TB} \\ \tilde{u}_{TB} - \tilde{u}_{TC} \\ \tilde{u}_{TC} - \tilde{u}_{TA} \end{bmatrix} = \sqrt{\frac{2}{3}} \cdot \begin{bmatrix} 1 & 0 \\ 1/2 & \sqrt{3}/2 \\ 1/2 & -\sqrt{3}/2 \end{bmatrix} \cdot \begin{bmatrix} \tilde{u}_\alpha \\ \tilde{u}_\beta \end{bmatrix} \tag{8.13}$$

instantaneous values of the voltage u_{Ca} and u_{Cb} are determined. As results from the comparison of Equation 8.8 and Equation 8.13, determination of the inserting voltages completely compensates the higher harmonics of terminal-supply voltages under balance and unbalance conditions in three-wire systems.

8.3.2.2 Balancing and Regulation of the Load-side Voltage

Let us consider application of a D-SSSC structure, which is presented in Figure 8.24, to balance and regulate of load-side voltages. On the basis of transformation

$$u_T^+ = \left[(u_{TA} + u_{Ca}) + a\cdot(u_{TB} + u_{Cb}) + a^2\cdot u_{TC}\right]/3 \tag{8.14a}$$

$$u_T^- = \left[\left(u_{TA} + u_{Ca}\right) + a^2 \cdot \left(u_{TB} + u_{Cb}\right) + a \cdot u_{TC}\right]/3 \tag{8.14b}$$

we identified positive (+) and negative (–) sequences of the terminal-supply voltages u_{TA}, u_{TB}, u_{TC}. From Equation 8.14b it also results that the negative sequence may be compensated by inserting a voltage u_{Ca}, determined on the basis of the equation

$$u_{Ca} = -u_{TA} - a^2 \cdot \left(u_{TB} + u_{Cb}\right) - a \cdot u_{TC} \tag{8.15}$$

After inserting a voltage u_{Ca}, positive and negative sequences of load-side voltages: u'_{La}, u'_{Lb}, u'_{Lc} take the following values

$$u_L'^- = \{u_{TA} + \underbrace{\left[-u_{TA} - a^2 \cdot \left(u_{TB} + u_{Cb}\right) - a u_{TC}\right]}_{u_{Ca}} + a^2 \cdot \left(u_{TB} + u_{Cb}\right) + a \cdot u_{TC}\}/3 = 0 \tag{8.16a}$$

$$\begin{aligned} u_L'^+ &= \{u_{TA} + \underbrace{\left[-u_{TA} - a^2 \cdot \left(u_{TB} + u_{Cb}\right) - a u_{TC}\right]}_{u_{Ca}} + a \cdot \left(u_{TB} + u_{Cb}\right) + a^2 \cdot u_{TC}\}/3 \\ &= \left(a - a^2\right) \cdot \left(u_{TB} + u_{Cb} - u_{TC}\right)/3 \end{aligned} \tag{8.16b}$$

As one can see from Equation 8.16b, the positive sequence of a load-side voltage may be changed by inserting a voltage u_{Cb} into the terminal voltage u_{Tb}. In this manner, by the simple structure of the D-SSSC, presented in Figure 8.24, we can not only filter higher harmonics but also balance and regulate load-side voltages. Balancing and regulation principles, described by Equations 8.15 and 8.16, illustrate voltage phasors, presented in Figure 8.25. In practical applications,

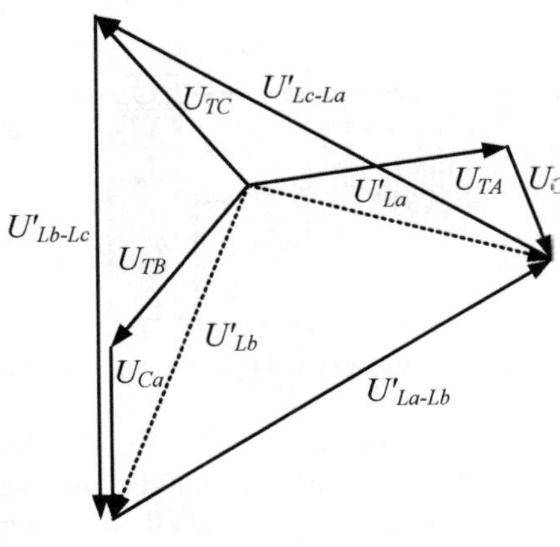

Figure 8.25. Voltage phasors of a D-SSSC (Figure 8.24) during balancing and regulation

we stress again D-SSSC (including the structure as in Figure 8.24), which are dedicated to balancing and regulation of the voltage in the broad sense and independently from a load type, must be additionally power supplied. Owing to the following functional opportunities, these arrangements perform their function under the generally accepted name DVR.

8.4 Dynamic Voltage Restorer

8.4.1 What is a DVR

A dynamic voltage restorer is a PE converter-based D-SSSC which can protect sensitive loads from all supply-side disturbances other than outages. It is connected in series with a distribution feeder and is also capable of generating or absorbing real and reactive power at its AC terminals. The basic principle of a DVR is simple: by inserting a voltage of the required magnitude and frequency, the DVR can restore the load-side voltage up to the desired amplitude and waveform even when the source voltage is either unbalanced or distorted. Usually, a DVR, as a cost-effective solution when compared to very costly UPS agreements, is connected in order to protect loads and can be implemented at both an LV level and an MV level, which gives an opportunity to protect high-power applications from voltage sags during faults in the supply system. A typical location in the distribution system and the operation principle of the DVR is shown in Figure 8.26 [35], where U_T – terminal supply voltage, U'_L – the load side voltage after restore, U_C – the inserted voltage by the DVR, I_L and I_S are the load and feeder currents, P_C – the real power generated or absorbed by the D-SSSC.

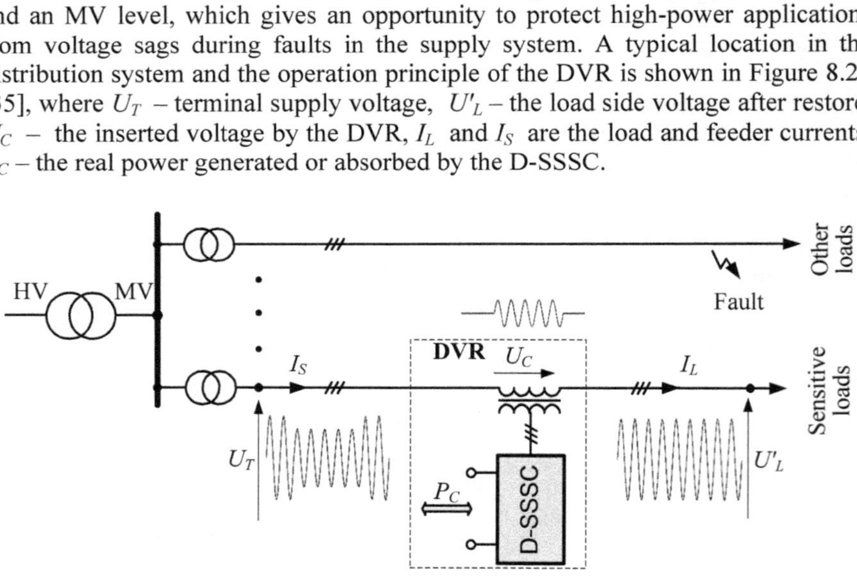

Figure 8.26. A typical location and operation principle of the DVR

DVR was commercially introduced in 1994 for the first time, and its first important installation was in North Carolina for the rug manufacturing industry [36]. Since then, the number of installed DVR has increased continuously. Obviously, it is implemented especially in those industry branches where supply-side disturbances can lead to dangerous situations for personnel or serious production losses.

8.4.2 Control Strategies of the DVR Arrangements

The following DVR control strategies can be applied:

- "*Presag*" strategy, where the load-side voltage U'_L after restoration is assumed to be in phase with this presag voltage [37];
- "*Inphase*" strategy, where the voltage U'_L after restoration is in phase with terminal-supply voltage U_T during sag [38];
- "*Minimal energy*" strategy, where the phase angle α of the voltage U'_L after restoration is determined on the basis of requirements P_C=0 or $dP_C/d\alpha$=0 [39].

These strategies and differences between them are presented in Figure 8.27, where dashed-line phasors are related to the presag condition, while solid-line phasors relate to restoration.

The "*presag*" strategy's characteristic (Figure 8.27b) is a significant magnitude of an inserted voltage U_C. Furthermore, a DVR controlled on the basis of this particular strategy should allow injection of heavy active power P_C, which especially impacts the required capacity of the energy storage or energy absorbed

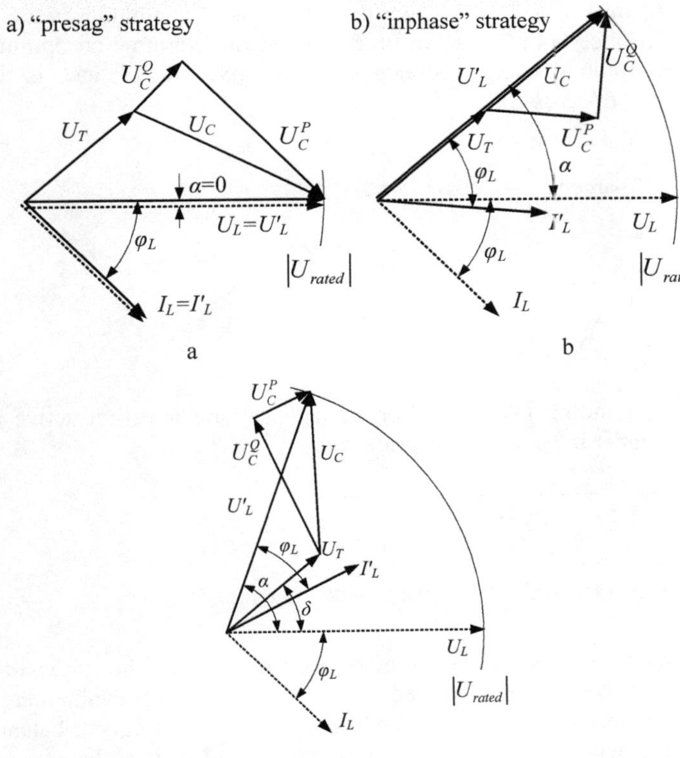

Figure 8.27. Phasor diagrams; illustration of the DVR control strategies: **a** "*presag*"; **b** "*in-phase*"; **c** "*minimal energy*"

from the grid supply during voltage sags. Therefore, the "*presag*" strategy is applied only in cases where the load is exceptionally sensitive to changes of supply-voltage phase angle α (for example some applications of thyristor converters). If the load is not sensitive to the changes of supply-voltage phase angle α, then the most often applied strategy is the "*in-phase*" strategy (Figure 8.27b). The main advantage of this solution is that the magnitude of an inserted voltage $U_C = U'_L - U_T$ is minimal. If the magnitude of the terminal-supply voltage is disturbed, the DVR generates the same voltage as the voltage drop. Therefore, the rated power of the DVR is minimized for the existing current and terminal-supply voltage [38]. In addition, application of the "*inphase*" strategy decreases the demand for injection of active power P_C; however, it does not minimize its value.

When the value of injected active power P_C is critical because of the capacity of the energy storage, it is advised to apply a "*minimal energy*" strategy (Figure 8.27c) instead. For this we distinguish two possible variations of this strategy. Its first variation refers to voltage sags, during which the magnitude of the terminal-supply voltage U_T is not dropped below the value $U'_L \cdot cos\varphi_L$, where φ_L is the load power factor angle. Then, suitably selecting the angle $\alpha = \alpha_{opt}$, we have the possibility of restoration of the load-side voltage U'_L at active power $P_C = 0$. In the second case, when $U_T < U'_L \cdot cos\varphi_L$, angle $\alpha = \alpha_{opt}$ should be determined on the basis of $dP_C/d\alpha = 0$. In this case, the active power P_C takes a minimum value, different from "*zero*". Since both cases can be combined to determine an optimum angle $\alpha = \alpha_{opt}$, the "minimal energy" strategy for any given disturbance is realized as follows if: $U_T \geq U'_L \cdot cos\varphi_L$, then

$$\alpha_{opt} = \varphi_L + \delta - \arccos\left(\frac{U'_L}{U_T} \cdot \cos\varphi_L\right) \tag{8.17a}$$

else

$$\alpha_{opt} = \varphi_L + \delta \tag{8.17b}$$

The corresponding DVR insertion voltage U_C and injection active power P_C requirement under the α_{opt} control strategy are

$$U_C = \sqrt{U_L'^2 + U_T^2 - 2 \cdot U'_L \cdot U_T \cdot \cos(\alpha_{opt} - \delta)} \tag{8.18}$$

$$P_C = S_L \cdot \cos\varphi_L - S_L \cdot (U_T / U'_L) \cdot \cos(\varphi_L - \alpha_{opt} + \delta) \tag{8.19}$$

Equations 8.17–8.19, which point out the essence of the "*minimal energy*" strategy, can be directly applied for the control and evaluation of DVR characteristics only in the case of one-phase distribution systems or balanced three-phase systems, with the assumption that only balanced voltage disturbances occur. Other cases of the "*minimal energy*" strategy, for example during single-phase or two-phase sags, with or without phase deviation, are presented in the literature [39].

Figure 8.28 shows the characteristics of DVR controlled on the basis of "*inphase*" and "*minimal energy*" strategies, in the case of restoration of the load-side voltage during single-phase voltage sags. As we can see, the injection active power P_C is lower by about 20% when applying the "*minimal energy*" strategy. However, at the start and end points of time interval Δt_{sag} of the sag, 15–25% short-time overvoltages also occur. This, however, is not advisable for numerous sensitive loads. Therefore, "*minimal energy*" strategy is sporadically realized in commercial DVRs.

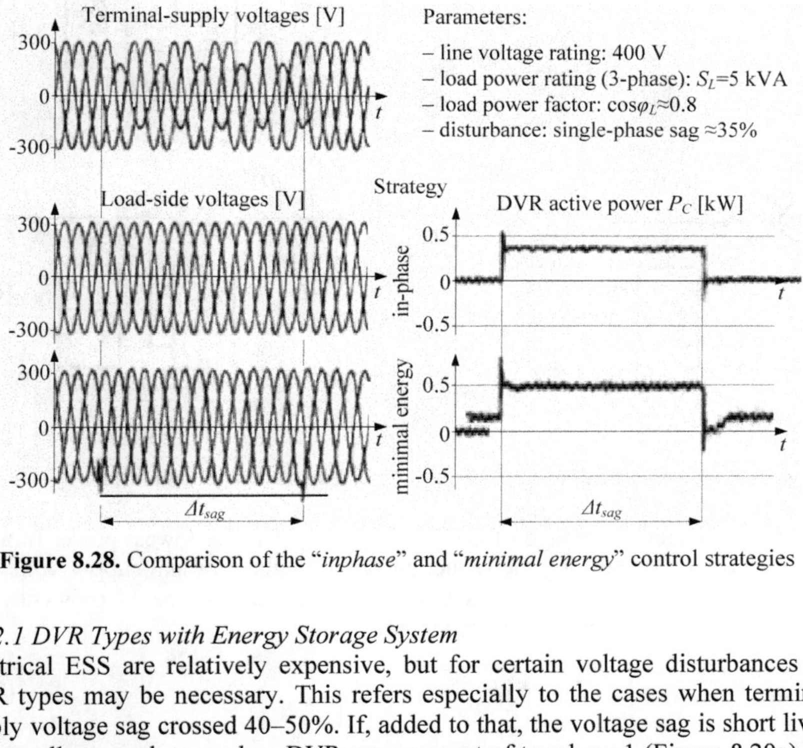

Figure 8.28. Comparison of the "*inphase*" and "*minimal energy*" control strategies

8.4.2.1 DVR Types with Energy Storage System

Electrical ESS are relatively expensive, but for certain voltage disturbances the DVR types may be necessary. This refers especially to the cases when terminal-supply voltage sag crossed 40–50%. If, added to that, the voltage sag is short lived, it is usually enough to apply a DVR arrangement of topology 1 (Figure 8.29a), for which a variable DC-link voltage is characteristic. In this case, the restoration of the load-side voltage may take as long as energy ΔE, delivered by DVR to the load, does not cause a change of the voltage U_{DC} in capacitor C_{DC}, below the assumed bottom level $U_{DC(\min)}$. Below this level, the D-SSSC is unable to insert the required voltage U_C. Thus, the energy ΔE is limited by the inequality

$$\Delta E \le C_{DC} \cdot (1/2) \cdot \left(U_{DC(0)}^2 - U_{DC(\min)}^2\right)$$

where $U_{DC(0)}$ is the rated DC-link voltage.

If we assume terminal-supply voltage sags, which take longer, then it is advisable to apply topology 2 (Figure 8.29a). A separated ESS and an additional DC/DC converter maintain a constant DC-link voltage. During the restoring process, energy is transferred from a large ESS to a smaller rated DC-link storage using a DC/DC converter.

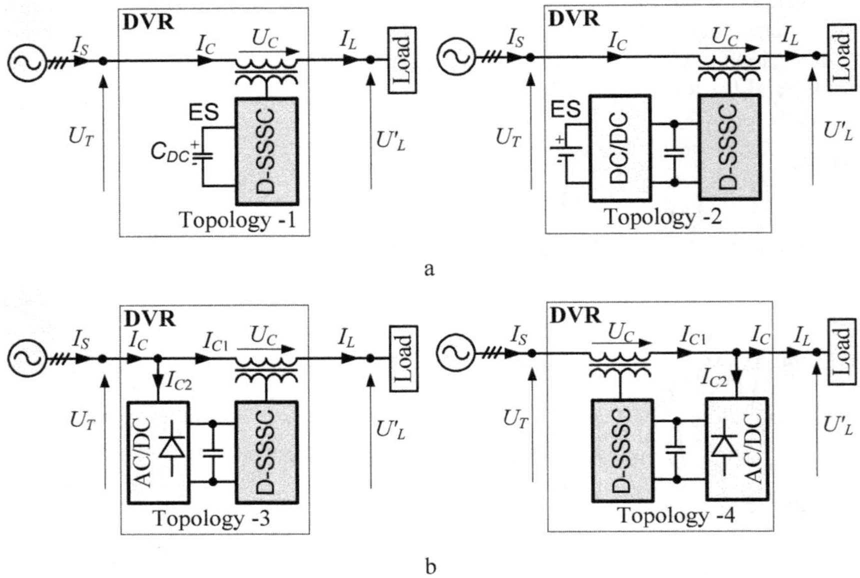

Figure 8.29. Topologies of the DVR: **a** with energy storage; **b** without energy storage

The undoubted advantage of DVR of topology 1 is the lowest power rating in the converters. On defining the reduction factor of the terminal-supply voltage as $\xi=(U_{rated}-U_T)/U_{rated}$ and considering that $U_{rated}=U'_L$, and because the only converter is D-SSSC, this power rating can by calculated as

$$S_{DVR(1)} = S_{D-SSSC} = U_C \cdot I_C = \xi \cdot S_L \tag{8.20}$$

where $U_C=\xi\cdot U_{rated}=\xi\cdot U'_L$ and $I_C=I_L=I_S$. However, for topology 2 we obtain

$$S_{DVR(2)} = S_{D-SSSC} + S_{DC/DC} = \xi \cdot S_L + S_{DC/DC} = 2 \cdot \xi \cdot S_L \tag{8.21}$$

where the power rating of the additional DC/DC converter $S_{DC/DC}=S_{D-SSSC} = \xi\cdot S_L$ is calculated for the "*worst*" case – resistance load. Then, the lowest possible value of S_{D-SSSC} equals the active power P_C , delivered from ESS to the load through a DC/DC converter and a D-SSSC arrangement.

8.4.2.2 DVR Types Without Energy Storage

DVR topologies without energy storage (Figure 8.29b) include an additional shunt AC/DC converter as a residual supply source. It is usually a diode rectifier, since only unidirectional power flow is assumed necessary and it is a cheap solution. In these topologies, saving is obtained on the energy-storage system, and the ability exists to compensate very much longer sags, which constitutes its great advantage. The main disadvantage is, however, that they draw more current from the feeder during the terminal-supply voltage drop. Therefore, DVRs without energy storage and with a shunt diode rectifier are not fit for soft-grid cases of deep voltage sags.

The limitation for maximal values of voltage sags refers especially to topology 3 (Figure 8.29b), being characterized by a supply-side-connected diode rectifier. For this topology, the uncontrollable DC-link voltage U_{DC} is proportional to the terminal-supply voltage. Hence, when load-side voltage $U'_L=U_{rated}$, we can conclude that

$$U_{DC} = k_1 \cdot (1-\xi) \cdot U_{rated} \text{ and } U_C = \xi \cdot U_{rated} \tag{8.22}$$

whereas the following inequality must always be satisfied

$$U_C \le n \cdot k_2 \cdot U_{DC} \tag{8.23}$$

where k_1 and k_2 are coefficients depending on the structure of the diode rectifier and the D-SSSC, and n is the transformer ratio. From Equations 8.22 and 8.23 we conclude that, in order to restore fully the load-side voltage, the transformer ratio must be selected in accordance with the following formula

$$n \ge k_1 \cdot k_2 \cdot \xi/(1-\xi) \tag{8.24}$$

For example, assuming $k_1 \cdot k_2=1$, for $\xi=0.5$ we obtain $n=1/2$. Such a transformer ratio, which is typical for commercial DVR, limits their ability for full restoration of the load-side voltage to the changes of factor ξ within the limits 0–0.3, which is shown in Figure 8.30 [40].

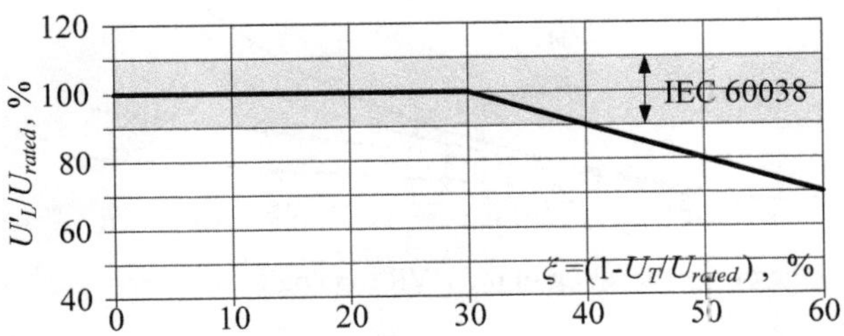

Figure 8.30. Typical output-voltage characteristic of a commercial DVR

As we can see, on the basis of the schemes presented in Figure 8.29b, current I_{C1} through D-SSSC in cases of DVR of topology 3 equals the load current $I_{C1}=I_L$ however, in cases of DVR of topology 4, being characterised by a load-side-connected diode rectifier, this current equals $I_{CI}=I_L/(1-\xi)$. Thus, since the voltages inserted by the DVRs $U_C=\xi\cdot U'_L$, the rated power of the D-SSSC structures are different and are given by

– for topology 3: $S_{D-SSSC} = \xi \cdot S_L$ (8.25)

– for topology 4: $S_{D-SSSC} = \xi \cdot S_L / (1-\xi)$ (8.26)

However, the rated powers of the shunt AC/DC converters (diode rectifiers), which are determined on the basis of the maximal values of currents and voltages, are the same and amount to [41]

$$S_{AC/DC} = \xi \cdot S_L / (1-\xi) \tag{8.27}$$

Taking into account Equations 8.25–8.27, the overall power ratings of the DVR topologies without energy storage $S_{DVR}=S_{D\text{-}SSSC}+S_{(AC/DC)}$ are calculated as

– for topology 3: $S_{DVR(3)} = (2-\xi)\cdot \xi \cdot S_L / (1-\xi)$ (8.28)

– for topology 4: $S_{DVR(4)} = 2 \cdot \xi \cdot S_L / (1-\xi)$ (8.29)

Figure 8.31 shows how the overall converter power rating varies with a change of factor "ξ" for the four DVR topologies considered. The DVR of topology 4, which is characterized by a load-side-connected AC/DC converter, requires the highest power rating when compared to other topologies DVR. The significant disadvantage of this topology is also larger currents to be handled by the D-SSSC.

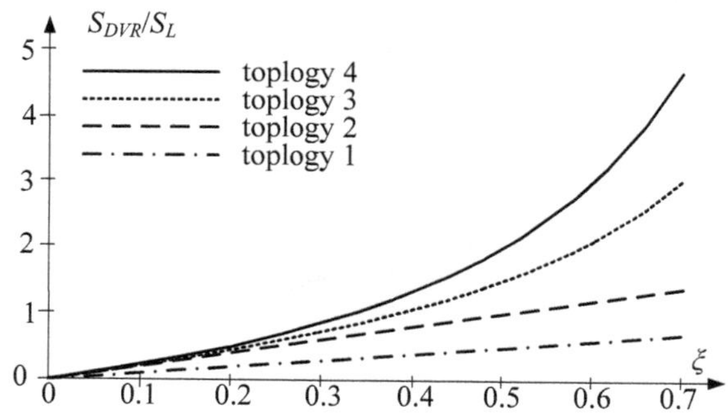

Figure 8.31. Overall power rating of the DVR topologies (only converters) *vs* factor ξ

In addition, the load can be disturbed by the non-linear currents drawn by the shunt diode rectifier (AC/DC converter). However, a DVR using topology 4 may be an efficient solution in terms of the shunt converter design, since the DC-link voltage can be held constant.

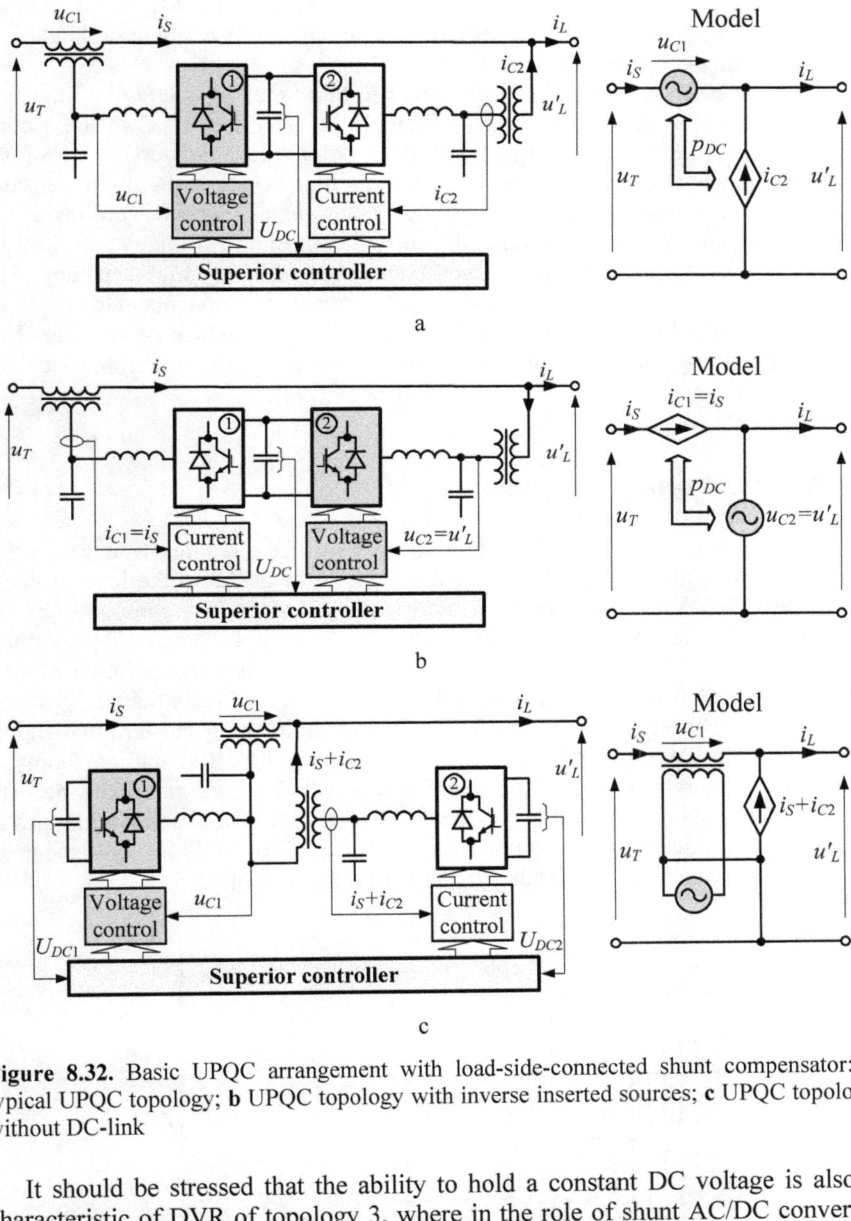

Figure 8.32. Basic UPQC arrangement with load-side-connected shunt compensator: **a** typical UPQC topology; **b** UPQC topology with inverse inserted sources; **c** UPQC topology without DC-link

It should be stressed that the ability to hold a constant DC voltage is also a characteristic of DVR of topology 3, where in the role of shunt AC/DC converter we apply an Active Front-end (AFE) rectifier. Moreover, the AFE rectifier can serve the function of a shunt active compensator. The same applies to DVR of topology 4; however, in this case we can unload the D-SSSC structure and series transformer from non-active components of the load current. In this way, DVR structures presented in Figure 8.29b, with a shunt APF rectifier instead of a diode rectifier, can also serve the function of shunt-series or series-shunt compensators, popularly called Active Power Line Conditioner (APLC) [42–44], or UPQC [45, 46]. Figure 8.32 presents three different topologies UPQC with a load-side-connected shunt compensator.

The most often researched and discussed in the literature is UPQC of the first topology – typical (Figure 8.32a), for example [47–49]. In UPQC of the second topology (Figure 8.32b), convergent with the topology delta conversion UPS [50], a series compensator forces a sinusoidal line current i_S, and a shunt compensator provides a sinusoidal load-side voltage u'_L. Therefore, since the waveforms of the reference signals are also sinusoidal, the current control and voltage control of the instantaneous voltages and currents may be realized with very high accuracy [51], and without the need for identification of separate components. However, an advantage of the third topology UPQC (Figure 8.33c) is the lack of DC-link. This is very important when applying shunt and series compensators realized on the basis of multi-level inverters of topology HB or FC [52].

8.5 AC/AC Voltage Regulators

Most voltage sags on the supply system have a significant retained voltage, so that energy is still available, but at too low a voltage to be useful to the load. In these cases, besides DVR arrangements without energy storage, as series boosters for restoration of the load-side voltage, we can also apply an AC/AC Voltage Regulator (VR) [53, 54]. The overall structure and principle of operation of such three-phase VRs is presented in Figure 8.33, where the inserted voltages U'_C are in phase, and the inserted voltages U''_C are in quadrature with the terminal-supply voltage U_T during sags. Also possible are relations of VRs, with a load-side connected primary winding of the transformers, which we do not focus on. The two inserted voltages U'_C and U''_C, are also needed only when the phase angles of the voltages U_T during sags are not balanced, which is actually a requirement for load-side voltages U'_L, or when the "*presag*" strategy is applied.

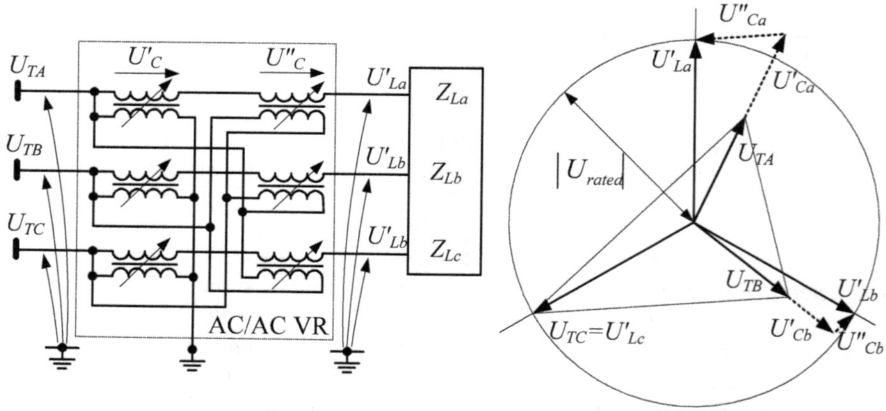

Figure 8.33. Overall structure and principle of operation of AC/AC voltage regulators

In practice, AC/AC VRs usually only allow insertion of the voltage U'_C in phase with the terminal-supply voltage U_T. If regulation of the magnitude and phase of the load-side voltage is required, then most often, even if considering the number of necessary series transformers (Figure 8.33), first of all DVR (Figure 8.29b) or UPQC (Figure 8.32) arrangements are applied. An alternative future for them may be VRs with matrix converters [55, 56]; however, so far solutions of this type have never been applied in practice. Furthermore, in this section we describe only three types of AC/AC VRs allowing only regulation of the magnitude of the load-side voltage, classified as follows:

- Electromechanical voltage regulators (as archetype modern solutions);
- Step voltage regulators (with multitaps transformers or autotransformers);
- Continuous voltage regulators (on the base PWM AC choppers).

An important point to note in the selection of a VR is that the chosen solution must solve the particular problem without creating additional problems.

8.5.1 Electromechanical Voltage Regulators

The principle of this type of VR is to control automatically an internal variable autotransformer to compensate for the variation of the magnitude of the terminal-supply voltage. The output of the autotransformer feeds the primary winding of a buck/boost transformer, of which the secondary winding is connected in series between the supply terminals and load terminals to inject an adding or opposing inserting voltage U_C into the supply line as shown in Figure 8.34.

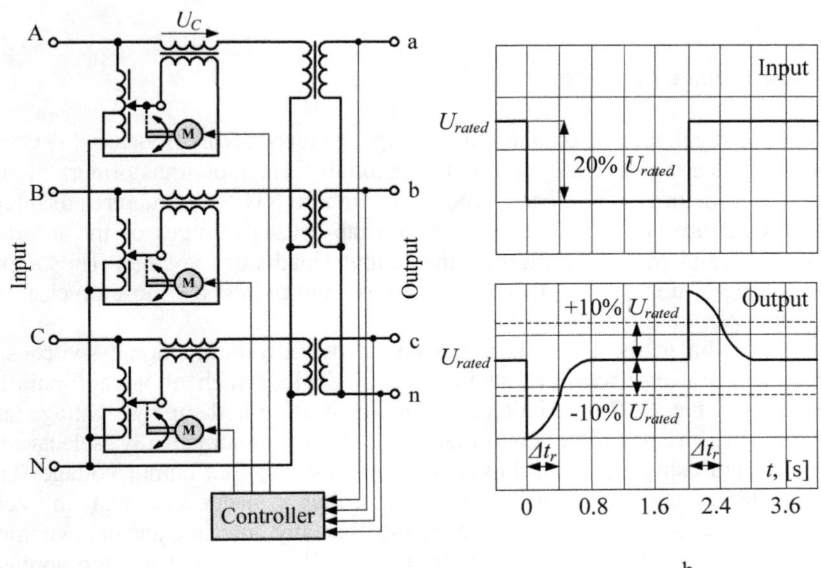

Figure 8.34. A typical electromechanical voltage regulator and its dynamic property: **a** circuit diagram; **b** changes of a output voltage

The three-phase electromechanical VRs are offered either as arrangements of operation common to each of the three phases of the feeder or as arrangements allowing individual regulation of the voltage magnitude independently in every phase. In the first arrangements one servomotor common to the three autotransformers is applied; however, in the second, each autotransformer must have its own servomotor, controlled independently. As presented in Figure 8.34a, an electromechanical VR realizes individual regulation. The load-side voltage of the electromechanical VR is monitored by individual controllers in each line. If this voltage deviates from the preset value due to a change in the terminal-supply voltage or the load current, the controller will drive an adequate servomotor, which then rotates the brush arm of the variable autotransformer in the required direction to boost or buck the input (terminal-supply) voltage until the correct preset value of the output (load-side) voltage is restored. This VR does not produce harmonics and therefore does not inject distortion into the incoming voltage supply.

The main advantages of electromechanical VRs are simple design, relatively low cost and noise immunity. The most important disadvantages, which show to some extent technical obsoleteness of these VRs, are moving parts, relatively large overall dimensions and weight, and what is more important, low regulation dynamics. For example, the dimensions, weight and regulation rate of a commercial three-phase electromechanical VR with individual regulation of power rating 45 kVA are: 134×93×63 cm, 340 kg and 45 ms/volt. As calculated during experimental research, the response time Δt_r of this VR, in the case of a 20% voltage change (U_{rated}=220 V), is approximately 400 ms (see Figure 8.34b). It is slower than step VRs and much slower than continuous VRs. Therefore, it is obvious that electromechanical VRs do not allow mitigation of rapid and short-time voltage sags and swells.

8.5.2 Step Voltage Regulators

Step-voltage regulators, also called Static Tap Changers Series Boosters (STCSB) operate by selecting separate taps of the autotransformer or transformer. Figure 8.35 shows structures of two basic types of one-phase STCSB (structures of three-phase STCSB are similar). A change of the tap causes changes of the inserted voltage U_C, and thus regulation of the output (load-side) voltage. The output voltage changes in steps, and therefore, the more tap points, the more precise can be the regulation of the output voltage.

Tap selection in the STCSBs is performed by fully bidirectional switches – contactors or power-electronics semiconductor devices such as an anti-parallel connection of the SCR. Variations in the input (terminal-supply) voltage are monitored by an electronic controller that in turn automatically selects the appropriate tap using a switch, thus maintaining the required output voltage. The instant of tap changing is phased by the controller to occur very near the zero crossing of the input voltage, thus ensuring that any interference or switching transients are reduced to a minimum. If, additionally, SCR switches are applied, then the response time of the STCSB with variations of the input voltage does not usually exceed one cycle (20 ms for the fundamental frequency, 50 Hz), and with

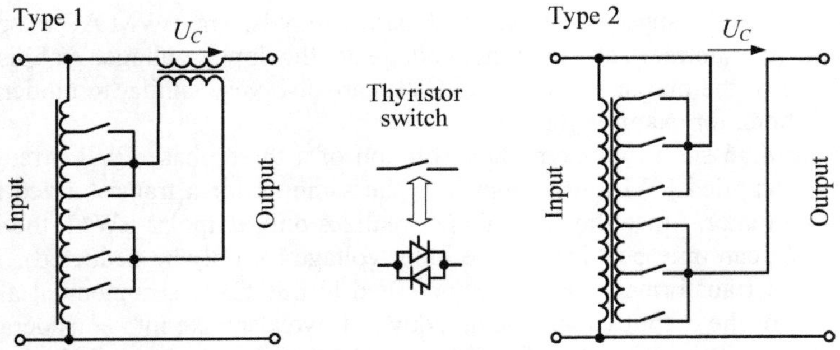

Figure 8.35. Basic types of static tap changer series boosters

special control algorithms, the time may even be reduced to about one half-cycle.

It should be stressed that an additional series-boost transformer, necessary in STCSB of type 1 (with autotransformer), is often used in STCSB of type 2 (Figure 8.35). In this case, tap-switching devices are insulated by a supply source and load, and this especially facilitates the working current and working voltage to fit the load parameters. Additionally, series-boost transformers are used in most of the improved precision of output-voltage regulation STCSBs arrangements. For example, in [57, 58], STCSBs are used, among others, for rapid voltage-magnitude regulation, *e.g.*, voltage-sag compensation. Modified STCSBs are also used as static phase shifters rated among FACTS [59, 60].

When summarizing STCSB arrangements, on the one hand we can see their advantages such as:

- Very high efficiency;
- Relatively small size and weight, and low cost;
- Relatively fast response, typically 1–1.5 cycles (20–30 ms).

On the other hand, they have some limitations in their application, resulting from the following characteristics:

- The voltage regulation (stabilization) is in steps;
- The output voltage tolerance is normally only ±3–5%.

These limitations are not valid for continuous-voltage regulators [61–63], which in addition characterize even faster response to input voltage changes.

8.5.3 Continuous-voltage Regulators

AC/AC Continuous Voltage Regulators (CVR), also called PWM (or self-commutated) AC voltage controllers (or stabilizers) or PWM AC boosters, are very fast and very tolerant power-electronics arrangements without moving parts and with no need for tap changing. The main component of CVRs is PWM AC chopper [64, 65]. The PWM AC chopper supplies a voltage with regulated magnitude to the primary winding of a series-boost transformer. The secondary winding of this transformer, similarly to former types of VRs, is connected in series between the supply terminals and load terminals. The PWM AC chopper can thus add (or subtract) an inserting voltage to the input voltage, stabilising the magnitude of the output voltage. The CVRs are also very similar to modern static phase shifters, for example [66].

Figure 8.36 shows an exemplary solution of a three-phase CVR arrangement, where the applied PWM AC chopper is the same as for a transistorized reactive current regulator. Since the AC chopper realizes only unipolar PWM, the inserted voltage U_{C1} can only be added to the input voltage (or only be deducted). The two series-boost transformers Tr_1 and Tr_2 applied to this CVR arrangement allow for regulation of the output voltage up and down. If we also take into consideration the transformer voltage ratio $1/n_1$ for Tr_1 and $1/n_2$ for Tr_2, it is not difficult to show that the output voltage can be expressed as follows

$$U_L' = U_T\left(1 - 1/n_2 + \delta/n_1\right) \tag{8.30}$$

where δ is the controllable duty cycle ($0 \le \delta \le 1$) of the switches S_{Aa}, S_{Bb}, and S_{Cc}. Hence, the control range of this CVR is the following

$$1 - 1/n_1 \le U_L'/U_T \le 1 - 1/n_1 + 1/n_2 \tag{8.31}$$

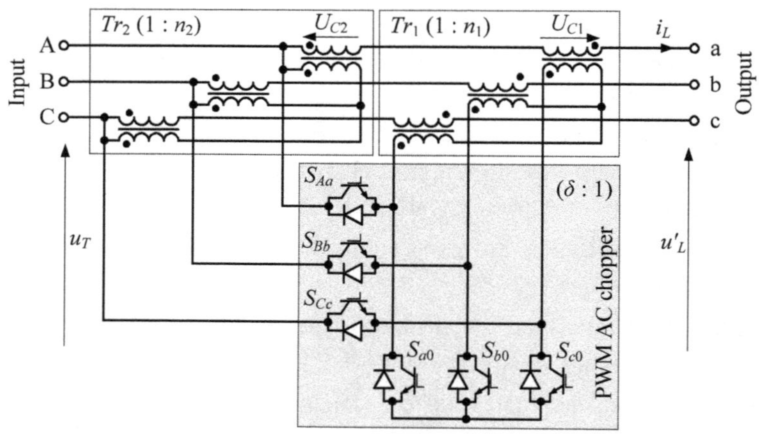

Figure 8.36. The structure of an example of the AC/AC transistored CVR

Figure 8.37 shows exemplary input (terminal-supply) and output (load-side) voltages and load-current waveforms for the CVR (Figure 8.36), in the case where the full control range of the output voltage is $0.7 \le U'_L/U_T \le 1.3$, *i.e.*, when the transformer voltage ratio are: $1/n_1=0.6$ and $1/n_2=0.3$. The presented waveforms apply to two values of the duty cycle: $\delta=0.2$ and $\delta=0.8$, with the switching frequency of the switches S_{Aa}, S_{Bb}, S_{Cc}, and S_{a0}, S_{b0}, S_{c0} only $f_i=2$ kHz. As regards high-harmonics disturbance, the switching frequency should be the highest possible.

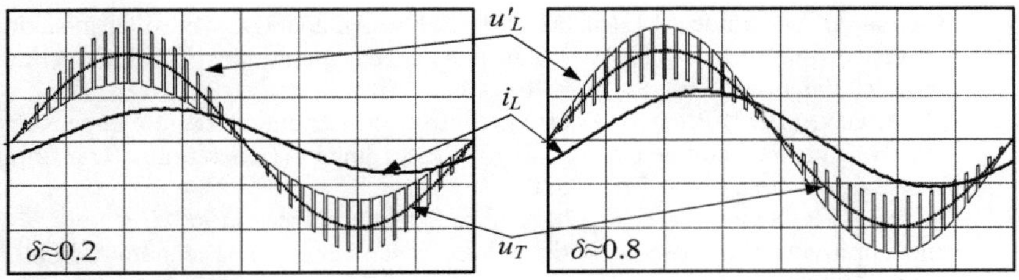

Figure 8.37. Input and output voltage and load-current waveform for the presented CVR

The significant disadvantage of CVR of the structure shown in Figure 8.36 is the inability of this arrangement to regulate independently the load-side phase voltages. This inability is associated with realization of three-phase switches, and thereby with the principle of operation applied to a PWM AC chopper [67, 68]. However, such a type of regulation allows CVR topologies with somewhat different PWM AC choppers applied; among them are also the topologies presented in Figure 8.38. In these CVRs, AC choppers are realized on the basis of single fully bidirectional switches with turnoff capability. We should note that a CVR of the topology shown in Figure 8.38a (without an additional series-boost transformer – see Figure 8.36) allows regulation of the phase output voltage only up and down, and a CVR of the topology shown in Figure 8.38b in both up as well as down. It depends on whether the three-phase PWM AC chopper allows either unipolar PWM or bipolar PWM [51, 62] to be applied.

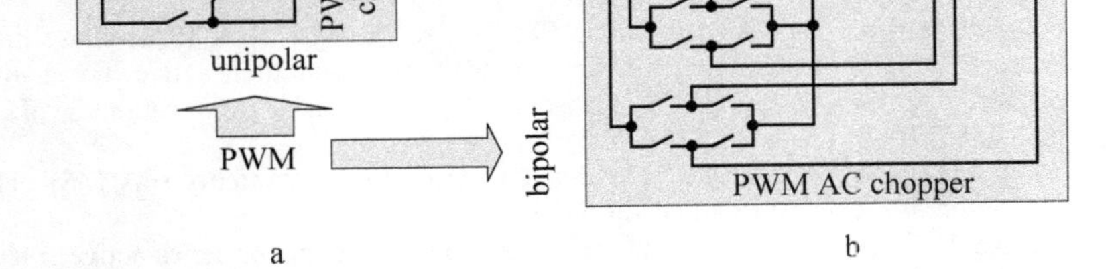

Figure 8.38. Examples of topology CVRs, allowing independent regulation in phases

References

[1] Blazic B, Papic I, (2006) Improved D-STATCOM control for operation with unbalanced currents and voltages. IEEE Transactions on Power Delivery, vol.21, no.1:225–233

[2] Escobar G, Stankovic AM, Mattavelli P, (2004) An adaptive controller in stationary reference frame for D-STATCOM in unbalanced operation. IEEE Transactions on Industry Electronics, vol.51, no.2:401–409
[3] Moon GW, (1999) Predictive current control of distribution static compensator for reactive power compensation. IEE Proceedings (Generation, Transmission, Distribution, vol.146, no.5:515–520
[4] Strzelecki R, Supronowicz H, (2000) Power factor correction in AC supply systems and improving methods. (in Polish) Warsaw University of Technology Publishing House
[5] Massoud AM, Finney SJ, Williams BW, (2004) Review of harmonic current extraction techniques for an active power filter. IEEE International Conference on Harmonics and Quality of Power:154–159
[6] Emadi A, Nasiri A, Bekiarov SB, (2005) Uninterruptible power supplies and active filters. CRC Press, USA
[7] Shin EC, Park SM, Oh WH, Kim DS, (2004) A novel hysteresis current controller to reduce the switching frequency and current error in D-STATCOM. Annual Conference of the IEEE Society:1144–1149
[8] Kincic S, Wan XT, McGillis DT, Chandra A, (2005) Voltage support by distributed static VAR systems (SVS). IEEE Transactions on Power Delivery, vol.20, no.2:1541–1549
[9] Haque MH, (2001) Compensation of distribution system voltage sag by DVR and D-STATCOM. IEEE Power Tech Conference, Porto, Portugal
[10] Lee SY, Wu CJ, (1997) Combined compensation structure of an SVC and an active filter for unbalanced three phase distribution feeders with harmonic distortion. Proceedings of the APSCOM International Conference:543–548
[11] Akagi H, (2004) Active filters and energy storage systems for power conditioning in Japan. Proceedings of International Conference on Power Electronics Systems and Application:80–88
[12] Ribeiro PF, Johnson BK, Crow ML, (2001) Energy storage systems for advanced power applications. IEEE Proceedings, vol.89, no.12:1744–1756
[13] Saminemi S, Johnson BK, Hess HL, (2006) Modelling and analysis of a flywheel energy storage system for voltage sag correction. IEEE Transactions on Industrial Applications, vol.42, no.1:42–52
[14] Schoenung SM, Burns C, (1996) Utility energy storage applications studies. IEEE Transactions on Energy Conversion, vol.11, no.3:658–665
[15] Oliveira da Silva SA, Donoso-Garcia PF, Cortizo PC, (2002) A Three-phase line-interactive UPS system implementation with series-parallel active power-line conditioning capabilities. IEEE Transactions on Industrial Applications, vol.38, no.6:1581–1590
[16] Song YH, Johns AT, (1999) Flexible AC transmission systems (FACTS). The Institution of Electrical Engineers, London, UK
[17] Tanaka T, Wada K, Akagi H, (1995) A new control scheme of series active filters. IPEC Conference:376–381
[18] Blaschke F, (1972) The principal of field orientation as applied to the new trans-vector close-loop control system for rotating-field machines. Siemens Review, vol.34:217–220
[19] Park RH, (1929) Two reaction theory of synchronous machines. Transactions AIEE, vol.48:716–730
[20] Soares V, Verdelho P, (1998) Instantaneous active and reactive current id-iq calculator suitable to active power filters. International PEMC Conference, vol.8:111–114

[21] Strzelecki R, Frąckowiak L, Benysek G, (1997) Hybrid filtration in conditions of asymmetric nonlinear load current pulsation. European Power Electronics and Applications Conference:1453–1458
[22] Aller JM, Bueno A, Paga T, (2002) Power system analysis using space-vector transformation. IEEE Transactions on Power Systems, vol.17, no.4:957–965
[23] Fortescue CL, (2002) Method of symmetrical coordinates applied to the solution of polyphase networks. Transactions AIEE, vol.37:1027–1140
[24] Clarke E, (1943) Circuit analysis of AC power systems. Wiley, New York
[25] Ferrero A. Leva S, Morando AP, (2004) A systematic, mathematically and physically sound approach to the energy balance in three-wire, three-phase systems. L'Energia Elettrica, vol.81:51–56
[26] Ferrero A, Leva S, Morando AP, (2000) About the role of the Park imaginary power on the three-phase line voltage drop. ETEP, vol.10, no.5:287–286
[27] Akagi H, Kanazawa Y, Nabae A, (1984) Instantaneous reactive power compensators comprising switching devices without energy storage components. IEEE Transactions on Industrial Applications, vol.IA-20, no.3:625–630
[28] Peng FZ, Lai JS, (1996) Generalized instantaneous reactive power theory for three-phase power systems. IEEE Transactions on Instrumentation and Measurement, vol.45, no.1:293–297
[29] Vilathgamuwa DM., Perera AADR., Choi SS, (2003) Voltage sag compensation with energy optimized dynamic voltage restorer. IEEE Transactions on Power Delivery, vol.18, no.3:928–936
[30] Akagi H, Ogasawara S, Kim H, (1999) The theory of instantaneous power in three-phase four-wire systems. A comprehensive approach. IEEE IAC Conference, vol.1:431–439
[31] Cardenas V, Moran L, Bahamondes A, Dixon J, (2003) Comparative analysis of real reference generation techniques for four-wire shunt active power filters. IEEE Power Electronics Specialist Conference:791–796
[32] Kim H, Blaabjerg F, Bak-Jensen B, Choi J, (2002) Instantaneous power compensation in three-phase systems by using p-q-r theory. IEEE Transactions on Power Electronics, vol.17, no.5:701–710
[33] Bhavaraju VB, Enjeti PN, (1996) An active line conditioner to balance voltages in a three-phase system. IEEE Transactions on Industrial Applications, vol.32, no.2:287–292
[34] Strzelecki R, Supronowicz H, (1997/1999) Filtration of the harmonic in AC supply systems. (in Polish) Adam Marszałek Publishing House, Poland
[35] Fitzer C, Arulampalm A, Barnes M, Zurowski R, (2002) Mitigation of saturation in dynamic voltage restorers connection transformers. IEEE Transactions on Power Electronics, vol.17, no.6:1058–1066
[36] Woodley NH, Morgan L, Sundaram A, (1999) Experience with an inverter-base dynamic voltage restorer. IEEE Transactions on Power Delivery, vol.14:1181–1185
[37] Jauch T, Kara A, Rahmani M, Westermann D, (1998) Power quality ensured by dynamic voltage correction. ABB rev., vol.4
[38] Kim H, (2002) Minimal energy control for a dynamic voltage restorer. IEEE PCC Conference, vol.2
[39] Vilathgamuwa DM, Perera AADR, Choi SS, (2003) Voltage sag compensation with energy optimized dynamic voltage restorer. IEEE Transactions on Power Delivery, vol.18, no.3:928–936
[40] Didden M, (2003) Techno-economic analysis of methods to reduce damage due to voltage dips. Ph.D. thesis, Catholic University of Leuven, Leuven, Belgium

[41] Nilsen JG, Blaabjerg F, (2005) A detailed comparison of system topologies for dynamic voltage restorers. IEEE Transactions on Industrial Applications, vol.41, no.5:1272–1280

[42] Aredes M, Heumann K, Watanabe EH, (1998) A universal active power line conditioner, IEEE Transactions on Power Delivery, vol.13, no.2:545–551

[43] Strzelecki R, Klytta M, Frąckowiak L, Rusiński J, (1999) Power flow in APLC topologies. Proceedings of the Electrical Power Quality Utilization Conference:391–398

[44] Strzelecki R, Kukluk J, Suproniwicz H, Tunia H, (1999) A universal symmetrical topologies for active power line conditioners. European Power Electronics and Applications Conference

[45] Fujita H, Akagi H, (1998) The unified power quality conditioner: the integration of series- and shunt-active filters. IEEE Transactions on Power Electronics, vol.13, no.2:315–322

[46] Emadi A, Nasiri A, Bekiarov SB, (2005) Uninterruptible power supplies and active filters. CRC Press, Boca Raton, USA

[47] Farrukh Kamran F, Habetler TG, (1998) Combined deadbeat control of a series-parallel converter combination used as a universal power filter. IEEE Transactions on Power Electronics, vol.13, no.1:160–168

[48] Strzelecki R, Benysek G, Rusiński J, Dębicki H, (2005) Modeling and experimental investigation of the small UPQC systems. IEEE Compatibility in Power Electronics Conference

[49] Han B, Bae B, Baek S, Jan G, (2006) New configuration of UPQC for medium-voltage application. IEEE Transactions on Power Delivery, vol.21, no.3:1438–1444

[50] Oliveira da Silva SA, Donoso-Garcia PF, Cortizo PC, Seixas PF, (2002) A three-phase line-interactive UPS system implementation with series-parallel active power-line conditioning capabilities. IEEE Transactions on Industrial Applications, vol.38, no.6:1581–1590

[51] Kaźmierkowski MP, Krishnan R, Blaabjerg F, (2002) Control in power converters. Selected problems. Academic Press, San Diego, USA

[52] Jin Wang J, Peng FZ, (2004) Unified power flow controller using the cascade multilevel inverter. IEEE Transactions on Power Electronics, vol.19, no. 4:1077–1084

[53] Dmowski A, (1983) AC voltage regulation. Selected systems. (in Polish), WNT, Warsaw, Poland

[54] Montenero-Hernandez OC, Enjeti PN, (2000) Application of a boost AC/AC converter to compensate for voltage sags in electric power distribution systems. Power Electronics Specialists Conference, vol.1:470–475

[55] Gyugyi L, Pelly BR, (1976) Static frequency changers. John Wiley, New York

[56] Strzelecki R, Noculak A, Tunia H, Sozański K, Fedyczak Z, (2001) UPFC with matrix converter. EPE Conference

[57] Demiric O, Torrey DA, Degeneff RC, Schaeffer FK, Frazer RH, (1988) A new approach to solid-state on load tap changing transformer. IEEE Transactions on Power Delivery, vol.13, no.3:952–961

[58] Lipkowski KA, (1983) Transformer-switches performance topologies of the AC/AC converters. (in Russian), Naukova Dumka, Kiev, Ukraine

[59] Hingorani NG, Gyugyi L, (1999) Understanding FACTS. Concepts and technology of flexible AC transmission systems. New York, IEEE Press

[60] Iravani MR, Maratukulam D, (1994) Review of semiconductor controlled (static) phase shifters for power systems applications. IEEE Transactions on Power Systems, vol.9, no.4:1833–1839

[61] Mozdzer AJ, Bose BK, (1976) Three-phase AC power control using power transistors. IEEE Transactions on Industrial Applications, vol.IA-12:499–505
[62] Hamed SA, (1990) Modeling and design of transistor-controlled AC voltage regulators. International Journal Electronics, vol.69, no.3:421–434
[63] Kwon BH, Jeong GY, Han SH, Lee DH, (2002) Novel line conditioner with voltage up/down capability. IEEE Transactions on Industrial Electronics, vol.49, no.5:1110–1119
[64] Fedyczak Z, Strzelecki R, (1997) Power electronics agreement for AC power control. (in Polish) Adam Marszałek Publishing House, Poland
[65] Fedyczak Z, Strzelecki R, Benysek G, (2002) Single phase PWM AC/AC semiconductor transformer topologies and applications. Power Electronics Specialists Conference, vol.2:1048–1053
[66] Lopes LAC, Jóos G, Ooi BT, (1998) A high-power PWM quadrature booster phase shifter based on a multimodule AC controller. IEEE Transactions on Power Electronics, vol.13, no.2:357–365
[67] Vincenti D, Jin H, Ziogas P, (1994) Design and implementation of a 25kVA three-phase PWM AC line conditioner. IEEE Transactions on Power Electronics, vol.9, no.4:384–389
[68] Strzelecki R, Fedyczak Z, (1996) Properties and structures of three-phase PWM AC power controllers. Power Electronics Specialists Conference, vol.1:740–746

9

Energy Storage Systems

Piotr Biczel

Warsaw University of Technology
Institute of Electrical Power Engineering
75 Koszykowa Street, 00-662 Warsaw, Poland.
Email: Biczel@ee.pw.edu.pl

9.1 Introduction

There is no doubt the near future will drive power generation towards DG and RES. Distributed generation units are not controlled and regulated by the power system operators. The power produced in DG units is optimized by the owner in order to maximize the profit, heat production or renewable sources utilization. This is done regardless of the power demand in the networks to which the sources are connected. If the DG utilization reaches more then dozen or so percent of the power production, the problem of power balancing will occur either in the local grid or in the whole system. The problem will become even bigger, if, as EU officially recommends, the power generation is based mainly on renewable sources, such as wind turbines or solar energy. Employment of renewable sources will cause not only decrease in the power quality, but also complications in balancing the power system.

It needs to be stressed that the power quality of both distributed and renewable generation is the strategic aim of EU in the area of sustainable development.

No reaction to the problems mentioned above will result in a decline of power quality or increased power losses. Voltage and frequency fluctuations can be particularly dangerous. If acceptable levels of voltage and current frequency are exceeded, this causes damage in the devices and the flicker effect occurs. Active and passive power flows of the intersystem energy will also increase. In extreme situations the power system may become unstable or even disintegrate.

The quality of energy and the stability of the system are further influenced by the widespread application of non-linear receivers. The components of higher harmonics generated in the network by such receivers cause voltage deformations of series impedances on the transmission lines, as well as increased losses in the cores of the transformers or other magnetic elements. Difficulties in power quality maintenance caused by local resonances may result in the failure of the passive power compensation units.

All the above-mentioned problems are caused by a similar phenomenon – the lack of correlation between the generated power and the load. The causes and effects of the problems depend on the time perspective:

- In the long-term perspective, *i.e.*, changes of the seasons of the year, what changes are the power demand the power plant productive power;
- In the medium-termed perspective (hourly changes);
- In the short-term perspective (seconds and milliseconds).

Considering that power systems have practically no storage capacity, the above-mentioned problems are nowadays solved by regulating production. The classic thermal systems have too small dynamics and power to satisfy the growing regulating needs. What is more, they do not support equal distribution of the remedial measures among the nodes of the power network. In the early 1970s it was already estimated that the U.S. power system would have operated most effectively if 6–12% of their generation capacity had been installed in storage systems. Only 3% were installed [1].

Energy storage may solve the problem. If the power system is incapable of storing energy, a storage device has to be employed. At present many techniques of energy storage are known. Each has its own properties, from high storage capacities, through high power, to the response time of milliseconds.

The paper [2] identifies areas of power storage utilities in power systems according to their generation, transmission and power utilization.

As far as power generation is concerned, power storage may be applied in the following areas:

- Rapid reserve – generation capacity that a utility holds in reserve to meet electric reliability requirements to prevent interruption of service to customers in the event of a failure of an operating generating station;
- Area control and frequency responsive reserve – the ability for grid-connected utilities to prevent unplanned transfer of power between themselves and neighboring utilities and the ability of isolated utilities to respond instantaneously to frequency deviations (frequency responsive reserve);
- Commodity storage – storage of inexpensive off-peak power for dispatch during relatively expensive on-peak hours. Commodity Storage refers to applications that require less than 4 h of storage.

As far as transmission and distribution of power is concerned, one can differentiate:

- Transmission system stability – ability to keep all components on a transmission line in synchronization with each other and prevent system collapse;
- Transmission voltage regulation – ability to maintain the voltages at the generation and load ends of a transmission line within 5% of each other;

- Transmission facility deferral – ability of a utility to postpone installation of new transmission lines and transformers by supplementing the existing facilities with another resource;
- Distribution facility deferral – ability of a utility to postpone installation of new distribution lines and transformers by supplementing the existing facilities with another resource.

As far as customer services are concerned, one should remember:

- Customer energy management – dispatching energy stored during off-peak or low cost times to manage demand on utility-sourced power;
- Renewable energy management – applications through which renewable power is available during peak utility demand (coincident peak) and available at a consistent level;
- Power quality and reliability – ability to prevent voltage spikes, voltage sags, and power outages that last for a few cycles (less than 1 s) to minutes from causing data and production loss for customers.

This part of the book presents some of the techniques of storing the electric power, as well as their possible applications.

9.2 The Structure of Power Storage Devices

As direct storage of electricity is impossible, it has to be converted to another kind of energy. Figure 9.1 presents the idea of energy storage.

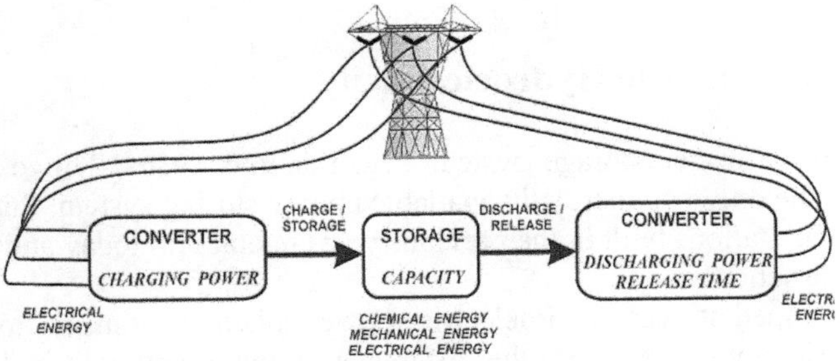

Figure 9.1. The idea of storing electricity [3]

Table 9.1 shows the known techniques and forms of storing electricity. Electricity may also be stored in the form of heat.

The following parameters are important for employment an energy storage device:

- Amount of the stored energy;
- Power;
- Storage period;
- Speed of charging and discharging.

Table 9.1. Options for storing electricity [4, 5]

Mechanical storage	Pumped water storage Compressed air storage Flywheels	Two converting stages
Electrochemical storage	Accumulators with internal storage Accumulators with external storage Hydrogen systems	Two converting stages
Electrical storage	Superconducting coils Capacitors (different technologies)	No converting stages
Thermal storage	Hot water PCM device	Three converting stages

The possible amount of stored energy depends on the kind of storage device employed as well as on its properties. The power input and the released power depend on the properties of energy converters. On the other hand, both employed storage device and the properties of energy converters are the decisive force of the dynamic properties of the stored energy.

What is important in practice is the number of energy conversion cycles necessary to store or receive the energy from the storage device. This number is dependent upon the kind of the employed storage device. Table 9.1 presents the number of conversions for various electricity-to-electricity storage techniques [4].

The most important storage techniques are briefly described in the following sections.

9.3 Pumped-storage Hydroelectricity

The history of water storage system began in the 1930s. Up to 1970 it was practically the only commercially available energy storing system. Pumped-storage hydroelectric stations built in the previous years operate till today and new ones are under construction.

The pumped-storage hydroelectric power plant is similar to the classic hydroelectric power plant with the difference of employing both hydro-generators and electric pumps. The water storage system of such a power plant consists of two large reservoirs located on two different elevations. The reservoirs are connected by shafts. In older stations two pipelines were built – the pumping system and hydro-generators. New pumped-storage hydroelectric power plants employ bi-directional shafts, turbines and electric machines. In consequence, in order to lower the investment costs, the number of shafts and pieces of equipment has to be reduced twice.

The principle of operation of a pumped-storage hydroelectric power plant involves changing the potential energy of the water into its kinetic energy, and than further into electricity. The peaking pumped-storage hydroelectric power plant consists:

- Upper reservoir;
- Lower reservoir;
- Turbines and hydro-generators;
- Pumps and electric motors.

The upper reservoir is situated higher than the lower reservoir. The vertical distance between the reservoirs which the water has to cover is called *"head"*.

Pumping the water from the lower into the upper reservoir, which takes place only when energy storage is needed, charges the system through increasing the potential energy of the water.

When water is released from the upper into the lower reservoir, its potential energy changes into kinetic energy under the influence of the gravitational force. The flowing water propels the turbine and hydro-generator, which results in electric power generation.

The peaking pumped-storage hydroelectric power plant usually operates in one of the following four modes:

- Power storage mode – water exclusively pumped into the upper reservoir;
- Power generation mode – water is exclusively released to the lower reservoir;
- Spinning reserve mode – the power plant is charged, *i.e.*, the water is stored in the upper reservoir; the power plant awaits the generation start-up command, while hydro-units rotate;
- Short-circuit mode – the water is pumped and released at the same time, the power is simultaneously stored and released.

The last two modes of operation are used to regulate and sustain the quality of power in the power system. The basic parameters of the peaking pumped-storage hydroelectric power plant are:

- Capacitance – the amount of possibly stored energy;
- Pumping power – the maximum possible power which the power plant impose on the power electronic system;
- Generation power – the maximum power which the power plant can transfer to the system.

The capacitance is closely related to the cubic capacity of both reservoirs, and the head. It is approximately described by the formula

$$E = \rho V g H \tag{9.1}$$

where H – head; g – gravitational acceleration; V – cubic capacity of the reservoir; ρ – water density.

Depending on the chosen solution, storing capacity is determined by the cubic capacity of either the upper or the lower reservoir. In practice, one should also take into account the efficiency of pumping, power generation, as well as output efficiency of the power plant. The overall efficiency of a peaking pumped-storage hydroelectric power plant amounts to around 70–80%.

The dynamic properties of the power plant depend on the following:

- Time span of the transition from stand-by to generation start-up;
- Time span of the transition from pumping to generation and *vice versa*.

The two decide the manner of employing the peaking pumped-storage hydroelectric power plant in a power electronic system.

In reality, the pumped-storage hydroelectric power plant has very good dynamic properties. Generators, if they are initially spinning in air, can reach the rated power in just a dozen or so seconds. If they start from standstill, generators can be brought to full power in 1 or 2 min.

The basic requirement for building a peaking pumped-storage hydroelectric power plant is the necessity for the appropriate head and adequate cubic capacity of the upper reservoir. In order to obtain the same capacitance and power storage efficiency, one needs proper head, or considerable cubic capacity of the upper reservoir. On this basis, three main types of peaking pumped-storage hydroelectric power plants can be differentiated:

- Power plants taking advantage of the natural elevation differences between the reservoirs with shafts connecting the two laid inside the mountain (Figure 9.2);
- Power plants taking advantage of the natural or artificial elevation differences between the reservoirs with pipes connecting the two laid on the surface (Figure 9.3);
- Power plant built at a river, with two dams creating the upper and lower reservoirs (Figure 9.4).

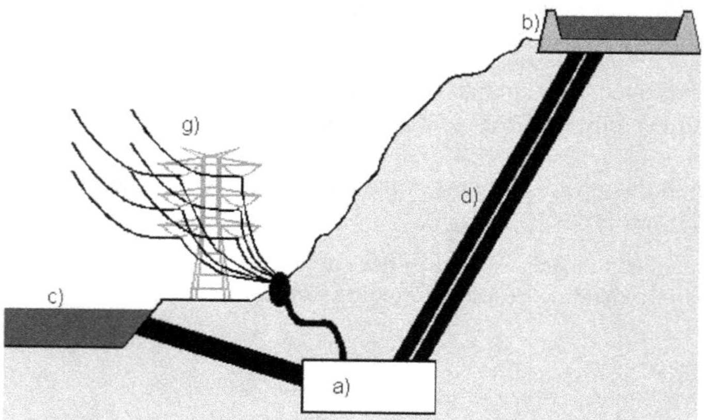

Figure 9.2. Schematic diagram of pumped-storage hydroelectric power plant with shafts connecting reservoirs laid inside the mountain: a) engine room with hydro-units, b) upper reservoir, c) lower reservoir, d) shafts connecting the reservoirs, g) power terminal

Older power plant designs used separate shafts and pumping machines for water release. Nowadays, reversible hydro-units are most commonly employed, which simplifies the structure of the power plant and lowers the construction costs.

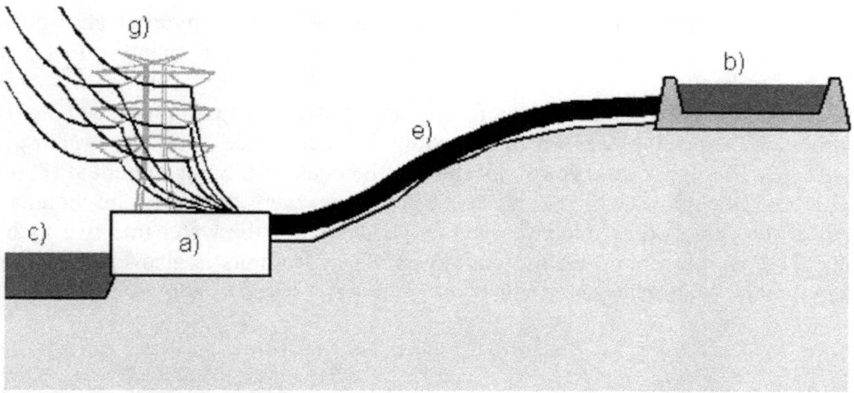

Figure 9.3. Schematic diagram of pumped-storage hydroelectric power plant with pipes connecting reservoirs laid on the surface: a) engine room with hydro-units, b) upper reservoir, c) lower reservoir, e) pipes connecting the reservoirs, g) power terminal

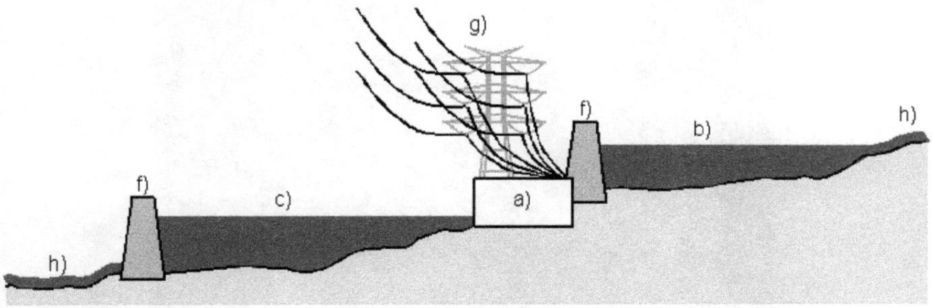

Figure 9.4. Schematic diagram of pumped-storage hydroelectric power plant built at a river: a) engine room with hydro-units, b) upper reservoir, c) lower reservoir, f) dam, g) power terminal, h) river

Figure 9.5. View of the upper and lower reservoirs of Porąbka-Żar power plant [7]

The list of almost all operating peaking pumped-storage hydroelectric power plants can be found in [6]. Some examples of typical power plant structure are presented below.

Figure 9.5 presents the view of the Porąbka-Żar peaking pumped-storage hydroelectric power plant, which was built in accordance with the structure in Figure 9.2 The upper reservoir is located on the peak of the Żar mountain (850 m ASL). The two shafts which connect the reservoirs were drilled in the mountain. Each of the shafts splits into two channels which carry the water into two hydro-units. The engine room with its four reversible hydro-units is also located inside the mountain. Man-made Lake Międzybrockie is employed as a lower reservoir.

Figure 9.6. View of the Żarnowiec power plant [8]

Figure 9.6 presents the view of the Żarnowiec peaking pumped-storage hydroelectric power plant, which was built in accordance with the structure in Figure 9.3. In the picture one can see the man-made upper reservoir, pipes connecting the two reservoirs, the engine room building, and the channel which carries the water away to the lower reservoir, Lake Żarnowiec. The Żarnowiec power plant employs four reversible hydro-units. It is the biggest peaking pumped-storage hydroelectric power plant in Poland.

Nidzica, and Solina-Myszkowice power plants serve as examples of power plants designed in accordance with Figure 9.4. This type of investment often concentrates on fire protection rather than power generation.

Figure 9.7. Solina power plant [9]

Figure 9.7 presents a view of the Solina power plant built at the River San. In the upper part of the picture one can see the upper reservoir – the man-made Lake Solina. Next to the lake is the engine room building. Man-made Lake Myczkowice performs the role of the lower reservoir. The power plant employs two hydro-units with generators and two reversible hydro-units. The overall generated power equals 200 MW. Due to the cubic capacity of the lower reservoir, the Solina power plant can operate in the power generation mode at full rating for up to 5–6 h.

Figure 9.8. Niedzica power plant (courtesy of ZEW Niedzica) [10]

Figure 9.8 presents a view of the Niedzica power plant, which is built on the River Dunajec. The upper reservoir is the man-made Lake Czorsztyńskie, the tail water reservoir – man-made Lake Sromowieckie Wyższe.

Pumped-storage systems are currently the most effective. They can store thousands of MWh for as long as six months. Due to their rapid response speed, hydroelectric storage systems are particularly useful as a backup in case of sudden demand changes.

Partly due to the large storage capacity and the relative simplicity of structure, the operating cost of a pumped-storage system is among the lowest for all storage systems. The cost of energy storage can be lower in pumped hydroelectric systems than in other systems. Unlike hydroelectric dams, a pumped hydroelectric system has minimal effect on the landscape. What is more, it produces no pollution or waste.

The main drawback of the pumped-storage system is its dependence on geological formations. There have to be two large-volume reservoirs along with a sufficient amount of head for the investment to be feasible. As a result, the location of the plant tends to be remote from human settlements, *e.g.* mountains, where construction is difficult and the distance to the nearest power grid huge.

There are almost 300 operating pumped hydroelectric power plants all over the world and new ones are under construction. Instead of using two different altitude over-ground reservoirs, underground pumped hydroelectric storage facilities are now being developed. Excavating the lower reservoirs out of solid rock, or employing old mines is taken into consideration. These new facilities can be completely isolated from natural water sources. Moreover, they can carry the same water up and down the huge vertical distance, thus giving greater energy per unit volume than a natural system.

9.4 Compressed Air Energy Storage System

The structure of Compressed Air Energy Storage (CAES) systems is similar to that of water pumped storage systems, with the difference that air, not water, is used as the storage medium. The structure of CAES is also simpler than that of water storage systems as only one tank has to be built. Off-peak power is taken from the grid to supply the electric motor which drives compressors. Compressors pump air into the sealed cavern, be it a naturally occurring aquifer, solution-mined salt cavern, or a constructed rock cavern. The cheapest solution is the aquifer storage. In case of peak demand, the turbine is driven by compressed air.

The air is compressed to about 75 bar in underground caverns in off-peak demand periods. During compression the air gets warm (in accordance with the laws of thermodynamics). That is why, for the best use of space, the air has to be cooled prior to injection.

During peak demand, electricity is produced by releasing the air from the mine. During decompression the air gets cooler, and the pressure produced in this process drives the turbine. However, the temperature might get too low and damage the devices. Hence, the whole process has several stages. First, the air is preheated in the recuperator, which uses the heat from the combustion exhaust. Then the compressed air is mixed with small quantities of natural gas or oil. The mixture is burnt in the combustor. The hot gas expands in the turbine and generates electricity.

In theory, the process should be isothermal; in practise, as was mentioned before, it is impossible due to the technical and thermo- dynamical limitations. The assumed maximal stored energy is described by Equation 9.3. The state of ideal gas is described by Equation 9.2. For isothermal process the value is constant

$$pV = nRT \tag{9.2}$$

where p – gas pressure; V – gas volume; n – number of gas moles; R –Boltzmann constant; T – gas temperature.

According to the definition, the work performed by the gas during isothermal process equals

$$W_{AB} = nRT \ln\left({p_A}/{p_B} \right) \tag{9.3}$$

keeping in mind the relation for isothermal processes

$$p_A V_A = p_B V_B \tag{9.4}$$

Equation 9.3 has to be reduced by the efficiency of individual devices and the fact that neither the compression nor the decompression is an isothermal process.

The economics of CAES is based on the time shift of the air compression. In the usual turbine assemblies the compressor absorbs up to two-third of the energy. If the compression does not take place at the time of power generation, the peak power which the gas turbine can generate is increased by the amount usually absorbed by the compressor. The profit is based on the price differences between the on-peak and off-peak demands. Typical CAES consists of five basic subsystems (Figure 9.9):

- Reversible electric motor/generator;
- Two or more stages of air compressor with intercooler and after-coolers;
- The recuperator and high and low pressure turbines;
- Sealed cavern;
- Control systems, fuel storage systems, grid connection, *etc.*

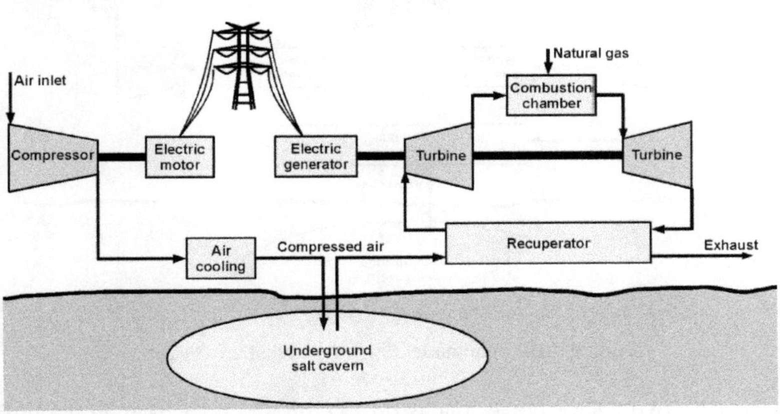

Figure 9.9. Schematic diagram of CAES

There are two main CAES assemblies, with one or two shafts. In a one-shaft assembly the electric device operates in reverse. In the charging mode (*i.e.*, air compression) the device works as an engine and drives the compressors. The shaft is divided into parts connected with clutches. While charging, the electric engine and the compressors are adjoined by the clutch. While discharging, compressors are disconnected from the electric device and the clutch joins the device and the gas turbine. The turbine drives the device, which then plays the role of a generator.

In a two-shaft assembly, two systems operate: one connects the electric engine and the compressors, the other the turbine and the generator. As far as the amount of stored energy and the storage period are concerned, CAES does not differ much from the peaking pumped-storage system. Both systems were designed to store great amounts of power for periods up to several months.

CAES also has good dynamic properties. Under normal conditions it can come to its rated power in about 12 min, and from an emergency start in about 9 min. Moreover, if natural geological formations are used, CAES does not require huge investment expenditure. Basic restrictions are the laws of thermodynamics and the issue of maintaining the isothermal process. As the latter is in reality impossible, heat has to be provided for the process of air decompression, and dissipated in the process of air compression. As a consequence, the whole system cannot be called a pure storage system. In order to release the stored energy, combustion in the gas turbine assembly has to be employed.

However, there is another idea of an adiabatic CAES system [11]. The structure of such a system is presented in Figure 9.10. After compression, the air is cooled and the energy generated in the process of cooling is not lost but stored in thermal storage. The stored heat is then used to warm up the air in the process of decompression. Such a solution does not require combustion for turbine-based power generation. There is no extra fuel consumption. Still, the structure of the system becomes particularly complicated.

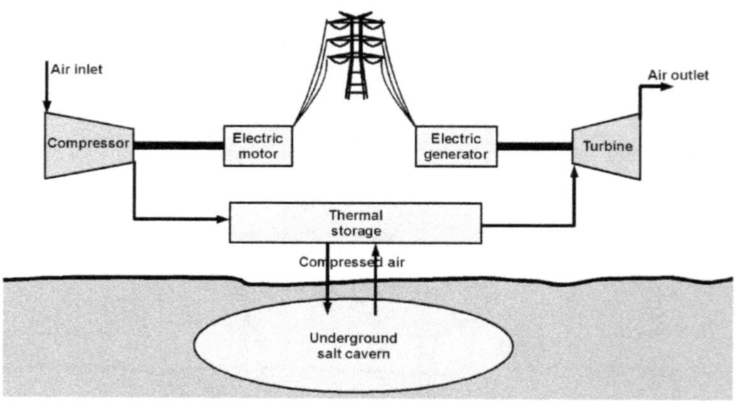

Figure 9.10. Schematic diagram of adiabatic CAES

At present, two CEAS power plants operate on a regular basis in the traffic of the power electronic system: Huntorf power plant in Germany and McIntosh power plant in Alabama, USA. The first commercial CAES power plant in the world is the German 290MW plant located in Huntorf (Figure 9.11), which has been operating since 1978 [12].

Figure 9.11. The view of CAES in Huntorf, Germany [12]

The Huntorf plant runs on a daily cycle, in which the air storage charges in 8 h cycles and power generation takes 2 h. During that process the pressure in the tank rises from 46 do 66 (max. 72) bar at the speed of 2.5 bar/h. In the 2 h discharge process the pressure falls back to 46 bar at a speed of 10 bar/h. The power plant has reported availability of 86% and a starting reliability of 98%. The Huntorf power plant employs a salt cavern as a storage tank.

Figure 9.12. The view of CAES in McIntosh, Alabama, USA [13]

The Alabama Electric Co-operative Inc. built the other commercial CAES power plant in McIntosh, AL, USA (Figure 9.12). It has been operating since 1991 and provides 110 MW of power generation. The McIntosh plant cavern, mined from a salt dome, supplies compressed air which supports the generation of 100 MW for as long as 26 h. Two more CAES plants are under construction – a 35 MW CAES unit in Japan and a 100 MW one in Israel.

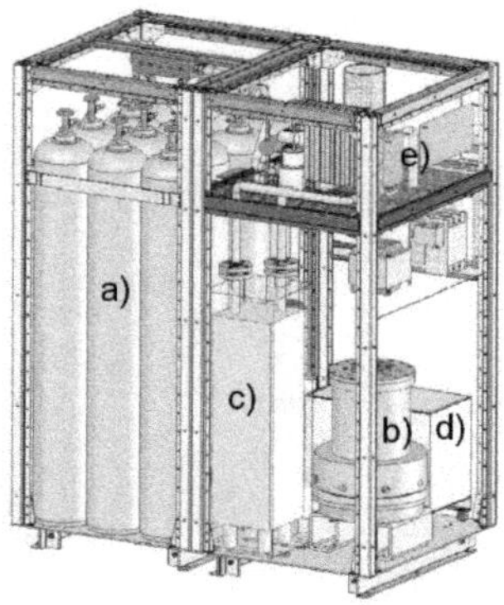

Figure 9.13. The view of an idea of a small CAES applied in UPS [14]: a) air tank, b) engine/generator, c) heat storage, d) compressor, e) power electronic devices and controllers

Compressed Air Energy Storage is mainly used to store great quantities of energy for long periods of time. Until now, CAES systems have been considered to be large installations with great significance for the huge power system. Nowadays, many researchers investigate small CAES systems' application in dispersed power generation and UPS, as an alternative to battery storage systems.

Figure 9.13 presents an idea of CAES applied in UPS [14]. Its structure is similar to that of an AA-CAES. It includes the thermal storage which stores the heat used to warm up the air before passing it through the turbine. The system is supposed to have following advantages over chemical batteries:

- Very long work time, *ca.*, 20 years;
- Possibility of operation in a standard cabinet, *e.g.*;
- Full power support time of up to 15 min for a 100 kVA UPC;
- Work temperature range 0–40°C;
- High infallibility;
- Non-toxic and recyclable parts.

Other authors [13, 15] postulate construction of units with power *ca.*, 1MW in a classic design. Such power is, according to estimates, adequate for distributed generation. Application of existing turbo-compressors for engines is also postulated.

To sum up, CAES are economically effective only on the large scale. As far as storage capacity is concerned, only peaking pumped-storage systems can compete with CAES. Due to very small losses, CAES can store energy for more than a year.

9.5 Flywheels

Flywheels are probably the oldest and most popular method of storing energy. First employed in windmills, they later found application in steam engines. Every modern car engines makes use of a flywheel, which stores energy to change torque level (caused by the engine cycles). We are not aware of how common flywheels are. Most of us encountered them for the first time in childhood, when we played with power drive toy cars.

The structure of a flywheel employed for electricity storage is distinct from the structure of the same element used for different purposes. In energy storage, designers aimed at maximizing efficiency, capacity, storage period, dynamic properties, and energy density of the system. As a result, the only similarity between flywheels employed for energy storage and those employed in, *e.g.*, steam engines is the principle of operation.

The principle of operation for flywheel in the energy storage system lies in setting the disk in rotation. The disk gains speed due to employment of an electric motor. The electric energy of the motor is then transformed into the kinetic rotational energy of the disk, which is further stored in the accelerated mass. When there is need to discharge the energy, the flywheel drives the electric generator. In the storing mode the flywheel inside the accumulator rotates. The maximum energy that theoretically can be stored in a rotating mass amounts is

$$E = \frac{J\omega^2}{2} \tag{9.5}$$

where J – moment of inertia of the flywheel, ω – rotational speed of the flywheel.

In order to increase the amount of stored energy, it is more effective to increase the rotational speed rather than the mass of a flywheel, as the former is in square power. A typical flywheel storage system consists of (Figure 9.14):

- Flywheel;
- Electric motor/generator;
- Power electronic devices;
- Bearings;
- Casing.

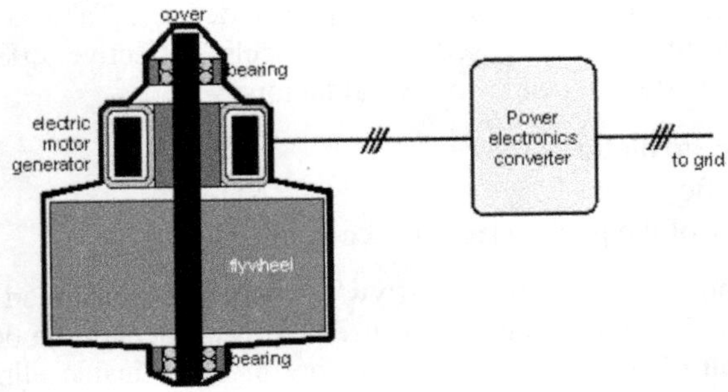

Figure 9.14. The flywheel's structure

In the past, an assembly of two electric machines was used; one drove the flywheel, the other played the role of a power generator. It was the consequence of different demands for generation mode and machine work. Nowadays, the application of heavy duty power electronic converters enable us to separate the amplitude and frequency of the voltage which drives the engine from the generated one, as well as from the voltage which is transmitted to the supply network. As a result, the rotational speed of the rotating mass can be increased. What is more, independence of the frequency of the supply network enables the increase in the range of the flywheel's rotational speed variations. In consequence, the flywheel can be discharged deeper. These days, the minimum to maximum speed ratio equals, *e.g.*, 3:1, which allows discharge of up to 90% of the stored energy.

Table 9.2. Properties of materials used for flywheels [16]

	Density [kg/m^3]	**Strength [MN/m^2]**	**Theoretical maximum specific energy [Wh/kg]**
Steel (AISI 4340)	7800	1800	32
Alloy (AlMnMg)	2700	600	31
Titanium (TiAl6Zr5)	4500	1200	37
GFRP Glass fiber reinforced polymer (60 vol% E-glass)	2000	1600	111
CFRP Carbon fiber reinforced polymer (60 vol% HT Carbon)	1500	2400	222

Another issue is the material the flywheel is made of. In the past, steel was used but it did not allow development of high speed, as the material did not withstand the load. Later on, alloys of, *e.g.*, aluminum or titanium alloys were used. The latest solution is composite materials, which allow the development of speeds up to 100,000 rpm and, at the same time, high power density. Table 9.2 presents the basic properties of materials used for flywheels. Effective efficiency of the flywheel storage system depends on several factors:

- Resistance to motion;
- Engine efficiency;
- Efficiency of the power electronic converters.

Resistance to motion decelerates the flywheel when it is charged and limits its maximal rotational speed. It is mainly caused by resistance on the bearings and air drag. Application of hermetic casing with low pressure inside eliminates the air drag.

The most difficult task related to the design of a flywheel storage system is employment of adequate bearings, which must hold the flywheel in a given position at minimal resistance to motion. It is not easy at the rotational speed of tens of thousands of rpm. Traditional ball bearing cannot be employed, as their resistance is too high. That is why magnetic bearings are employed. Flywheel energy storage systems have many advantages:

- Relatively simple structure;
- Long life;
- Low maintenance requirements;
- High energy density;
- Possibility of building small units.

The dynamic properties of flywheels are very good. They enable significant amounts of energy to be released in seconds. Although such a system has the disadvantage of not being able to store energy for too long, it can be a good supplement to the assemblage of short- and long-term energy storage systems.

Flywheel energy storage systems are employed in electrical vehicles and power electronic devices. As far as electrical vehicles are concerned, flywheel energy storage systems are employed mainly in public transport vehicles, which stop and accelerate frequently. Employment of flywheel storage systems may significantly reduce energy consumption due to the recovery of the braking energy. Application of flywheels in public transport vehicle will not be thoroughly described in this paper.

The intention of flywheels is to regulate the medium-termed processes of high dynamics in power electronic systems, *e.g.*, quick frequency regulation. The need to regulate the frequency stems from the actual changes of real input and produced power. When consumption is higher than production, the frequency decreases. If production exceeds the consumption, the frequency increases. Changes in frequency can occur very quickly, even in several minutes, which makes them hard to regulate. Nowadays, peaking pumped-storage hydroelectric power plants are used, but flywheel storage may prove better, as its dynamics is better. The problem of frequency regulation will be further exacerbated by increased application of renewable sources and distributed generation.

Figure 9.15. The view of Beacon Power's Smart Energy Matrix™ [17]

Technology of modular flywheel energy storage for power electronic systems has been developed by Beacon Power™. In order to increase the power or capacity of such a systems, the modules are combined into an assembly called Smart Energy Matrix™. Such systems operate in Amsterdam, NY and San Ramon (CA, USA). Figure 9.15 shows the interior of a container with utilities of total power of 100 kW.

Beacon Power™ intends to construct a 20 MW Smart Energy Matrix™ with 5 MWh capacity, and 4 s full power starting time [17].

Another system produced by Rosseta Technik GmbH and presented in Figure 9.16 found application in, *e.g.*, power supply for tram-car traction networks [10]. The basic technical data of such a system are:

- Weight 1.5 tonnes;
- Voltage 450–1,000 V;
- Current ±1,000 A;
- Power 500 kW;
- Max. rotational speed 25,000 rpm;
- Power capacity 6 kWh.

Flywheel energy storage systems develop rapidly. Their good dynamic properties, relatively small dimensions and low operating costs make them ideal for application in the transport and power industries. In the power industry flywheel storage will be applied mostly as short-term compensation for generated power changes in renewable sources, *e.g.*, windmills or solar energy power plants, as time-constants of flywheels are similar to those of wind or insulation changes.

Figure 9.16. The view of T1 system produced by Rosseta Technik GmbH [18]

9.6 Battery Storage

Electrochemical secondary cells are converters for reversible energy. They can convert electricity into chemical energy and the other way round. Energy is stored in the form of chemical ingredients, which can be stored inside the converter or in an outside container.

Battery energy storage systems have several advantages. Battery storage systems provide spinning reserve, area frequency and voltage control. Moreover, they can be located near the loads, with a consequential reduction in system losses thanks to their small sizes. The high initial cost of battery storage systems proves to be a disadvantage. Furthermore, at the present time, batteries require replacement every 8–10 years.

Nowadays, the only type of battery available for large energy storage applications is the lead-acid battery. Its typical efficiencies, made up of battery efficiencies of about 78% and power conditioning system efficiencies of about 94%, amount to around 72%. Examples of large battery installations in operation are:

- 17 MW, 14 MWh in Germany;
- 21 MW, 14 MWh in Puerto Rico;
- 10 MW, 40 MWh in United States of America.

Accumulator with Internal Storage
Chemical batteries, *e.g.*, lead-acid, nickel-cadmium, nickel-metal-hydride, lithium-ion and lithium-ion-polymer batteries, belong to the group of accumulators with internal storage. In such batteries, the energy converter and storage container cannot be physically separated. The converter is replaced by a barrier between the active material and the electrolyte.

The principle of operation is similar to all batteries. A battery consists of cells connected in series. Each cell is composed of two active material electrodes immersed in an electrolyte. One can differentiate the batteries on the basis of the material the electrodes are made of, the kind of active material employed and the type of electrolyte in which the electrodes are immersed. The charge process can be understood in general as accumulation of active material around the electrodes. Energy is retrieved in the discharge process.

Depending on the employed material, the cells differ in the number of cycles possible, energy density, charging time, internal resistance, price, *etc*. The kind of battery to be installed at a specific place should be chosen in the course of an accurate analysis of the working and economical conditions.

Lead-acid. Lead-acid batteries have been commercially used for storing electrical energy for more than 100 years. Since decades it has been, and still is, the most widespread electrical energy storage system Lead-acid batteries cover a wide range of applications. To cover various requirements, different battery designs had to be developed. It is also the cheapest battery type in comparison to all other readily available storage systems with appropriate characteristics. A major drawback of the lead-acid battery is the low specific gravimetric energy content resulting from the high molecular weight of lead.

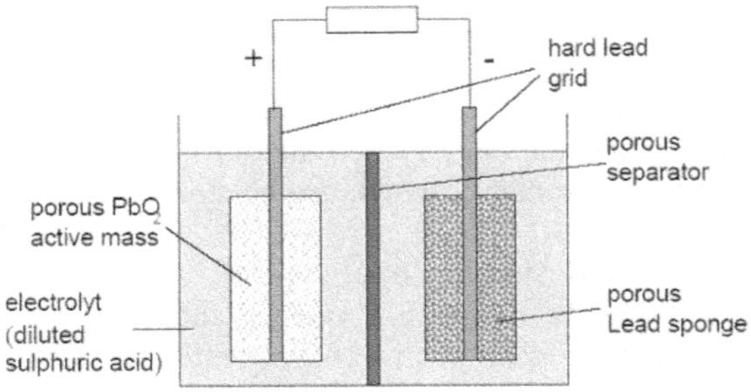

Figure 9.17. Structure of a lead-acid cell [5]

In the charged state, lead-acid batteries consist of a positive electrode with lead dioxide (PbO_2) and a negative electrode with lead (Pb) as the active material (Figure 9.17). Both electrodes contain a support grid, which is made of hard lead alloys. Sulfuric acid (H_2SO_4) is used as the electrolyte. The nominal voltage of a lead-acid battery is 2 V, the open-circuit voltage of a charged battery, depending on the concentration of the electrolyte, amounts to *ca.*, 2.1 V. The open circuit potential of the positive electrode in a full charged battery is approx. +1.75 V.

Lead-acid batteries are inexpensive (approx. 150 €/kWh), readily available, reliable, and highly recyclable. Although inexpensive, their usable energy is limited by their weight. The consequence of the battery's low specific energy – 25–35 Wh/kg – and poor energy density are the considerable dimensions and weight of the battery pack. Another disadvantage is low cycle life – average less than 1000 cycles [4]. Moreover, lead-acid batteries should not be discharged below 80% of their rated capacity, also called Depth of Discharge (DOD). Exceeding the 80% DOD shortens the life of the battery.

Further research on improving these batteries is being conducted. A gelled lead-acid battery uses an electrolyte paste instead of a liquid. Such batteries do not have to operate in an upright position, as no electrolyte will spill in an accident. Non-aqueous lead-acid batteries do not usually have as long a cycle life and are more expensive than the deep flooded ones. A new type of lead-acid batteries with higher specific energy is under development.

There are two main concepts: film electrode battery or bipolar design. Bipolar design allows reducing internal connections' resistance about 10^6 times. Such a battery can reach up to 50 kWh/kg [4].

Nickel-cadmium. NiCd batteries have been available as a commercial product for many decades and are highly renowned. Their lifetime and number of cycles are satisfactory. Moreover, a standard NiCd battery can be employed at temperatures of –20°C. On the other hand, cadmium employed in NiCd batteries is known as a threat to the environment.

Several types of NiCd batteries are on the market, with differences in applied technologies and the handling of gases. The electrolyte is potassium hydroxide (KOH), which does not significantly change its concentration or density during charging or discharging. Only water participates in the reaction. NiCd batteries available on the market contain liquid electrolyte. They are all sealed.

The rated voltage of NiCd cells amounts to 1.2 V and can be discharged at higher rates without the accessible capacity falling significantly below the rated capacity. Even for discharge rates of 5 C5 a high-performance NiCd battery can supply 60–80% of the rated capacity. What is more, compared to lead-acid batteries, the influence of the temperature on the battery capacity is rather small. Nevertheless, high temperatures, in the region of 40°C and more, should be avoided, as at high temperatures the charging efficiency is very low and the self-discharge rate increases significantly. At 20°C the self-discharge rate amounts to 20% per month. The energy efficiency compared with lead-acid batteries is significantly lower, *i.e.*, in the range of 60–70%

NiCd cell has a low internal resistance with typical values in the region of a few mΩ for the DC resistance. Under normal operating conditions, a NiCd battery can reach up to 2,000 100% DOD cycles. Its lifetime can reach 8–25 years. Despite their good electrical properties, the market share of NiCd batteries in autonomous power supply systems is not very high due to the high costs of their employment. NiCd batteries are *ca.* three times more expensive than lead-acid batteries.

Nickel-metal-hydride. The active material of the positive electrodes of a nickel-metal-hydride (NiMeH or NiMH) battery in its charged state is NiOOH, the same material as in a NiCd battery. The negative active material in the charged state is hydrogen, the component of hydride. In the cell, oxygen can be transported from the positive to the negative electrode and recombined there with hydrogen to form water. Thus, the cells can be used as dry cells and installed in any desired position. The open circuit voltage reaches between 1.25 and 1.35 V/cell, and the nominal voltage amounts to 1.2 V. The efficiency of a NiMH battery is rather high and reaches about 80–90%. The maximum power available for such a battery is lower than in NiCd batteries and averages 120–170 W/kg, but there are also high power batteries with specific power up to 9,000 W/kg. The specific energy is 50 – 80 Wh/kg. Self discharge at 25°C amounts to 20% per month but at 45°C it is as high as 60% per month. The battery cycle lifetime is higher than 2,400, and increases due to DOD reduction. In the case of 4% DOD it can reach a 100,000 [4].

Due to their better environmental compatibility and their higher gravimetric energy density, NiMH batteries replace NiCd batteries in the market of portable appliances. But, owing to the fact that their costs are approximately five times higher than that of lead-acid batteries (the cost varies from 300 to 1,000 €/kWh), they are not commercially available in larger capacities.

Lithium-ion and lithium-ion-polymer. Lithium batteries are the most rapidly developing battery technology of the last few years. Nowadays, lithium ion and lithium polymer batteries capture the market for portable applications, *i.e.*, mobile phones. Despite the fact that at present they are not used in power supply systems, they are worth a closer look. Their efficiency and charge–discharge characteristics are very well suited for their applications. Unfortunately, at the moment, lithium batteries are too expensive for applications where the high energy density is not so important.

Li-ion batteries of the modern type have a nominal voltage of 3.6 V. The electrolyte here is an organic solvent with dissolved lithium salts. Compared to NiCd or nickel-metal-hydride batteries, they are less tolerant to operation with high currents, which makes discharge at high currents difficult. Neither do they currently achieve the same cycle life as NiCd or nickel-metal-hydride batteries – about 1,000 and can reach 300,000 for 5% DOD. The battery typical specific energy is 80–180 Wh/kg and specific power 100–400 W/kg [4].

Lithium batteries require constant current or constant voltage charging. Their recharge behavior is very good. Full charging of the battery to achieve adequate lifetimes is not as important as with lead-acid batteries. But the voltage limit must be observed very accurately. The end limit of the charge voltage of 4.1 V must not be exceeded by more than 50 mV. The discharge of lithium batteries is also restricted by the material specific voltage. The end-of-discharge voltage is 2.3 V/cell for the cobalt and 2.7 V/cell for the manganese. Maximum discharge currents are in the region of 2 C5.

The main types of batteries are compared in Table 9.3. Supercapacitors are added for the sake of comparison.

Table 9.3. Comparison of the properties of batteries [5]

Battery technology	Energy density	Energy density	Efficiency	Life time	Cycle life time	Temperature for operation	
						Charging	Discharging
	[Wh/kg]	[Wh/l]	[%]	[a]	[cycles]	[°C]	[°C]
Lead-acid	20–40	50–120	80–90	3–20	250–500	–10 to +40	–15 to +50
NiCd	30–50	100–150	60–70	3–25	300–700	–20 to +50	–45 to +50
NiMH	40–90	150–320	80–90	2–5	300–600	0 to +45	–20 to +60
Li-ion, Li-polymer	90–150	230–330	95–95		500–1,000	0 to +40	–20 to +60
Supercaps	1–10	2–15	90–95	~ 10	500,000	–25 to +75	–25 to +75

Accumulator with External Storage

The secondary batteries with internal storage use their electrodes both as a part of the electron transfer process and as an energy storage utility. As a consequence, both energy storage capacity and the power rating are related to the electrodes' size

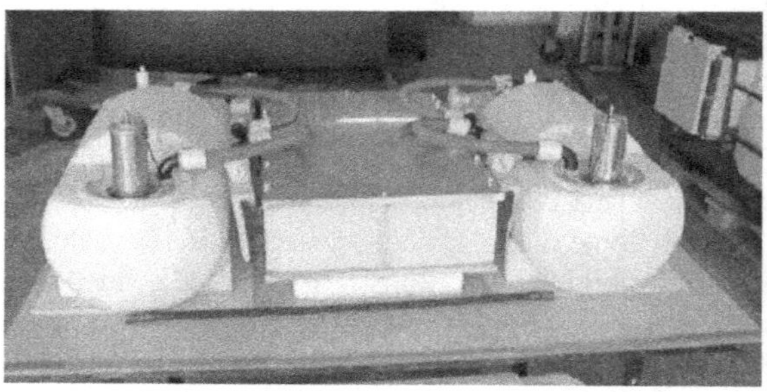

Figure 9.18. An example of a secondary battery with external storage [5]

and shape. This drawback is solved by electrochemical batteries with external storage. The reaction occurs within the electrochemical cell, but the energy is stored in two tanks separated from the cell (Figure 9.18). The electrochemical cell has two compartments, one for each storage medium, physically separated by an ion-exchange membrane, which allows designing the battery power and the energy content separately.

There are several methods to use chemical energy for energy storage. One of the most commercially advanced is regenerative fuel cell technology, also known as redox flow cell technology. It converts electrical energy into chemical energy by *charging* two liquid electrolyte solutions. The chemical energy is then converted back to electrical energy during discharge.

Energy is stored and released by means of a reversible electrochemical reaction between two salt solutions, which occurs within an electrochemical cell. The cell has two compartments physically separated by an ion-exchange membrane. Unlike in a typical battery system, the electrolytes flow into and out of the cell through separate manifolds and are electrochemically transformed inside the cell. The Regenesys™ system is an example of the commercial application of the regenerative fuel cell technology. Due to its high response speed capability of storing huge amounts of energy, it is a very good solution for most cases of power systems. The first Regenesys™ system was built in Little Barford. The plant was designed to store 120 MWh of energy and discharge it at a nominal power rating of 15 MW.

9.7 Hydrogen Storage

The idea of storing electric energy in the form of hydrogen is connected to modern solutions in the field of electric energy generation, and especially renewable energy sources. Storing energy in the form of hydrogen has the following advantages:

- Easy energy transmission;
- Unlimited storage period;
- Easy installation scaling;
- Numerous technologies of hydrogen producing and its transmission into electric energy.

All hydrogen electric energy storage technologies consist of three elements:

- Primary power-to-hydrogen conversion system;
- Hydrogen storage system;
- Hydrogen-to-electricity conversion system.

Hydrogen is produced in the following technological processes:

- Reforming hydrocarbons;
- Coal or coke gasification;
- Plasmatic process;
- Electrolysis or photo electrolysis;
- Biomass gasification.

Table 9.4 presents the basic properties of hydrogen.

Table 9.4. Basic properties of hydrogen

relative atomic mass	1.00794 u
density	0.0899 kg/Nm^3
specific HHV energy	142.0 MJ/kg
specific LHV energy	120.0 MJ/kg
HHV energy density	11.7 MJ/Nm^3
LHV energy density	9.9 MJ/Nm^3
boiling point	20.268K(–252.88°C)
melting point	14.025K (–259.13°C)
critical point	33.250K (–239.90°C)

In order to transform hydrogen into electric energy one employs mostly fuel cells, turbines, and hydrogen-fed engines. Due to their high efficiency, fuel cells are believed to be the best solution. Thorough discussion of transforming electric energy to hydrogen, and the other way round, is not the concern of this publication.

The biggest problem of employing hydrogen in energy storage is storing the hydrogen itself. As the lightest known chemical element [19], hydrogen has a very low density, which results in very low volumetric energy density, in spite of very high mass specific energy. For that simple reason, the biggest problem is storing adequate amounts of hydrogen in a technically reasonable volume.

The amount of energy possible to be stored in the form of hydrogen depends on the quantity of the stored hydrogen itself. The bigger the quantity of the stored hydrogen, the bigger the amount of the retrieved energy. The energy stored in hydrogen can be described with the following formula

$$E = mE_{sp} = \rho VE_{ed} \tag{9.6}$$

where m – mass of the stored hydrogen; E_{sp} – hydrogen specific energy; ρ – hydrogen density in normal conditions; V – stored volume of hydrogen; E_{ed} – hydrogen energy density.

Table 9.5. Comparison of hydrogen with other fuels

Substance	**Specific energy (LHV)**		**Energy density kWh/Nm^3**
	MJ/kg	kWh/kg	
Hydrogen	120	33.33	
Lignite	27		
Hard coal	8		
Petroleum		11.60	
Petrol	44.75	12.00	
Methanol		5.47	
Methane		13.9	
Natural gas		10.6–13.1	36,000MJ/kg
Propane	46.33	12.88	

Unfortunately, due to its low volumetric power density, the storage and transport of great amounts of hydrogen are rather problematic. Table 9.5 presents the comparison of hydrogen with other fuels.

Compressed Hydrogen
At present this is the best known and the most commercially wide-spread technology. The hydrogen is stored in pressure tanks, usually at the pressure of 200–800 bar. The tanks are usually made of either composite or steel. As the first are considerably lighter than the latter, they are commonly employed in transport.

One of the basic problems of pressure installation is the necessity to compress the hydrogen, which absorbs great amounts of energy. Figure 9.19 presents the cost of energy employed to compress hydrogen [19]. The graph shows both the ideal processes (broken line) and the multistage process employed in reality (full line).

Liquefied Hydrogen
Liquid hydrogen is stored in cryogenic tanks at 21.2 K at ambient pressure. Because the hydrogen critical temperature is very low (33K), the liquid hydrogen must be stored only in open tanks. If the storage container was closed, the pressure inside at room temperature (RT) could increase to about 104 bar. This is possible due to unavoidable heat leakage, and must be permitted for safety reasons. Hydrogen losses depend on the size of the tank. The losses could be significant for tanks used in vehicles, and may amount to 3–4% a day [19].

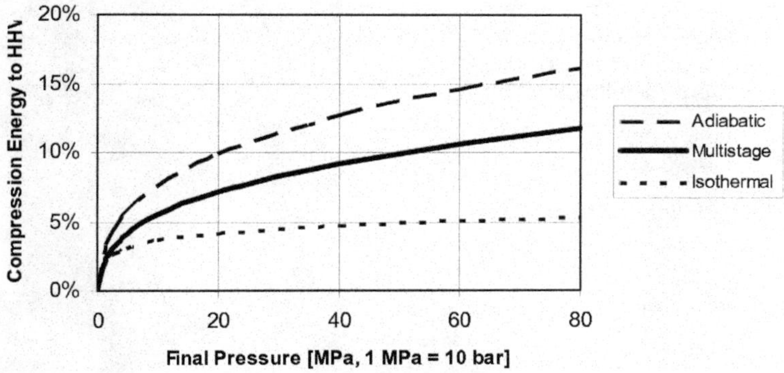

Figure 9.19. Energy costs of hydrogen compression [19]

The simplest liquefaction cycle is the Joule-Thompson cycle, also called the Linde cycle. First, the gas is compressed and then cooled in a heat exchanger. It flows through a throttle valve where it undergoes an isenthalpic Joule-Thomson expansion, producing some liquid. The cooled gas is separated from the liquid. Then it returns to the compressor *via* the heat exchanger. Before the first expansion step occurs, hydrogen is usually precooled using liquid nitrogen (78K) [20]. A huge amount of energy is necessary for liquefaction. The energy cost of liquefaction can be as high as about 30% of the energy stored in liquid hydrogen.

Metal Hydrides
One of the promising technologies is storing hydrogen in the form of metal hydrides, which can store and release gaseous hydrogen in the same way as a sponge absorbs and releases water. This process is already employed in some prototypes in the portable systems area (*e.g.*, camcorders, laptops, *etc.*,).

A hydride is the name given to the compounds of hydrogen with other elements. Such compounds are unstable, when pressure or temperature changes the dissociation and releases hydrogen in the gaseous form. Depending on the compound, the pressure used to charge, or discharge the hydrogen out of the hydride amounts to 0.1–1 MPa, and the temperature cannot exceed 150–300°C. The concentration of the hydrogen mass in the mass of the whole tank can be as high as 7%. What is more, the capital and operating costs are moderate, and the system's maintenance is unproblematic. At present the researchers look for alloys which would allow higher concentration and closer packing of hydrogen atoms in a hydride at relatively low costs and low charging/discharging temperatures.

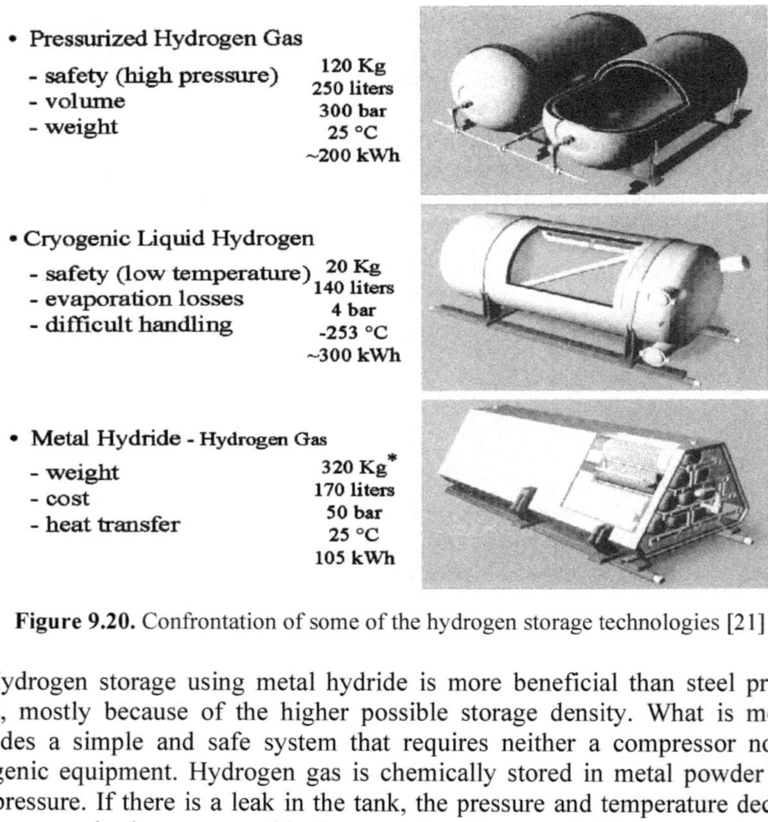

Figure 9.20. Confrontation of some of the hydrogen storage technologies [21]

Hydrogen storage using metal hydride is more beneficial than steel pressure tanks, mostly because of the higher possible storage density. What is more, it provides a simple and safe system that requires neither a compressor nor any cryogenic equipment. Hydrogen gas is chemically stored in metal powder under low pressure. If there is a leak in the tank, the pressure and temperature decrease, and, as a result, the process of hydrogen release is paused. Figure 9.20 shows the views and basic parameters of the above-mentioned devices [21].

Hydrogen Storage in Carbon Nanotubes

Carbon surfaces absorb hydrogen, which allows the carbon material to achieve a power density similar to that of liquid hydrogen. However, in order to achieve satisfactory storage results, great surfaces of carbon materials have to be employed. Various kinds of carbon materials with developed surfaces are known, *e.g.*, fullerenes, activated carbon, nanotubes.

Activated carbon energy storage demands low temperatures, *ca.* 100K, but the storage density is rather low. In the case of nanotubes the density is higher, from 6.5%. However, both technologies are still being researched.

Figure 9.21 presents a comparison of the capacities of the modern technologies in hydrogen storage [21].

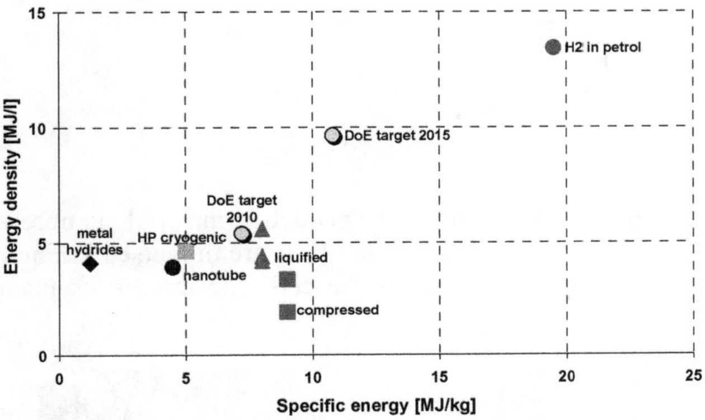

Figure 9.21. Comparison of the capacities of hydrogen storage technologies [21]

Hydrogen is also considered for a medium energy transport and storage for large power system. The idea of storing great amounts of hydrogen is though of as a compensation for energy flows in power electronic systems. Rashid [22] describes the idea of storing great amounts of low pressurised hydrogen in salt caverns. This technology employs underground reservoirs (caverns and porous rocks formerly exploited in natural gas or petroleum industry or salt caverns) which are sealed and filled with gas. Hydrogen losses from such tanks amount to 1–3% per year. It is the cheapest of the modern hydrogen storage technologies.

9.8 Superconducting Magnet Energy Storage

Superconducting Magnetic Energy Storage was proposed among others by Lasseter [23]. The idea was to meet the daily power variation in the French power system. It is interesting that in the 1970s it was easier to make a superconducting coil than power electronic device. The first operating SMES system with three-phase inverter was investigated in Los Alamos National Laboratory in 1974 [1].

The principle of operation of a SMES employs the storage of energy in the magnetic field around the coil carrying direct current; see Figure 9.22. The possible stored energy is described by Equation 9.7

$$E = \frac{LI^2}{2} \tag{9.7}$$

where L – inductance of the coil; I – current flowing through the coil.

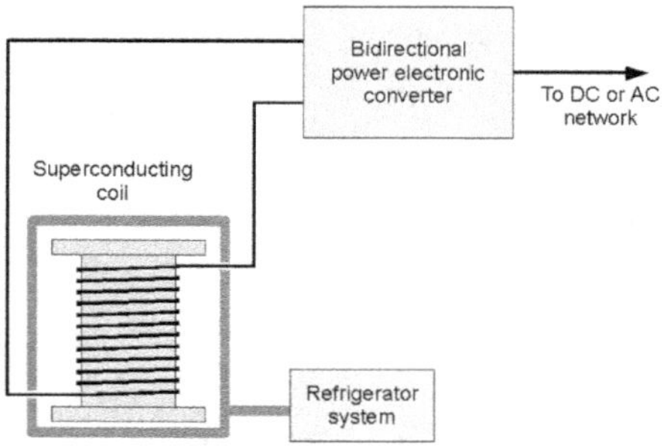

Figure 9.22. Pictorial diagram of a SMES

Employment of coils made of superconductive materials is necessary in order to reduce the energy loses. In practice, the coils are immersed in liquid nitrogen or helium. The basic technical issues for the coil's design are its manufacture and cooling.

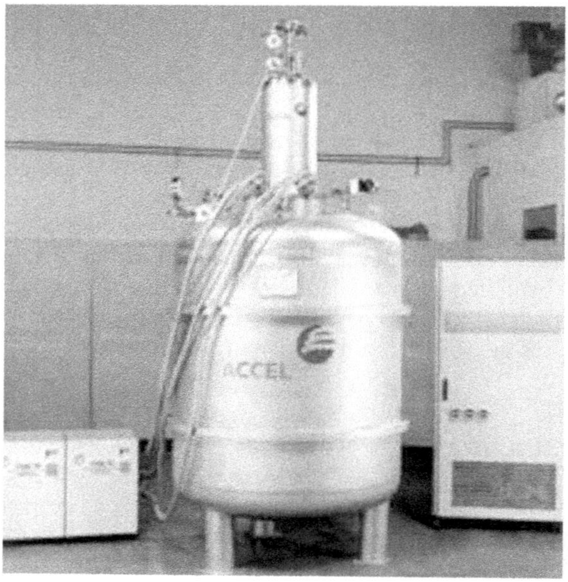

Figure 9.23. The view of the SMES device produced by ACCEL Instruments GmbH [24]

As there is no need to convert energy, SMES has very good dynamics, which is impossible in the case of devices based on mechanical energy storage. The response time of SMES amounts to seconds. What is more, as a result of practically no resistance on the windings, SMES can generate very high power. It also results in its high efficiency. However, even a small AC component in a coil's current can cause losses. SMES is perfect for situations where great power is needed for a short period of time, as the basic disadvantages of SMES are its low capacity and short storage period.

Figure 9.23 presents an example of a SMES available for commerce. It is a device produced by ACCEL Instruments GmbH. Basic parameters of the device are as follows:

- SMES current 1,000 A;
- Stored energy 0.56 kWh;
- Average power 200 kW;
- Maximum power 800 kW.

Other known solutions are presented in [25].

9.9 Supercapacitors

The supercapacitor (ultracapacitor or double layer capacitor) is a new energy storage device which, in its operation principle, is similar to a traditional capacitor. However, its capacity and discharge current is very much higher than that of a traditional one. That is why supercapacitors are used as storage units in power system, vehicles, *etc.* In the near future, supercapacitors may replace chemical batteries, especially if long-time storage is not required, as, at the moment, the scale of their size is comparable to that of batteries, from small ones used in cellular phones to large ones found in cars.

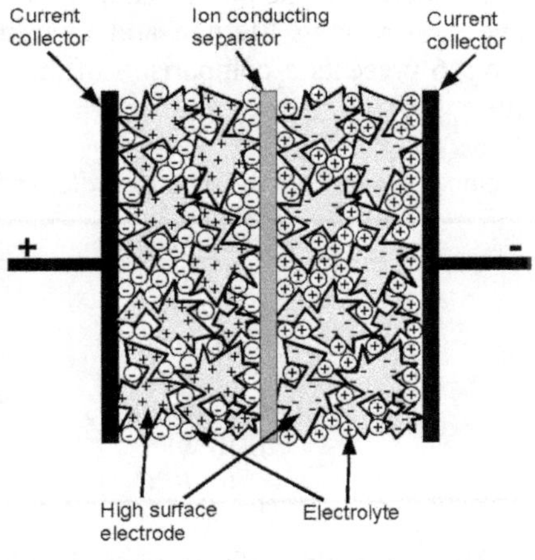

Figure 9.24. The structure of ultracapacitor

Figure 9.24 shows a double layer supercapacitor structure. It consists of two carbon nanotubes or activated carbon electrodes immersed in electrolyte. There is a membrane immersed in the electrolyte between the electrodes. The membrane conducts ions but not electrons, so electrons cannot flow through the device. The electrolyte consists of ions arranged in a very thin layer on the surface of the electrodes, which behaves as a dielectric. No chemical reaction takes place in the device. If the voltage is lower than a few volts, the layer does not conduct electrons. Therefore, till this voltage level is needed the structure works like a capacitor. Dielectric dipoles layers have a very small thickness, *i.e.*, about 10 A. The capacity of a capacitor is proportional to the surface area of the electrode and inversely proportional to the thickness of the dielectric. The area of the electrode can reach as high as 2,000 m^2 per gram of the capacitor's mass. Hence, a supercapacitor's capacitance can reach the region of thousands of Faradays. However, it has to operate at low frequency. The supercapacitor behaves like pure capacitance at frequencies below 100 Hz [26].

The possible stored energy for supercapacitor is described by the same formula as in the case of a regular capacitor, the only difference being the much higher capacity of the ultracapacitor.

The energy density typical for an ultracapacitor amounts to 5 Wh/kg, which is much higher than in the case of a chemical battery – 40 Wh/kg. Although supercapacitors have a lower energy density than batteries, they avoid many of the battery's disadvantages. First, batteries have a limited number of charge – discharge cycles. Moreover, the charge and discharge process takes a long time because it involves chemical reactions with non-instantaneous rates. Finally, batteries have a limited cycle life with a degrading performance. In contrast to batteries, supercapacitors can be charged and discharged almost an unlimited number of times; they can discharge in matters of milliseconds and are capable of producing enormous currents. In face of all the above facts, the supercapacitors are applied to load leveling and in situations where a sudden boost of power is needed in a fraction of a second. Furthermore, the maintenance costs of supercapacitors are rather low, as they have a very long lifetime and their performance does not degrade with time. Table 9.6 presents a comparison of the properties of a typical supercapacitor and a battery [26].

Table 9.6. Comparison of a typical supercapacitor and a battery

	Supercapacitor	**Battery**
Discharge time	1–30 s	0.3–3 h
Charge time	1–30 s	1–5 h
Cycle life	>500,000	500–2,000
Efficiency	90–95%	70–85%
Power density	1,000–2,000 W/kg	50–200 W/kg
Energy density	1–10 Wh/kg	20–100 Wh/kg
Operating temp.	–40–70°C	0–60°C

Supercapacitors are an extremely safe storage device as they are easily discharged. They have low internal resistances, even if many of them are coupled together. The voltage is proportional to the current state of charge, which allows the control system to estimate the actual level of stored energy. Even though they have a lower energy density, due to their readiness in releasing power, supercapacitors have already replaced batteries in many fields. The main advantages of supercapacitors are:

- Possibility to generate great currents;
- Very short charging and discharging time;

- Very high number of cycles without deterioration of the properties;
- Simple structure without the employment of toxic substances.

The disadvantages of supercapacitors are mainly:

- High voltage to charge tractability;
- Low single ultracapacitor voltage;
- Low specific energy.

The high tractability is a serious obstacle in the application of supercapacitors. It requires the employment of complicated bidirectional systems of power electronic converters capable of working in a significantly wide range of voltages, and great currents. Most modern research on supercapacitors focuses on power electronic converters.

As far as low single ultracapacitor voltage is concerned, in order to increase it, series connection is employed. However, similarly to chemical batteries, such a connection causes the problem of voltage dispersion to separate capacitors. The problem may be solved by application of small balancing system. However, such a solution raises the costs of the capacitor significantly and complicates its structure.

Supercapacitors were initially used by the US military to start the engines of tanks and submarines. Nowadays, they are employed mostly in the field of hybrid vehicles and handheld electronic devices. The ultracapacitor has recently become commercially available. A typical commercially available ultracapacitor can hold 2500 farads, and release 300 A of peak current with a peak voltage handling of about 400 V. Supercapacitors found application in areas where fast and efficient, yet short-term, storage devices are needed, especially in:

- Electric car drives;
- UPS;
- The systems improving the quality of energy (the shape and value of the voltage) in distribution networks;
- Energy storage devices in traction networks;
- Driving systems with frequent start-ups and braking, *e.g.*, elevators;
- Non-overload generating systems, *e.g.*, fuel cells, as sources of great current.

9.10 Application of Energy Storage Devices

Energy storage systems vary in principle of operation, capacity, and dynamic properties. Each type of the system should be utilized in a different way. A proper system can be chosen on the basis of the following factors:

- Type of the compensated fluctuations;
- Properties of the power system;
- Required storage capacity;
- Minimum storage period;
- Charging and discharging time;
- Available location and environmental factors;

- Required energy density and lifetime;
- Minimum number of storage cycles.

The range of power storage application can traditionally be divided into power quality maintenance and balancing the energy in the system. In order to improve the energy quality (*i.e.*, set value of the voltage, its frequency and shape) fast storage devices should be employed. Such devices do not have high capacitance, but can generate high power electricity. Balancing the system, on the other hand, requires a high capacitance storage device with not so good dynamics.

Table 9.7 presents selected properties of the described power storage systems. In order to choose an appropriate storage system, one should confront the properties of the storage device with the demands of the power system in which it will be employed; see Figure 9.25. As can be seen, all the described systems should be developed, as each of them meets different needs and finds application in different areas.

Table 9.7. Comparison of the storage systems

Storage system	Capacity	Power	Start up time	Storing time
Pumped water	~ 10^3 MWh	10^1–10^3 MW	2–5 min	Months
CAES	~ 10^2 MWh	10^2 MW	12 min	Months
Flywheels	kWh	10^2 kW	Seconds	Hours
Batteries	kWh to MWh	kW to MW	Milliseconds	Days, weeks
SMES	~ 1 MW	~ 0.5 MWh	Seconds	Minutes
Supercapacitor	10^3F		Milliseconds	Minutes

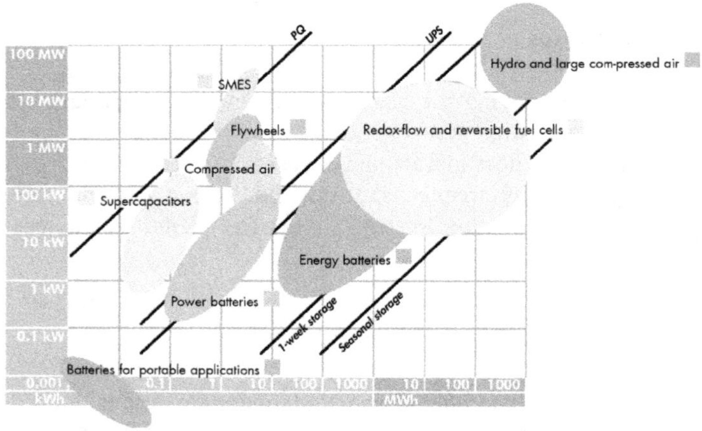

Figure 9.25. The areas of application of the particular types of energy storage systems [27]

It is necessary to emphasize that the demand for energy storage systems will increase in line with the development of renewable power sources and distributed generation. Apart from the classic huge (water) systems, new ones will be necessary. They will have to be flexible in power and capacity leveling (as *e.g.*, chemical batteries, supercapacitors, fuel cells). The bigger ones will be employed close to wind farms, the smaller ones in distribution grids near to the middle-sized and large renewable power plants.

The main aim of a storage device is to limit the negative impact of the source on the public grid. Secondly, it optimises the utilization of the primary carriers. The most interesting case of a storage utility is a stand-alone system supplied by unpredictable sources such as sun and wind, which can obviously be balanced without a storage device [2, 5, 28].

The storage system employed for renewable sources should have very good dynamic properties, *i.e.*, it should start generation or consumption with the rated power, and change the direction of the energy flow in milliseconds. In the case of wind turbines, the only solution is to employ electrochemical systems or supercapacitors. However, from the economical point of view, lead-acid batteries are still the best choice.

References

[1] Walling RA, Miller NW, (2002) Distribution generation islanding – implications on power system dynamic performance. IEEE Conference Publications:92–96

[2] Kim JE, Hwang JS, (2001) Islanding detection method of distributed generation units connected to power distribution system. Proceedings of the IEEE Summer Meeting:643–647

[3] Willoughby R, (2001) Integration of distributed generation in a typical USA distribution system. IEE CIRED2001 Conference Publication, no.482:18–21

[4] Brahma S, (2000) Effect of distributed generation on protective device coordination in distribution system. IEEE PES Summer Meeting:115–1119

[5] Salles MBC, Freitas W, Morelato A, (2004) Comparative analysis between, SVC and DSTATCOM devices for improvement of induction generator stability. IEEE MELECON, Dubrovnik, Croatia

[6] Gyugyi L, (1989) Solid-state control of electronic power in AC transmission systems. International Symposium on Electric Energy Converters in Power Systems, T-IP.4, Capri, Italy

[7] Mcdermott T, Dugan RC, (2003) PQ reliability, and DG. IEEE Industrial Application Magazine:17–23

[8] Simons G, Sethi P, Davis R, (2001) The role of renewable distributed generation in California's electricity system. IEEE Power Engineering Society Summer Meeting, vol.1:546–547

[9] Kojovic L, (2002) Impact DG on voltage regulation. IEEE Power Engineering Society Summer Meeting, vol.1:97–102

[10] Wang HF, Jazaeri M, Cao YJ, (2005) Operating M and control interaction analysis of unified power flow controllers. IEE Proceedings (Generation, Transmission, Distribution), vol.152, no.2:264–270

[11] Usta O, Redfem M, (2000) Protection of dispersed storage and generation units against islanding. IEEE PES Summer Meeting 2000:976–979

[12] Barker P, (2000) Determining the impact of distributed generation on power systems. Part 1: radial distribution systems. IEEE:1645-1656

[13] Mathur RM, Varma KR, (2002) Thyristor – based FACTS controllers for electrical transmission systems. IEEE Press, New York

[14] Gyugyi L, (1992) Unified power flow control concept for flexible AC transmission systems. IEE Proceedings, part-C, vol.139, no.4

[15] Sannino A, Svensson J, Larsson T, (2003) Power electronic solution to power quality problems. Electric Power Systems Research, vol.66:71–82

[16] Gyugyi L, Hingorani NG, (1990) Advance static VAR compensator using gate-tran-off thyristor for utility applications. CIGRE, no.23-203

[17] Hingorani NG, Gyugyi L, (2000) Understanding FACTS: concepts and technology for flexible AC transmission systems. Piscataway, IEEE Press

[18] Marks M, (2000) Distributed generation: CEQA review and permit streamlining. California Energy Commission, report 700-00-019

[19] IEEE Standard 1547-2004, (2004) 1547 IEEE standard for interconnecting distributed resources with electric power systems. Institute of Electrical and Electronics Engineers, Piscataway, New Jersey

[20] Bhowmik A, Maitra A, Schatz SM, (2003) Determination of allowable penetration levels of distributed generation resources based on harmonic limit considerations. IEEE Transactions on Power Delivery, vol.18, no.2:619–624

[21] Al-Mawsawi SA, (2003) Comparing and evaluating the voltage regulation of a UPFC and STATCOM. Electric Power and Energy System, vol.257:735–740

[22] Rashid MH, (2003) Flexible AC transmission. Chapter13: power electronics circuit devices and applications. Pearson Prentice Hall, third edition

[23] Lasseter R, (2002) Microgrids. Proceedings of the Power Engineering Society Winter Meeting, vol.1:27–31

[24] Ackermann T, Andersson G, Soder L, (2001) Distributed generation: a definition. Electric Power System Research, vol.57:195–204

[25] Friedman NR, (2002) Distributed energy resources interconnection systems: technology review and research needs. National Renewable Energy Laboratory, report SR-560-32459

[26] Dugan RC, Price SK, (2002) Issues for distributed generation in the US. Power Engineering Society Winter Meeting, vol.1:121–126

[27] Rifaat RM, (1995) Critical considerations for utility/cogeneration inter-tie protection scheme configuration. IEEE Transactions on Industry Applications, vol.31, no.5:973–977

[28] Daly PA, Morrison J, (2001) Understanding the potential benefits of distributed generation on power delivery systems. Rural Electric Power Conference:A211–A213

10

Variable and Adjustable Speed Generation Systems

Włodzimierz Koczara

Institute of Control and Industrial Electronics,
Warsaw University of Technology,
Koszykowa 75 Street
00-662 Warszawa, Poland.
Email: Koczara@isep.pw.edu.pl

10.1 Introduction

10.1.1 Conventional Generation Systems

Variable speed generation is an emerging technology. At the present time, electrical power systems are based on generation standard AC electricity by Wound Rotor Synchronous Generator (WRSG). Figure 10.1 shows a simple generation power system, where the wound rotor synchronous generator WRSG is driven by a prime mover DE producing torque T_d. The WRSG operates with the fixed speed ω_{gm} and supplies a load. The generator produces an opposite torque T_g.

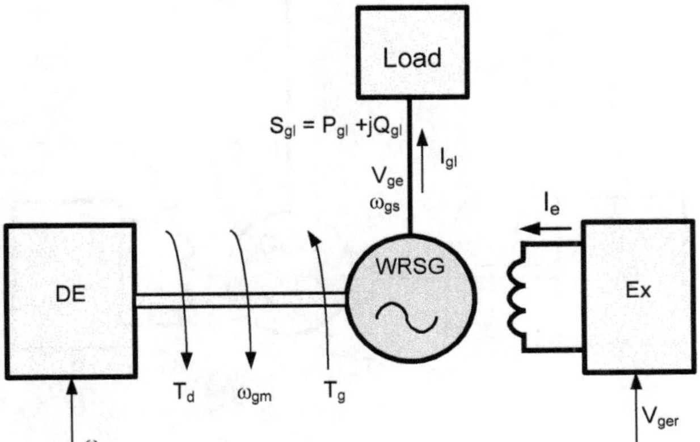

Figure 10.1. The classical generation system with WRSG

The WRSG produces an AC e_{ge} which is the function of the excitation current I_e time t, rotor shaft rotational frequency (*i.e.*, angular speed) ω_{gm} and the number of poles p

$$e_{ge} = E_{ge\max} \sin(\omega_{gm}(p/2)t) \tag{10.1}$$

where E_{gemx} stands for the maximum amplitude.

The WRSG output voltage frequency is

$$f_{ge} = \omega_{gm}(p/2)/2\pi \tag{10.2}$$

Therefore, any changes of the rotor shaft rotational frequency ω_{gm} will result in changes of the output voltage frequency. For a given standardized frequency the series of speeds ω_{gm} and the number of poles p are strictly coupled. For instance, for 50 Hz the two poles generator has to operate with the speed of 314.16 rad/s, for four poles the speed is 157.08 rad/s, while for six poles it falls to 104.72 rad/s. Hence, the prime mover DE has to be sized to only one speed and the choice of speeds is very limited. Moreover, when the speed is not precisely maintained then the voltage frequency follows the rotor shaft rotational frequency and it results in the degradation of quality of the output voltage V_{ge}. The generator load depends on the demanded volt-ampere power S that is the function of demanded active P and reactive Q powers.

In the case of connection of the WRSG to the stiff grid, as is shown in Figure 10.2, the rotor shaft rotational frequency and then stator frequency are supported by the power from the grid.

The generator output voltage is equal to the power grid voltage and the generating unit is the source of active power P_{ge} and reactive power Q_{ge}. The active

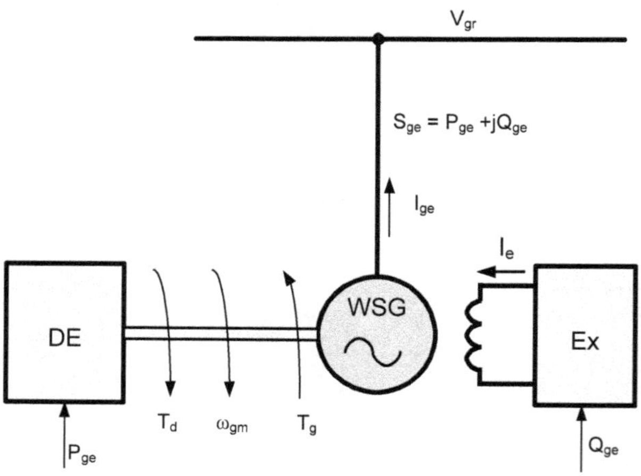

Figure 10.2. The generation system with a wound rotor synchronous generator connected to the power grid

power depends on load angle δ_g and reactive power is controlled by E_{ge} produced by the generator. Figure 10.3 shows an equivalent diagram of the grid connected generation system (Figure 10.3a) and simplified vector diagram (Figure 10.3b). The active power, provided by the generator, is the function of load angle δ_g

$$P_{ge} = E_{ge} V_{gr} \sin(\delta_g) / x_g \tag{10.3}$$

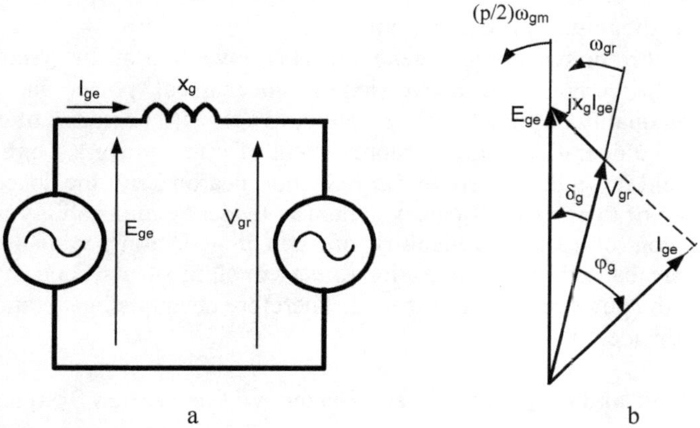

Figure 10.3. The wound rotor synchronous generator connected to power grid: **a** an equivalent diagram; **b** simplified vector diagram

and reactive power

$$Q_{ge} = E_{ge} V_{gr} \sin(\delta_g) / x_g - V_{gr}^2 / x_g \tag{10.4}$$

where voltages are expressed as rms values.

The load torque produced by the generator

$$T_g = \left(E_{ge} V_{gr} \sin(\delta_g) / x_g\right) / \omega_{gm} \tag{10.5}$$

is also load angle δ_g dependent. This torque may be the result not only of the driving torque but also, as is shown in Equation 10.5 and vector diagram shown in Figure 10.3, of instantaneous changes of vector V_{gr}. Then the load angle

$$\delta_g = \int_{t_1}^{t_2} \left(\frac{p}{2} \omega_{gm} - \omega_s \right) dt \tag{10.6}$$

The unwanted changes of the load angle above 90° may result in a significant pole slip and the generator has to be instantly disconnected, even when the generation system is in a healthy state. A control of load angle, by adjusting the driving torque T_d, is very slow and temporarily not practically possible. Therefore the generator speed change ω_{gm}, caused by the driving torque (Equation 10.5), is slow. Moreover, the rotating parts inertia (inertia of the prime mover and the generator) J_{gs} is high, which makes a delay in speed (and angle δ_g) change and unwanted oscillations

$$\omega_{gm2} = \omega_{gm1} + \frac{1}{J_{gs}} \int_{t_1}^{t_2} \left[T_d(t) - T_g(\delta_g) \right] dt \qquad (10.7)$$

where ω_{gm1} is the initial speed at the time t_1.

The wound rotor synchronous generator is an invention of the electromagnetic era because it combines a conversion of mechanical power into electrical depending on quality of the AC voltage. However, the strict demand of given fixed speed limits the operation and the improvement of prime movers. High speed gas turbines should drive the generator *via* reduction gearbox and the speed of diesel engines is out of their high efficiency area. The tendency to instability and lack of load angle control reduces reliability of such a generating system. All these drawbacks are caused by the need for direct coupling of a wound synchronous generator to the power grid or to the load. Therefore decoupled generation systems have to be considered.

10.1.2 Variable and Adjustable Speed Decoupled Generation Systems

Power electronic converters are commonly used in drives and UPS supply systems. Power electronic systems, used in electrical drives, convert AC power to demanded AC or DC power according to the reference speed, supplying the motor by pulse width modulated (PWM) voltage or current mode. In UPS systems DC/AC converter, supplied from DC voltage, delivered by a battery, produces the AC sinusoidal voltage dedicated to a given load [1]. The application of power electronic converters to generation systems brings additional degree of freedom in speed [2–7]. This additional degree of freedom results in a number of new performances of generation systems that are not available in conventional wound rotor generators connected directly to the load. Figure 10.4 shows the principle of application of power electronic converters in power generation systems. The prime mover DE, operating with speed ω_{gm}, drives a generator G which converts mechanical power into electrical. The generator G produces AC voltage V_{ge} of which the angular frequency is

$$\omega_{ge} = p/2 \; \omega_{gm} \qquad (10.8)$$

The operation of power electronic converter PEC results in changes of the generator angular frequency ω_{ge} to the demanded standardized angular frequency according to reference signal ω_{gs}. In steady state the converter frequency ω_{pc} is equal to the demanded output frequency

$$\omega_{pc} = \omega_{gs} \tag{10.9}$$

The power electronic converter produces single or multi-phase voltage

$$v_{gs} = V_{gsm} \sin\left(\omega_{pc} t\right) \tag{10.10}$$

where V_{gsm} is the amplitude of the AC voltage.

The v_{gs} voltage is completely independent from generator angular frequency which means that the output frequency is decoupled from the generator frequency and speed.

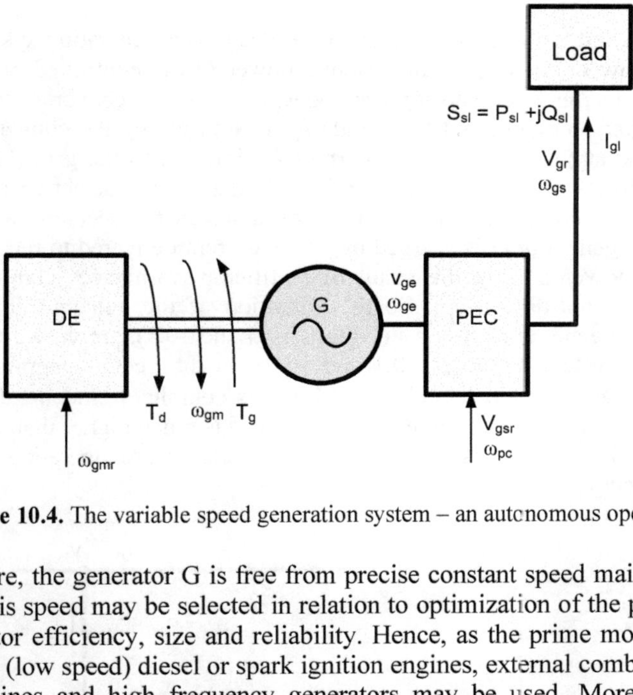

Figure 10.4. The variable speed generation system – an autonomous operation

Therefore, the generator G is free from precise constant speed maintenance [3, 4, 7] and this speed may be selected in relation to optimization of the prime mover and generator efficiency, size and reliability. Hence, as the prime mover, internal combustion (low speed) diesel or spark ignition engines, external combustion (high speed) turbines and high frequency generators may be used. Moreover, steam turbines or other prime movers such as Sterling motors may be applied. In practice all existing prime movers may be used. Figure 10.5 shows an example of equivalent block diagram of the generation system shown in Figure 10.4. The power electronic converter PEC is made with an intermediate DC link system. The generator produces AC voltage E_{ge} which supplies a rectifier Re. The output of the rectifier is connected to the DC link system DCS equipped with energy storage. The rectified voltage is converted to AC voltage by an inverter IN.

The intermediate DC link DCS is equipped with energy storage, which decouples the output AC voltage from instantaneous changes of generator voltage. As the energy storage, a capacitor ("*voltage source*") or inductor ("*current source*") is used.

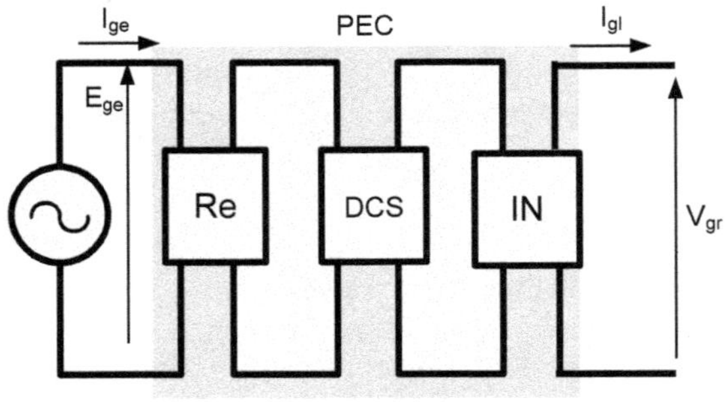

Figure 10.5. Equivalent diagram of the variable speed generation system

In the case of the grid connection of the decoupled generation system (Figure 10.6), the flow of active P_{gs} and reactive power Q_{sa} is controlled by the power electronic converter PEC. The angular frequency ω_{pc} of the converter PEC is equal to the grid frequency ω_{gs} and load angle δ_{pc} is adjusted by the converter PEC by phase φ_{pc} and amplitude I_{mgs} of the current I_{gs} injected to the grid (Figure 10.7). Hence the power electronic converter PEC is the source of active and reactive power. The driving engine speed has to be adjusted to meet this active power demand. The generator G is charged by active current delivered to rectifier Re and its reactive power is only the result of rectifier performances coming from its topology and control concept. So the condition of the generator operation, in variable speed generation systems (Figures 10.4 and 10.5), are very different from conventional systems (Figures 10.1 and 10.2). In the vector diagram, shown in Figure 10.7, the generator angular speed ω_{gm} is completely independent and not coupled to any other vector and usually this speed is much higher than grid voltage angular speed. This means that now new generators are operating with much higher frequency.

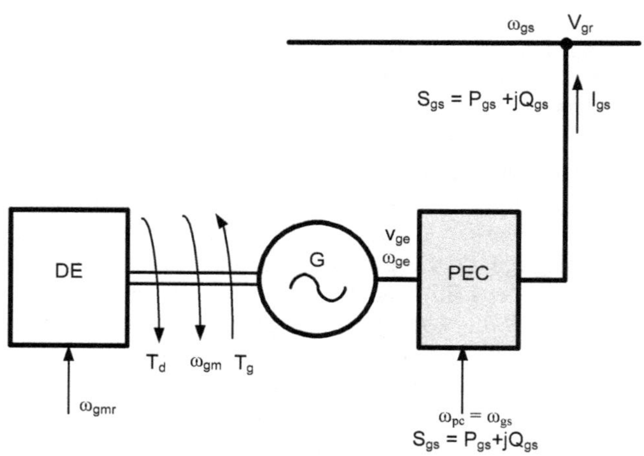

Figure 10.6. The variable speed generation system – grid connected operation

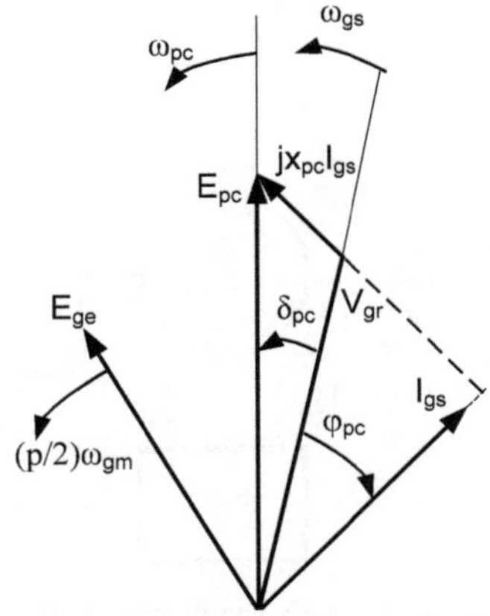

Figure 10.7. Vector diagram – case of connection of variable speed generation system to power grid

The position of the current vector (angle φ_{pc} and the module) is created by the power electronic converter PEC almost instantly (by high frequency PWM switching method). This is almost instant control of load current and power.

The active power, provided by the converter, is a function of load angle δ_{pc}

$$P_{gs} = E_{pc} V_{gr} \sin(\delta_{pc}) / x_{pc} \tag{10.11}$$

and reactive power

$$Q_{gs} = E_{pc} V_{gr} \sin(\delta_{pc}) / x_g - V_{gr}^2 / x_{pc} \tag{10.12}$$

where voltages are expressed as RMS values.

The driving prime mover DE has to deliver active power consumed by load P_l and to cover losses of the generator ΔP_G and converter ΔP_{PEC}

$$P_{DE} = P_1 + \Delta P_G + \Delta P_{PEC} \tag{10.13}$$

The power P_{DE} produced by the prime mover is the function of its speed ω_{gm} and produced torque T_d

$$P_{DE} = \omega_{gm} T_d \tag{10.14}$$

Therefore the speed of the engine has to follow power demand. A block diagram of the speed adjustment is shown in Figure 10.8.

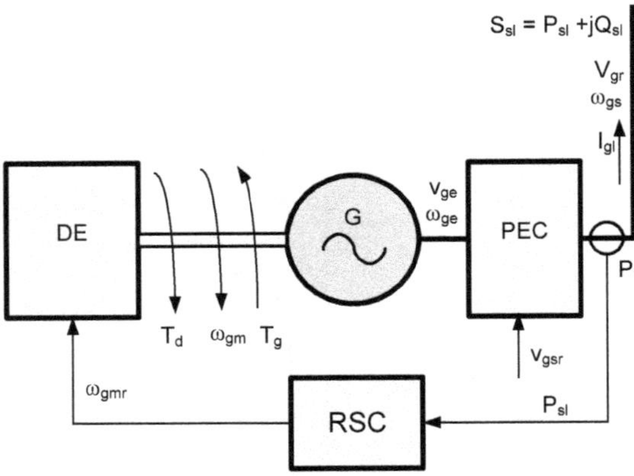

Figure 10.8. An adjustable/variable speed generation system

A power sensor PS is producing signal P_{sl} proportional to load which is delivered to reference speed controller RSC. Therefore, the engine speed follows the active power demand according to static and dynamic performances of the RSC controller, the prime mover and rotating parts inertia J_{dg}. In many prime movers the developed torque T_d depends on actual speed and then in the simple equation of acceleration

$$d\omega_{gm}/dt = 1/J_{dg}\left(T_d - T_g\right) \tag{10.15}$$

more precise data should be considered

$$\omega_{gm}(t) = f\left\{T_d\left(t, \omega_{gm}\right), T_g(t), J_{dg}\right\} \tag{10.16}$$

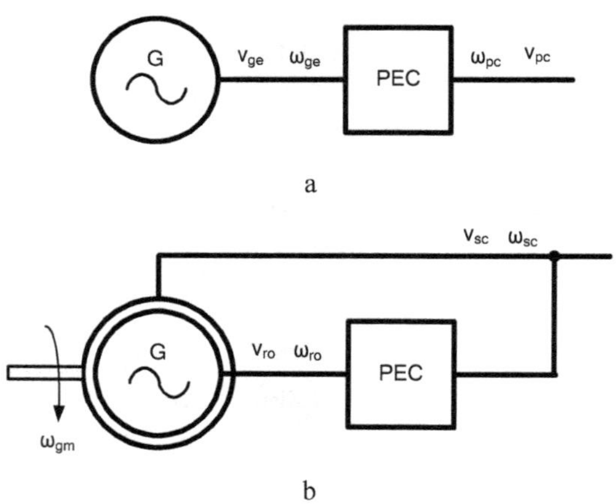

Figure 10.9. Decoupled generation systems: **a** decoupled by topology; **b** decoupled by control

The speed controlled generation system operates as an adjustable speed unit. However, they are commonly called, similar to drives, variable speed generation systems, and this name will be used in later sections.

Decoupled generation systems have two basic architectures. The first, presented in Figure 10.9a, is decoupled by topology which means that the output frequency is produced only by the power electronic converter PEC and such a system may be dedicated to the most commonly known generators. The second architecture, shown in Figure 10.9b, is dedicated only to slip-ring induction generators. In this system output AC voltage is produced by stator of the generator G as the function excitation current delivered to the rotor by the converter PEC. The stator voltage frequency ω_{sc} is a function of the generator angular mechanical speed ω_{gm}, the number of poles p and the rotor current angular frequency ω_{ro}

$$\omega_{sc} = \omega_{gm} + \omega_{rc} \tag{10.17}$$

Therefore the rotor angular mechanical speed has to be precisely followed by the angular speed of the rotor current produced by the converter PEC. In such a system the decoupled operation is achieved only by the control of the rotor current.

10.2 Electrical Power Systems

10.2.1 Introduction

The topology of the power electronic converters used in power generation depends on an applied generator, a speed range and the kind of energy storage in an intermediate DC link. In a decoupled (by topology) generation system different electrical machines — brushless permanent magnet, cage induction, wound rotor synchronous, DC with mechanical commutator or reluctance generator — may be applied as generator. Recent development of high quality magnets results in wide application of permanent magnet generators with radial or axial flux construction.

The speed range depends on the prime mover used. In the case of high speed turbines the speed is fixed around their best efficiency and small changes of speed may be caused by load power or driving torque variation. The generation unit operates, with regard to classical generators, with non-conventional speed. As their speed is not constant they will also be called "*variable speed generation systems*". Internal combustion engines provide a wide speed range and their specific fuel consumption depends on torque and speed. Therefore, the adjustment to high efficiency points of operation is the key issue to energy saving operation.

The variable speed generation systems have different construction and control strategy depending on the specification related to the kind of operation. There are three basic systems:

- Operating only in the presence of grid voltage;
- Operating only autonomously [4, 8];

- Operating autonomously and in grid connection mode [7].

At the present time the DC link storage is mostly made by DC capacitors and voltage source converters with internal commutated devices are used as power electronic systems. However, with very high power, transistor and thyristor converters may be applied with DC inductors as DC current source.

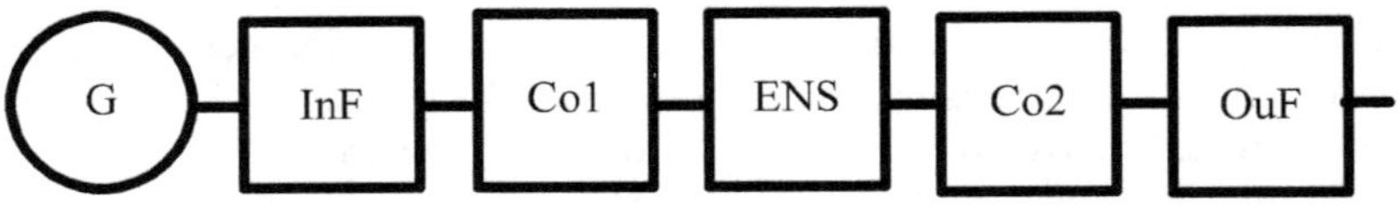

Figure 10.10. Basic components of electrical part of the variable speed generation system

The electrical power system of the variable speed decoupled generation has main sub-blocks which are shown in Figure 10.10. The generator G delivers power to converter Co1 and energy storage ENS *via* input filter InF. The converter Co2 produces the required output delivering power to the load through an output power filter. The input filter InF may be not applied when the inductance of the generator is significant.

10.2.2 Autonomous Generation Systems with Permanent Magnet Generators

The fixed excitation of the permanent magnet results in a simple construction. Figure 10.11 shows a schematic diagram of a single-phase generation unit supplied from the permanent magnet generator PMG. A three-phase rectifier Re provides DC link voltage V_{dc} which is converted to a single-phase AC voltage.

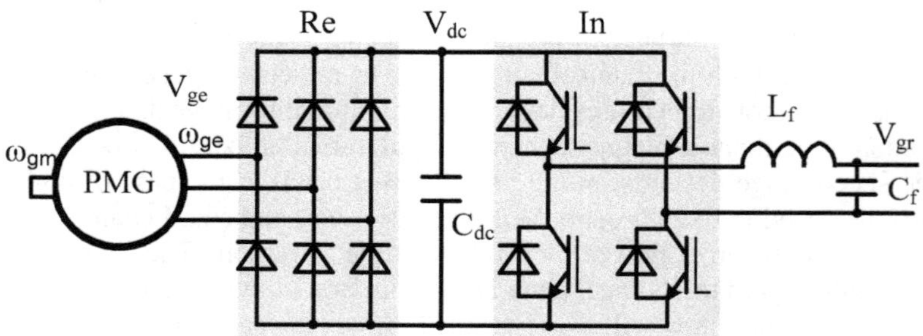

Figure 10.11. An autonomous single-phase variable speed generation for fixed speed or for the low voltage and narrow speed range

The DC link voltage has to fulfill the condition $V_{dc}>V_{gr}$, *i.e.*, it has to be greater than the amplitude of the AC output voltage The inverter In operates as a pulse width modulator and its output voltage consists of a series of pulses having a high content of the demanded fundamental. The output low pass filter (L_f–C_f) reduces voltage high harmonics.

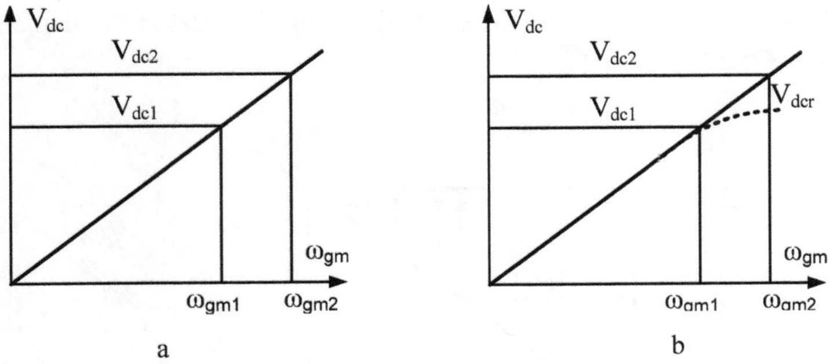

Figure 10.12. The rectified generator voltage as a function of speed: **a** voltage proportional to the rotor speed; **b** not stiff voltage regulation

The V_{dc} voltage is proportional to the rotor speed (Figure 10.12a) and then the speed has to be higher than given ω_{gm1} responding to $V_{dc1}=V_{gr}$. Maximum DC link voltage is limited by content of harmonics, increasing amplitude of fundamental and by transistor and DC link capacitor voltage ratings.

The reduction of harmonics content in output voltage at high speed may be achieved when the generator voltage regulation is not stiff (Figure 10.12b). Usually the load current increases when the speed is higher. The high current produces a DC voltage drop below V_{dc2} what may permit one to get high speed without the increase of harmonics of the output voltage. However, when the generation system is not charged (for instance in the case of a sudden disconnection of load), then the output voltage harmonics increases.

High quality of the output AC voltage provides the variable speed generation system when the DC link voltage V_{dc} is stabilized. In the three-phase generation system, shown in Figure 10.13, the generator rectified DC voltage V_{gedc} is boosted by a DC/DC converter (made from inductor L_{dc}, transistor T_{dc} and a diode D_{dc}) to V_{dc}.

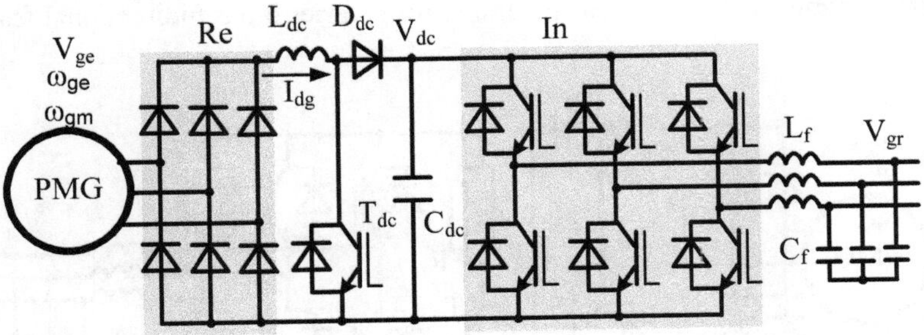

Figure 10.13. Three-phase variable speed generation system with stabilized DC link voltage

Figure 10.14 presents the rectified generator voltage V_{gedc} as a function of speed and boosted–stabilized DC link voltage V_{dc}. At the low speed ω_{gm1}, boosting factor described as $b_f=(V_{dc}/V_{gedc1})$ is the biggest. Theoretically the boosting factor may be high but in practice is not much higher than 2 [8]. In this case the speed range is as

2 to 1. At speed ω_{gm2} the generator rectified voltage is equal to the demanded DC link voltage and booster is off.

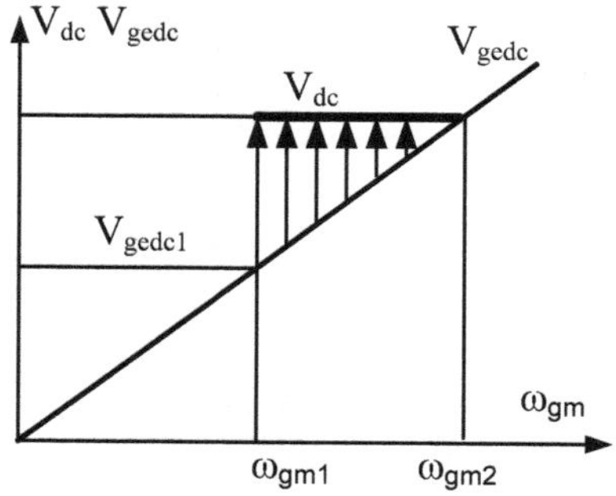

Figure 10.14. The step-up chopper operation and V_{DC} voltage stabilization

The booster also controls the generator current. The load torque produced by the generator is proportional to the rectified generator current I_{dg} (Figure 10.13)

$$T_g = k_T I_{dg} \tag{10.18}$$

where k_T is equivalent linkage flux of the permanent magnet generator.

Therefore, by control of the rectified current it is possible to control maximum load torque and the power delivered by the generator at given speed

$$P_{dg} = V_{gedc} I_{dg} \tag{10.19}$$

However, the control of the rectified DC current I_{dg} is limited when a short circuit appears or in case of the maximum speed, *i.e.*, at ω_{gm2}.

The rectifier and the booster transfer the power P_{dg} in one direction from the generator to the DC link whereas the transistor inverter It has bidirectional features

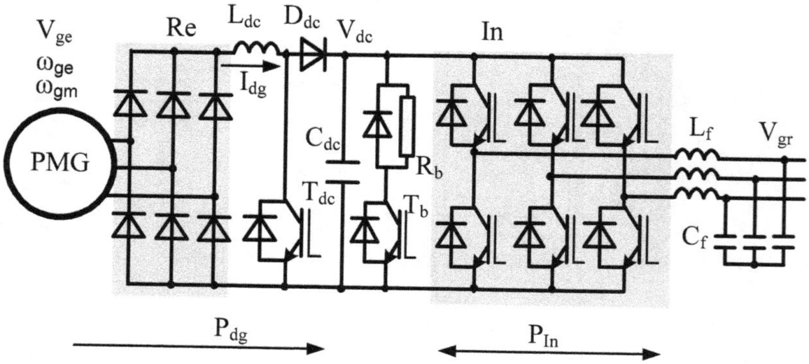

Figure 10.15. The variable speed generating system with DC overvoltage protection by resistor R_b

so it delivers power P_{In} to the load. It may also draw power from the load. Therefore, as the load is the motor in state of energy recovery braking (generator braking) the inverter absorbs this power and transfer it to the DC link. Moreover the DC link capacitor C_{dc} capacitance is low (region of tens of mF) and then its voltage grows quickly to dangerous overvoltage. To protect the generation system against this overvoltage the recovered energy may be dissipated by a braking resistor controlled by a transistor. Figure 10.15 shows the generation system with braking resistor R_b controlled by series connected transistor T_b. The resistor discharges DC link capacitor C_{dc} until the DC link voltage drops to the reference level.

The rectifier, connected to the rotor of the permanent magnet generator, (Figure 10.16) draws a square or trapezoidal waveform current [8]. The permanent magnet generator phase inductance results in an overlapping phenomenon causing the delay of the fundamental component of the current *vs* to induced voltage.

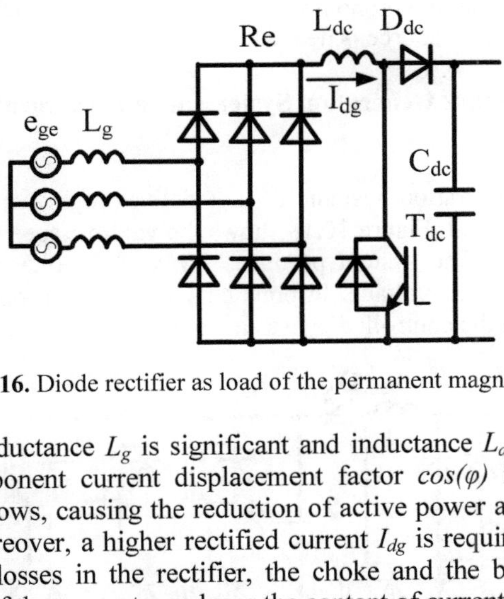

Figure 10.16. Diode rectifier as load of the permanent magnet generator

If the stator inductance L_g is significant and inductance L_{dc} removed then the fundamental component current displacement factor $cos(\varphi)$ is reduced and the reactive current grows, causing the reduction of active power additional heating of the generator. Moreover, a higher rectified current I_{dg} is required which results in additional power losses in the rectifier, the choke and the booster. However, a small inductance of the generator reduces the content of current harmonics.

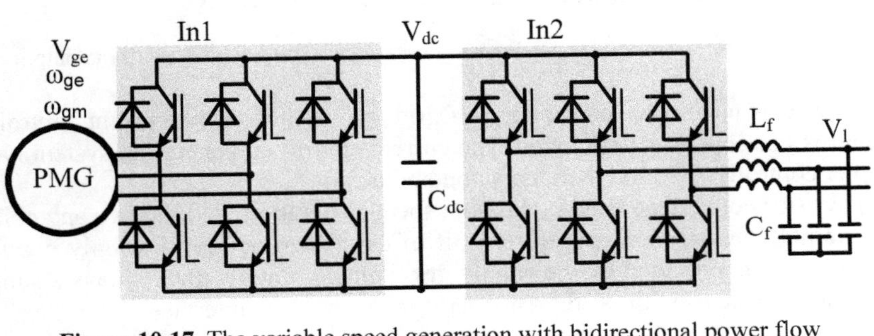

Figure 10.17. The variable speed generation with bidirectional power flow

The converter connected to the permanent magnet generator may have several different topologies, resulting in different impacts on the current drawn from the generator. Figure 10.16 shows the generating system with two fully controlled three-phase transistor converters. The converter In1, connected to the generator, assures bidirectional flow of power, actively creating the waveform of the current and fundamental current displacement.

Therefore, in the case of zero displacement of the current, the generator and converter may perform at minimum volt-ampere power. Moreover, within a change of current phase to π, the generator changes its state to motoring and absorbs power from the DC link protecting it against overvoltage events. At low speed the converter In1 operates as a booster controlling the DC voltage as shown in Figure 10.14.

The generation system shown in Figure 10.17 may be modified using the permanent magnet generator of a cage induction generator instead. The cage induction machine control is well developed in the drive system. However, in the case of demanded reliable autonomous operation, a self excitation or preliminary excitation by an external source is needed.

10.2.3 Non-autonomous Generation Systems with Permanent Magnet Generators

Non-autonomous generation systems do not deliver high quality power without connection to a stiff grid. Figure 10.18 shows the variable speed generation system with L output filter. The system produces PWM AC voltage which cannot be accepted as standard AC voltage. In practice such a system operates as a voltage source inverter with the controlled phase and amplitude of the current delivered to

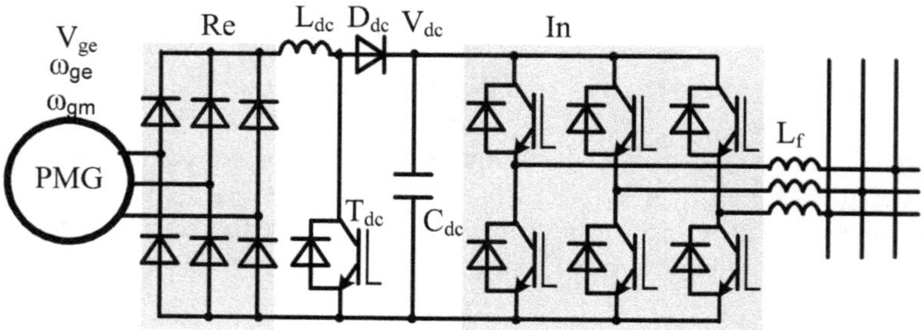

Figure 10.18. The variable speed generation system with L type filter in the output

the grid. This means that in the case of grid power failure the current controlled system is not able to supply loads. The current controlled generation system has a simpler controller solution than the autonomous.

Thyristor converters, used in variable speed generation, do not have any ability for internal commutation caused turn-off of conducted device. The only possible commutation is provided by the supply grid voltage. Figure 10.19 shows a simple generation system with a thyristor bridge as a grid interface. The thyristor converter TyIn operates in an inverter mode with firing angle α range $(\pi/2)<\alpha<\pi$. The converter produces DC voltage

$$V_{dgr} = V_{dgr\max}\cos(\alpha) \tag{10.20}$$

where V_{dgrmax} is output DC voltage for α=0.

Thus converter balances rectified generator voltage V_{dge}

$$V_{dge} = V_{d\max}\omega_{gm} / \omega_{gmx} \tag{10.21}$$

where V_{dmax} is rectified voltage for ω_{gmx}.

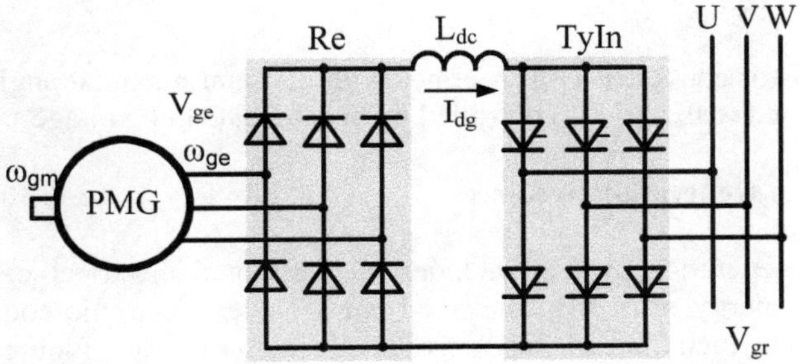

Figure 10.19. The variable speed generation system with thyristor grid interface

For state of balance

$$V_{d\max}\omega_{gm} / \omega_{gmx} = V_{dgr\max}\cos(\alpha) \tag{10.22}$$

The cos(α) is a function of speed

$$\cos(\alpha) = (V_{d\max} / V_{dgr\max})(\omega_{gm} / \omega_{gmx}) \tag{10.23}$$

The power factor (displacement factor) cos(φ) is close to cos(α)

$$\cos(\varphi) = \cos(\alpha) \tag{10.24}$$

which means that at the low speed the cos(φ) is very low and the power delivered to grid needs high content of reactive power.

Moreover, even in the case of short disturbance cuts in the supply grid voltage — just for tens of milliseconds — the thyristor converter operating in inverter mode passes to a state of short circuit and should be disconnected. Hence such a system performs with very low reliability.

The cos(φ) of the thyristor converter may be improved when the rectified generator voltage is boosted and stabilized in the system shown in Figure 10.20.

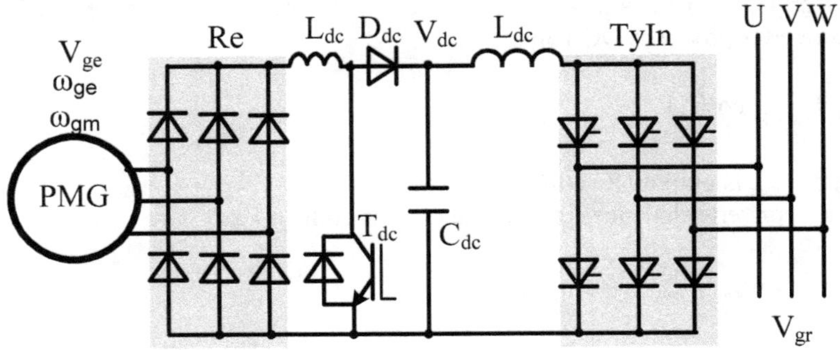

Figure 10.20. The improvement of converter *cos(φ)* by the application of the booster of rectified DC voltage

The thyristor converter TyIn operates with constant maximal angle $\alpha_{max}<\pi$ and power delivered to the grid is controlled by boosted DC link voltage.

10.2.4 Hybrid Generation Systems

The hybrid generation system includes an additional electrical energy source. Usually this energy source is connected *via* a power electronic converter to DC link voltage in which there already exists small energy storage. Figure 10.21 shows a block diagram of the hybrid variable speed generation system where this additional energy storage ENS2 is connected through DC/DC converter Co3 to DC link energy storage ENS1.

An example of variable speed hybrid generation with an electrochemical battery bank as energy storage is shown in Figure 10.22. The battery bank B_b and smoothing inductor L_{bf} are connected to DC link voltage circuit *via* bidirectional DC/DC converter made from transistors T_{bc} and T_{dc}. The battery energy is used to supply the inverter In when the engine is off or when demand for load power is higher than the engine can deliver. At low speed the power of the engine is low and the battery adds its power and then improves the quality of power produced by the variable speed generating unit.

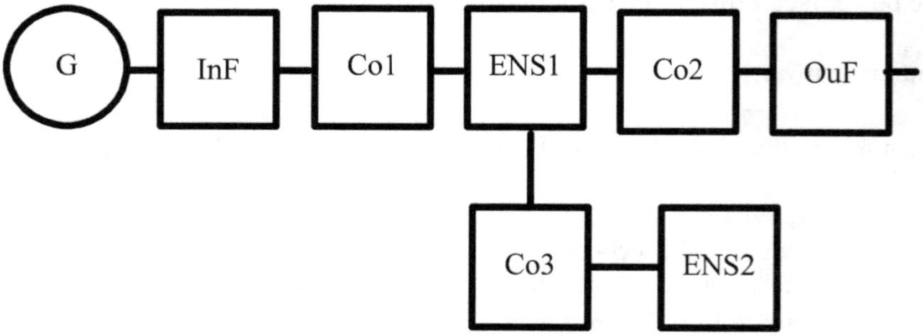

Figure 10.21. The block diagram of the hybrid generation system

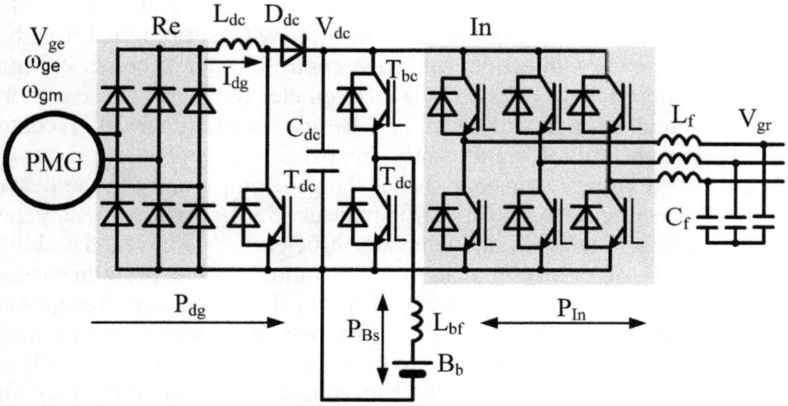

Figure 10.22. The hybrid variable speed generation system with battery bank

The battery is charged in a state of low load operation. Instead of a battery bank, a supercapacitor bank may be applied as shown in Figure 10.23. In this case stored energy is much less than in battery and such a storage system is designed for a short time. This time is usually sized to the time of engine acceleration from low to high speed.

10.2.5 Engine Starting in Power Electronic Generation Systems

Internal combustion engines have their own electrical motor called "*starter*" which drives the engine until it produces torque sufficient to keep the operation stable. The starter is supplied from a special battery bank. In hybrid variable generation systems, equipped with battery or supercapacitor, it is possible to use this storage to start the engine. In this case the permanent magnet generator is used as the

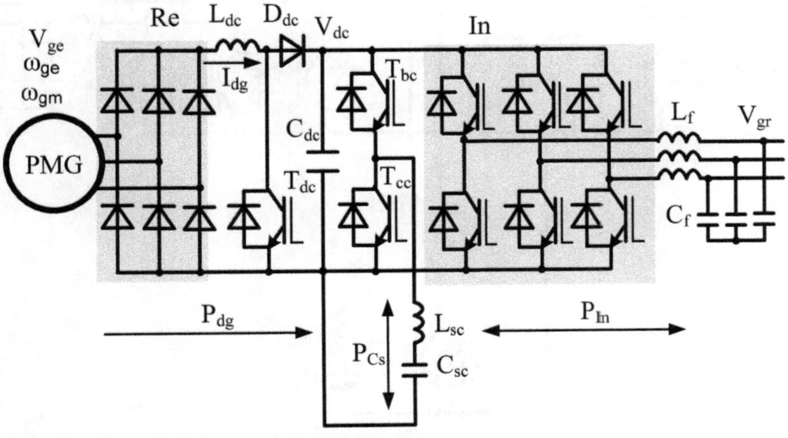

Figure 10.23. The hybrid variable speed generation system with supercapacitor bank

motor. Figure 10.24 shows the hybrid generation system with fully controlled transistor converter In1 connected to the generator PMG and supplied from battery B_b. To start in the right direction, the drive control system requires information about pole position. It may be done by a position electromechanical sensor or by a sensorless method such as PIPCRM [9]. Besides starting the fully controlled transistor converter In1 it significantly improves the generation system. The current controlled voltage converter shapes the generator current to the demanded waveform, for example, to sine wave. It reduces the harmonics of the generator current, increasing in this way the generator efficiency. Moreover, the ability for bidirectional flow of energy extends the generation system performances. This means that during regenerative braking of the load the DC link voltage may be stabilized by transfer of energy to PMG which then operates as a motor with rising speed and changing electrical energy into kinetic mechanical energy. Parallel to the generator, energy may be stored by the battery bank. Such an ability may also be used to damp grid voltage oscillations.

External combustion engines, just like high speed turbines, require starting systems as well. Figure 10.25 shows an example of a high fixed speed generation system equipped with a special starting supply system. The starting system is connected to the generator only during starting by the switch SS. The low voltage battery is boosted by DC/DC converter SB and supplies converter SIn which, with the generator, operates as a brushless DC motor. After the drive reaches minimum speed, ensuring safe driving operation of the turbine, the starting system is disconnected. The turbine reaches its high speed and the permanent magnet generator produces high voltage but disconnected starting system rated voltage may be much lower than the main power system.

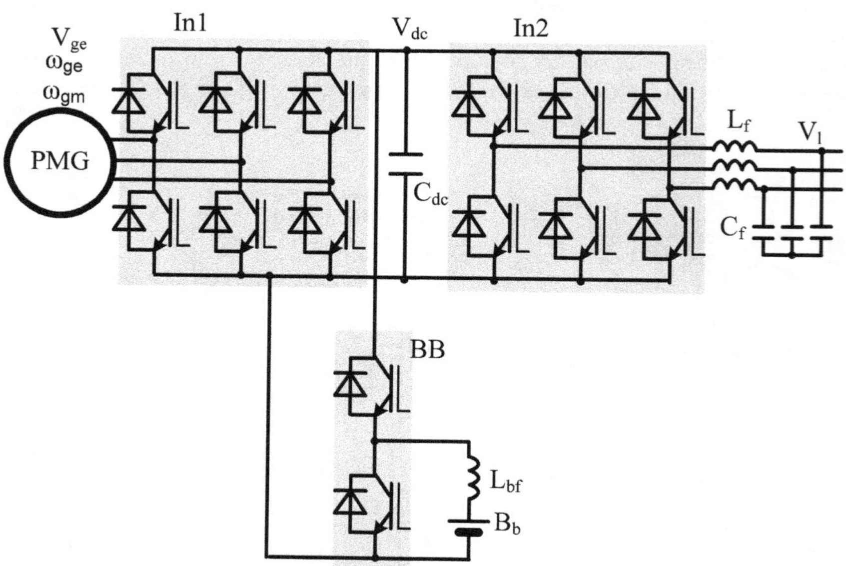

Figure 10.24. The variable speed generation system and starting from external battery

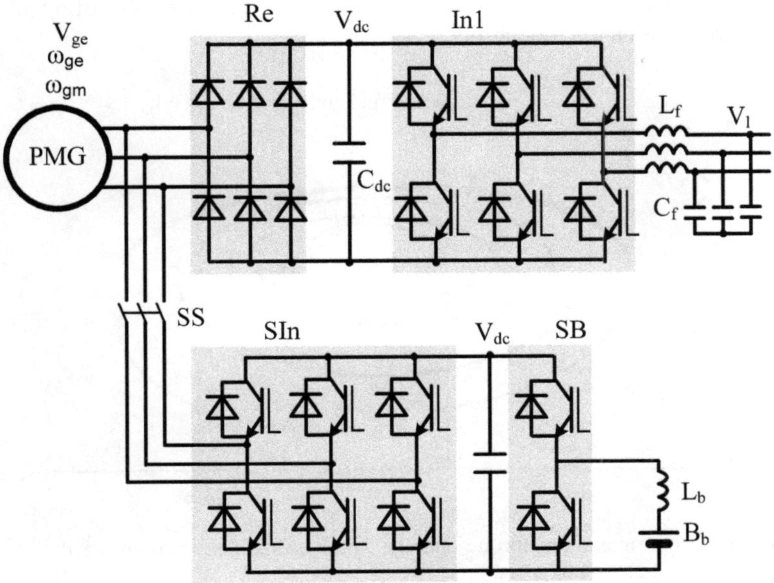

Figure 10.25. The variable speed generation system – starting from battery through external converter

10.3 Prime Movers and the Control System

10.3.1 Prime Movers

In low power, *i.e.*, below 2 MW, generation units the most popular prime mover is the diesel engine. This engine performs over a wide range of speeds with a very flat maximum torque curve but its efficiency depends on the load torque. A simplified example of an industrial diesel engine maximum torque T_{dmax} and power P_{dmax} as a function of angular speed are shown in Figures 10.26 and 10.27. The figures also present specific fuel consumption g_f representing grams of fuel per 1 kWh produced energy. In the map of the specific fuel consumption the most efficient areas of operation (*i.e.*, the lowest specific consumption) are marked as g_{min}. The map indicates that, for low load torque, the specific fuel consumption is high. Additionally, in Figures 10.26 and 10.27 the dashed line shows torque and power of fixed speed ω_{d50} related to conventional wound rotor synchronous generator directly producing 50 Hz voltage. This is usually ω_{d50}=157.08 rad/s (1,500 rpm).

Each type of diesel engine has its own different fuel consumption map but most engines have their area of minimum specific fuel consumption for speeds higher than 157.08 rad/s. This means that the change strategy of generation technique from fixed speed to variable speed may result in the speed adjusting to more efficient points of engine operation.

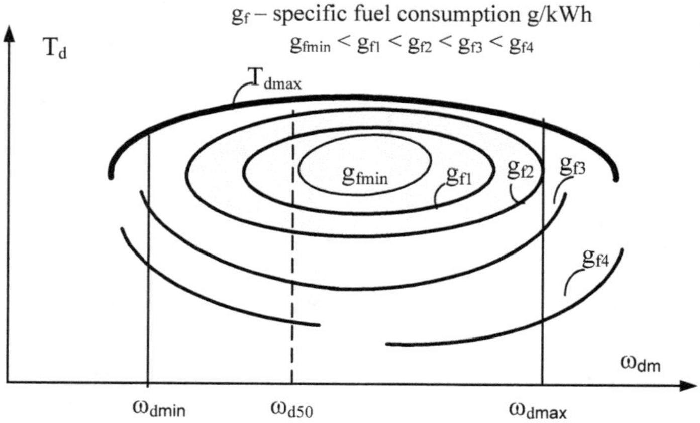

Figure 10.26. The maximum torque and the specific fuel consumption of the industrial diesel engine as a function of speed

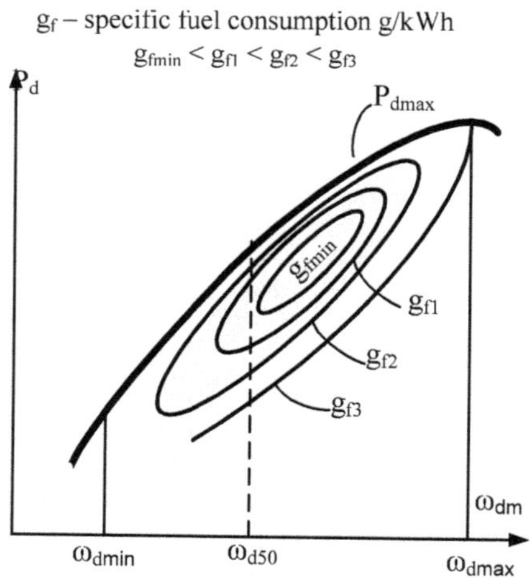

Figure 10.27. The maximum power and the specific fuel consumption of the industrial diesel engine as a function of speed

10.3.2 Speed Control Strategies

The criteria of control of the variable speed generation system are based on fuel savings and/or quality of the output AC voltage in the case of the step load. The generator speed controller GSC (Figure 10.28) provides a reference speed signal S_{De} on the basis of signals received from power electronic converter PEC, generator G and driving engine DE. The speed control systems use the method of DC link voltage stabilization [4], output power [7], rectified generator current, and rectified generator power.

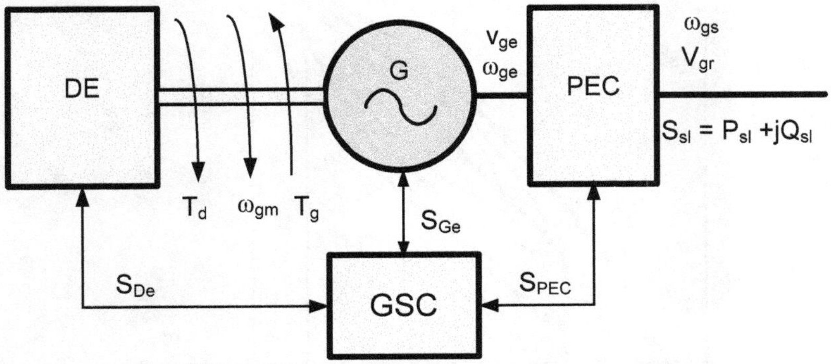

Figure 10.28. Control of the variable speed generation system

The received signal, used to determine reference speed, is filtered and corrected using, for instance, look-up table. Figures 10.29 and 10.30 show torque and power representing the two different methods of speed control.

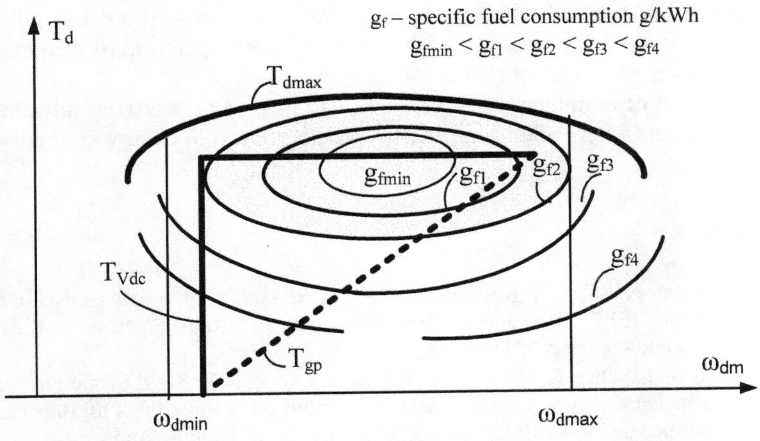

Figure 10.29. Load torque of the variable speed generation systems

The first method (torque T_{Vdc}) is based on maintenance of constant voltage in the DC link of the power electronic converter. Assuming limited rectified current [4] the DC voltage is controlled by generator speed. In the case of load increase the DC link voltage drops. Therefore for higher power demand (Equation 10.14) higher speed is adjusted. This method results in very efficient generation system operation and derived power P_{Vdc} (Figure 10.30) is very close to the maximum power of the engine.

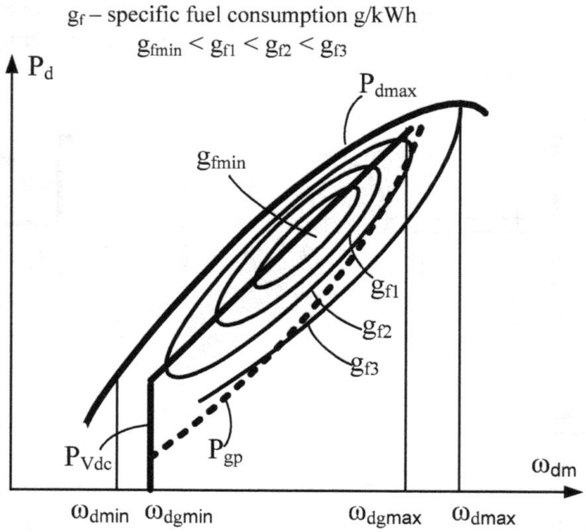

Figure 10.30. Load power of the variable speed generation systems

However, the margins between maximum available torque (power) and load torque are small which results in low power rate in transients. In contrast to this method the dashed line represents the torque (Figure 10.29) and the power (Figure 10.30) based on the method of rectified or output power without any correction from look-up tables. This method leads to less efficient operation of the generation set but assures a much higher power rate in transients.

The method of maintenance of constant DC link voltage is very advantageous in hybrid systems where additional energy is delivered from energy storage.

References

[1] Al-Khayat N, Seliga R, Koczara W, (2002) DSP control of variable speed integrated generator. IEEE International Symposium on Industrial Electronics IEEE-ISIE, IEEE catalog number 02TH8606C, L'Aquila, Italy

[2] Al-Tayie J, Seliga R, Al-Khayat N, Koczara W, (2003) Steady state and transient performances of new variable speed generating set. Proceedings of 10th European Conference on Power Electronics and Application, Toulouse, France

[3] Bolognani S, Venturo A, Zigliotto M, (2000) Novel control technique for high-performance diesel-driven generator-sets. Conference Proceedings on Power Electronics and Variable Speed Drives, no.475

[4] Da Ponte M, Grzesiak L, Koczara W, Hybryd generator apparatus. Patent WO 9828832, Patent US6175217

[5] Grzesiak L, Koczara W, Da Ponte M, (1998) Novel hybrid load-adaptive variable-speed generating system. Proceedings of the IEEE International Symposium on Industrial Electronics:271–276

[6] Grzesiak L, Koczara W, Da Ponte M, (1999) Application of permanent magnet machine in the novel hygen adjustable-speed load-adaptive electricity generating

system. Proceedings of the IEEE International Electric Machines and Drives Conference, Seattle, USA:398–400

[7] Koczara W, Ernest E, Al-Khayat N, Tayie JA, (2004) Smart and decoupled power electronic generation system. Conference Proceedings IEEE PESC, Aachen, Germany

[8] Koczara W, Leonarski J, Dziuba R, (2001) Variable speed three phase power generation set. Proceedings of 9th European Conference on Power Electronics and Application, Graz, Austria

[9] Koczara W, Wisniewski J, Jakubowski P, (2007) Pole position identification of permanent magnet axial flux motor using PIPCRM sensorless method. Proceeding of the 12th European Conference on Power Electronics and Applications, Aalborg, Denmark

11

Grid Integration of Wind Energy Systems

Detlef Schulz

Department of Electrical Engineering,
Electrical Power Engineering,
Helmut-Schmidt-University,
Holstenhofweg 85, D-22043 Hamburg, Germany.
Email: Detlef.Schulz@hsu-hh.de

11.1 Introduction

Considering all renewable sources, wind energy has a global share of 42%. The number of Wind Energy Converters (WECs) rapidly increases world-wide. By the end of 2007 a global wind power of nearly 94 GW had been installed [1]. Looking at country-dependent models for improved feed-in tariffs or quota system models the economical position of this energy conversion type has improved significantly over the last ten years. At the same time these different financial conditions has led to diverse installation numbers. As a result Europe has 75% of the installed wind power, America 15%, Asia 10%, Africa 1% and Australia/Pacific 1.5%. Beacuse of this growth, mainly Europe has become a high technology wind energy area with new drive train concepts, high generator powers and high rotor blade lengths. Unit powers increased rapidly from 180 kW in 1992 to 6 MW in 2007, whereas the higher unit powers above 3 MW are exclusively European technology. These high power WECs are mainly designed for offshore installations, which is a European speciality due to the necessary geological and economical conditions. The highest country share of global wind power belongs to Germany with 28%, followed by Spain, USA, India and Denmark. High potential for future years is forecast for USA, Spain, France and Poland. The United States installed 5.2 GW in 2007, the highest share of new wind power [1]. Onshore and offshore installations require different technologies and therefore they are discussed separately.

11.2 System Overview

Wind energy converters refers to the technical system for the conversion of wind power into electrical power. A description of wind energy conversion into usable electrical energy requires an explanation of the combined mechanical and electrical components, which are balanced together in control circuits. Relating to the scope of this book, the electrical functionality of the WECs and their interaction with the power system will be in the foreground. In Figure 11.1 an overview of WEC construction types with upwind rotor is given. Figure 11.1a shows a WEC type with gearbox. The rotor speed, which is often in the range of 8–22 rpm, is the input value for the gearbox that usually has a transmission ratio of 1:90. On the output side the speed is around 1,500 rpm, which is the synchronous speed of four-pole generators. The use of standard four-pole induction machines is favored in this WEC type. A drawback of this system is the need of a fast rotation gear that can be a source of mechanical outages over the expected operation time of 20 years. This WEC type can be visual by identified by the long horizontal dimensions of the drive train, resulting in a long nacelle shape. Figure 11.1b shows an other solution without a gearbox.

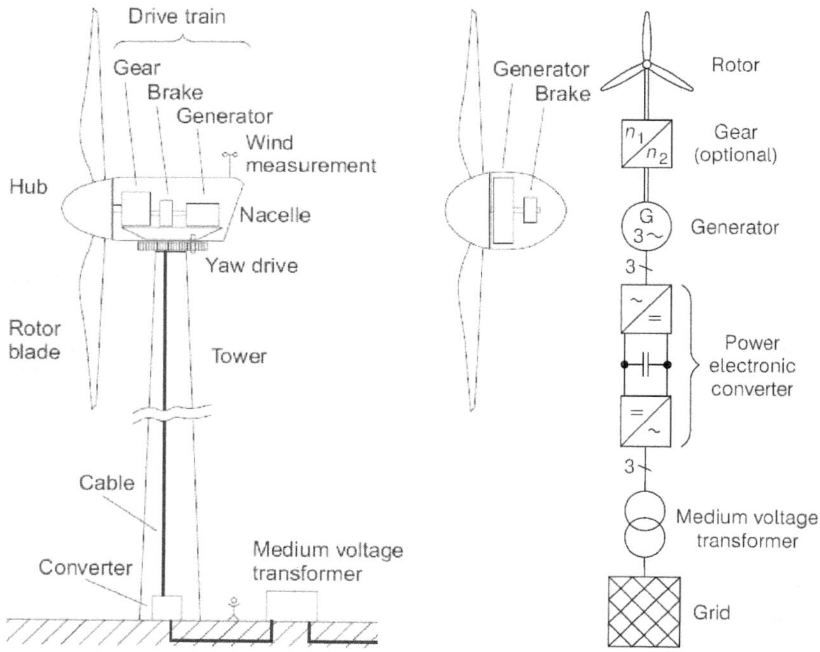

Figure 11.1. Principle of a wind energy converter: **a** WEC with upwind rotor, gearbox and four-pole machine; **b** tower head of a gearless type with multi-pole generator; **c** overview of a grid connected WEC

There the pole number of the generator has to be higher for a lower synchronous speed in the range of the mechanical rotor speed. A higher pole number requires more coils in the stator circumference and therefore the diameter of the generator is greater compared to the WEC type with gearbox. The gearless type can be equipped with synchronous machines with electrical or permanent magnet excitation. An advantage of the system is the absence of a high speed gear, in contrast to the disadvantage of the higher weight of the nacelle due to the bigger iron and copper parts of the generator. This type is identifiable from the nacelle shape with its high diameter and short length. Figure 11.1c gives the system overview with focus on the electrical part and grid connection. In this case the generator is connected to a full-size power electronic Voltage Source Converter (VSC), consisting of two back-to-back Voltage Source Inverters (VSI). The converter output is grid connected *via* a medium voltage transformer. Even though a three-phase generator system is depicted, multi-phase systems are also available on the market. For the generator and inverter, low voltage components are used, that require high currents in high power ranges, but reduce the isolation and construction efforts of the components.

Figure 11.2 shows respectively one practical construction example of the two described WEC types. In Figure 11.2a the long nacelle dimension is visible Figure 11.2b shows the gearless type with a higher nacelle diameter. Both described WEC types are in use in power ranges from some hundred kW up to 5 MW. Each type can also be used for higher power ranges in future.

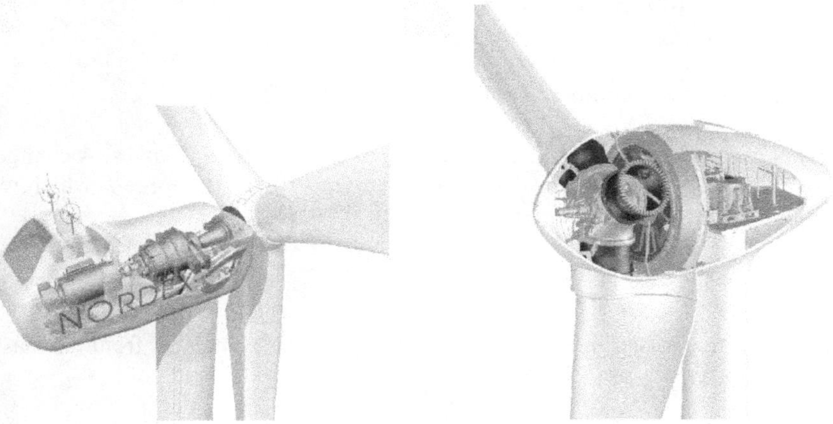

a b

Figure 11.2. Exemplary main types of WECs: **a** Nordex N80, fast rotating four-pole generator with gearbox, nominal power 2.5 MW, rotor diameter 80 m (source: Nordex); **b** Enercon E-82, multi-pole generator without gearbox, nominal power 2 MW, rotor diameter 82 m (source: Enercon)

Another type of specially constructed WEC is shown in Figure 11.3. It is totally different from the two systems described so far. The WEC uses a combination of low-speed gearbox and medium voltage multi-pole synchronous generator with permanent excitation. A gear with low speed is not as sensitive as high speed components. Permanent magnet excitated generators have up to 30–40% less volume and weight compared to electrical excitated machines. The medium voltage level allows smaller cross sections for the machine windings and cables. For the grid connection a medium voltage power electronic converter is necessary.

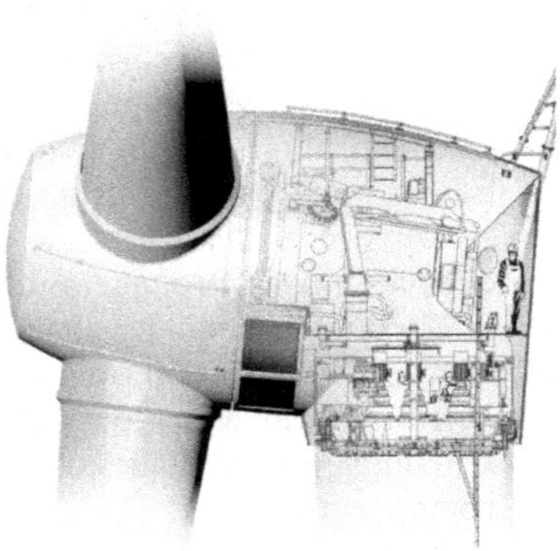

Figure 11.3. Half-section drawing of a 5 MW WEC (Multibrid M5000) with permanent magnet excitated synchronous generator and slowly rotating gearbox (source: Multibrid)

11.3 Wind Energy Converters

11.3.1 Energy Conversion

Wind is caused by temperature and pressure differences in air masses. A change of these parameters alters the air density ρ. A volume V causes a force F which, with of g the acceleration of gravity is

$$F = V \cdot g \cdot \Delta\rho \tag{11.1}$$

When this force arises, kinetic energy is produced. It is calculated from the mass m and the velocity v

$$E = \frac{1}{2} \cdot m \cdot v^2 \tag{11.2}$$

If an air mass flow $\dot{m}$ instead of a constant mass and a constant velocity is assumed, the wind power P_w results

$$P_W = \dot{E} = \frac{1}{2} \cdot \dot{m} \cdot v^2 \tag{11.3}$$

The air flow $\dot{m}$ can be calculated from the density ρ, wind velocity v and the rotor area A, which is defined by the rotating blades

$$\dot{m} = \rho \cdot \dot{V} = \rho \cdot A \cdot v \tag{11.4}$$

By insertion of Equation 11.4 into Equation 11.3, the theoretical wind power P_0 results

$$P_0 = \frac{1}{2} \cdot \rho \cdot A \cdot v^3 \quad (11.5)$$

This means that power P depends on the third power of the wind speed, and this is mainly influenced by the installation site and the tower height. Of course there is a limit to the convertable wind power. This limit is expressed by the power coefficient c_p, also known as the Betz-factor [2]

$$P = \frac{1}{2} \cdot \rho \cdot A \cdot v^3 \cdot c_P(v) \quad (11.6)$$

The wind speed dependent power coefficient describes the amount of energy converted by the wind turbine. Its theoretical maximum value is c_p=0.593. In practice WEC power coefficients occur in the range of c_p=0.4–0.5; see Figure 11.4. Below the nominal operation range the factor is smaller. It is highly influenced by the rotor blade profile [3–9].

11.3.2 Tip Speed Ratio and Power Curve

The technical applications considered include horizontal axis Wind Turbines (WT) with three blades with electrical powers above 1 MW, which are connected to the electrical transmission system. In almost all of them the rotors are located upwind; only some older devices have downwind rotors. Depending on the nacelle weight the tower is constructed from steel or steel/concrete. Standard tower heights are up to 100 m. Due to height dependent wind speeds, some new devices have lattice towers up to 160 m in height.

Commonly, WT have a high tip speed; the tip speed ratio λ is calculated using the tip speed v_{TS} and the speed of the rotor plane v_{RP}

$$\lambda = \frac{v_{\mathrm{TS}}}{v_{\mathrm{RP}}} = \frac{\Omega \cdot r_{\mathrm{R}}}{v_{\mathrm{RP}}}; \quad \Omega = 2 \cdot \pi \cdot n . \quad (11.7)$$

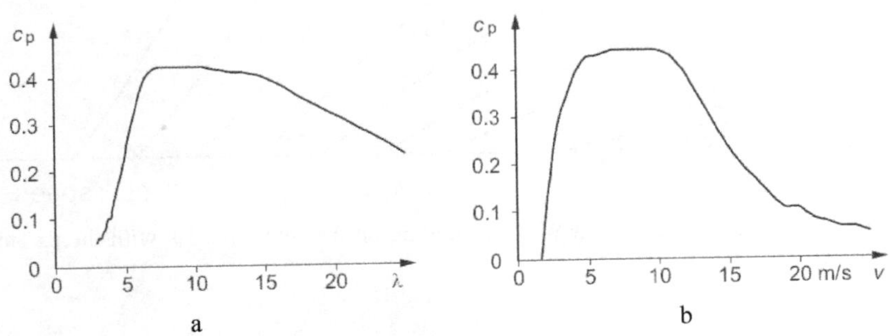

Figure 11.4. Power coefficient c_p of a WEC depending on: **a** the tip speed ratio λ; **b** the wind speed v

The rotor speed Ω is given in rpm, the rotor radius r in m and the speed n in s^{-1}. Typical values of the tip speed ratio of three-blade WT are in the range of 8–10. The tip speed ratio influences the power coefficient. Figure 11.4 shows the dependencies of the power coefficient on the tip speed ratio and on the wind speed. Both diagrams deliver a non-linear relation between the depicted values.

Applying the curve of $c_p(v)$ from Figure 11.4 to Equation 11.6, the power-speed curves of a WEC occur as shown in Figure 11.5. These curves show the rotor power over the rotational speed depending on the parameter wind speed. As expressed by Equation 11.6, the power increases non-linearly with increasing wind speed. Obviously the power maxima are dependent on wind speed the rotor speed. For an optimal power yield at variable wind speeds a variable rotor speed is necessary. Therefore variable speed WECs deliver in particular at strong changing wind speeds a higher energy output compared to fixed speed devices.

Attention should be paid to the fact that, due to the wind speed dependency of the power coefficient c_p, the tip speed ratio λ is also speed dependent; compare with Figure 11.4.

For adaptation of the generator curve to the mechanical trait of the WEC a torque-speed graph is required. Generally the rotor torque T can be derived from the rotor power P, the angular frequency ω in s^{-1} and the rotor speed n

$$T = \frac{P}{\omega}; \quad \omega = \frac{2 \cdot \pi \cdot n}{60} \tag{11.8}$$

with n in rpm.

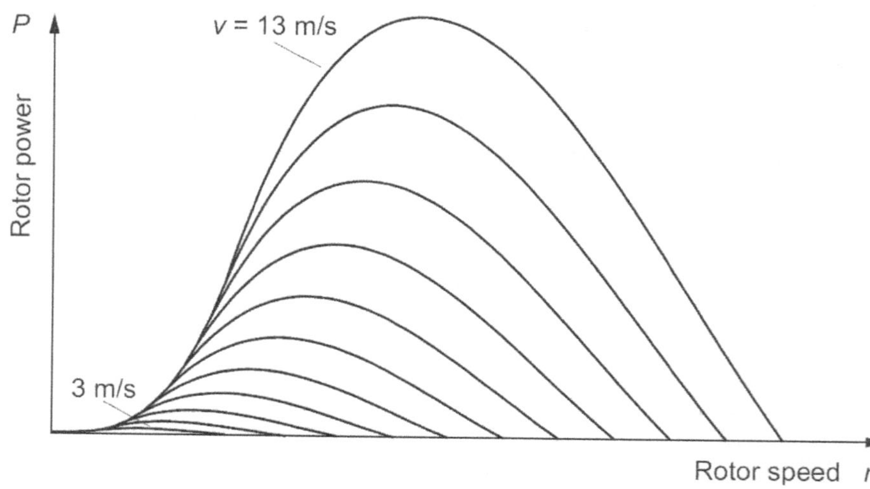

Figure 11.5. Rotor power P of a WEC depending on the rotor speed n with the parameter wind speed v

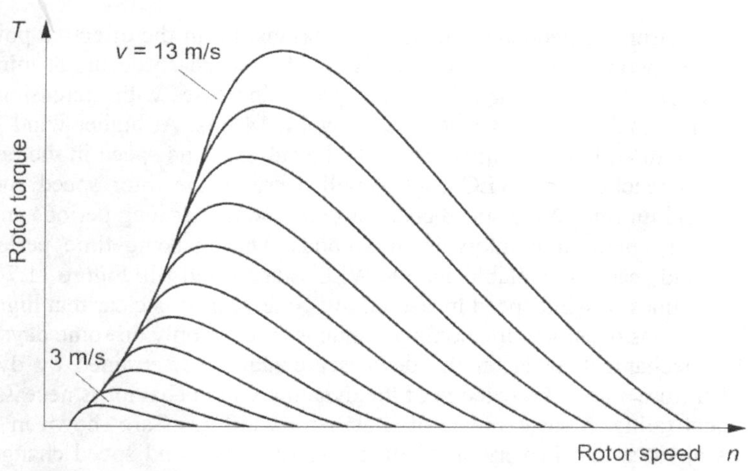

Figure 11.6. Rotor torque T of a WEC depending on the rotor speed n with the parameter wind speed v

Using Equation 11.8, the set of curves from Figure 11.5 can be converted to the new array of curves in Figure 11.6. These graphs are useful to determine the working curve of the generators and, based on this, to develop the control circuits and the control strategy of the WEC.

11.3.3 Operation Modes

Unlike conventional fossil energy conversion, WECs produce a wind speed dependent output power. The power is calculated according to Equation 11.6, which

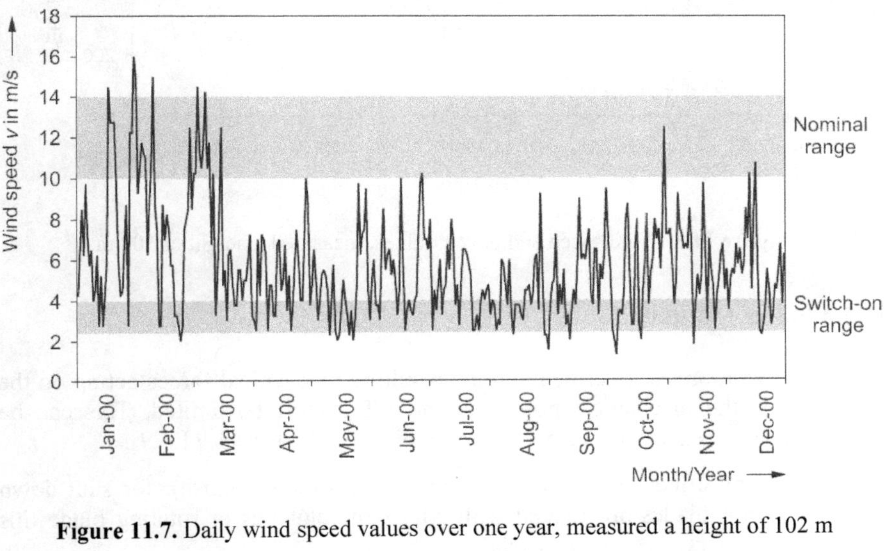

Figure 11.7. Daily wind speed values over one year, measured a height of 102 m

describes the strong dependency on wind conditions. From the effective power of the WEC the operation modes can be derived. The switch-on occurs at minimum wind speeds of 2.4–4 m/s. Then the output power increases with increasing wind speed. The nominal WEC power is reached at 11–14 m/s. At higher wind speeds the power is limited to the nominal value. If the cut-off wind speed in the range of 18–25 m/s is reached, the WEC control will decrease the rotor speed and shut down the wind turbine. Wind speed changes occur both over long periods and very quickly within small time intervals of seconds. During some time periods no minimum wind speed is available and the WEC is at a standstill. Figure 11.7 shows daily mean values of wind speed measured over one year. It is clear that high wind speed fluctuations occur and the nominal range is reached only on some days of the year. The mechanical loads on the devices are mainly determined by dynamic forces. Therefore a good knowledge of the dynamic wind behavior is necessary for a component rating. Measurements of short-term wind gusts are shown in Figure 11.8. It is obvious that in very small time intervals fast wind speed changes can arise. Such big speed changes cause strong variations of the output power. They also stress the mechanical components of the WEC. Therefore the mechanical power acting on the device has to be limited in order to protect the WEC.

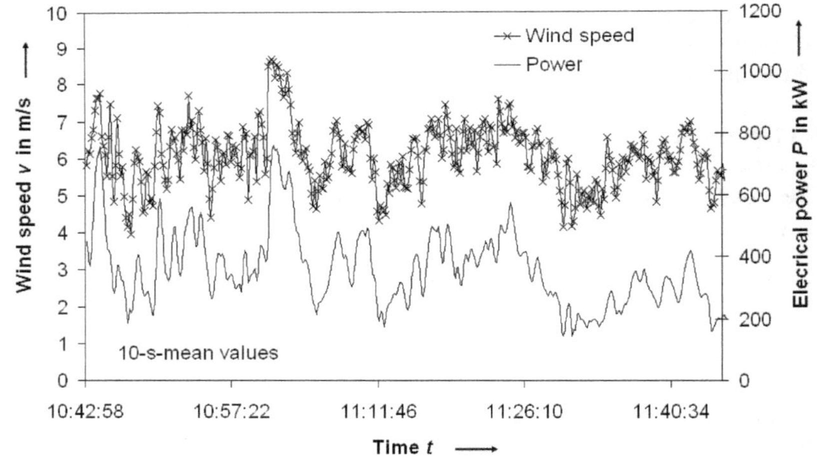

Figure 11.8. Wind speed and power values, measured a height of 102 m

11.3.4 Power Limitation

Due to the wind gusts described and the resulting mechanical forces acting on the rotor blades, the mechanical power on the WEC must be limited. This can be realised with three approaches, which are depicted in Figure 11.9 [3–9]:

- The stall-effect on fix mounted rotor blades (stall-control): for shut down the rotor blades are braked with adjustable clutches or rotating blade tips (tip spoilers); see Figure 11.9a;

- An adjustable stall-effect, realised with turnable rotor blades (active-stall-control): the blades turn the thin side in the wind direction; see Figure 11.9b;
- The reduction of the effective rotor area by axial rotation of the rotor blade in the wind direction (pitch-control): the turn is opposite to the active-stall-control; it has a higher rotation angle and a higher rotation speed; see Figure 11.9c.

Stall-controlled WECs are limited to power ranges below 2 MW, and active-stall-control is realised by only one manufacturer. In higher power classes all devices are equipped with pitch-control, which enables a fast reaction on high mechanical power spikes and makes exact mechanical dimensioning possible. The power limitation is independent of speed regulation.

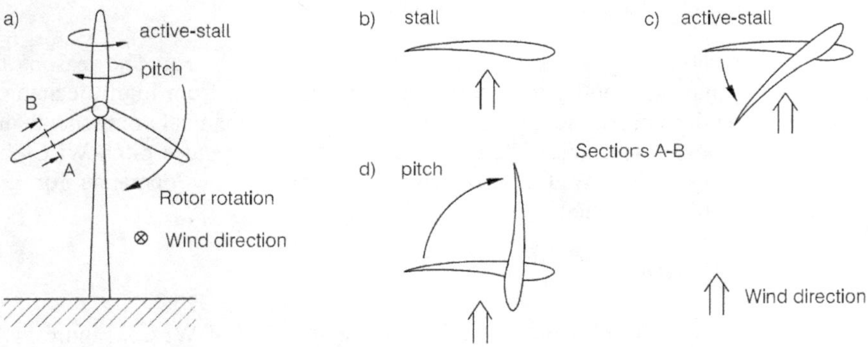

Figure 11.9. Aerodynamic limitation of the mechanical power: **a** overview of the WEC; **b** stall-; **c** active-stall-; **d** pitch-control

Depending on the type of power limitation, different power-speed curves of WECs can be observed. Figure 11.10 shows a comparison of the power-speed graphs of three WECs each with 1.5 MW of nominal power. Despite different speed ratings the typical behavior is visible. The stall-controlled WEC in Figure 11.10a shows a small overshoot of the nominal power, followed by a decrease of the power. WECs with active-stall-control do not exceed the nominal power and no power increase arises; see Figure 11.10b.

Pitch-controlled WECs have the same power-speed curve as the active-stall-controlled types; see Figure 11.10c, but they rotate the blade ends in the wind direction and so in the opposite direction compared to the active-stall-control. Another important difference is the faster blade adjustment of the pitch devices. This allows a quick reaction on wind speed changes. A bigger moving angle of the blades makes force reduction on the blades possible up to high wind speed ranges of 25 m/s and so the nominal electric power can be delivered over a wider operation range.

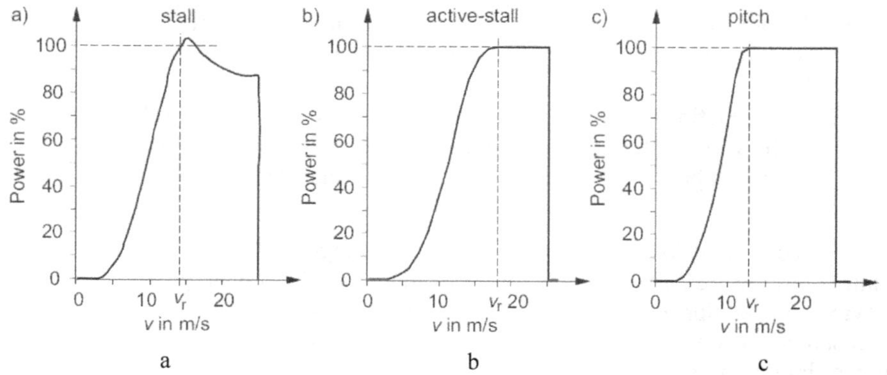

Figure 11.10. Power over wind speed of WECs with different power limitation strategies: **a** stall-control; **b** active-stall-control; **c** pitch-control

Basically the energy yield is influenced by the speed control. The reasons for the use of active-stall- and pitch-control are for protection from high mechanical loads. Stall devices make the highest demands on the mechanical components and cause higher noise emissions. Therefore in higher power ranges pitch-WECs are preferably in use. Stall-WECs also cause problems if new demands for grid integration should be applied.

11.3.5 Speed Control

The type of speed control usually decides the energy yield of WECs. Figure 11.11 shows the working curves of fixed speed and variable speed generators in combination with the power-speed curves of WECs.

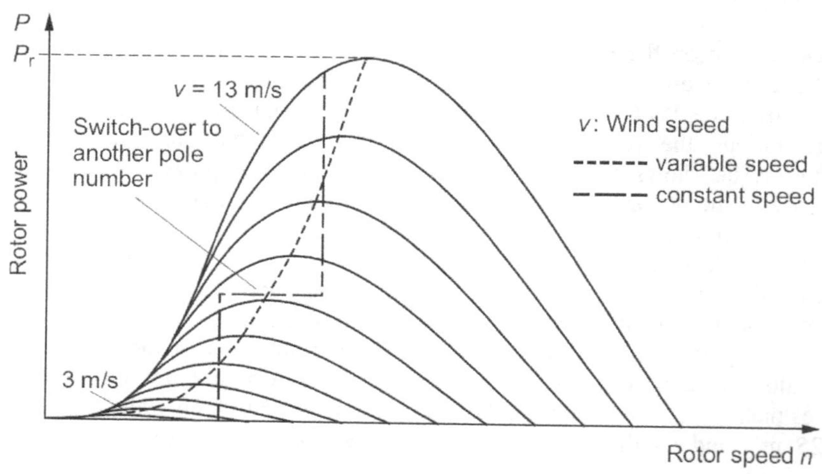

Figure 11.11. Working curves of fixed speed and variable speed generators in combination with the power-speed curves of WECs

Generally one can decide easily between single-stage or two-stage fixed speed and variable speed control. Single-stage fixed speed controlled devices cannot adapt their rotor speed to the wind speed and such WECs cannot deliver an optimal energy yield. Two-stage fixed speed WECs are more flexible; they work mostly with two-speed pole-changing generators.

Variable speed generators adapt their speed to the actual power conditions. Their optimal working point is adjusted by using a look-up table or a maximum power tracking method. Variable speed operation enables an optimal energy conversion within the whole speed range. Particularly in time intervals with frequently changing wind speeds; a higher energy yield compared to fixed speed devices is possible. Speed adaptation is realized by the application of power electronic converters.

11.3.6 Power Curves of WECs

Depending on the applied power limitation and speed control, resulting power curves can be obtained. Figure 11.12 shows the active power curve given by the manufacturer and the measured power over the wind speed of a WEC with variable speed control as shown in Figure 11.11 [10]. The measured curve lies above the guaranteed power values. Such deviations can result from tolerances of the wind speed measurement, which influence the power curve considerably; see Equation 11.6.

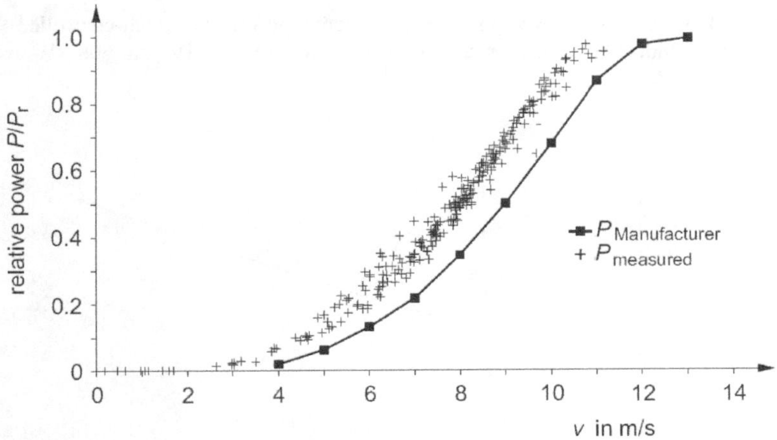

Figure 11.12. Relative power of a variable speed WEC given as manufacturer curve and measured curve over wind speed

Figure 11.13 shows the measured apparent power and reactive power of pitch-controlled WECs with double-fed induction generator and nominal powers of 1.5 MW and 600 kW over the relative power. It can be seen that the apparent power increases with increase of the active power, whereas the reactive power is almost constant. The reactive power has to be controlled to a minimum value; the visible part of it is necessary for the magnetization of the generator. If the WEC has a fixed speed control and a generator with two different pole pairs, the reactive power consumption differs compared to a variable speed device [11].

In Figure 11.14 the apparent and reactive power of a fixed speed WEC is depicted. The reactive power consumption increases with increasing apparent power. For all values two curves exist, which represent the WEC behavior depending on the pole pair number used in the two operation modes.

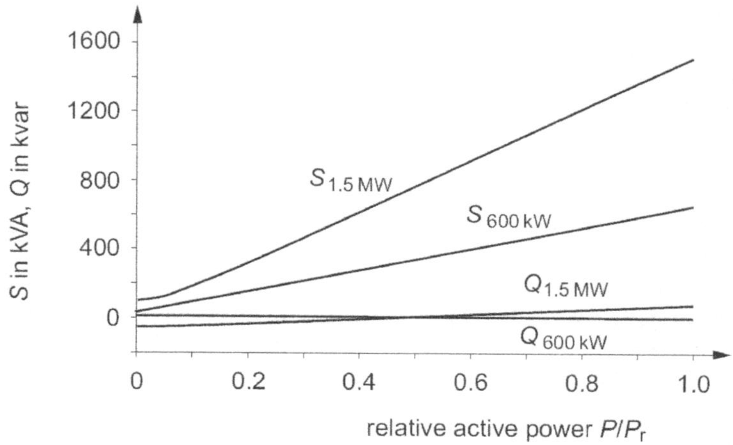

Figure 11.13. Measured apparent power S and reactive power Q of pitch-controlled WECs with double-fed induction generator and nominal powers of 1.5 MW and 600 kW over the relative power P/P_r

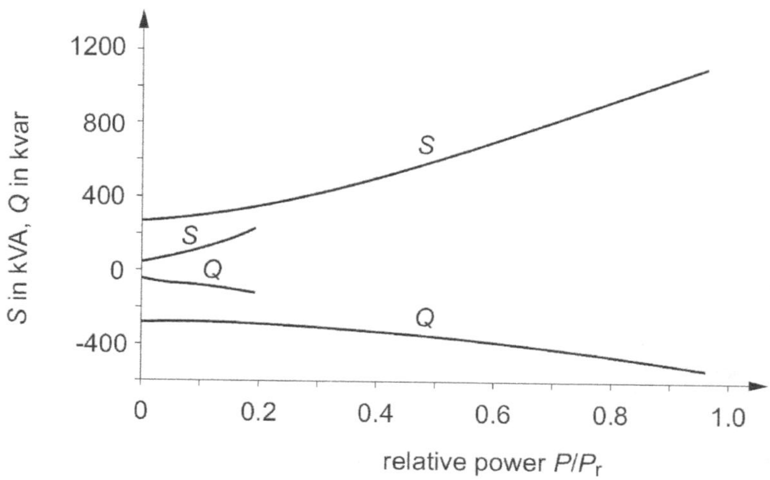

Figure 11.14. Apparent and reactive power of a stall-regulated, fixed-speed WT with a nominal power of 1 MW

Because of the mostly simple grid connection of fixed speed WECs (see Section 11.4.3), no improved reactive power control is possible.

11.4 Grid Integration

11.4.1 Generator Types

Due to the variable wind conditions, the power flow from the WEC to the mains is also discontinuous. The derived operation modes were described in Section 11.3.3. For the conversion of mechanical to electrical power, different generator types are usable; see Figure 11.15 [12]. Their electrical properties strongly influence the behavior of the whole WEC and determine the operation possibilities on the grid. In addition to this, the system features can be improved by using a VSI rated for the full apparent power (full-size) or only the rotor power.

Direct Coupled Induction Machine (IM)
In WECs direct coupled induction machines operate mostly as four-pole types. The stator terminals are directly connected to the grid; see Figure 11.15a. A gearbox transforms the mechanical rotor speed into a higher speed for the generator operation in oversynchronous mode. The induction machine consumes reactive power for its magnetization from the grid.

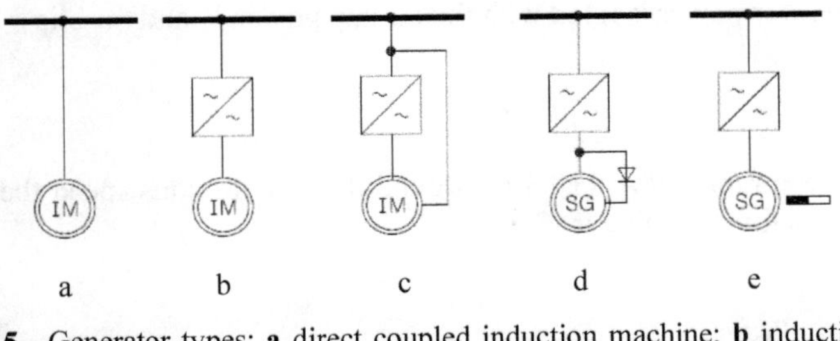

Figure 11.15. Generator types: **a** direct coupled induction machine; **b** induction machine with full-size VSI; **c** double-fed induction machine; **d** electrical excited synchronous generator with VSI; **e** Synchronous Generator (SG) with permanent magnet excitation

Therefore this machine type cannot work in islanding mode or build its own grid, *e.g.*, after a blackout. Due to the direct grid coupling, wind dependent power spikes directly produce voltage drops and flicker in the grid; see Section 11.5. The switch-on to the grid is realized with thyristor switches, which are short-circuited after the start. Because the direct grid coupling gives no control of the possible power factor, it depends only on the actual operation point of the generator. Recent grid requirements cannot be met with this system; its use is limited to older WEC types.

Induction Machine (IM) with Full-size Converter

The induction machine is connected to the grid with two back-to-back voltage source inverters; see Figure 11.15b. Because of the inverter rating for the full apparent power, the power electronics involve considerably costs. The converter decouples the generator electrically from the mains, and wind dependent power spikes are damped by the DC link. Due to the limited switching frequency of the grid side inverter, harmonics in current and voltage occur; See section 11.5. Because of the high inverter costs this system is not as used much as the next type.

Double-fed Induction Machine (DFIM)

In this type the stator terminals of the generator are directly connected to the grid and the rotor is connected with a voltage source converter to the mains; see Figure 11.15c. The energy flow over the rotor converter is bidirectional; in subsynchronous mode energy flows to the rotor and in oversynchronous mode energy flows from rotor to the grid. Over the stator terminals always energy flows to the grid. Because of the shared energy transport the rating of the rotor converter is smaller than for a full-size system. It depends on the speed range of the WEC and is mostly a third of the synchronous value in both directions; that produces a converter rating of a third of the nominal power. This power rating can be calculated as follows. The total power P is the sum of the stator P_s and rotor P_r powers

$$P = P_s \pm P_r \tag{11.9}$$

whereas the rotor power is rated with the air gap power P_a and the slip s

$$P_r = P_a \cdot s \tag{11.10}$$

In the speed range of $n=1\pm0.3\cdot n_s$, around the synchronous speed the slip s of a four-pole induction machine is

$$s_{-30\%} = \frac{n_s - \frac{2}{3}n_s}{n_s} = 0.\bar{3} \tag{11.11}$$

$$s_{+30\%} = \frac{n_s - \frac{4}{3}n_s}{n_s} = -0.\bar{3} \tag{11.12}$$

The calculations show that, at a speed variation of a third around the synchronous speed, a third of the total power must be transmitted over the VSI. The variable speed control of the machine is realized as shown earlier in Figure 11.11. For an explanation of the DFIM operation, the variable-speed-curve of Figure 11.11 is transformed according to Equation 11.8 into a torque-speed-curve as shown in

Figure 11.16. This curve is tipped over to the left side and added to the torque-speed-curve of the DFIM; see Figure 11.17. Working points result as intersection points of the graphs [13]. The standard operation curve of an induction machine includes the synchronous speed n_s at the synchronous point SP. This point is determined by the grid frequency f_1 and the pole number p

$$n_s = \frac{f_1 \cdot 60}{p} \tag{11.13}$$

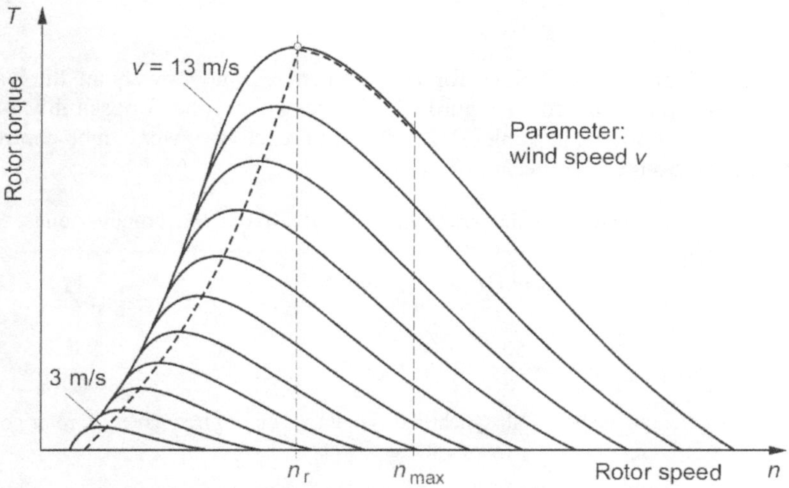

Figure 11.16. Variable speed operation curve of a WEC

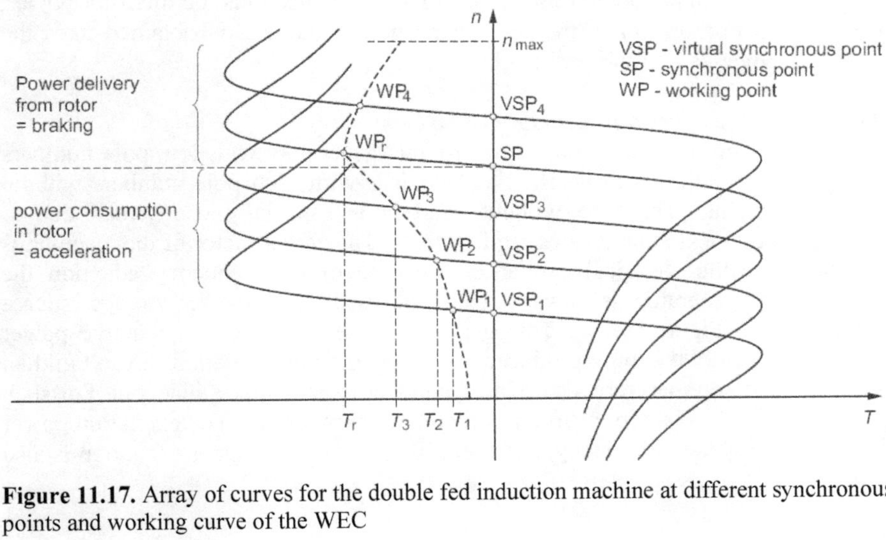

Figure 11.17. Array of curves for the double fed induction machine at different synchronous points and working curve of the WEC

Because of the double connection of the stator and the rotor, Equation (11.13) is not applicable; the grid frequency is replaced for further considerations by the virtual frequency f_v

$$f_v = f_r \pm f_s \tag{11.14}$$

A positive sign arises for different electrical rotation directions. This virtual frequency is used for the calculation of the synchronous speed

$$n_s = \frac{f_r \pm f}{p} \cdot 60 \tag{11.15}$$

Table 11.1 shows the values for the frequencies and powers in the virtual synchronous points marked in Figure 11.17. Using the control possibility of the rotor converter as shown in Table 11.1, active and reactive power can be controlled independently from each other.

Table 11.1. Frequencies and power directions at different synchronous points

Synchronous point	f_v in Hz	f_r in Hz	P_r	P_s
VSP1	40	10	+	–
SP	50	0	0	0
VSP2	60	10	–	+

VSP virtual synchronous point, SP synchronous point, f_v virtual frequency, P_r rotor power, P_s stator power

The generator itself has a higher rotor winding number compared to standard machines, which allows the operation of the grid side inverter without transformer. Figure 11.18 shows a measured curve of the stator power of a DFIM, which is compared to the manufacturer curve [11]. The difference must be the rotor power, which is fed to the rotor in the sub-synchronous mode and is obtained from the rotor in the over-synchronous operation range.

SG with Electrical Excitation and Full-size Converter

Two systems are on the market: synchronous generators with low pole numbers and speed adaptation with gearbox and devices with high pole numbers without need for gearing. The type without gearbox is especially widespread due to reduced mechanical maintenance requirements. The power factor of the machine is adjustable by the electrical excitation. Because of the frequency adaption the synchronous generator is always equipped with a full-size voltage source converter; see Figure 11.15d. This generator type consumes no reactive power from the grid side; it is only produced with the DC rotor excitation. It can build an islanding grid and therefore also rebuild an electric grid after a blackout. For such purposes it needs a battery charged excitation current source. The excitation power decreases the system efficiency, and therefore permanent magnet excitation is also used.

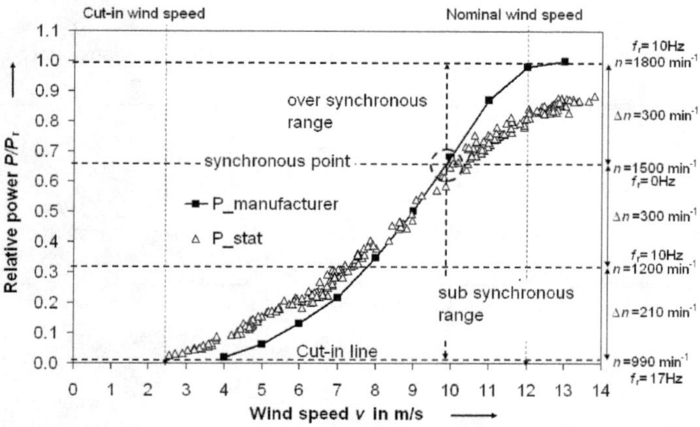

Figure 11.18. Array of curves for the double fed induction machine at different synchronous points and working curve of the WEC

SG with Permanent Magnet Excitation and Full-size Converter

For this type, permanent magnet excitation is applied. No excitation losses and excitation control exist. Control of the power factor is realized with the grid side inverter. Magnets are mounted on the rotor for the DC excitation. In WECs various hard magnetic materials such as SmCo or NdFeB with high energy production are used. Ferrites only have a low energy density. The magnetic energy density determines the generator volume and weight, higher values causing more magnet costs. If permanent magnets are used, the generator volume increases from 20–40% compared to electrical excitation. The Multibrid-WEC described in Section 11.2 is a special design of a permanent magnet excited generator type.

11.4.2 Types of Common Grid Coupling

If the common generator types are combined with the common power limitation methods, three main classes of WEC types result for the power class above 1 MW; see Figure 11.19 [13]. Older systems as shown in Figure 11.19a work with direct grid coupling and a thyristor start switch. They are not more acceptable because of their bad power quality parameters; see Section 11.5. The WEC types in Figure 11.19b,c are widely spread, the latter technology having the highest installation numbers. Mostly two back-to-back inverters are used, whereas the generator inverter controls the generators in an optimum working point with high efficiency and the grid side inverter controls the power quality properties of the WEC. Table 11.2 shows the electrical properties of some power electronic switches which are used in WECs. The most important values are the blocking voltage and kW-standard-units allow a flexible power adaptation. Only one WEC-application exist in the medium voltage range; see Section 11.2. Mostly low voltage converters are installed, and therefore high currents arise in the megawatt-range. A subdivision into 600 voltage applications are known from other technologies, such as oil and gas extraction or railway drive trains.

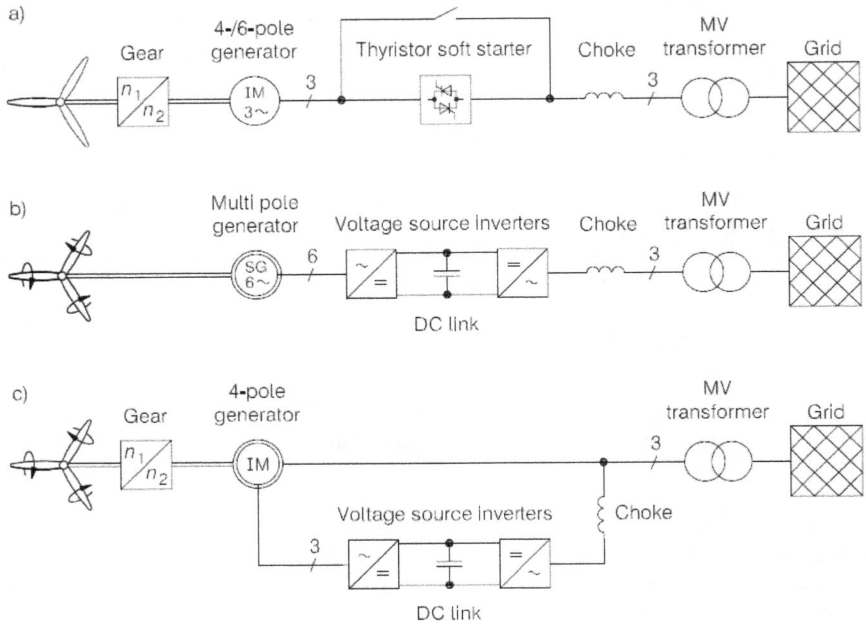

Figure 11.19. WEC types in the power range above 1 MW: **a** stall-regulated, constant-speed controlled WEC with direct grid connection; **b** pitch-regulated, gearless WEC with variable speed control with multi-pole synchronous generator and voltage-source converter; **c** pitch-regulated, variable speed controlled WEC with double-fed induction machine and voltage-source converter

Table 11.2. Power electronic switches for the use in WECs

Switch	Blocking voltage in kV	Maximum current in A	Pulse frequency used in WECs in Hz
Thyristor	12	5,000 (at 5 kV)	50
GTO	6	6,000	200–1,000
IEGT	6.5	1,200 (at 6.5 kV) [14]	2,000–20,000
IGBT	6.5	1,200 (at 3.3 kV) [15]	2,000–20,000
IGCT	10	3,000 (at 6.5 kV] [16]	150–500

GTO gate turn-off thyristor, IEGT injection enhancement gate transistor, IGBT insulated gate bipolar transistor, IGCT insulated gate commutated thyristor

11.4.3 Wind Park Design and Energy Management

Wind energy converters operate mostly concentrated in wind parks. This allows the joint use of local wind conditions and of the medium voltage structure on the electrical side. Almost all WECs work with low voltage generators and voltage-source power electronic converters. The wind turbines are connected to a medium voltage transformer. The medium voltage level of the wind park power is then connected to a high voltage transformer in a transformer station; see Figure 11.20.

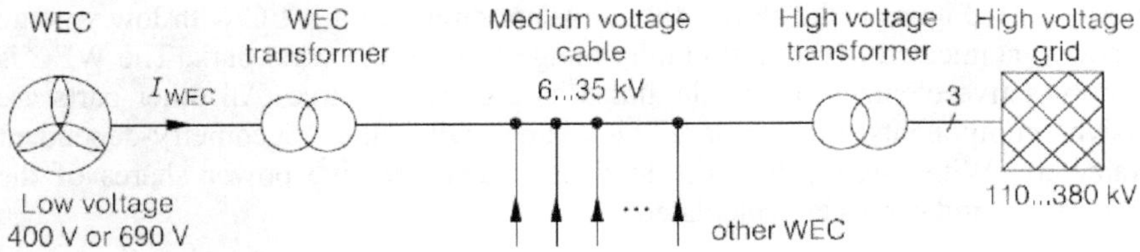

Figure 11.20. Principle of the grid connection of WECs in a wind park

Figure 11.21 shows an example structure of a wind park with 56 wind turbines with a total power of 95.2 MW. Beside the WECs, transformers and cables are installed, which contribute to the reactive power consumption.

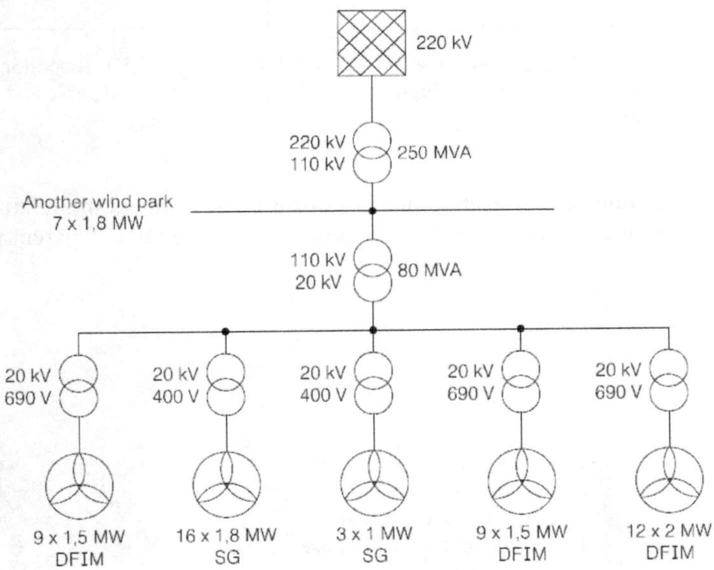

Figure 11.21. Example structure of a wind park with low voltage WECs DFIM double-fed induction machine, SG synchronous generator

11.4.4 Reactive Power Management in Wind Parks

WECs are equipped with a standard power control that adjusts the power factor λ to unity, because only the active power feed-in into the grid is remunerated. This is a proper solution for stand-alone WECs, but not for wind parks. There the Power Factor (PF) must be controlled to unity at the PCC. Reactive power management can be realised if the characteristics of all transmission elements are known.

Passive components, such as transformers and cables, also contribute to the reactive power balance of wind parks. Therefore they have to be considered in the calculation of the total power share. Based on the given wind park structure of Figure 11.21, Figure 11.22 shows an equivalent circuit of a WEC with low voltage generator, connected to the ultra high voltage level of the wind park. The WEC is the only active element and is simulated as a current source. All other parts are passive components, which are simulated with their geometry-dependent parameters. With these values the load dependent reactive power shares of the transformers and cables are calculated.

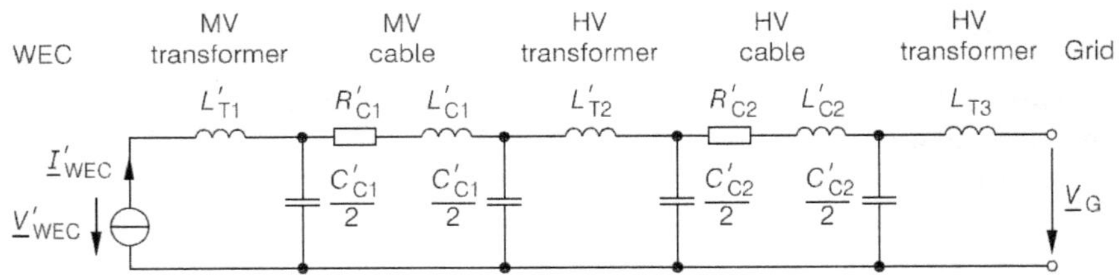

Figure 11.22. Equivalent circuit of a WEC with low voltage (LV) generator, medium voltage (MV) transformer and cable, high voltage (HV) transformer and cable and ultra high voltage (UHV) transformer

Transformers

For the calculation of the reactive power Q of the transformers the relative short-circuit voltage v_{sc}, the transmitted active power P and the rated apparent power S_r is used

$$Q_T = v_{sc} \frac{P^2}{S_r} \qquad (11.16)$$

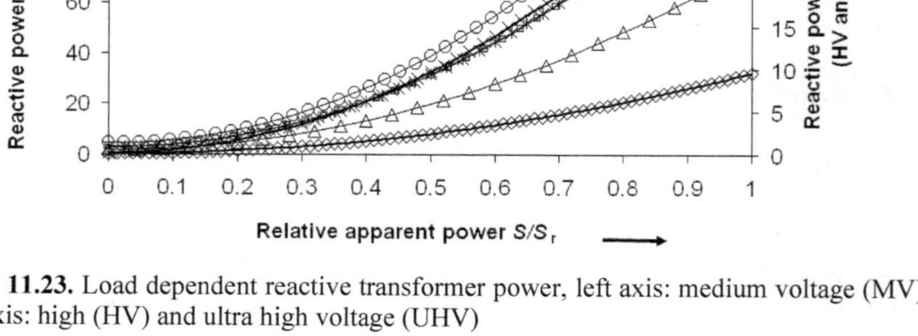

Figure 11.23. Load dependent reactive transformer power, left axis: medium voltage (MV); right axis: high (HV) and ultra high voltage (UHV)

With Equation 11.16 the reactive power consumption over the whole power range of the wind park is calculated for all medium voltage (MV) device transformers, the high voltage (HV) substation transformer and the ultra high voltage (UHV) transformer at the PCC. Between the HV and UHV transformer is an additional feeding from another wind park.

The calculation results are shown in Figure 11.23. It is indent, that the derived curves always show a square dependency over the relative apparent power [17, 18].

Medium and High Voltage Cables

Cables have both inductive and capacitive components; resistive parts can be neglected. The inductive $Q_{C,\,ind}$ and capacitive $Q_{C,\,ca}$ reactive parts are calculated from the cable length l and the inductance layer $L^{'}$ or capacitive layer $C^{'}$ per length unit

$$Q_{C,ind} = 3 \cdot I^2 \cdot 2\pi f \cdot L^{'} \cdot l \tag{11.17}$$

$$Q_{C,cap} = V_{LL}^{\ 2} \cdot 2\pi f \cdot C^{'} \cdot l \tag{11.18}$$

The total reactive power of the cable is then calculated

$$Q_C = \left|Q_{C,ind}\right| - \left|Q_{C,cap}\right| \tag{11.19}$$

For the calculations of the reactive powers, manufacturer equivalent circuit cable data are used as summarised in Table 11.3. The reactive power consumption of medium and high voltage wind park cables over the transmitted active power is shown in Figure 11.24. It is seen, that the lowest total reactive power occurs at full load. The capacitive reactive power of the high voltage cable is higher compared to

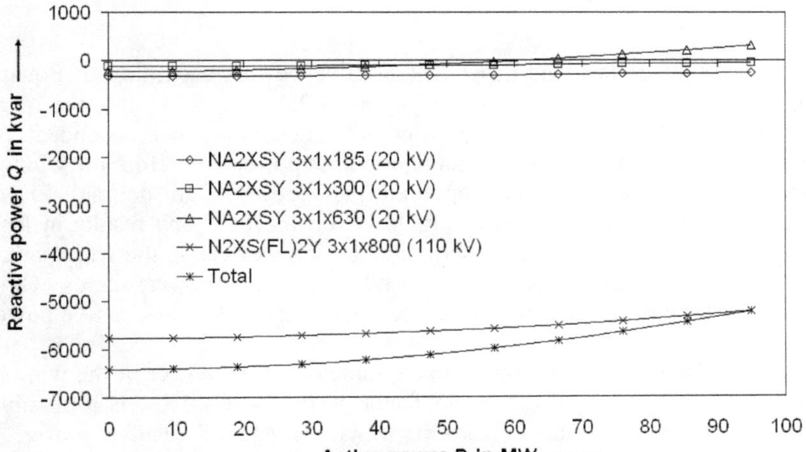

Figure 11.24. Reactive power consumption of medium and high voltage cables

that of medium voltage cables and determines the total cable characteristic in the wind park.

Table 11.3. Manufacturer equivalent circuit data of medium voltage and high voltage cables

Voltage in kV	Cable description	Cross section in mm²	*R'* in Ω/km	*L'* in mH/km	*C'* in µF/km
20	NA2XSY	3×1×185	0.410	0.37	0.27
20	NA2XSY	3×1×300	0.195	0.35	0.325
20	NA2XSY	3×1×630	0.063	0.315	0.43
110	N2XS(FL)2Y	3×1×800	0.03	0.4	0.22

Wind Park Power Balance

An optimised active power transmission between WEC and PCC in a wind park requires equal inductive and capacitive reactive power parts. This is not easy to realise because the reactive power share is not constant, but load dependent, and results from the summation of the reactive powers of the passive transmission elements that are represented by the transformer curves and cable curves in Figures 11.23 and 11.24. The wind park reactive power balance is calculated according to the equivalent circuit of Figure 11.22 with the reactive powers of the transformers (index T) and cables (index C). For the wind park as shown in Figure 11.21 the total balance is calculated

$$Q_{\mathrm{WP}} = Q_{\mathrm{WT}} + Q_{\mathrm{T1}} + \left|Q_{\mathrm{C1,ind}}\right| - \left|Q_{\mathrm{C1,cap}}\right| + Q_{\mathrm{T3}} + \left|Q_{\mathrm{C2,ind}}\right| - \left|Q_{\mathrm{C2,cap}}\right| + Q_{\mathrm{T3}} \qquad (11.20)$$

The complete wind park transmission system includes three transformers and two cables

$$Q_{\mathrm{TL}}(P) = \sum_{1}^{3} Q_{\mathrm{T}n}(P) + \sum_{1}^{2} Q_{\mathrm{C}n}(P) \qquad (11.21)$$

A calculation of the load-dependent reactive powers according to Equations 11.20 and 11.21 leads to the curve in Figure 11.25. It is shown that the reactive power consumption and the PF of the wind park are strongly load-dependent. A PF of unity is only reached at 50% of nominal output power. Therefore the wind park works over a wide range not optimally and reaches an optimal PF only accidentally. This leads to an unfavorable PF on the PCC and results in loss of revenue. The WECs are the only flexibly adjustable devices in the wind park. All transmission elements deliver load-dependent reactive power shares. Only a permanent PF adjustment of the WECs can guarantee the highest active power at the PCC.

Figure 11.26 shows as an example the consumed reactive power of the wind park for a flexible adjustment of the power factor on the WECs. Now is it possible to control the reactive power to zero in a wide range of relative power. The effectiveness of this control depends on the control range of the power factor of the single WECs, which may vary for different WEC technologies and manufacturers.

In Figure 11.27 the calculated control range of some WEC types with an adjustable power factor of cosφ=±0.966 is applied to the considered wind park. In this case, the reactive power can be held at zero between 34% and 84% of the rated wind park power [11].

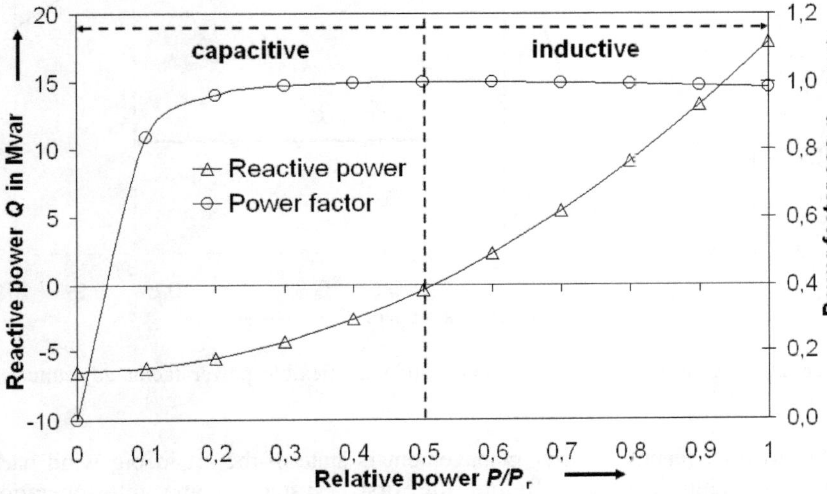

Figure 11.25. Reactive power consumption and power factor of the investigated wind park over relative power P/P_r

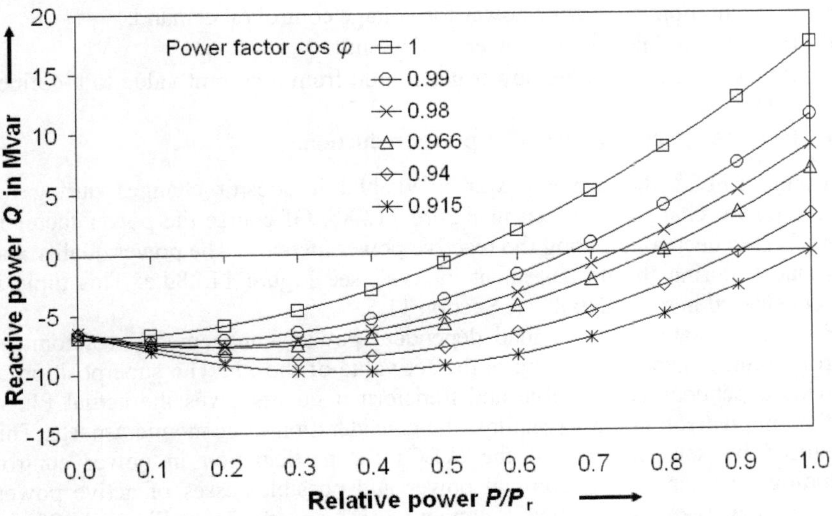

Figure 11.26. Calculated reactive power consumption of the wind park at variable power factor of the WECs

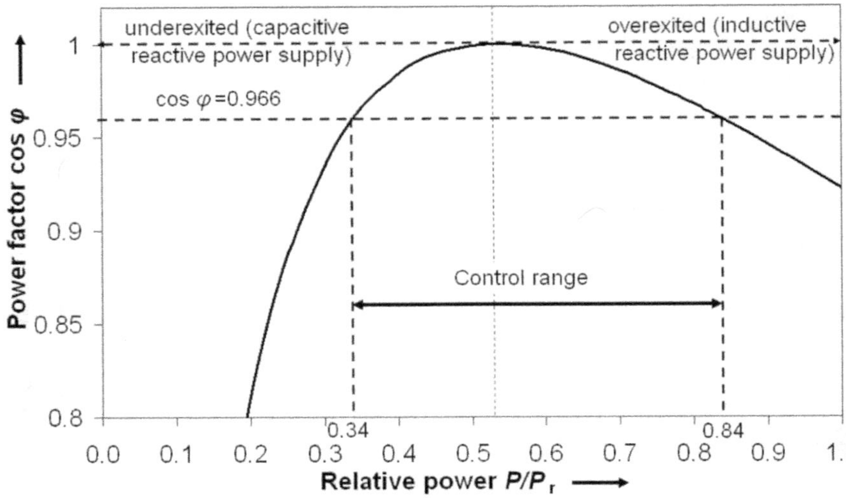

Figure 11.27. Control range of the wind park with flexible power factor adjustment of WECs

Meanwhile, reactive power management is state-of-the art in big wind parks with long cable distances. Before the first systems come into operation, measurements must remove all technical doubts. The results of such joint test measurements between wind park operator and utility are shown in Figure 11.28 [11]. This test should prove some defined tasks:

- Consumption of reactive power for voltage control on demand;
- Bisection of the reactive power on demand;
- Reduction of the active power generation from a current value to a defined value;
- Increase of active power after power reduction.

In Figure 11.28 the active power is visible; it doesn't changes during the reactive power change as shown in Figure 11.28b. Of course the power factor in Figure 11.28c decreases during the reactive power increase. The power quality was not reduced during the measurement interval; see Figure 11.28d,e. This topic is later considered in more detail; see Section 11.5.

For the permanent use of load dependent power management an automated control circuit is necessary. Input is the set-value of the PF. The superposition of the wind-dependent WEC, cable and transformer curves gives the actual PF. A practical implementation also requires the consideration of economic aspects. This is realised by selective use of the WECs for participation in power control, depending on their current nominal power and possible losses of active power. This idea is expressed by the box "*Economic WEC selection*"; see Figure 11.29.

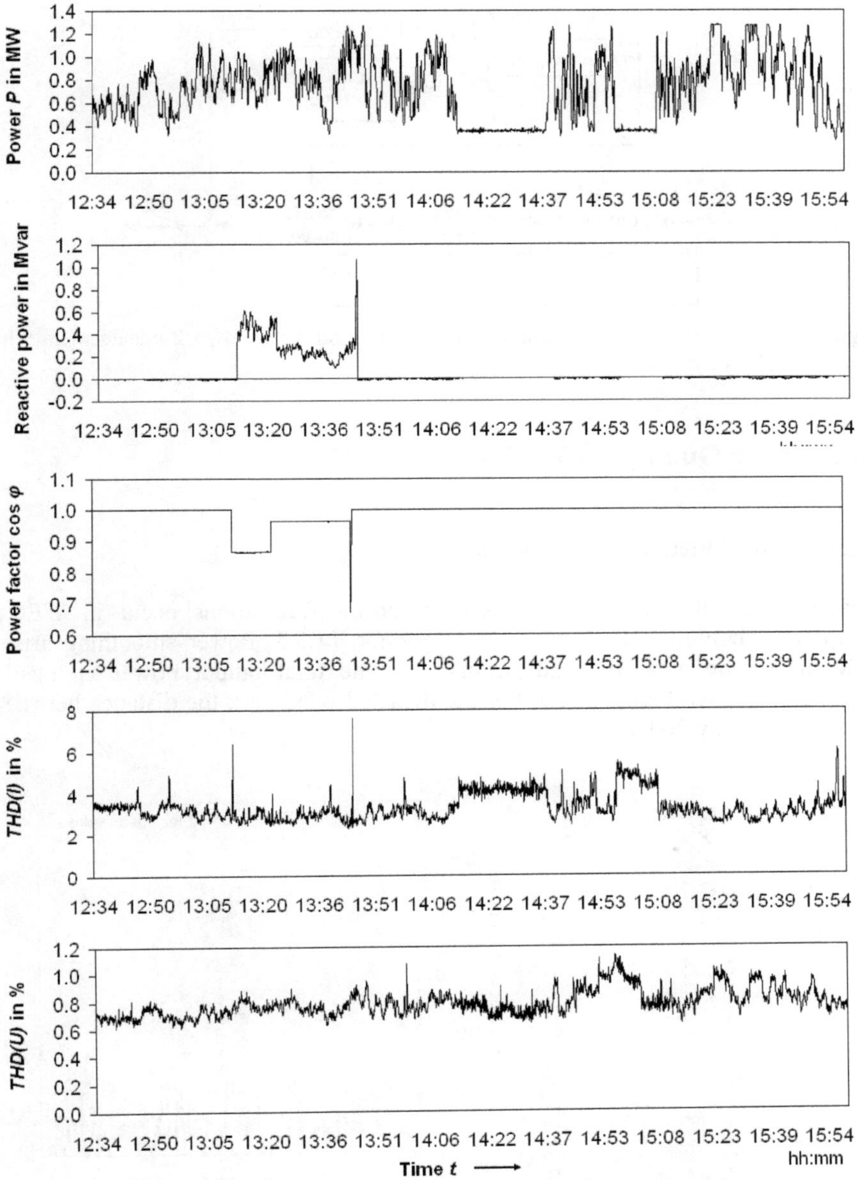

Figure 11.28. Measurements of active and reactive power adjustment of a WEC with a rated power of 1.5 MW. From above: active power; reactive power; power factor; total harmonic distortion of current; total harmonic distortion of voltage

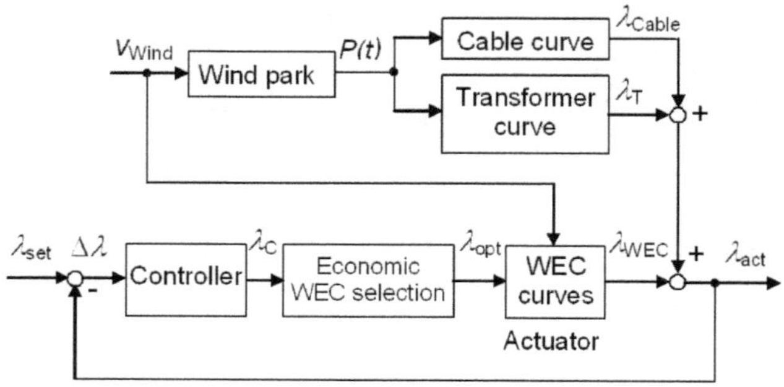

Figure 11.29. Structure of a load-dependent reactive power wind park management with flexible WEC power factor

11.5 Power Quality on WECs

11.5.1 Power Fluctuations and Flicker

Depending on the wind speed variation, power fluctuations occur in WECs; compare with Figures 11.7 and 11.8 in Section 11.3.3. Power smoothing arises only in a wide area. In small wind parks the total output power is mostly discontinuous; see Figure 11.30. For the depicted wind park the distance between the WECs is only 200 m.

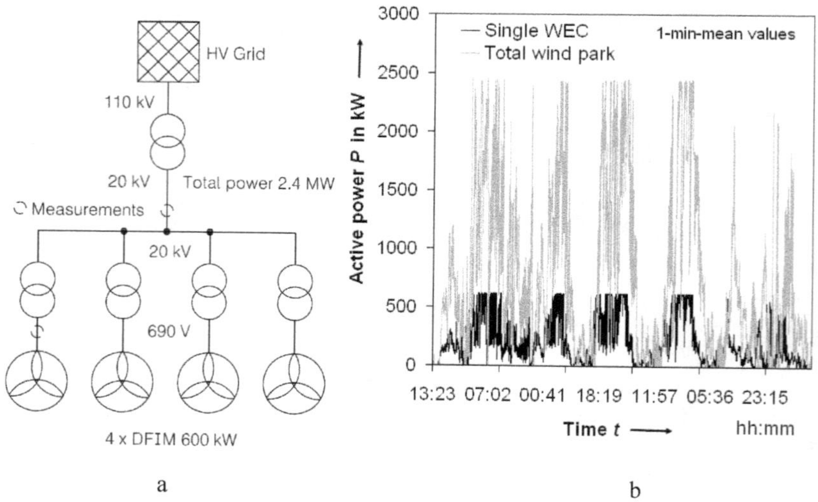

Figure 11.30. Power fluctuations in a small wind park: **a** wind park structure with measurement points; **b** power of a pitch-controlled 600-kW-WEC with double-fed asynchronous generator and total power of all four devices over 13 days

If the power changes, the variable current flow causes voltage drops on the grid impedance. These voltage changes cause, *e.g.*, light density changes on bulbs, which is called flicker. Hence, the magnitude of the voltage change depends on the values of the current and the grid impedance. Large power fluctuations cause large current changes; the flicker effect is strongly influenced by the power class of WECs. When a load with power P is connected to the grid, current I flows

$$I = \frac{P}{V \cdot \cos\varphi} \tag{11.22}$$

Every power change per time unit is equivalent to a current change ΔI

$$\Delta I = \frac{\Delta P}{V \cdot \cos\varphi} \tag{11.23}$$

The resulting current through the grid impedance causes a voltage drop, which reduces the grid voltage and causes several effects on the consumer, such as power decrease or malfunction. Such voltage drops are easily visible in light sources.

Evaluation of Flicker

Flicker is a voltage change in the frequency range between 0.005 Hz and 35 Hz, which is evaluated with the light density changes of a 60-W-bulb visible to the human eye [19]. The light density of bulbs is proportional to the square of the supply voltage. Hence, the flicker intensity depends on the frequency and amplitude of the voltage change. It is evaluated according to the standards EN 61000-3-3, -4-7, -4-15 and IEC 1000-3-5 [20–23]. In order to understand in the next pages the most important parameters are discussed here.

Short voltage variations are detected by short-term flicker P_{st}, which is calculated with the flicker causing time t_f in the calculation period T_p of 10 min; the exponent is a standardization factor

$$P_{st} = \left(\sum t_f / T_p \right)^{1/3.2} \tag{11.24}$$

Measurement devices use weighting factors a_i for the cumulative distribution of the power quantils $P_{i\%}$

$$P_{st} = \sqrt{\sum_i a_i \cdot P_{i\%}} \tag{11.25}$$

Corresponding to the standard EN 61000 the short-term flicker must not exceed the value of P_{lt}=1.

The long-term flicker P_{lt} is calculated by a cubical smoothing of the short-term flicker over N=12 intervals of 10 min; it must be in line with the standard EN 61000 and not exceed the value of P_{lt}=0.65

$$P_{\mathrm{lt}} = \sqrt[3]{\frac{1}{N}\sum_{i=1}^{N} P_{\mathrm{st},i}^{3}} \tag{11.26}$$

Some countries, *e.g.*, Germany, have additional local regulations for flicker evaluation. There exists a limit for the long-term flicker of grid connected power plants on the low and medium voltage level of P_{lt}=0.46, considering the superposition effect from multiple flicker sources [24]. Corresponding to the standard EN 50160 [25] the long-term flicker during a weekly interval must be, for 95% of the time, maximal P_{lt}=1. Utilities are responsible for limiting the long- term flicker in their grid sections below this value independent of the number of flicker sources.

For the practical evaluation of flicker limits, the curves according to EN 61000-3-3 as shown in Figure 11.31 are applied. The flicker limits depend on the value and the repetition rate and, in addition to this, on the shape of the voltage changes. Rectangular changes cause more visible flicker effects than sinusoidal voltage deviations. Generally the curves have the same trend; this is caused by the characteristical sensitivity of the human eye. The most disturbing frequency of light density changes is at 8.8 Hz; therefore this frequency range has the lowest flicker. This flicker frequency is equivalent to a repetition rate per half period of the grid voltage of r=1056 min^{-1}

$$r = 2 \cdot f \cdot 60 \frac{s}{\min} = 2 \cdot 8.8 \frac{1}{s} \cdot 60 \frac{s}{\min} = 1056 \ \min^{-1}$$

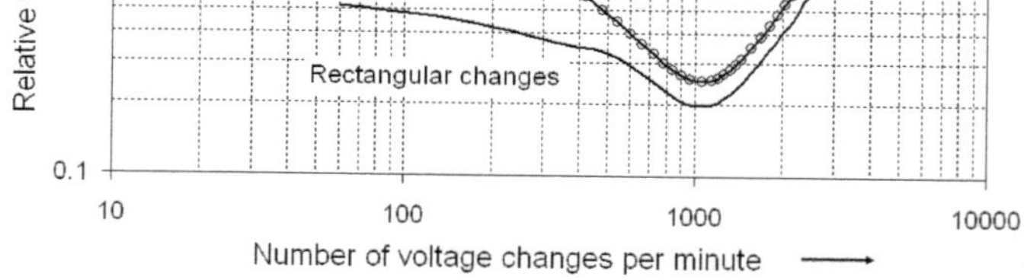

Figure 11.31. Flicker tolerance curve according to EN 61000-3-3, IEC 1000-3-3 for periodic rectangular and sinusoidal voltage changes (CENELEC-curve)

Flicker relevant frequencies f_r, caused by the tower shadow effect of a WEC with b=three blades at a speed of n_r, are calculated according to

$$f_r = b \cdot n_r \tag{11.27}$$

To minimize the risk of financial losses for WECs a pre-evaluation of the possible flicker emissions is required. This procedure follows the standard IEC 61400-21 [26]. It will not be amplified here; a remarkable fact is the simulation of the WEC on a fictive grid with defined properties. A variation of these properties is realized by the choice of the resistive and inductive impedance parts. Also, for WEC power quality testing, local regulations exist, *e.g.*, the FGW-standards [27]. Some of these different standards are unified to a MEASNET standard for the power quality evaluation of WECs [28].

Flicker Characteristic of WECs

Basically the flicker effects of WECs are influenced by the device power, the generator type and its kind of grid coupling. Flicker is generally caused by load changes due to wind speed variations, wind shear between lower and upper rotor blade tip and tower shadow effects. The last two items are only important for high rotor diameters in the megawatt power class. In the 5-MW-class the WECs are equipped with a single blade control to minimize the tower shadow effect.

Stall-regulated, constant-speed devices (see Figure 11.19a) transmit all wind changes as power changes to the grid. Therefore this WEC type causes the highest flicker emissions, which do not exceed the flicker limit values only in small power classes below 1 MW. Modern WECs with pitch-regulated power limitation and variable speed control have typical curves of voltage changes as shown in Figure 11.32 for a 600-kW WEC.

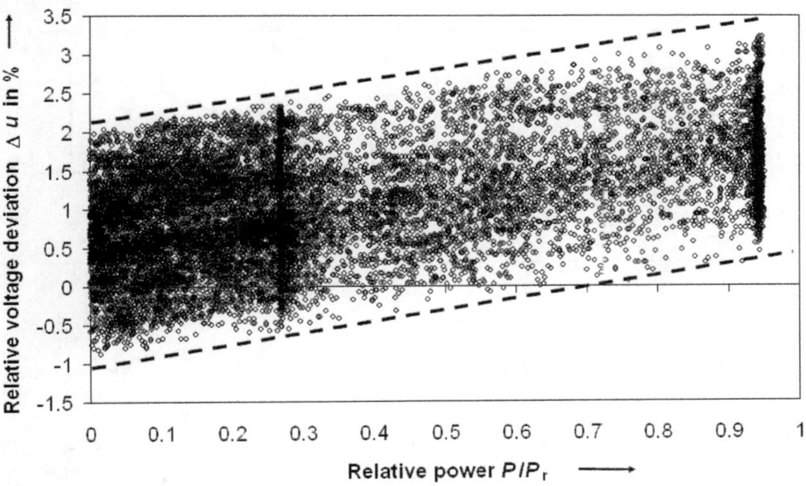

Figure 11.32. Typical curve of the relative voltage deviation over the relative power, measured on the 690 V level of a pitch-controlled 600-kW WEC with double-fed induction machine, 1-min mean values, measured over 13 days

Other WEC types in the megawatt power classes show similar trend curves to the measured wind turbine. With increasing relative power increases the relative voltage deviation ΔU

$$\Delta u = \frac{u(t) - u_{\mathrm{rG}}}{100} \tag{11.28}$$

For an estimation of the grid disturbance potential the k-factor is used, which is calculated from the maximal current I_{max} and the rating current I_r

$$k = \frac{I_{\max}}{I_{\mathrm{r}}} \tag{11.29}$$

With this factor the stress on the transmission elements and also the potential flicker load can be estimated. The k-factor is set to one if the generator is a synchronous type or the WEC is grid connected with a power electronic converter. For directly grid connected induction machines with speeds of 95–105% of the synchronous speed, k is set to four. The value is k=8, if the starting current is unknown; for a known starting current k is calculated from the ratio of the starting and the rating current.

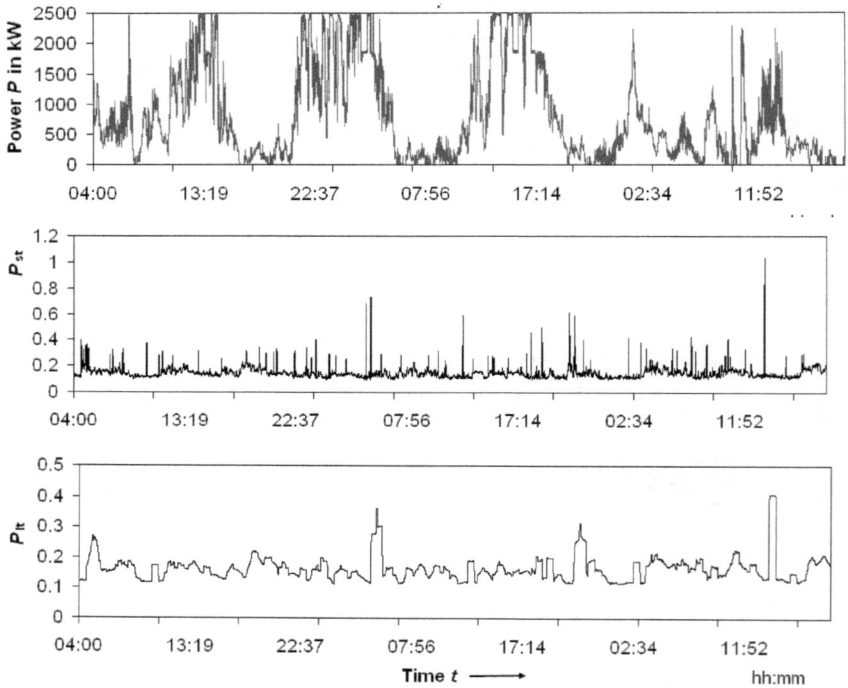

Figure 11.33. Measurements over 10 days on the 20 kV-level of a wind park with a total power of 2.4 MW, see Figure 11.29a, 1-min-mean values. *From the top*: power; short-term flicker (limit value P_{st}=1); long-term flicker (limit value P_{lt}=0.46)

The flicker generation is also influenced by the grid impedance, the value of which depends strongly on the voltage level. The grid impedance decreases with increasing voltage level. A guiding value of the ratio between the relative impedances of the low, medium and high voltage level is 100:10:1, which may vary depending on the installed components. Compared to the low voltage connection, the flicker values of WECs are always smaller on the medium voltage side because of the lower grid impedance and the low currents. Figure 11.33 shows measurement results of a 10-day measurement interval on the 20-kV level of a wind park with 2.4 MW power. In Figure 11.33a the power over time is depicted. This curve shows some strong fluctuations, which result in short-term flicker values according to Figure 11.33b. It can bee seen that the long-term flicker in Figure 11.33c is always smaller than the short-term values because of the cubical smoothing in its calculation according to Equation 11.27. The measured wind park does not exceed the flicker limits. In practical grid measurements it is mostly difficult to assign the measured flicker values clearly to their sources. A promising method is the measurement of the existing flicker level in the grid with disconnected WEC. If a power quality assessment is necessary, such pre-measurements should be carried out over the full planned time interval to avoid later misinterpretations of the results and proposals for supposed improvements. A possible feature for the detection of the flicker direction is currently not applied in the existing measurement devices.

Reduction of Flicker

There are possible measures on the grid side and on the consumer side to reduce the flicker level within technical limits. On the PCC the following conditions should be considered during the planning of a wind park project:

- Sufficient short-circuit power on the PCC;
- Distribution of fluctuating loads to different PCCs;
- No simultaneous load deviations on the same PCC;
- Limitation of power oscillations between wind parks.

Often the grid conditions are fixed due to limited installation area of a wind park. Then the only choice is to find a more distant PCC with a higher short-circuit power. This will result in higher costs. A split of the wind park to different PCCs is also applicable to limit the simultaneous power changes.

Basically the flicker is reduced by the decrease of the grid impedance and a decreasing current change. A lower grid impedance can be realized by:

- Increased grid meshing: not used practically because of high costs;
- Use of transformers with higher short-circuit power: used if the operation losses do not increase;
- Applying a higher voltage level: frequently used for WECs in high power classes or wind parks.

On the side of the (active) consumers, such as WECs, the flicker effect in the grid can be reduced by limitation of the current fluctuations during starting events and the power fluctuations caused by wind changes. These demands require their consideration in the technical concepts of the WECs, which therefore can be equipped with:

- Soft starters for WECs with constant speed control: they use mostly thyristor circuits; see Figure 11.19a;
- Voltage-source power electronic converters with high capacity values in the DC link for the damping of power fluctuations;
- Zero-current synchronisation control during the grid switch-on;
- Dynamic compensation of load changes with several kinds of active filters; this methods is technical possible, but not applied practically because of the high costs.

A high flicker level is often a reason for the non-approval of wind parks by the utilities. Sometimes they give only a time-limited grid connection allowance until a measurement certificate exists.

11.5.2 Harmonics

Non-sinusoidal functions can be split with the FOURIER-transformation into parts of sine and cosine functions. These shares are multiples of the basic oscillation and are called harmonics. In the European transmission system the fundamental frequency of 50 Hz is the first harmonic. The *n*-th integer multiple of the basic frequency is the *n*-th harmonic. If the multiplier is a fraction, the result is an interharmonic. The interharmonic parts below the basic frequency are called subharmonics. Even though all kinds of harmonics are only results of a mathematical split time function, in practice "*measured*" harmonics are referred to because of device internal calculations. The harmonic content describes the deviation from the ideal sinusoidal function.

Cause of Harmonics

Harmonics occur if the current or voltage is non-sinusoidal. Basically non-linear loads, pulsed voltages of converters with PWM or switching events in general are the reason for such function shapes. Because of the direct relation between some kinds of harmonic types and their causes, it is possible to draw conclusions about the cause from the measured harmonics.

Even numbered harmonics arise only if the time function is half-wave unsymmetrical. Such asymmetries appear as transient stages at fast load changes or as permanent effects at unsymmetrical control of power electronic converters. Faulty current measurements can also generate such problems. All these described possibilities can arise in WECs. Subharmonics are produced by periodical switching events with variable frequency. Interharmonics are generated by periodical switching processes, when the frequency is not synchronized to the fundamental frequency. This happens at low and high frequency switching, asynchronous switching of power electronic converters, matrix converters or devices with burst firing control.

Some relations between the signal properties and the result of the harmonic analysis can be seen:

- The steeper the signal the higher the generated frequencies;
- Periodic signals deliver discrete spectra;
- Non-periodic signals deliver continuous spectra.

Evaluation of Harmonics

Current and voltage harmonics are evaluated according to IEC 61000-3-4 [29]. The most important IEEE power quality standards are also listed in the references [30–34]. Relevant harmonic parameters are:

- Values of the single harmonics;
- THD of current or voltage;
- Partial Weighted Harmonic Distortion (PWHD) of current or voltage.

Beside these parameters, sum parameters for the separated evaluation of the even and uneven harmonic parts also exist, which are not considered here because of their absence in the power quality standards for WECs.

Whereas the THD of current shows the influence of the whole-numbered current multiples of the base frequency, the PWHD of current evaluates the influence of the current harmonics of higher orders. These sum parameters are calculated with the single harmonic current parts I_n and the first harmonic current I_1, the same calculations are possible for the voltage

$$THD(I) = \frac{\sqrt{\sum_{n=2}^{40} I_n^2}}{I_1} \tag{11.30}$$

$$PWHD(I) = \frac{\sqrt{\sum_{n=14}^{40} n \cdot I_n^2}}{I_1} \tag{11.31}$$

Up to now, no international standard exists with defined sum parameters for interharmonic evaluation; they are only considered as flicker relevant voltages in IEC 61000-2-4 and -2-12 [35, 36]. Only some countries have local standards for the limitation of interharmonic currents and voltages.

If measurements are applied, attention should be paid to the maximum frequency resolution of the device and the current and voltage probes, which should be at least twice as high as the fastest measurement value. Corresponding to IEC 61000-4-30 [37], the measurement interval must be over 10 periods of the grid voltage.

The internal Fast Fourier Transformation calculates the multiples of the device basic frequency. This basic frequency f_b is derived from the sampling frequency f_{sa} and the storage depth d. For a typical storage depth of 1920 points and a sampling frequency of 9.6 kHz we obtain a frequency resolution of 5 Hz

$$f_{\mathrm{b}} = \frac{f_{\mathrm{sa}}}{d} = \frac{9.6\,\mathrm{kHz}}{1920} = 5\,\mathrm{Hz} \tag{11.32}$$

A frequency resolution of a tenth of the grid frequency is useful because of the required 10-period-interval. If the measurement channels use a joined analog/digital converter, the frequency resolution decreases with increasing channel use. For long-term measurements the storage capacity of the device is important, depending on the selected frequency resolution and the measurement time.

Harmonic Generation Rules and WEC Specialities
The generation of harmonics follows known basic rules:

- Characteristic harmonics as multiples of the pulse frequency of PWM converters;
- Grid commutated power electronic converters produce current harmonics depending on their pulse number p;

$$I_{\mathrm{n}} = m \cdot (p \pm 1) \quad m = 1, 2, 3, \ldots \tag{11.33}$$

- Harmonic side bands by amplitude modulation of basic and pulse frequency in self and grid commutated power converters;
- Interharmonics f_μ arise as side bands of characteristic harmonics of PWM converters, they are calculated with the pulse number p, the load frequency f_l and the harmonic order n;

$$f_{\mu} = n \cdot 50\,\mathrm{Hz} \pm k \cdot p \cdot f_{\mathrm{l}}, \quad k = 1, 2, 3, \ldots \tag{11.34}$$

- Even harmonics due to asymmetries, caused by control faults;
- Interharmonics due to control actions;
- Interharmonics with the frequency $f_{n,m}$ due to the back-to-back configuration of two converters, calculated in line with IEC 61000-2-4 with the power converter pulse numbers p_1 and p_2 and the input f and output F basic frequency, for k_2=0 is $f_{n,m}$=f_n and only harmonics arise;

$$f_{n,m} = \left[(p_1 \cdot k_1) \pm 1\right] \cdot f \pm (p_2 \cdot k_2) \cdot F \quad k_1, k_2 = 0, 1, 2, 3, \ldots \tag{11.35}$$

- 5/7 harmonics as compensation effect of standard three-phase PWM converters, if the grid voltage is distorted [38].

In addition to the basic rules, specific harmonics often arise, which are typical, but not limited to WECs:

- Interharmonics due to generator pole switching, if generators with two pole pairs are applied;
- Interharmonics due to the speed-depending frequency conversion between rotor and stator of DFIM [39];
- Harmonics due to the slot numbers of the generators, if one part of the generator is directly connected to the grid, depending on the speed; 1.5

MW-generators have, *e.g.*, slot numbers of N_s=72 on the stator and N_r=60 on the rotor;

- Harmonics due to resonance effects [40];
- Non-characteristic harmonics due to grid impedance unbalance [41].

Harmonic Characteristic of WECs

The limited switching frequencies of the frequency converters causes distortions on the output currents of the WECs. These distortions are evaluated with Equation 11.28. Its height is proportional to the ratio of the grid impedance values R_G/X_G and inversly proportional to the PMW pulse frequency. If the THD of current values of a PMW converter are allocated to the relative output power, we obtain a curve as depicted in Figure 11.34. It shows a characteristic graph with the function

$$THD(I) \sim \frac{1}{P} \tag{11.36}$$

This typical curve shape occurs independently of the voltage level and is valid for single WECs and also for wind parks. It is caused by the structure of the calculation formula in Equation 11.30, because the harmonic content is always referred to the fundamental wave, which increases with increasing output power. For the mathematical description of the typical measured curve shape of the THD of current over the relative power of PWM converters, the parameters a and m can be used [11]

$$THD(I) = f(P) = ax^{-m} \text{ with } x = \frac{P}{P_r} \tag{11.37}$$

The parameters a and m are variable; they are influenced by the value of the grid impedance, the PWM converter switching frequency, installed grid chokes and filters. Table 11.4 shows the calculated parameters of five WECs. The allocation of such trend functions is only possible if the THD of current varies only slightly.

Table 11.4. Parameters a and m according to Equation 11.37 obtained from measured THD(I) curves of 1.5 MW WECs

WEC type	Generator type	THD(I)$_{max}$ in %	THD(I)$_{min}$ in %	a	m
1	DFIM	90	4	2.29	0.97
2	DFIM	50	1	0.99	1.10
3	DFIM	100	3	2.80	0.89
4	DFIM	450	8	4.46	1.12
5	SG	130	5	3.93	0.75

DFIM double-fed induction machine, SG synchronous generator, Total Harmonic Distortion of current

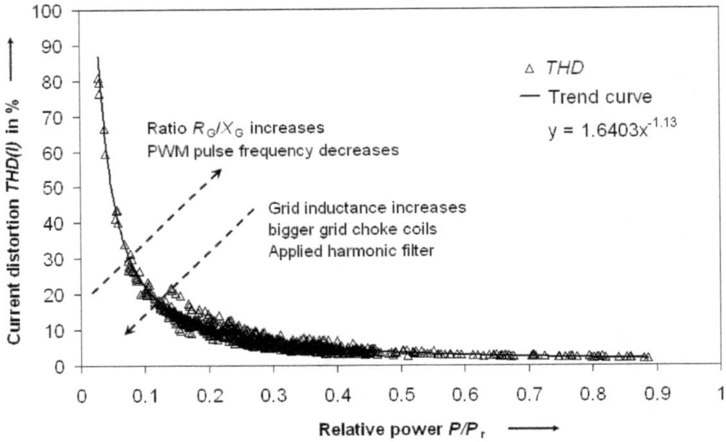

Figure 11.34. Total harmonic distortion of the current THD(*I*) on the 20 kV-level of a wind park

All described effects are visible in Figure 11.34, which also shows the calculated trend curve of the measured values. If the trend curves of WECs are known, countermeasures can be applied more easily, *e.g.*, the rating of passive harmonic filters or the design and programming of active filters.

Use of Harmonic Superposition Effects

Harmonics can be minimised by using the superposition effects. If harmonics are shifted to each other, their sum may be reduced or reinforced. Therefore the superposition of two sinusoidal functions with the same normalised amplitude of one is considered

$$f_\Sigma = y_1 \sin(\omega t) + y_2(\sin \omega t + \varphi) \tag{11.38}$$

These functions can be shifted against each other in small time or angular steps. Depending on the shifting angle φ, the sum of the two single amplitudes is calculated. This is realised for the fundamental wave and also for their harmonic frequencies. Figure 11.35 shows the calculation results for the fundamental frequency and the harmonics of order n=3, 5, and 7. It is very clear that the sum of the two fundamental waves from angles between zero and 120° is higher than the single wave amplitude and from 120° to 180° it decreases. As a general result, the calculated sum function has spans with decreased and increased amplitude, compared to the amplitudes of the single functions. The range, in which the amplitude sum of the two normalised functions increases, can be described using the periods T_n of the harmonics [42]

$$m \cdot \left(-\frac{T_n}{3}\right) \leq f_{\Sigma n} \leq m \cdot \left(\frac{T_n}{3}\right) \text{ with } n \in \Re, m = 1, 3, \ldots . \tag{11.39}$$

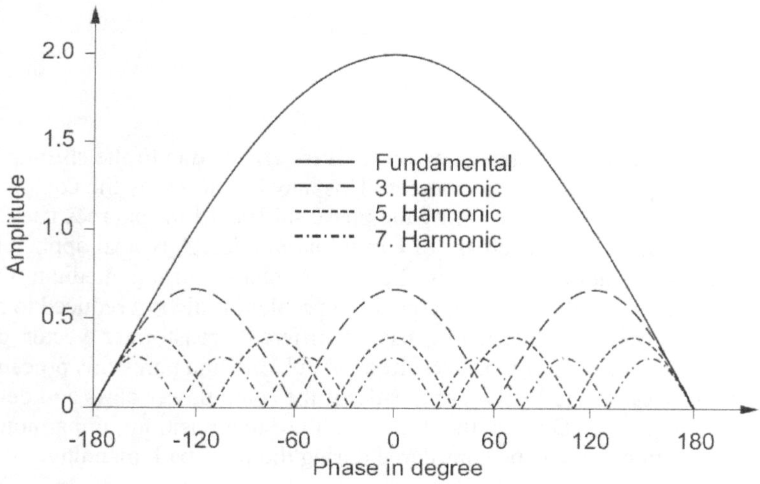

Figure 11.35. Amplitude sum of respectively two functions with normalised amplitudes *vs* shifting angle φ, results for the 1st, 3rd, 5th and 7th harmonic

A decrease of the sum amplitude of the respectively two amplitudes occurs in the remaining ranges

$$f_{\sum_n} \leq \pm\, m \cdot \left(\frac{T_n}{3} \ldots \frac{2T_n}{3} \right) \text{ with } n \in \Re,\ m = 1,3,\ldots. \tag{11.40}$$

The resulting superposition rule of two functions with the same period is obvious

$$f_{\sum_n}(t) = 2 \cdot \left| \sin\left((0.5 \cdot \omega \cdot t) \cdot n \right) \right|,\ n \in \Re \tag{11.41}$$

This function is valid for the superposition of two equal functions of any frequency. From Figure 35 it can be concluded that the attenuation of harmonics at small shifting angles is more probable if the order and therefore the frequency is higher. For that reason the third or fifth harmonic will not be so attenuated at small shifting angles as high order harmonics.

If the number of harmonic sources is higher than two, the problem becomes more complex. The variance increases with the number of connected harmonic sources. Whereas the sum of two independent functions delivers definite results, the sum of more magnitudes with independent shifting angles brings scattered results. The reasons for different shifting angles and different power factors cosφ of more functions can be, *e.g.*, two-point current control with different bandwidth, different control modes or load changes. To obtain values for these cases, the probability of increase and decrease must be calculated.

A popular empiric relation used for the superposition of multiple harmonic sources of WECs is given by [43]

$$THD(I)_{\Sigma} = THD(I)_n \cdot \frac{1}{\sqrt{n}} \tag{11.42}$$

Generally the harmonic superposition delivers effects due to the shifting angle arising between harmonic current sources. This may be caused by the complex grid impedance between two sources or by the phase shifting of the sources itself. Both effects can be used for the decrease of current harmonics. A typical application of the shifting of harmonic sources is the use of phase-shifting medium voltage transformers in WECs. Then the superposition problem is always reduced to a two-source problem with the alternating use of different transformer vector groups Dyn5 and Dzn6. This causes a phase shift of φ=30°. In wind parks this procedure is often applied because the WECs are mostly of the same power class and deliver a similar output power. Due to the fact that the superposition of harmonics is described algebraicly, it can be considered during the wind park planning.

TSO Requirements

Because of the increasing power share of the wind energy, more and more Transmission System Operators (TSO) demand a participation of the WECs in additional grid services. These requirements are country dependent; only the standard IEC 61400-21 gives consistant frame of technical demands [26]. In addition to this, the local utilities are free to issue other additional technical guidelines. Some basic features are:

- Frequency dependent control of the active power flow;
- External control possibility of the active power by the TSO;
- Grid voltage control by adjustable reactive power flow;
- Low-voltage ride-through during grid short-circuit events.

All these demands are fixed in diagrams which are part of the technical guidelines. Figure 11.36 shows an example for some grid system services demanded by German transmission system operators. In Figure 11.36a is the power curve over the frequency shown. WECs must not disconnect from the grid within the frequency range between 47.5 Hz and 51 Hz. The minimum duration of the power generation during grid frequency deviations is depicted in Figure 11.36b; Figure 11.36c shows the voltage dependent reactive power supply.

Also the WEC-behavior during grid short-circuit events is clearly defined in the local TSO guidelines. Figure 11.37 shows the low-voltage ride-through curves as required for the passing of grid failure events. The reason for such demands is the desired grid assistance by the WECs. At falling frequency a certain output power must supplied by the devices; see Figure 11.37a. Depending on whether the grid short-circuit is close to the generator terminals or remote at the wind turbine, the reaction on the resulting voltage drop must be different; see Figure 11.37b,c.

Such demands anticipate the necessary properties of completely decentralized energy systems with a high supply share from renewable systems in the future. Especially the possible voltage control is a useful feature of the decentralised power generation with WECs [47, 48], because the voltage cannot be controlled centralised. In future the participation on the primary grid frequency control is also foreseen.

All additional grid service requirements cause additional technical efforts for the WEC manufacturer. If a voltage control is applied, the necessary reactive power must flow over the grid side frequency converter. This means that this converter has to be rated higher compared to only the active power supply. Therefore the amount of reactive power is given exactly for each operation point of the WEC; see Figure 11.38.

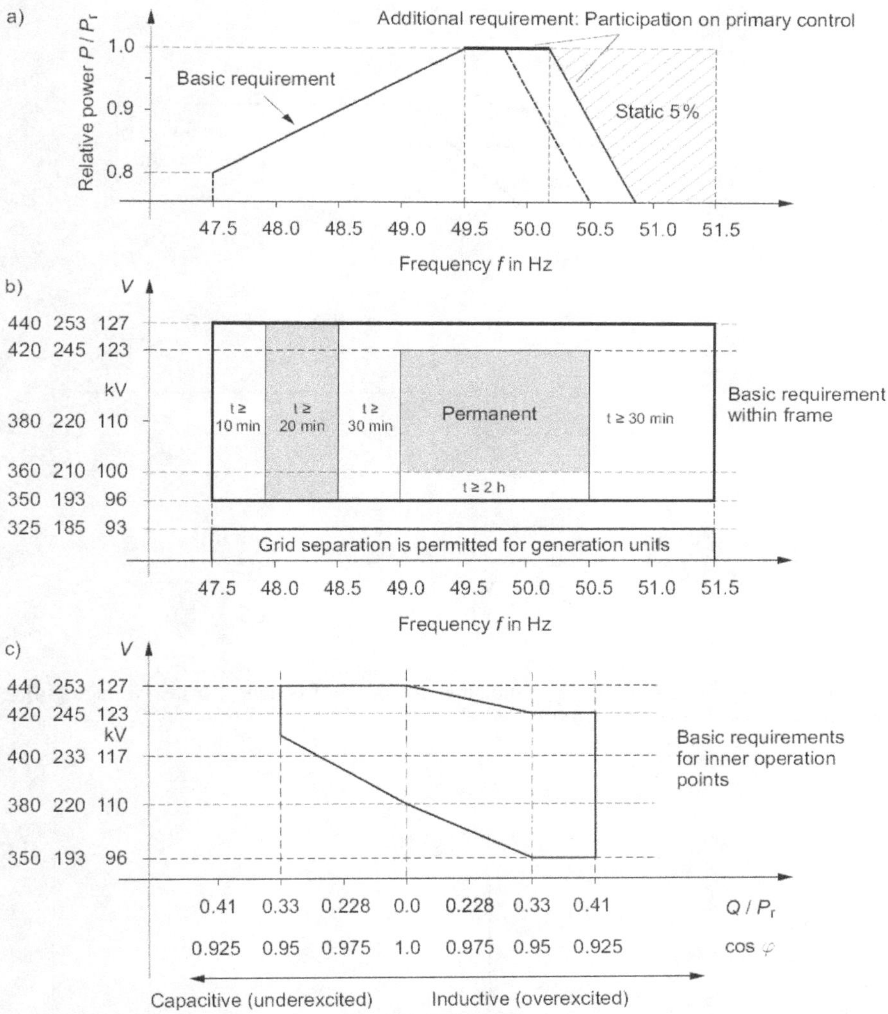

Figure 11.36. TSO requirements for WECs as applied in Germany [44]: **a** power curve; **b** supply duration; **c** reactive power supply

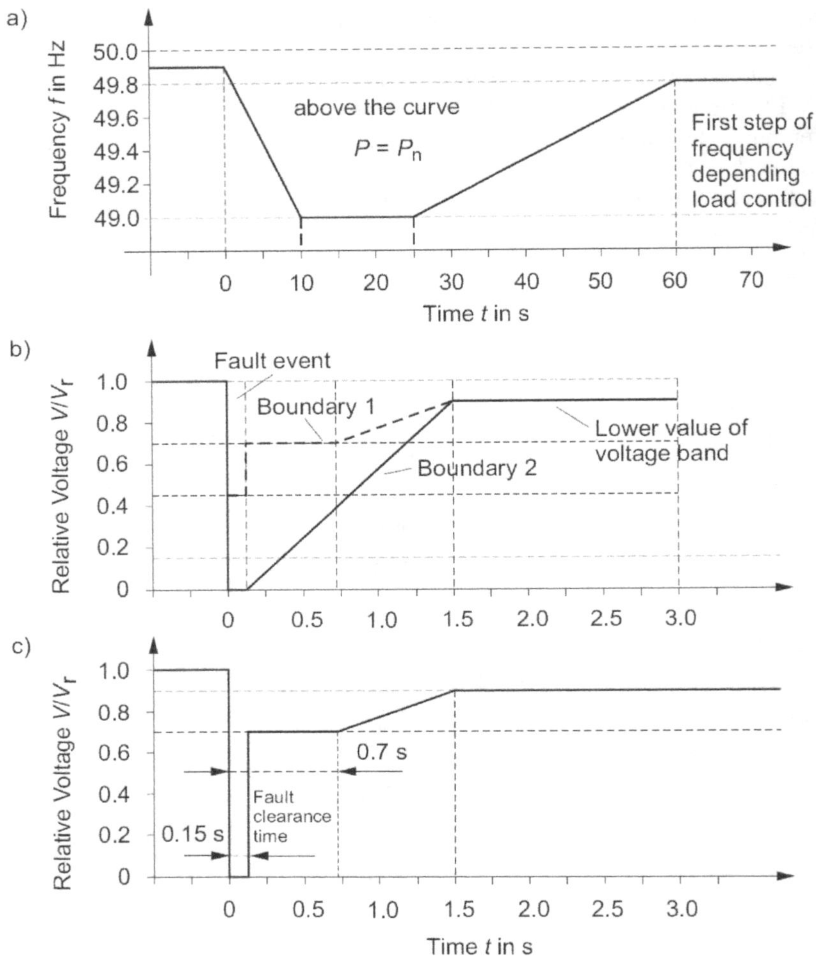

Figure 11.37. German ride-through guidelines for: **a** low frequency; **b** low-voltage [44]; **c** short-circuit close to generator [45]

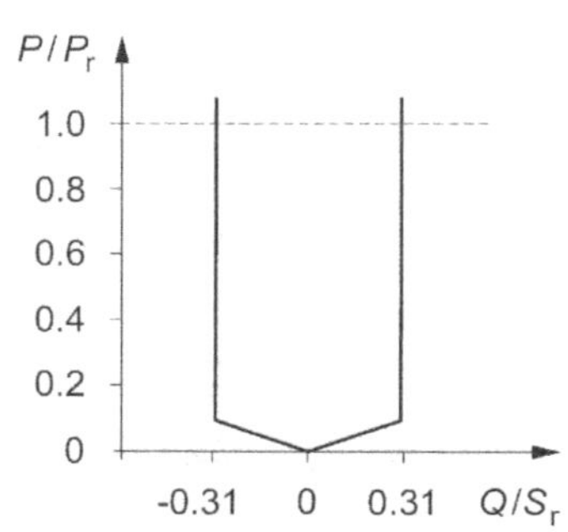

Figure 11.38. Reactive power range of a WEC [46]

11.6 Offshore Wind Energy

11.6.1 Installation Numbers and Conditions

In 2007 a global capacity of over 827 MW was installed in offshore wind parks, most of them in Europe. More than 45 GW are planned in Europe, thereof over 27 GW in Germany and over 10 GW in Great Britain. Also in North America a capacity of 1 GW is planned. These numbers show that the offshore market can be one of the most successful energy stories of the current century. A reason for the present global offshore power distribution with a European preponderance is the necessary occurrence of some good offshore conditions: good wind speeds, flat shore grounds, near grid connection points and adequate feed-in tariffs.

The high economic interest in WEC offshore technology is triggered by the good wind conditions on sea, which allows a doubling of the yearly energy generation compared to onshore wind parks and also the use of new installation areas. In contrast the installation effort is much higher and requires, unlike onshore WECs, special sea cranes.

11.6.2 Wind Park Design

Depending on the installation place, different technical requirements for the WECs occur. Offshore devices are evaluated corresponding to their:

- Reliability: failure frequency, maintenance efforts;
- Life time: erosion, corrosion resistance, load limitation;
- Technical risk: use of approved concepts;
- Installation costs: possibly low nacelle weights for low grounding efforts, low technical complexity;
- Energy yield: use of pitch-controlled wind turbines;
- Control possibilities: active and reactive power, grid services;
- Power quality properties: harmonics, flicker.

There are diverse possibilities for internal grid structures in wind parks. The design is influenced by the voltage type and level of the existing PCC and the installed generator type and voltage level of the WECs. Usable combinations of these three technical degrees of freedom result in the following five systems:

- Low voltage (LV) generators, connected to a medium voltage (MV) wind park grid with fixed grid frequency, high voltage (HV) AC onshore connection;
- As the first bullet with an HVDC land connection;
- As the first bullet with MV generator; for the generator see Figure 11.3;
- As the first bullet with MV generator;
- As the first bullet with a variable frequency in the internal wind park grid.

Also the speed control can be varied for fixed or variable speed system and it can be realised as a single or group control with to relatively smooth wind conditions. Radial or ring distribution systems similar to land installations are used as internal wind park grids. Up to a wind park power of 120 MW, two-winding transformers are used; above 120 MW up to 300 MW three-winding types are applied.

11.6.3 Transmission Types

For the energy transmission between offshore wind park and land substation, three systems are usable: the three-phase HVAC, a multi-phase HVAC or the HVDC connection. Their technical properties are listed in Table 11.5. Figure 11.39 shows the principle drawing of the three transmission systems. With the idea of six-phase HVAC systems the limited transmission capacity can be doubled [49]. This is only possible up to a maximum transmission distance due to the reactive cable power.

Table 11.5. Parameters of the transmission systems [50]

Parameter	HVAC three-phase	HVDC
Power per cable in MW	180–250	300–350
Maximum distance in km	80–120	Unlimited
Economical transmission voltage in kV	150	150
Cable available for voltage in kV	170	150
Cable diameter in mm	200 for three conductors	90 for one conductor

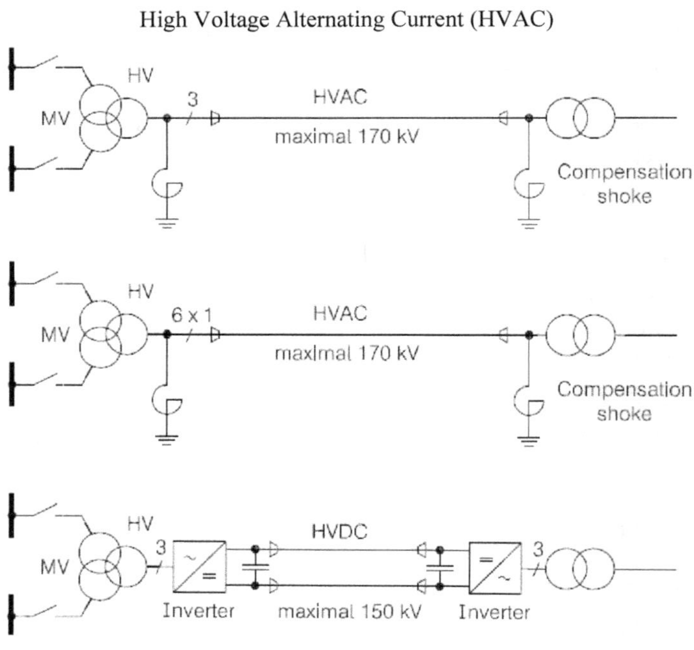

Figure 11.39. Transmission systems between offshore wind park and land substation. *From the top*: three-phase HVAC; six-phase HVAC; HVDC

Table 11.6. Advantages and disadvantages of high voltage AC and DC transmission systems

System	Advantage	Disadvantage
HVAC	• Proven technology • Low space required • No cost intensive power electronics • High reliability	• High cable costs • Reactive power of cable causes losses, requires compensation • Ground heating • Harmonics transmission • Limited length
HVDC	• Unlimited length • Low cable costs • No compensation required • Usable as static compensator • Electrical decoupling of systems • Free choice of wind park frequency • Independent active and reactive power control • High power quality • Black start (grid building) capability	• No reference for wind parks • High costs for power electronics • Power electronic losses • No electrical coupling between wind park and grid, no contribution to grid stability possible • A lot of space necessary

In Table 11.6 the advantages and disadvantages of the HVAC and HVDC systems are compared. Both transmission types are planned for future offshore wind parks, depending on their distance to the land substations. A wind park project mostly starts with a small pilot project with some WECs and will then be finished with full power in a second stage if the pilot project was successful. This makes the decision for the suitable transmission system difficult at the beginning.

11.7 Future Requirements and Developments

11.7.1 WEC Types

Considering future developments, voltage level, generator design, power electronic design and energy conversion strategy can be discussed. Of course the low voltage WEC types existing today will be further developed in future. This means novel generators and power electronic converters with higher output power will be on the market.

If the unit power increases up to 8 MW or 10 MW either the current or the voltage may increase. A conventional solution is the system development at the low voltage level by upscaling the generators and PWM converters. Medium voltage solutions bring decreasing currents, a smaller generator design and a lower nacelle weight which is important mainly for offshore wind parks. Currently the available medium voltage equipment for WECs is limited to only some supplies.

This hinders the development from the low voltage to medium voltage system approach.

In addition to this, special designed generators are mostly used in WECs. Double-fed induction generators or synchronous generators with electrical or permanent magnet excitation dominate the market. Whether costly special generators should be used or standard generators in combination with costly power electronics, see Section 11.4.1. The use of completely new generator designs, like transversal flux machines, has to now not been convincing in terms of manufacturing costs and efficiency. WEC concepts equipped with multi generators have also been discussed, but not realised yet.

Considering the energy conversion chain, only one conversion step between mechanical and electrical energy is necessary in WECs. Following this idea, the combination of a hydraulic gearbox with a simple standard synchronous generator seems to be a promising concept for the future. In this case no special designed generator or power electronic is necessary, but the gearbox efficiency has to be as high as for conventional gears.

11.7.2 Energy Management, Storage and Communication

The aim of a large-area Energy Management System (EMS) is to optimize power flow in the transmission system. This includes the control of power generation, power consumption, import and export of power and energy storage. This can only work properly if the different generation units cooperate within the electrical grid. With increasing power generation from renewable units a transformation of the transmission system is necessary. Beside the construction and modernization of the grid the implementation of "intelligent" properties is also required to reach the "*SmartGrid*" status. All system services have to be carried out by the decentralized power units.

If the EMS cannot adapt the generated and consumed power, energy storage is required. For use in transmission systems, high storage capacities are necessary. Such energy amounts can be handled today only from pumped-storage power stations or compressed air storage. High storage potential can be made available if consumer control is applied. This is mainly usable with industrial cooling in cold storage houses. By controlled disconnection and connection of such units, a storage capacity of some 10 GW is disposable, *e.g.*, only in Germany 40 GW [49]. This method should be used if the storage requirements hinder the further development of a necessary transfer of the energy supply [51].

In a future grid every unit should be able to work in four states: generator, load, flushing storage or charging storage. Between all units there must be an information exchange to secure the system stability. WECs are able to meet external control requirements and participate in the grid control; see Figure 11.28 in Section 11.4.4. For the participation in the grid services a controllable interface is necessary. In addition to this, the external control requirements may call for a communication interface. Both have to be consistent for all generation units. The communication standard IEC 61850 [52] for substations is not sufficient for WECs; because of this a separate standard IEC 61400-25 [53] and a draft for the control and monitoring of decentralised generation units IEC 57/660/NP are in cause of preparation.

11.8 Economics and Reimbursement

Two economic strategies for the remuneration of energy feed-in compete; first, the increased feed-in tariff based on certain laws as in Germany, Spain or France and, second, the quota system model for energy projects as applied in Great Britain. The most important value is the realised feed-in payment per kWh. Law-based tariffs generate a high increase of the installed wind power capacity. Because of a higher technical and financial risk the offshore installations stay behind expectations. Within the last 15 years the costs per kilowatt have been reduced by 50% from 1,600 to 800 €/kW. This concerns standard wind turbines in the megawatt power range. New developed devices, especially offshore WECs, are more costly because of their improved technology and more auxiliary components. Generally, within the last two years, a cost increase due to a higher market demand and increasing material costs has been observed. Currently the energy generation costs of WECs are in the range of 8–9 €cent/kWh for onshore and 6–8 €cent/kWh for offshore installations. At the same time, the European feed-in tariffs differ significantly for onshore and exceptionally for offshore wind parks. Figure 11.40 shows a comparison of the feed-in tariffs of different European countries. In the newest German amendment of the renewable energy law an additional bonus is foreseen with 0.7 €cent/kWh on the medium voltage level and 1.4 €cent/kWh on the high voltage level for system services like participation in grid voltage control.

The difference between the feed-in tariffs and the energy generation costs, reduced by interest payments and maintenance costs, is the estimated net profit. Naturally wind park projects with higher net profit are preferably installed. In the offshore market there is also a high financial imbalance. According to a study by a

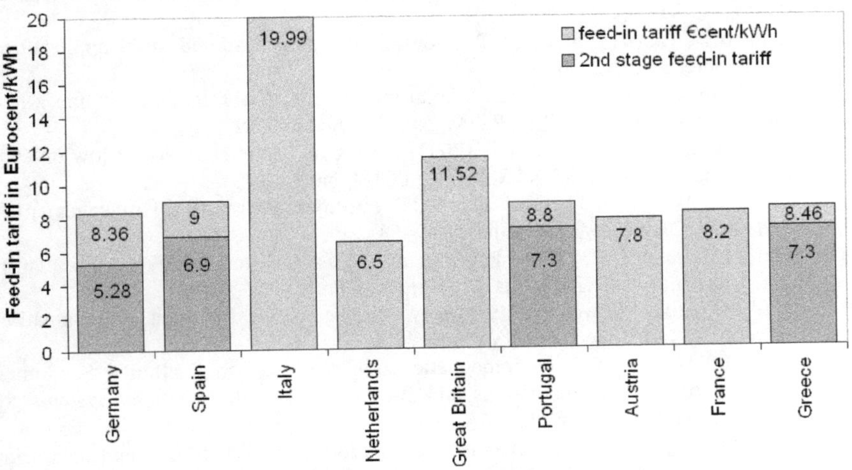

Figure 11.40. European feed-in tariffs of onshore WECs

European accounting company, is the imbalance for future offshore projects is very high with an average internal rate of return of 1.2% in Germany and 19.5% in Great Britain [54]. Such financial imbalances lead to supplier concentration in countries with advantageous conditions. Due to limited manufacturer capacities this can hinder offshore development in countries with lower rates of return.

References

[1] Global Wind Energy Council, (2007) Continuing boom in wind energy – 20 GW of new capacity in 2007

[2] Betz A, (1926) Wind energy and its use by wind mills. (in German), Göttingen, Bandenhoeck & Ruprecht

[3] Bennert W, Werner U, (1989) Wind energy. (in German), Berlin, Verlag Technik

[4] Molly JP, (1990) Wind energy. (in German), 2nd Edition, Karlsruhe, CF Müller

[5] Hau E, (2005) Wind turbines. Fundamentals, technologies, application, economics. 2nd Edition, Berlin, Heidelberg, Springer

[6] Luque A, (2003) Handbook of photovoltaic science and engineering. John Wiley & Sons

[7] Gasch R, Twele J, (2002) Wind power plants. Fundamentals, design, construction and operation. Berlin, Solarpraxis

[8] Burton T, Sharpe D, Jenkins N, (2002) Wind energy handbook. Chichester, New York, John Wiley & Sons

[9] Manwell JF, McGowan JG, Rogers AL, (2002) Wind energy explained. Theory, design and application. New York, John Wiley & Sons

[10] Schulz D, (2002) Investigation of power quality parameters of grid-connected PV and wind energy systems. (in German), dissertation thesis, TU Berlin

[11] Schulz D, (2006) Integration of wind energy converters into electrical grids. (in German), habilitation thesis, TU Berlin

[12] Heier S, (2006) Grid integration of wind energy conversion systems. 2nd Edition, New York, John Wiley & Sons

[13] Heuck K, Dettmann KD, Schulz D, (2007) Electrical power systems. (in German), Wiesbaden, Vieweg

[14] Yamano N, (2004) 6.5kV IEGT module development for industrial applications. PCIM, Nuremberg:326–329

[15] Rahimo M, (2004) 2.5kV – 6.5kV industry standard IGBT modules setting a new benchmark in SOA capability. PCIM, Nuremberg:314–319

[16] Tschirley S, Bernet S, Carroll E, (2004) Desing and characteristics of low on-state voltage and fast switching 10kV IGCTs. PCIM, Nuremberg:281–287

[17] Schulz D, Wendt O, Hanitsch R, (2005) Improved power factor management in wind parks. DEWI – Magazine no.27:49–58

[18] Schulz D, Hanitsch R, (2006) Power factor adjustment and energy management in large grid-connected wind parks. IEEE IECON Paris:4201–4206

[19] Schulz D, (2004) Power quality: theory, simulation, measurement and evaluation. (in German), Offenbach, VDE-Verlag

[20] IEC 1000-3-3, (1996) Electromagnetic compatibility. Part 3: limits. Section 3: limitation of voltage fluctuations and flicker in low-voltage supply systems for equipment with rated current ≤ 16A

[21] IEC1000-4-7, (1991) Electromagnetic compatibility. Part 4: testing and measuring techniques. Section 7: general guide on harmonics and interharmonics

measurements and instrumentation, for power supply systems and equipment connected thereto
[22] IEC 61000-4-15, (1998) Electromagnetic compatibility. Part 4: testing and measuring techniques. Section 15: flicker-meter, functional description
[23] IEC 1000-3-5, (1994) Technical Report. Electromagnetic compatibility. Part 3: limits. Section 5: limitation of voltage fluctuations and flicker in low-voltage power supply systems for equipment with rated current greater than 16 A
[24] VDEW, (1998) Generation units connected to the medium voltage grid. (in German), 2nd Edition
[25] EN 50160, (2000) Voltage characteristics of electricity supplied by public distribution systems
[26] IEC 61400-21, (2001) Wind turbine generator systems. Part 21: measurement and assessment of power quality characteristics of grid connected wind turbines
[27] FGW, (2006) Technical guideline for wind energy converters. Part 3: determination of the electrical properties – power quality. (in German), rev.18
[28] Measnet, (2007) Power quality measurement procedure, ver.3
[29] IEC 61000-3-4, (1998) Technical Report. Limits. Limitation of emission of harmonic currents in low-voltage power supply systems for equipment with rated current greater than 16 A
[30] IEEE standard 519-1992, (1992) Recommended practices and requirements for harmonic control in electric power systems
[31] IEEE standard 1001-1988, (1988) Guide for interfacing dispersed storage and generation facilities with electric facility systems
[32] IEEE standard 1159-1995, (1995) Recommended practice for monitoring electric power quality
[33] IEEE standard 1250-1995, (1995) Guide for service to equipment sensitive to momentary voltage disturbances
[34] IEEE standard 1547-2003, (2003) Standard for interconnecting distributed resources with electric power systems
[35] IEC 61000-2-4, (1994) Electromagnetic compatibility. Part 2: environment. Section 4: compatibility levels in industrial plants for low-frequency conducted disturbances
[36] IEC 61000-2-12, (2003) Electromagnetic compatibility. Part 2: environment. Section 12: compatibility levels for low-frequency conducted disturbances and signalling in public medium-voltage power supply systems
[37] IEC 61000-4-30, (2003): Electromagnetic compatibility. Part 4: testing and measurement techniques. Section 30: power quality measurement methods
[38] Eggert B, (1998) Grid influence of an ideal three-phase inverter connected to the mains. (in German), ETG-Report 72, Berlin, Offenbach, VDE-Verlag:301–315
[39] Schulz D, Tognon E, Hanitsch R, (2003) Investigation of the harmonic transformation properties of double fed induction generators in wind energy converters. PCIM Power Quality Conference, Nuremberg:207–212
[40] Schostan S, Dettmann KD, Schulz D, (2007) Investigation of an atypical sixth harmonic current level of a 5MW wind turbine configuration. IEEE EUROCON Conference
[41] Dettmann KD, Schostan S, Schulz D, (2007) Wind turbine harmonics caused by unbalanced grid currents. Electrical Power Quality and Utilisation Journal, vol.XIII, no.2:49–55
[42] Schulz D, Lchamsuren B, Hanitsch R, (2001) Power quality improvements of solar systems: simultaneous inverter operation results in reduction of distortions. PCIM Power Quality Conference, Nuremberg:183–188

[43] Gerdes G, Santjer F, (1996) Reduction of grid interferences in wind farms. European Union Wind Energy Conference, Göteborg:440–454
[44] VDN, (2007) TransmissionCode. Grid and system regulations of the German grid utilities. (in German)
[45] E.ON, (2006) GridCode. High and extra high voltage
[46] VDN, (2004) Renewable generation units on the high and extra high voltage. (in German)
[47] Burges K, DeBroe A, Feijoo A, (2003) Advanced power control in a wind farm network. IEEE Power Tech Conference, Bologna
[48] Schulz D, Schostan S, Hanitsch R, (2005) Participation of distributed renewable generation on the grid control and grid support. PCIM, Nuremberg:414–419
[49] Brakelmann H, (2006) New six-phase system with high transmission power for VPE-isolated HVAC sea and land cables. (in German), Energiewirtschaft, vol.105, no.4:34-43
[50] Wensky D, (2002) Grid connection of wind parks with high power. (in German), Dena-tagung Perspektiven für die Stromversorgung der Zukunft, Berlin
[51] Stadler I, (2007) Optimisation of electricity transport in energy systems with high renewable energy penetration by demand response. 4th Conference on Sustainable Development of Energy, Water and Environment Systems, Dubrovnik
[52] IEC 61850, Communication networks and systems in substations, Part 1…10
[53] IEC 88/179/CD, (2004) Committee draft IEC 61400-25. Edition 1: wind turbines. Part 25: communications for monitoring and control of wind power plants
[54] KPMG, (2007) Offshore wind parks in Europe. Market study

12

Grid Integration of Photovoltaics and Fuel Cells

Detlef Schulz[1], Matthias Jahn[2] and Thomas Pfeifer[2]

[1]Department of Electrical Engineering,
Electrical Power Engineering,
Helmut-Schmidt-University,
Holstenhofweg 85, D-22043 Hamburg, Germany.
Email: Detlef.Schulz@hsu-hh.de

[2]Fraunhofer Institut Keramische Technologien und Systeme
Winterbergstraße 28, D-01277 Dresden, Germany.
Email: Matthias.Jahn@ikts.fraunhofer.de; Thomas.Pfeifer@ikts.fraunhofer.de

12.1 Introduction

Photovoltaics (PV) is a renewable technology with the highest theoretical potential. World-wide 5.86 GW_p were installed in 2006. Thereof 2,860 MW_p are in Germany; 1,709 MW_p in Japan and 624 MW_p in the United States [1]. These three countries have installed over 89% of the total world PV power. The index p of the power means peak values for a standardized installation of 1,000 W/m^2 and a temperature of 25°C. Most of the PV plants work grid connected because of increased tariffs for energy feed-in. Because of their similar output behavior, Fuel Cell (FC) systems are also discussed here. They generate DC output power like PV generators, which must be converted into AC with power electronic converters. FC are usable in different power system applications and for automotive and other mobile technologies. An important advantage is the lack of dangerous emissions. In 2006, 85 MW electrical power from FC was global installed, with 40.8 MW in the United States, 17.1 MW in the European Union (mostly Germany) and 9.9 MW in Japan [2, 3].

12.2 Photovoltaic Power Plants

12.2.1 System Overview

A grid connected PV system always consist of a PV generator, an inverter and a grid transformer; see Figure 12.1. In the PV generator DC power is generated, which is transformed into AC power with the power electronic inverter. The output transformer adapts the constant AC voltage of the inverter to the grid voltage level.

To avoid an undesired islanding mode of the system or an undesired energy flow from the grid to the PV system, an islanding detection monitoring is connected between the inverter output and the mains.

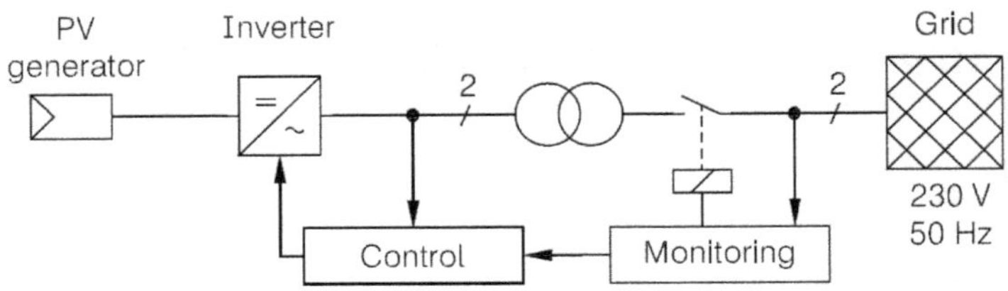

Figure 12.1. Principle of a single-phase grid connected PV system

The details of the energy conversion, the derivation of the current-voltage curves, the working principle of the inverter and the plant design are explained in the following sections.

12.2.2 Energy Conversion

In the PV generator occurs the conversion of the incident radiation into electrical energy by means of the photo-electric effect [4]. This occurs when doped semiconductors of main group IV of the periodic table of elements generate charge carriers when they are irradiated with light. The most used basic material is silicon which is doped with elements of the main group III or V. The doping result is either an n-conducting or a p-conducting semiconductor. Between n- and p-semiconducting layers exist a p-n-junction which causes an electric field and generates a DC voltage. Contacts on the upper and lower side of the layers allow a current flow to an external electric circuit. These contacts are designed as metal fingers on the upper side and as metal film on the lower side of the PV cell; see Figure 12.2. PV cells are manufactured as 5-inch cells (12.7×12.7 cm) or 6-inch cells (15.2×15.2cm). Former cells had a 4-inch (10×10 cm) format. One PV cell delivers a no-load voltage in the range 0.5–0.7 V and a short-circuit current of 2.5 to almost 7 A. In PV modules the cells are connected in series for voltage increase.

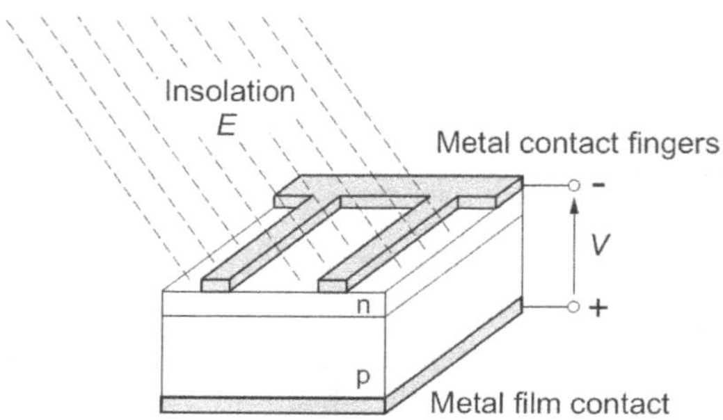

Figure 12.2. Principle of a photovoltaic cell

Depending on the application, these modules are connected in series or parallel circuits; the resulting system is called a PV generator.

12.2.3 Cell Types

PV cells can be classified into crystalline and thin film types; see Figure 12.3 [5]. Crystalline cells are sawn from silicon blocks, which are manufactured with different technologies [4]. With the Czochralski method silicon is pulled in a column; the Bridgeman method involves the casting of silicon ingots. The round columns or square ingots are sawn into wafers with diamond saws and the wafer edges are ground. Among the silicon cells three internal crystal classes exist. Mono or single crystalline blocks obtain a continuous and unbroken crystal lattice of the entire sample. In amorphous structures the crystal lattice is highly discontinuous. Between these two extremes exists the polycrystalline structure. Crystalline cells

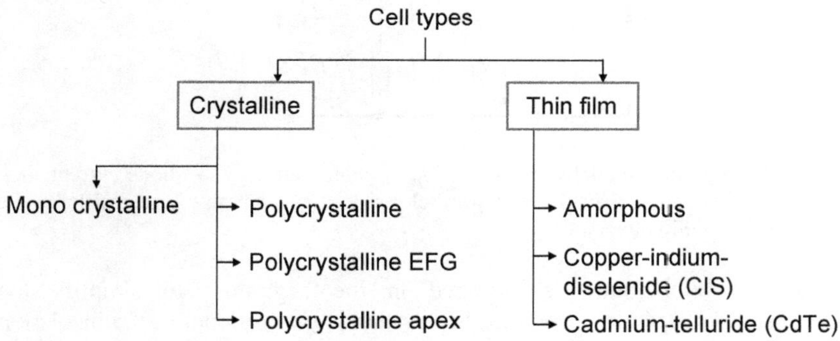

Figure 12.3. Photovoltaic cell types

Table 12.1. Required space and efficiency of different PV cell types [5]

Cell type	Required space in m^2/kW_p	Module efficiency in %
mono crystalline	7–9	17
polycrystalline (EFG)	8–9	13
polycrystalline	9–11	13
thin film	11–13	9
amorphous	16–20	7

EFG edge-defined film-fed growth

due to technological factors, need a minimum thickness of 200 micrometers; their production is costly because of the sawing waste and the grinding. The surface is chemical treated for antireflex coating and contacts are printed on it.

Thin film cells save material compared to crystalline cells. They are manufactured with a total semiconductor thickness of 10 micrometers or lower. A carrier substrate made of glass, plastics or metal is coated with the semiconductor material within a gas phase. Typical thin film types are amorphous, copper-indium diselenide (CIS) or cadmium telluride (CdTe) cells; see Figure 12.3. The required space and the efficiency of different PV cell types are shown in Table 12.1.

12.2.4 Modeling of PV Cells

For the understanding of the operation behavior of PV systems, equivalent circuit models are used. Solar cells are like diodes of a p-n junction. Therefore a one- or two-diode model is used for the simulation of PV cells. Such models are valid for the single cells and also for modules [6]. The one-diode model as shown in Figure 12.4 includes a light dependent current source, a diode, a parallel and a series resistor. By means of a variable resistor the output load is adjustable.

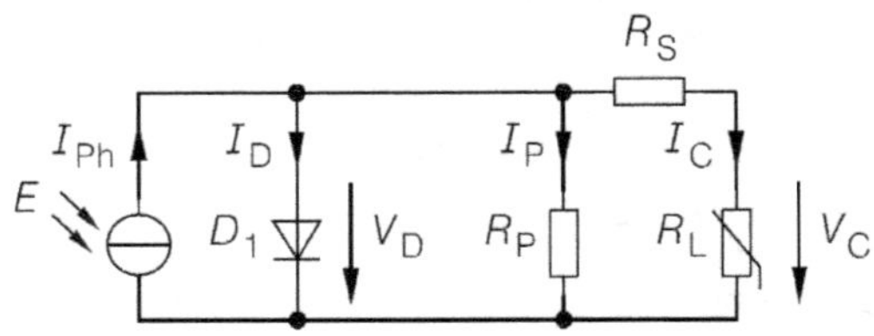

Figure 12.4. Single-diode-model of a PV cell, I_{ph} – photo current, I_D – diode current, I_P – parallel current, R_S – series resistor, R_P – parallel resistor, R_L – variable load resistor, V_D – diode voltage, V_C – cell voltage

A constant temperature is assumed in the beginning to simplify the explanations. Then a photo current I_{Ph} occurs that is proportional to the insolation E

$$I_{\mathrm{Ph}} \sim E = c \cdot E \tag{12.1}$$

The photo current is modeled by a controlled current source. According to the equivalent circuit in Figure 12.4 the output voltage and the cell current of the PV cell are calculated

$$V_C = V_{\mathrm{D}} - I_{\mathrm{C}} \cdot R_{\mathrm{S}} \tag{12.2}$$

$$I_{\mathrm{C}} = I_{\mathrm{Ph}} - I_{\mathrm{D}} - I_{\mathrm{P}} \tag{12.3}$$

A big parallel resistance R_P causes a small current I_P. The resistor models the leakage current on the grounded edges of the *p-n* junction. Ideally the parallel resistor is infinitive and the current is zero. Practically the resistor has an average value of 1 kΩ for crystalline cells. Thin film cells and cells in laminated modules have a lower leakage resistance that leads to a higher current. Applying a resistor value of $R_P \neq 0$ to the commonly used diode the expression for the cell current is

$$I_C = I_{Ph} - I_S \cdot \left\{ \exp\left(\frac{V_D}{m \cdot V_T} \right) - 1 \right\} - \frac{V_D}{R_P} \text{ with } V_T = \frac{k \cdot T}{e} \tag{12.4}$$

where I_S – saturation current, V_T – temperature voltage, m – diode factor, k – Boltzmann constant, T – absolute temperature, e – electron charge.

The series resistor simulates the contact resistors that should be as little as possible. A common average value as used in simulations is R_S=20 mΩ. Using Equation 12.2, the cell current can be calculated

$$I_C = I_{Ph}(E) - I_S \cdot \left\{ \exp\left(\frac{V_C + I_C \cdot R_S}{m \cdot V_T} \right) - 1 \right\} - \frac{V_C + I_C \cdot R_S}{R_P} \tag{12.5}$$

In the simulations a temperature voltage of V_T=25 mV was applied. The diode factor is m=1 for an ideal junction without recombination and m=2, if a complete recombination is assumed. Although from the point of physical modeling only integer values for the diode factor exist, in the electrical PV simulations a mean value of m=1.5 is applicable. The one- diode model allows quite an accurate simulation of the PV cell characteristic.

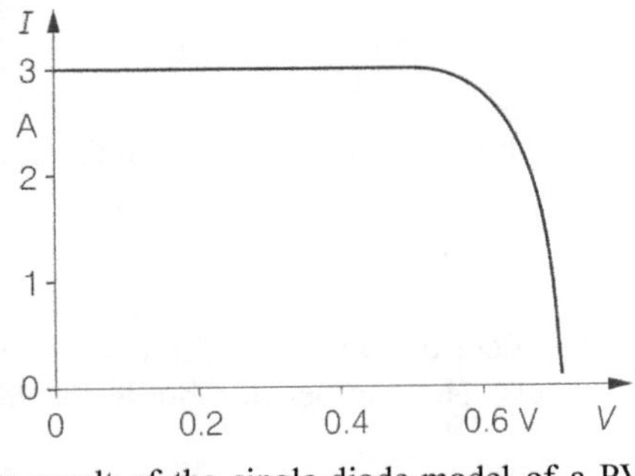

Figure 12.5. Simulation result of the single-diode-model of a PV cell with variable load $R_L(t)$, R_S=20 mΩ, R_P=1 kΩ, R_L=100 Ω/s·t

Figure 12.5 shows the simulation result for a PV cell at a temperature of T=25°C and an insolation of E=1000 W/m^2. The value of the resistor R_L changes during the simulation time t with the relation R_L=100 Ω/s·t. Hence, the cell behavior for different loads is obtained.

A refinement in the simulation of PV cells is possible by using the two-diode model. Therefore an additional diode is connected in parallel to the existing diode. This allows a physical correct simulation with two separate diode factors of m=1 and m=2 and also enables the use of two different temperature voltages. The basic equations in Equations 12.1 and 12.2 are also valid for the two-diode model, but for the calculation of the cell current the second parallel branch has to be considered

$$I_C = I_{Ph} - I_{D1} - I_{D2} - I_P \tag{12.6}$$

Now the cell current can be expressed by using the saturation current, the temperature voltage and the diode factors

$$\begin{aligned} I_C = I_{Ph}(E) - I_{S1} \cdot \left\{ \exp\left(\frac{V_C + I_C \cdot R_S}{m_1 \cdot V_{T1}} \right) - 1 \right\} \\ - I_{S2} \cdot \left\{ \exp\left(\frac{V_C + I_C \cdot R_S}{m_2 \cdot V_{T2}} \right) - 1 \right\} - \frac{V_D}{R_P} \end{aligned} \tag{12.7}$$

A simulation of a PV cell with the two diode model delivers the same result as for the one diode model. Because of this, in the following considerations the one diode model is used for an extended modeling of the PV modules.

12.2.5 Modeling of PV Modules

PV modules consist of series connected cells. The module voltage V_M is the sum of the cell voltages V_C. It is calculated using the cell number n

$$V_M = \sum_1^n V_C = n \cdot V_C \tag{12.8}$$

In a series connection the module current I_M is equal to the cell current I_C

$$I_M = I_{C1} = I_{C2} = \cdots = I_{Cn}$$

Figure 12.6 shows the equivalent circuit of a PV module consisting of two series connected single-diode models. The number of modules is choosen acording to the desired module voltage.

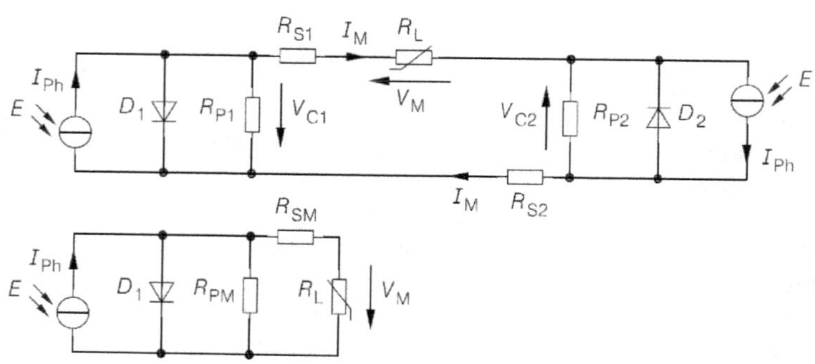

Figure 12.6. Equivalent circuit of a PV module. *From the top*: consisting of two series connected single-diode-models; with a single-diode-model and equivalent parameters

For the simulation of PV models with the one diode model the equivalent parameters for the module series resistance $R_{MS}=n \cdot R_S$, the module parallel resistance $R_{MP}=n \cdot R_P$ and the module temperature voltage $V_{MT}=n \cdot V_T$ are used [7].

12.2.6 Operation Behaviour

With increasing incident light raising the short-circuit current of a PV module, increasing temperature reduces the no-load voltage; see Figure 12.7. The optimal power generation is reached if the product of short-circuit current and no-load voltage is maximal. This point is called the Maximum Power Point (MPP), which is controlled by adjustment of the module output current. Modules can be connected in series or in parallel. Their practical connection is determined by the available space and direction of the installation.

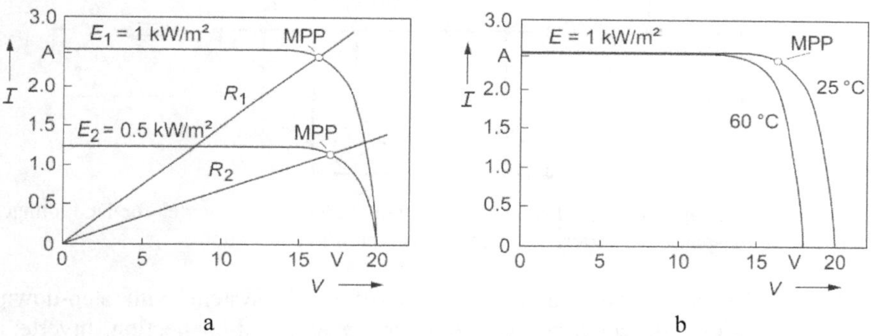

Figure 12.7. Current-voltage-characteristics of a PV module: **a** influence of the incident light; **b** influence of temperature

If a large generator is installed, the modules within similar directed areas are often connected in series. Different directed modules are conneted in parallel because of possible different shading conditions. The values for the current and voltage of the PV generator are specified by the rating of the PV inverter.

Most of the PV systems work grid connected; only in some special applications like parking automats or the supply of remote loads do they work without coupling to the mains. If energy should be fed into the grid, the DC values have to be transformed into an AC system with a frequency of 50 Hz by a power electronic inverter.

12.2.7 Inverter Types

Modern grid connected PV systems use self-commutated PWM inverters with pulse frequencies from 10–25 kHz. In the power electronic circuit all kinds of transistors are used; often MOSFET or IGBT are applied. Older grid-commutated inverter types often cause power quality problems and therefore disappear from the market. In PV systems different inverter topologies may be used depending on the technical boundary conditions:

- Inverters with a 50 Hz transformer: simple topology, high reliability, high volume and weight, maximum efficiency of 95%;
- Inverters with a high frequency transformer: costly concept, low volume and weight, maximum efficiency of 91%;
- Transformerless inverter: low weight, voltage transmission ratio up to 1:3, maximum efficiency of 95%;
- Cuk- or Zeta-inverter: transmission ratio up to 1:5, maximum efficiency of 91%;
- Resonance inverter: complex control, maximum efficiency of 95%.

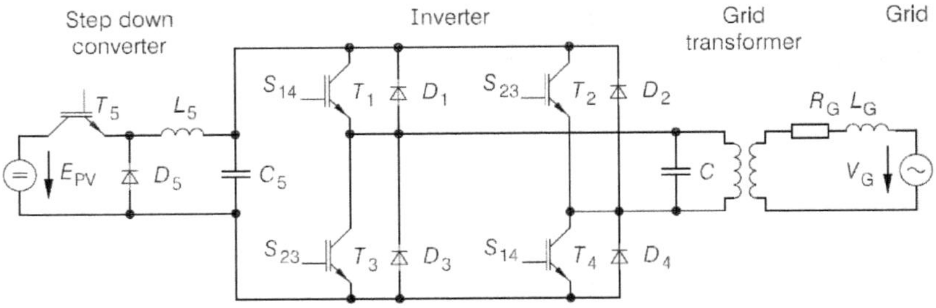

Figure 12.8. Equivalent circuit of a PV system with step-down converter for DC voltage adaptation, single-phase converter, grid transformer and grid connection

Figure 12.8 shows an equivalent circuit of a PV system with step-down converter, single-phase converter, grid transformer and grid connection. Inverters without transformers use a step-up converter for the conversion of the PV input voltage to the AC grid voltage. In all inverter types the DC voltage value is adapted by step-up or step-down converters to the AC values if required.

The inverter efficiency depends on the output power, which is a function of the incident light. In most European regions solar radiation seldom occurs above 800 W/m^2; the PV systems often work under partial load. Therefore it is the inverter efficiency compared by standardised European efficiency which influences the typical load conditions [5]

$$\eta_{Euro} = 0.03 \cdot \eta_{5\%} + 0.06 \cdot \eta_{10\%} + 0.13 \cdot \eta_{20\%} + 0.1 \cdot \eta_{30\%} + 0.48 \cdot \eta_{50\%} + 0.2 \cdot \eta_{100\%} \tag{12.9}$$

12.2.8 Plant Design

PV modules may be connected to the grid with module inverters, string inverters or central inverters; see Figure 12.9. Module inverters with small power ratings are fixed on the back side of every module. They can adjust an optimal MPP per device that results in a high total energy yield of the PV system. This decentral concept necessitates high effort if a monitoring system should be applied. String inverters convert the DC power of a whole module string. Compared to the module inverter, the MPP control is less optimal if the incident light is unevenly distributed or shading arises on some modules. However, a monitoring system is easier to implement. Central inverters offer the best monitoring possibility because only one data interface and one processing unit are necessary. However, no individual MPP tracking is possible. The application of the inverter types depends on the monitoring needs, the incident light distribution, the shading and the module direction. PV units up to 4.6 kVA are single-phase connected; above this power level

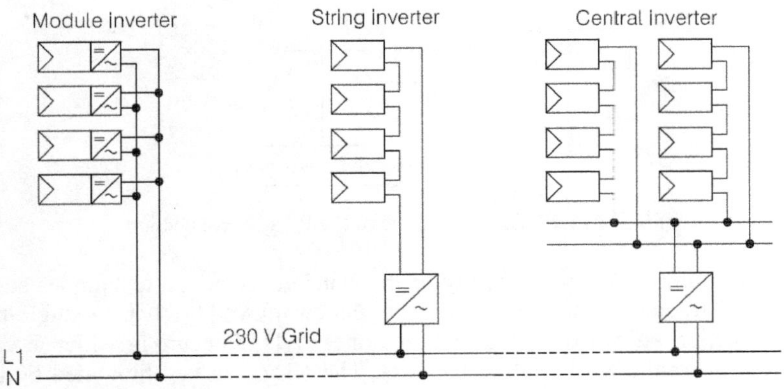

Figure 12.9. Classification of PV inverters according to [8]

they are always designed as three-phase plants. PV systems are mostly installed widely distributed in the low voltage grid with unit powers up to 30 kVA, but there are also medium voltage solar parks with powers up to 4 MVA. The type of grid connection determines the possibilities for the choice of the islanding detection monitoring.

12.2.9 Grid Interfacing and Islanding Detection

A connection to the mains requires failure monitoring for the islanding case. If the mains are switched off, the inverter output should shut down quickly. When the inverter operates despite the grid outage, islanding operation occurs. That means the output of the active and reactive power is equal to the required power of the still connected consumer loads

$$(P + \mathrm{j}Q)_{\mathrm{PV}} = (P + \mathrm{j}Q)_{\mathrm{C}} \tag{12.10}$$

The probability of islanding is not very high, because the power balance requires a balance between consumer load and inverter output power and additionally constant incident light and sparse disturbances from the Maximum Power Point Tracking (MPPT). Investigations of the islanding probability resulted in a maximum probability of 0.2 if ideal islanding conditions are assumed [9, 10]. But this event can occur randomly, as was shown in laboratory tests. Because of this, the use of detection devices is prescribed independently from the practical likelihood of inverter islanding. So far no international standard for the handling of this unintentional stand-alone operation exists. National standards diverge widely. Thereof some general rules for the handling of the islanding mode can be extracted. A consistent demand of all existing national standards is the immediate grid separation of the PV system by means of a switch; see Figure 12.10. This always requires a shut down of the inverter. The detection method of the islanding itself is a matter of dispute.

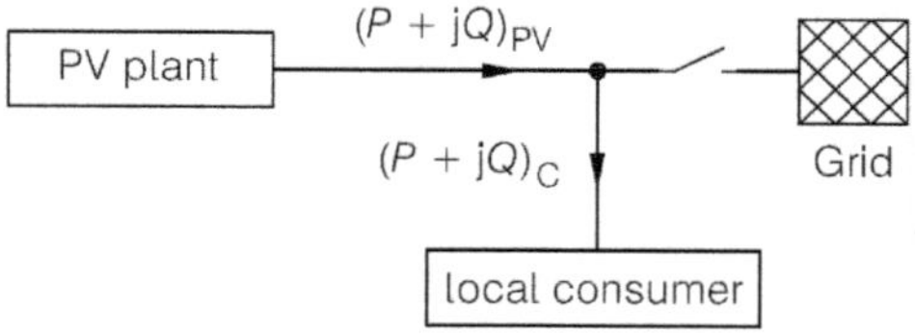

Figure 12.10. Basic principle of the PV grid connection

A main question is the reliability of the standard detection techniques such as voltage and frequency monitoring with protection relays [11, 12]. In single-phase grid connection so-called Non-detection Zones (NDZ) occur, in which no relay tripping arises although islanding happens. The NDZ can be eliminated by using additional passive or active grid monitoring methods; see Figure 12.11. Passive techniques evaluate only grid parameters, cause active methods disturbances on the mains and analyse the reaction on these perturbations. Passive methods don't influence the grid, but they can't avoid either a NDZ or partly faulty activations. Active monitoring methods have no NDZ but influence the grid [7, 13]. The measurement itself can be applied either before or behind the grid separation switch.

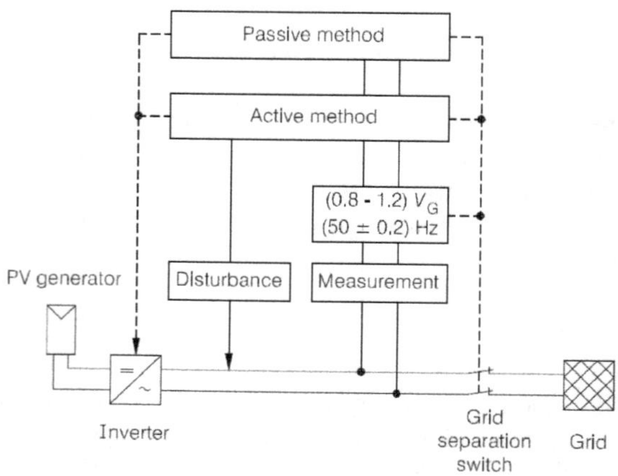

Figure 12.11. Active and passive islanding detection methods

Different passive monitoring methods exist [7]:

- Voltage and frequency monitoring with standard relays;
- Unstable phase synchronisation;
- Voltage harmonic monitoring;
- Monitoring of a voltage phase jump;
- Use of a phase-frequency graph;
- Monitoring of the power-frequency ratio.

Some of them are used in countries that do not agree to the application of one of the following active monitoring techniques:

- Variation of active and reactive power;
- Active change of the inverter output frequency;
- Active voltage phase shift;
- Monitoring of the grid impedance.

An international approach is still under discussion; no international solution exists [14]. Table 12.2 shows examples of the approaches of different countries and lists the national standards for islanding monitoring. The monitoring of voltage and frequency limits is demanded in all countries.

Table 12.2. International standards for islanding monitoring (examples)

Country	Monitoring condition	Standard
Australia	An accepted active method	Guidelines for grid connection, 2000; Draft AS/NZS 2000
Austria	IDF, impedance value of 1 Ω	CLC/BTTF 86-2(SEC)38
Belgium	No additional requirements	BFE CCLB 06, 2000
Canada	One active method without grid distortions	IEEE 929-2000
France	No additional requirements	No special standard
Germany	IDF with impedance measurement, limit value 1 Ω	VDEW regulations, draft DIN VDE 0126
Italy	No additional requirements	CEI 11-20
Japan	One active method	MITI guideline1998
Netherlands	Free choice	Netcode DTE, Supplementary conditions
Poland	No additional requirements	No special standard
Spain	No additional requirements	Condiciones administrativas y tecnicas
Switzerland	Free choice, function test	Eidgenössisches Starkstrominspektorat
USA	One active method without grid distortions	IEEE 929-2000

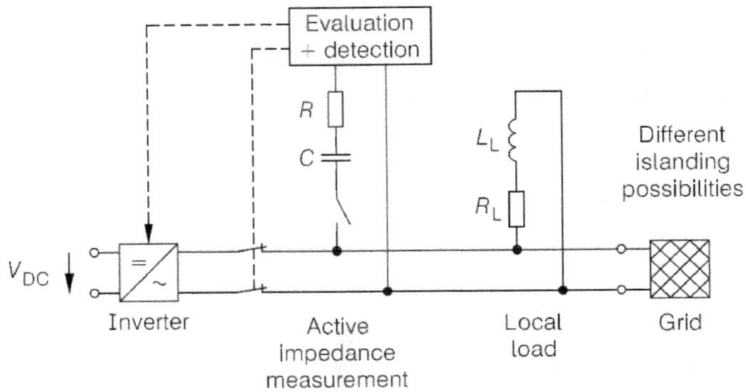

Figure 12.12. Principle of the active impedance measurement

Figure 12.12 shows the principle of an active impedance measurement, which causes distortion signals as shown in Figure 12.13. Such grid perturbations may influence the power quality [7, 15].

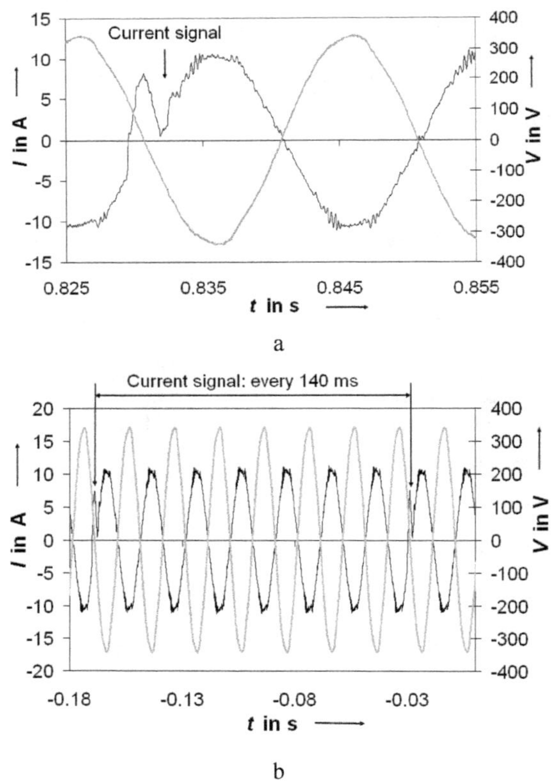

Figure 12.13. Islanding detection signal of an active impedance measurement: **a** single signal; **b** time-lag of the signals

Because of this, a monitoring device with the active grid impedance measurement was investigated in more detail in order to get real disturbance values of such methods. An evaluation of the charging event of a capacitor allows the determination of the grid impedance value. This value can be calculated from the calculated time constant during the charging or by using two measured complex voltage and current pairs. With the complex voltages and currents results for the complex impedance $\underline{Z}_G$

$$\underline{Z}_{\mathrm{G}} = \frac{\underline{V}_{\mathrm{G}}}{\underline{I}} = \frac{V_{\mathrm{G}} \cdot \mathrm{e}^{\mathrm{j}\varphi_V}}{I \cdot \mathrm{e}^{\mathrm{j}\varphi_I}} = \frac{V_{\mathrm{G}}}{I} \cdot \mathrm{e}^{\mathrm{j}(\varphi_V - \varphi_I)} \tag{12.11}$$

For two measurement points the impedance can be calculated in a simple way by using the difference values

$$\underline{Z}_G = \frac{\Delta \underline{V}}{-\Delta \underline{I}} = \frac{\underline{V}_1 - \underline{V}_2}{\underline{I}_2 - \underline{I}_1} \tag{12.12}$$

12.2.10 Power Quality

All power quality issues mentioned here are considered in addition to the effects on wind energy converters as discussed in Section 11.5. Generally, the harmonic emissions of PV systems may be lower compared to wind turbines, caused by the higher PWM pulse frequencies in the range of 10–25 kHz. Only sometimes is the grid impedance on low voltage PCCs so high that the voltage harmonics may break the limit values [16]. In grid connected PV systems, specific harmonics arise due to

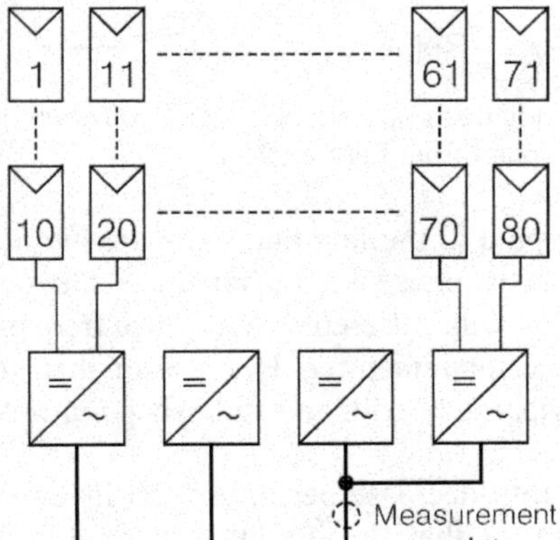

Figure 12.14. Topology of the measured PV plant

the MPPT and active islanding monitoring. Also the flicker emissions are very small compared to wind energy converters because of the lower unit powers and the slow incident lighting changes. Flicker occurs only in systems with a high output power in the upper kVA range connected to low voltage grids with high impedance values or in systems in the MVA power range. In parallel installed systems of the same power class harmonic superposition also arises as described in Section 11.5.2 [17]. If the PWM pulse frequencies differ strongly, this effect can be neglected.

In the following, a typical method for the detection of a power quality problem of a PV system is described. The PV plant as shown in Figure 12.14 suddenly delivered a decreased output power. Therefore the output power and the total harmonic distortion values of the three-phases were measured with a power quality analyser. The calculation of the THD values was explained in Section 11.5. In phase L3 two inverters of the same type are connected. Hence, the diagram in Figure 12.15 contains two inverter curves.

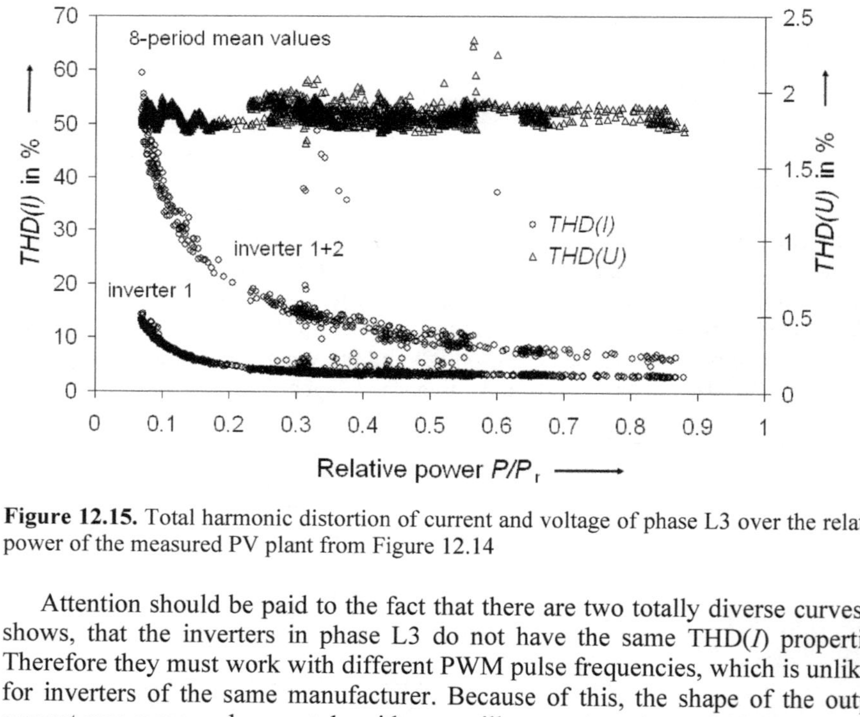

Figure 12.15. Total harmonic distortion of current and voltage of phase L3 over the relative power of the measured PV plant from Figure 12.14

Attention should be paid to the fact that there are two totally diverse curves. It shows, that the inverters in phase L3 do not have the same THD(*I*) properties. Therefore they must work with different PWM pulse frequencies, which is unlikely for inverters of the same manufacturer. Because of this, the shape of the output current was measured separately with an oscilloscope in order to find the source of disturbances.

A typical power quality decrease because of an inverter malfunction is shown in Figure 12.16. It is seen that the output current differs in distorted inverter operation more from the sinusoidal shape compared to the regular operation. The Total Harmonic Distortion of the current (see Section 11.5.2) increases significantly from 4.9–7.1%. Such a result is directly obtained from the defect inverter [7]. In the beginning of the analysis the failure has to be detected in the whole system with a more generalised measurement method.

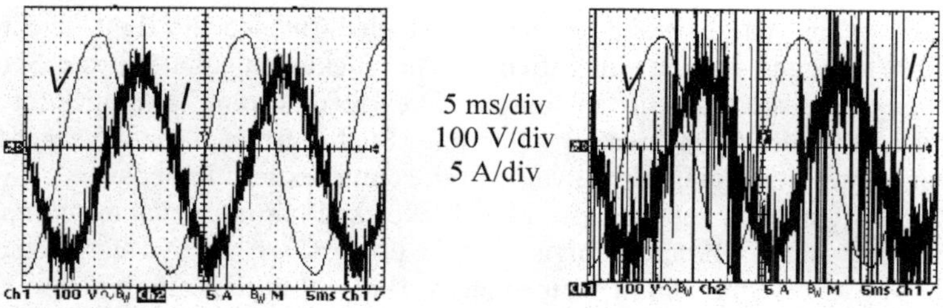

THD(I) = 4.9 % THD(I) = 7.1 %

a b

Figure 12.16. Voltage and current during: **a** regular; **b** distorted operation

Influence of the Grid Impedance

As also discussed in Section 11.5, the grid impedance influences both the flicker emissions and the harmonic values. The output current of PV plants flows over the

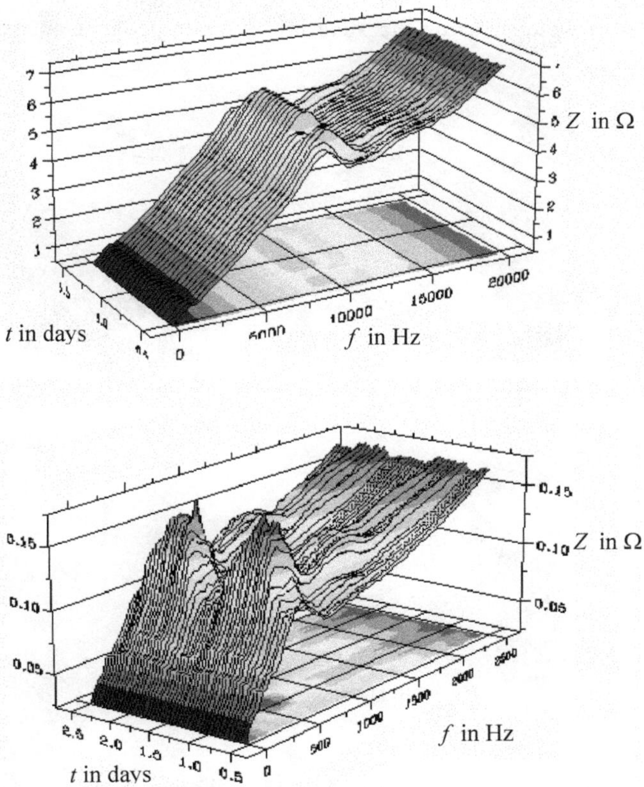

Figure 12.17. Measured frequency depending low-voltage grid impedance. *From the top*: laboratory grid, depicted up to 20 kHz; industrial grid, depicted up to 2 kHz

grid impedance and causes voltage changes which influence the light density of bulbs. These changes can be quantified by the flicker calculation; see Section 11.5.1. PV systems work as current sources. They feed current harmonics into the grid because of their PWM inverter switching. Such current harmonics generate voltage harmonics depending on the value of the complex grid impedance.

High grid impedance values cause high flicker and harmonic values compared to lower impedances. Therefore high electric powers are always connected on higher voltage levels with smaller impedances. But within the same voltage level, diverse impedances due to different grid structures also exist. An example of this fact is given in Figure 12.17 that compares the impedance behavior of two connection points over frequency and over time. The properties of a PCC in a laboratory grid are depicted in Figure 12.17a. A nearly linear impedance increase over frequency is visible up to 10 kHz. Then a resonance area follows in the frequency range between 10–13 kHz. From 13–20 kHz another linear range is visible. The high absolute impedance value of over 6 Ω at a frequency of 20 kHz is remarkable. Compared to this, the impedance of the industrial grid as depicted in Figure 12.17b is very low with an absolute value of 0.15 Ω at 2 kHz. But a clear resonance range appears above 500 Hz. In addition to this, a time dependent change of the impedance graph arises due to a switch-over in the grid supply. This changes the grid structure that determines the impedance behaviour. Resonance

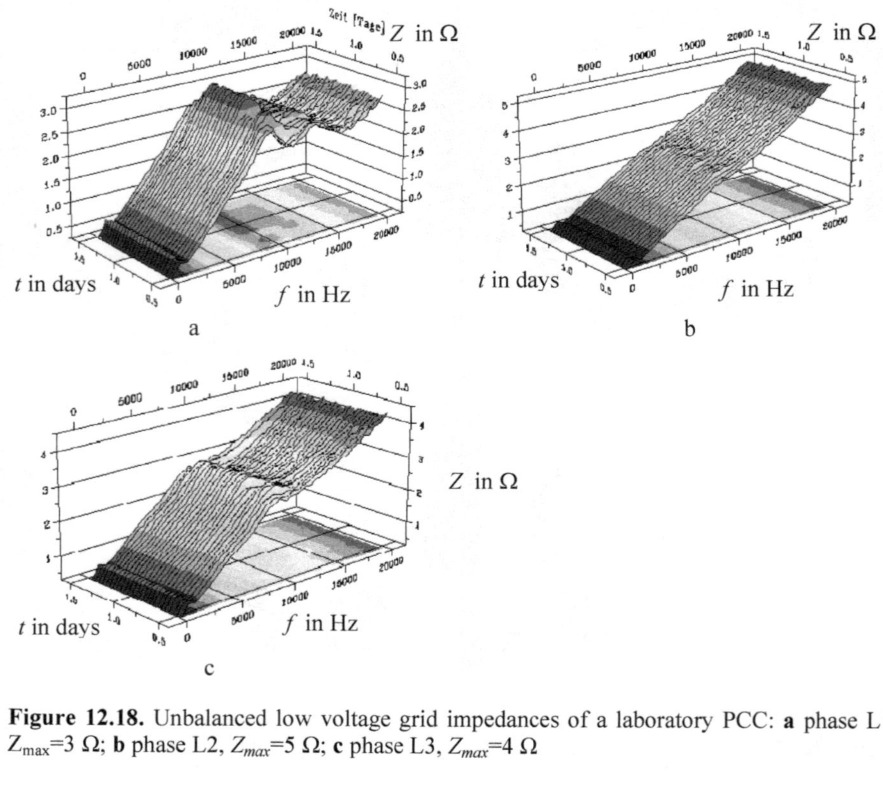

Figure 12.18. Unbalanced low voltage grid impedances of a laboratory PCC: **a** phase L1, Z_{max}=3 Ω; **b** phase L2, Z_{max}=5 Ω; **c** phase L3, Z_{max}=4 Ω

effects occur if capacitive loads are connected to the grid that has standard inductive behavior. In low voltage grids such combinations often occur if large lighting circuits with fluorescent lamps are connected [7]. Then compensation circuits should be applied to avoid undesired resonance effects on the grid.

Also within one PCC the absolute impedances value and the frequency behavior of the three-phases may be different due to uneven distributed single-phase loads or an imbalance of three-phase consumers; see Figure 12.18. The maximum values of the phases are different with 3 Ω for L1, 5 Ω for L2 and 4 Ω for L3.

12.2.11 Future Development

All future developments concentrate on the cost reduction of the whole system. Due to the high cost share the focus is on the specific production costs in €/kW$_P$ of cells and modules. Therefore two strategies are combined: first the optimisation of the production process and second the increase of the cell efficiency, both by using new technologies. An optimised cell production may mean less sawing waste for crystalline cells or a better thin-film coating from the gas phase. This results in a better use of the costly silicon material. Also, the application of tandem cells with a higher energy transformation rate due to two or more overlapping active absorption layers may increase the cell efficiency considerably.

From the grid point of view, PV systems will contribute more and more to a fundamental change of the energy system towards a non-polluting and decentralised supply unit with new control demands and high availability. The disadvantages of discontinuous power generation must be balanced by using other renewable technologies and improved storage technologies.

12.2.12 Economics

Market prices of complete PV systems vary between 4,500–5,500 €/kW$_P$ [1]. They are split into 50% for the modules, 13% for the inverter (*circa* 75 €cent/W$_P$), 7% for the support frame, 10% for cabling and 20% for planning and incident light. The energy generation costs of PV systems depends on the local insolation, they are — in 2007 in Central Europe at 30–50 €cent/kWh — far above the standard energy costs for private consumers of around 22 €cent/kWh. A so-called "*grid parity*" that means PV generation costs are just as high as the costs for private consumers should be a commonly accepted aim in Europe by 2020. An independent study of the journal "Photon" forecasts generation costs of 15 €cent/kWh until 2010 if the whole production chain is in one hand [18].

PV systems provide an energy harvest factor of four; that means that in the lifetime of the device four times the production energy amount is generated. This is below the energy harvest factor of eight as reached with wind turbines, but the total PV potential is much higher compared to wind energy.

As long as the grid parity is not achieved, a PV systems need an increased feed-in tariff to be a profitable investment. The stimulation instruments are different in the European countries and the rest of the world. Quota system models compete with guaranteed increased feed-in tariffs as first applied in Germany with the first version of the renewable energy law in 2000, which is now prepared for the second amendment.

Another energy conversion technology with a DC output is realised with fuel cells, which allow a continuous operation with a so-called "cold combustion" process. Similar to PV systems, they have to be connected *via* power electronic converters to the grid. In contrast to PV technology, they need fuel. Due to their harmless combustion outputs, fuel cells are among the environmental-friendly energies. Details of the technical cycle and possible applications are introduced in the next sections.

12.3 Fuel Cell Power Plants

12.3.1 Fuel Cell Types

In contrast to the indirect energy conversion in conventional power plants based on the combustion process and a generator, a direct conversion of chemical energy into electrical energy takes place in fuel cells [19], see Figure 12.19.

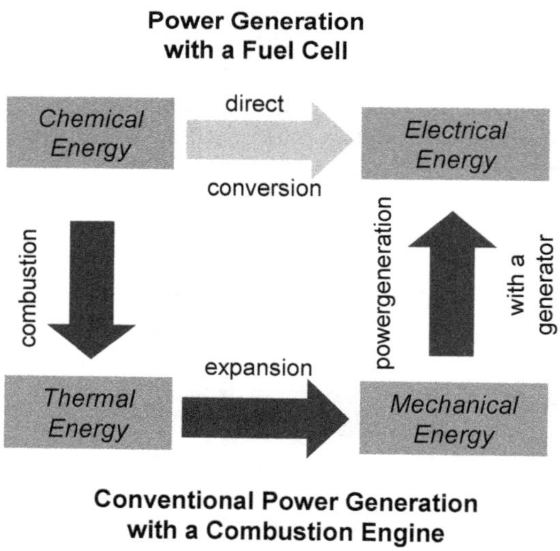

Figure 12.19. Power generation with an engine and a fuel cell

The energy conversion in a fuel cell (also referred to as "*cold combustion*") may be described as an inversion of the electrolysis of water. The overall reaction of the fuel (*e.g.*, hydrogen, H_2) and oxygen (*e.g.*, from air) is separated in two chemical reactions, where in the oxidation reaction on the anode side electrons are released and in the reduction reaction on the cathode side electrons are consumed, thus generating a usable electrical current. The anode reaction is described with

$$H_2 \rightarrow 2e^- + 2H^+ \tag{12.13}$$

The cathode reaction is

$$2\,H^+ + 2e^- + \frac{1}{2}O_2 \rightarrow +\; H_2O \tag{12.14}$$

The overall reaction is

$$H_2 + \frac{1}{2}O_2 \rightarrow +\; H_2O \tag{12.15}$$

The different fuel cell types vary in the ion conducting electrolyte that is used, as well as in the kind and flow direction of the charge carriers (Table 12.3). For a better understanding of the main processes in this chapter, we focus on two fuel cell types: the Polymer Electrolyte Membrane Fuel Cell (PEMFC) and the Solid Oxide Fuel Cell (SOFC); see Figure 12.20.

In the SOFC the water is produced on the anode side and the O^{2-} - ions move through the solid ceramic electrolyte (YSZ=Yttrium Stabilized Zirconiumoxide) while in the PEMFC the H^+ - ions move through the Nafion membrane and the water is produced on the cathode (air-) side of the fuel cell. Table 12.3 shows a short summary of the major fuel cell types.

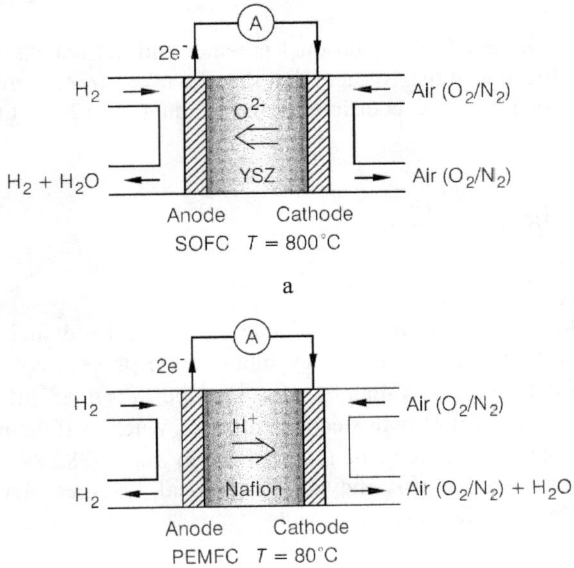

a

b

Figure 12.20. Production of water in: **a** PEMFC; **b** SOFC

Due to the nature of the fuel cell process, for an idealized conversion reaction the maximum amount of utilizable work (in terms of electricity) is given by the Gibbs free energy of formation

$$\Delta G_{\mathrm{f}} = G_{\mathrm{f,\,products}} - G_{\mathrm{f,\,reactants}} \tag{12.16}$$

Table 12.3. Major fuel cell types

FC Type	Electrolyte	Charge carrier	Operating temperature	Cell components	Fuels
PEMFC	Polymer membrane	H^+	80–120°C	Carbon based	H_2
DMFC	Polymer membrane	H^+	60–80°C	Carbon based	Methanol
PAFC	Liquid H_3PO_4 (immobilized)	H^+	200°C	Carbon based	H_2
AFC	Liquid KOH (immobilized)	OH^-	60–220°C	Carbon based	H_2
MCFC	Molten carbonate	CO_3^{2-}	650°C	Stainless based	H_2, CH_4
SOFC	Ceramic	O^{2-}	600–1,000°C	Ceramic based	H_2, CH_4, CO

PEMFC Proton Exchange Membrane Fuel Cell, DMFC Direct Methanol Fuel Cell, PAFC Phosphoric Acid Fuel Cell, AFC Alkaline Fuel Cell, MCFC Molten Carbonide Fuel Cell, SOFC Solide Oxide Fuel Cell

where in case of the PEMFC the product is water and the reactants are hydrogen and oxygen. Reduced to molar values (that may be taken from common literature) and using the stoichiometric coefficients from Equation 12.15, this relationship becomes

$$\Delta \bar{g}_{\mathrm{f}} = \left(\bar{g}_{\mathrm{f}}\right)_{\mathrm{H_2O}} - \left(\bar{g}_{\mathrm{f}}\right)_{\mathrm{H_2}} - \frac{1}{2}\left(\bar{g}_{\mathrm{f}}\right)_{\mathrm{O_2}} \tag{12.17}$$

with units of $\lfloor \Delta \bar{g}_{\mathrm{f}} \rfloor = \mathrm{kJ/mol}$.

By undertaking a general energy balance on an idealized fuel cell it may be deduced that the total (chemical) energy input to the process corresponds to the difference of the formation enthalpies Δh_{f}, For the "*burning*" of hydrogen this value is $\Delta h_{\mathrm{f}} = -241.83$ kJ/mol with steam as product, whereas if the produced water is condensed back to liquid the value is calculated to $\Delta h_{\mathrm{f}} = -285.84$ kJ/mol.

The maximum possible efficiency of a theoretical, loss-free fuel cell process is therefore given by the following expression

$$\mathit{Efficiency} = \frac{\Delta \bar{g}_{\mathrm{f}}}{\Delta \bar{h}_{\mathrm{f}}} 100\% \tag{12.18}$$

Due to the thermophysical fundamentals of the enthalpy and the Gibbs free energy, the maximum efficiency of the fuel cell process is temperature(and pressure) dependent. In Table 12.4 the theoretical values of the efficiency limit are summarized for the operating temperatures of the PEMFC and the SOFC, respectively.

Table 12.4. Temperature dependent maximum fuel cell efficiency

Temp. in °C	Δg_f in (kJ/mol)	EMF_{Max} in V	Efficiency limit
80	–228.2	1.18	80 %
800	–188.6	0.98	76 %

Δg_f molar Gibbs free energy, EMF_{Max} Electromagnetic force

In Figure 12.21 the maximum efficiency of a fuel cell is compared with the efficiency of the Carnot process which is calculated by

$$\eta_{\text{Carnot}} = 1 - \frac{T_0}{T} \tag{12.19}$$

with T_0=298K (ambient temperature) and T as the mean process temperature of the right-hand cycle considered.

As seen in Figure 12.21, for mean process temperatures below 1150K the fuel cell process has a higher efficiency potential compared to conventional (combustion based) power plant cycles. This is one of the reasons for the increasing interest in the development of fuel cell systems for electricity generation

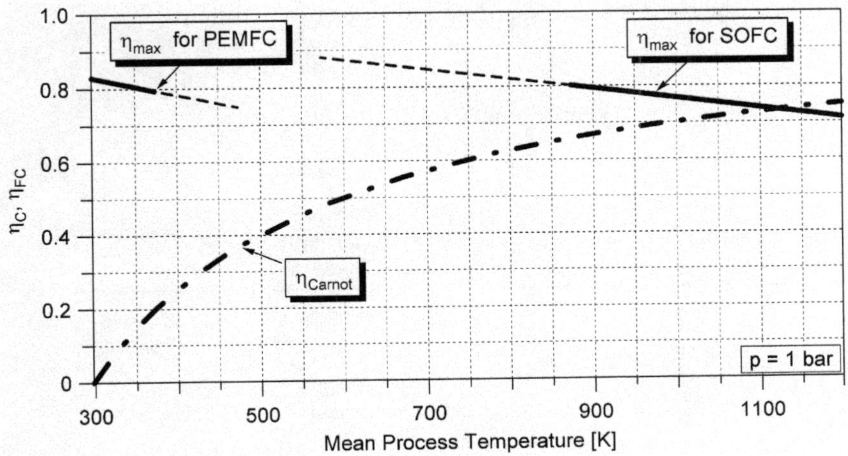

Figure 12.21. Maximum fuel cell efficiency compared to the Carnot cycle (based on pure H_2/O_2 reaction and the formation of H_2O liquid and vapor)

and Combined Heat and Power generation (CHP). However, in practical operation and with the irreversible losses of the real processes, the performance of fuel cell systems is considerably below those theoretical maximum values.

12.3.2 Energy Conversion

Depending on the fuel cell type, a specific fuel composition is required for the electrochemical current generation, *i.e.* a fuel feed rich of hydrogen, carbon monoxide, methane, methanol, ethanol or ammonia at a certain pressure, temperature and humidity level has to be provided. By the electrochemical conversion of those fuels using air (or pure oxygen) a direct current is generated at a specific voltage level that is given by the number of fuel cells which are electrically connected in series, shaping the fuel cell Stack. For practical use, a further DC to AC power conversion and voltage setting is usually required and the internal power demand of the fuel cell system also has to be covered.

The irreversible losses within the fuel cell appear as stack thermal power output (at a certain temperature level) that is usually transferred to the cathode gas stream or to a separate cooling cycle. In addition, the anode exhaust gas of every fuel cell will still contain some unreacted fuel, which can be recycled to the anode inlet or may simply be burned for additional heat generation. In order to meet all the process parameters, several internal heat sinks and sources are usually required to be connected within a fuel cell system. Depending on the process design and on the heat losses to the surroundings, excess heat may be used for heating or cooling purposes (the latter using a separate absorption chiller).

All those energy conversion steps within a fuel cell system may be simplified to a general energy balance schematic, which is shown in Figure 12.22 for a hydrogen consuming fuel cell. Other reformate components may be considered by their molar flow rate $\dot{n}$ and the corresponding enthalpy of combustion h_c.

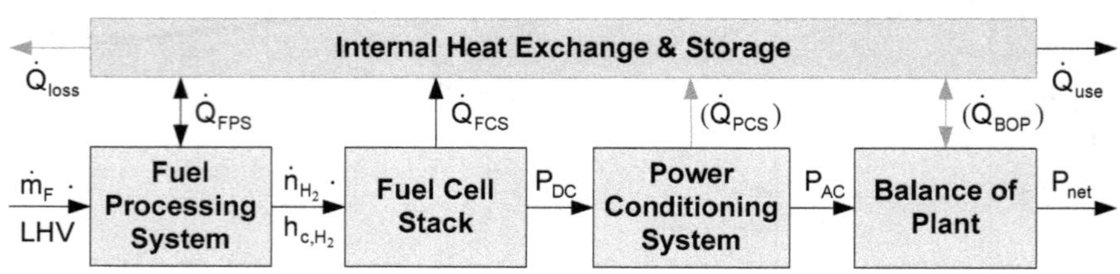

$$\eta_{FPS} = \frac{\dot{n}_{H_2} \cdot h_{c,H_2}}{\dot{m}_F \cdot LHV} \qquad \eta_{FCS} = \frac{P_{DC}}{\dot{n}_{H_2} \cdot h_{c,H_2}} \qquad \eta_{PCS} = \frac{P_{AC}}{P_{DC}} \qquad \eta_{BOP} = \frac{P_{net}}{P_{AC}}$$

Figure 12.22. General energy balance schematic for fuel cell systems

The energy input to the system is represented by the fuel mass flow and its lower heating value. By multiplying the denoted efficiency terms of the consecutive energy conversion steps the following relations for the system's electrical, thermal and total efficiency are achieved

$$\eta_{el} = \frac{P_{net}}{\dot{m}_F \cdot LHV} \qquad \eta_{th} = \frac{\dot{Q}_{use}}{\dot{m}_F \cdot LHV} \qquad \eta_{tot} = \frac{P_{net} + \dot{Q}_{use}}{\dot{m}_F \cdot LHV} \tag{12.20}$$

Considering those efficiency terms, it may be seen that the energy balance equations for a fuel cell system are identical to those of any conventional power plant cycle. However, depending on the process design, a fuel cell system at rated power output may have a higher nominal electrical efficiency than, *e.g.*, internal combustion engines or gas turbines, and – even more important in practical applications – the fuel cell system's efficiency decreases much less in part load operation.

A fuel cell system itself may not be viewed as a Regenerative Energy Source. However, fuel cells may become one of the key components within energy supply chains that are based on renewable energy. In the concept of the so called "*Hydrogen Economy*" fuel cells are expected to generate electricity from "*Solar Hydrogen*" in widespread mobile and stationary applications, where hydrogen from electro-, bio- or thermochemical processes is used as the energy storage and carriage medium. (By definition solar radiation, biomass, hydro and wind power are forms of solar energy – so the term "*Solar Hydrogen*" is supposed to cover a variety of renewable energy sources.) Whether and whenever this future concept may truly be implemented as the preferred energy supply structure is quite heavily discussed at present. But apart from that, even without a hydrogen infrastructure, fuel cells may be used for electricity and heat generation from renewable energy sources. With an appropriate fuel conditioning system the use of biomass gasification, fermentation or liquefaction processes for power generation is possible, where the medium and high temperature fuel cell types are especially useful conversion technologies.

12.3.3 Grid-connected Applications

At present, there are three main development pathways of fuel cell technology which results in the classification of portable, mobile and stationary applications; see Figure 12.23. In this chapter the focus is on stationary fuel cell systems, where the following system concepts may be defined in terms of grid-connected applications:

- Micro-CHP fuel cell systems for residential applications with 1–10 kW_{el};
- Block-type fuel cell systems for district heating and industrial applications with 50kW_{el}–1MW_{el};
- Fuel cell power plants and combined cycles (FC and gas and/or steam turbine) for large scale electricity generation (5 MW_{el}–1 GW_{el}).

In Figure 12.24 a schematic of possible fuel cell locations in the power distribution network is given. Depending on the rated power output of the system and the local codes and standards, different inverter concepts must be considered for the current feed to the power distribution grid. Micro-CHP systems in the low

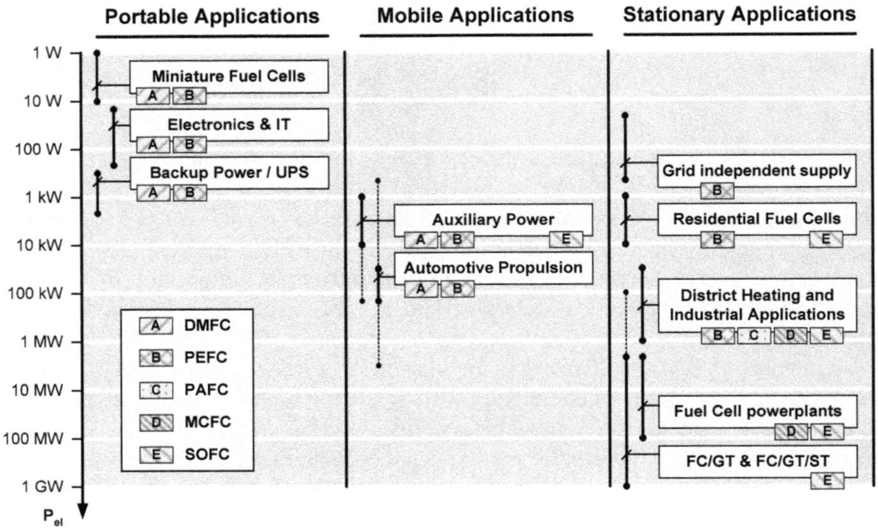

Figure 12.23. Development pathways of fuel cell technology

voltage network sections may be equipped with single-phase inverters (up to 4.6 kVA), whereas larger units are connected symmetrically to all three-phases [20].

Especially at low voltage level, the system operation may be heavily influenced by the power quality in the upstream distribution grid. In the event of a short-circuit or power outage, fuel cell systems have to be disconnected immediately by an automatic circuit breaker in order to prevent the network section from islanding.

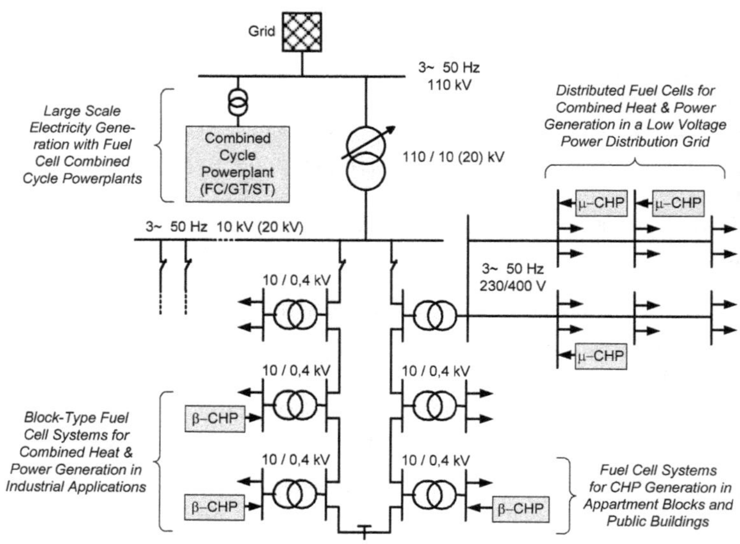

Figure 12.24. Exemplary fuel cell applications in a simplified power distribution grid

However, the necessity of detecting those faults using very sensitive voltage, frequency and/or impedance monitoring may also lead to a number of unnecessary disconnections from the grid in case of poor power quality parameters. In contrast to fluid-free energy converters like PV-modules or batteries, those unscheduled disconnections from the grid cause very difficult transitions of the fuel cell unit to idle operation with a reduced fuel input to the system, where the operating temperatures in all the relevant process stages have to be maintained.

Most of the current fuel cell system developments are intended for CHP, where the energy demand at a certain customer site is (partially) covered by cogeneration of electricity and heat, usually leading to a higher overall efficiency of the system compared to separated generation. The thermal output of the fuel cell unit may be used for room heating, water heating, steam generation and industrial processes as well as for cooling and air conditioning purposes (where an additional absorption chiller is driven by the waste heat).

Due to the particular characteristics of the end-user energy demand, some special operating conditions arise from the fact that the load-profiles of the various energy forms usually differ in the time domain. As the electrical and the thermal output of the CHP-unit are coupled directly, the following operational schemes have to be considered:

- Power-controlled operation: the rated power output of the fuel cell unit is optimized for the customers electricity demand and the mode of operation follows the electrical load-profile. Very steep gradients and peak loads may be buffered by additional devices or by the grid. The generated heat is usually supplied to a high-capacity thermal storage;
- Heat-controlled operation: the design of the fuel cell unit is optimized for the expected heat demand and the mode of operation follows the thermal load-profile. For the full coverage of the user-requirements (including peak loads) an additional burner and thermal storage devices are usually added to the system. The power distribution grid acts as a virtual electrical storage, collecting the excessive electricity from the CHP-unit as well as delivering the remaining power demand to the consumer.

For technical and economical reasons, most of the stationary fuel cell systems are designed for heat-controlled operation at present. If use of distributed CHP generators becomes more widespread in the future, some special requirements may arise in the design and operation of the power distribution grid. In a period of low power consumption, heat-controlled CHP-operation may cause an excess power generation in a certain network section, which must be transferred to the upstream voltage level. On the other hand, during a period with lower heat demand (*e.g.*, in summer) the local power consumption has to be covered almost completely by the grid. These changes to the conventional load-flow concept of hierarchical structured power distribution grids are leading to new technical requirements, especially in transformer control, fault detection and grid monitoring at low voltage level.

12.3.4 Plant Design

The design of a fuel cell power plant depends on the system operating conditions. In Figure 12.25 the fuel cell is used in an isolated network together with a wind turbine. In this example, hydrogen is used for the storage of the excess electricity produced by the wind turbine. Using the fuel cell, it can be converted into electrical power again if weather conditions do not allow operation of the wind turbine.

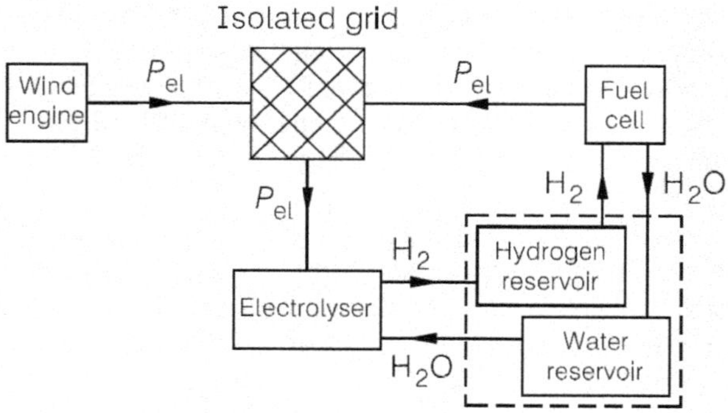

Figure 12.25. Fuel cell operation in an isolated network

For local power generation in household applications a conventionally available fuel must be used. The most common fuel supplied to residential buildings is natural gas. After the removal of sulfur compounds and reforming into hydrogen and carbon monoxide the gas may be used in a fuel cell to produce heat and electric energy. In this case heat-controlled operation is the preferred mode. Figure 12.26 shows the design of such a CHP unit.

Because operation is in heat-controlled mode the electric efficiency does not need to be as high as for centralized power generation. In many countries a politically motivated increased price for the network supply is received. It can be regarded as a heating system with additional power generation.

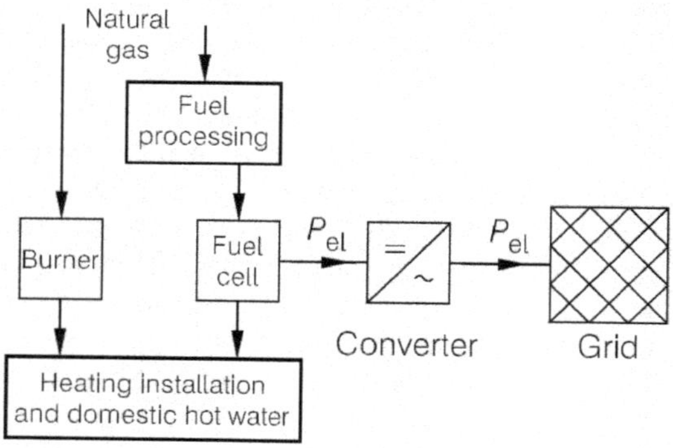

Figure 12.26. Local power generation for household applications

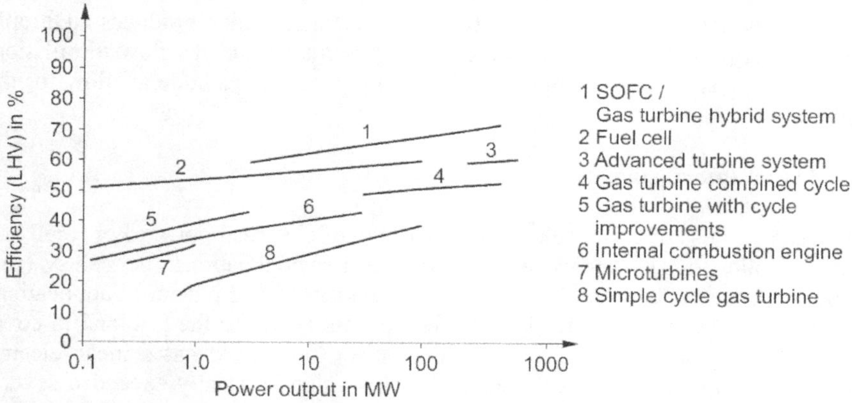

Figure 12.27. Electric efficiency of Power Generation Systems according to [21]

For centralized power generation the electric efficiency is important. Although it is possible to use hydrogen produced from renewable energy sources, the most important advantage of the fuel cell is efficiency. To recover additional costs in purchasing a fuel cell the efficiency must be higher than that of conventional plants.

As shown in Figure 12.27, the highest efficiency (related to the lower heating value of the fuel) is achieved by combining the Solid Oxide Fuel Cell (SOFC) with a gas turbine. For profitable operation this application needs to have a power of P_{el}>4 MW. For the integration of renewable resources the system can also be operated with biogas that is produced by anaerobic digestion of biomass, *e.g.*, from agricultural products or bio-waste.
Figure 12.28 shows the design of a fuel Cell power plant with a pressurized SOFC. With cell pressure increases the cell voltage, fuel and air may be compressed to approximately p=8–9 bar. The system was developed by Siemens Westinghouse and is operated on natural gas.

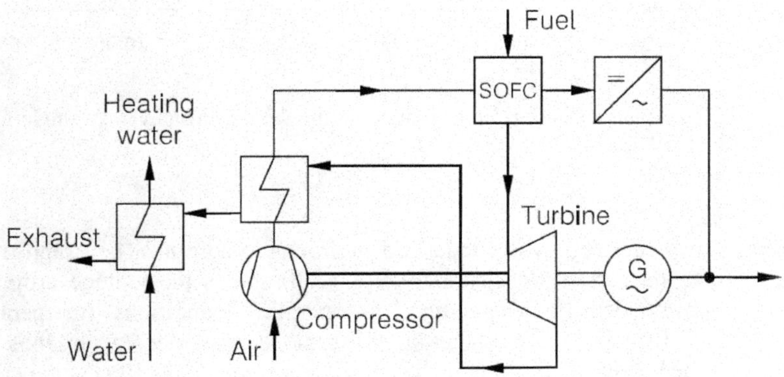

Figure 12.28. Centralized power generation by a pressurized SOFC according to [21]

On the outlet of the SOFC a gas turbine is installed which produces additional electric energy and also powers the compressors for the air gas flow installation. The exhaust gas flow of the turbine is used to preheat the cathode air flow for the SOFC.

12.3.5 Grid Interfacing

The basic purpose of a fuel cell is the generation of electrical power. For small and mobile applications in cars and trucks or remote area applications, fuel cell systems work in isolated operation. The systems for residential and industrial applications in towns use the electricity network to feed excess power to the grid and to cover peak loads. For example, an average four person household has a mean electric power consumption of about 0.4–0.5 kW. This value is usually exceeded several hundred percent during peak load times in the mornings and evenings; see Figure 12.29 [22].

For grid interfacing of Fuel Cell units there are consequently some important considerations. The electrical problems can be solved with standard technologies already available for photovoltaics. For the grid interfacing of fuel cells two additional technologies are necessary [23].

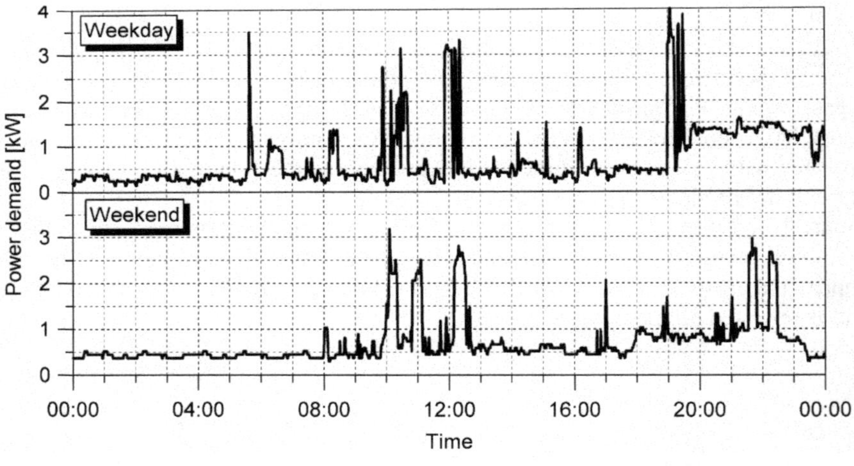

Figure 12.29. Exemplary electricity consumption during the day in an average single-family home (according to VDI 4655)

System Control

The voltage of a fuel cell is not independent from the current. The graph of the voltage against the current density in Figure 12.30 shows the voltage drop with increasing current density. Depending on operating conditions (temperature, pressure, *etc.*,) different cell voltages at the same current density are possible. Furthermore, the behaviour of a fuel cell stack changes during its life cycle because of degradation processes within the stack. To control and shift the fuel cell voltage to a fixed value, voltage regulators, DC/DC converters, and chopper circuits are used. The main types of electronic switches used in modern power electronic equipment are thyristor, MOSFET and IGBT.

Inverters

Fuel cells generate a DC power output. This might be advantageous in small systems but for grid connection a DC/AC conversion is necessary. For residential applications in households a single-phase AC inverter may suffice. For larger systems used for industrial applications the fuel cells must be connected to the grid *via* three-phase inverters. The efficiency of the DC/AC conversion influences the overall system efficiency.

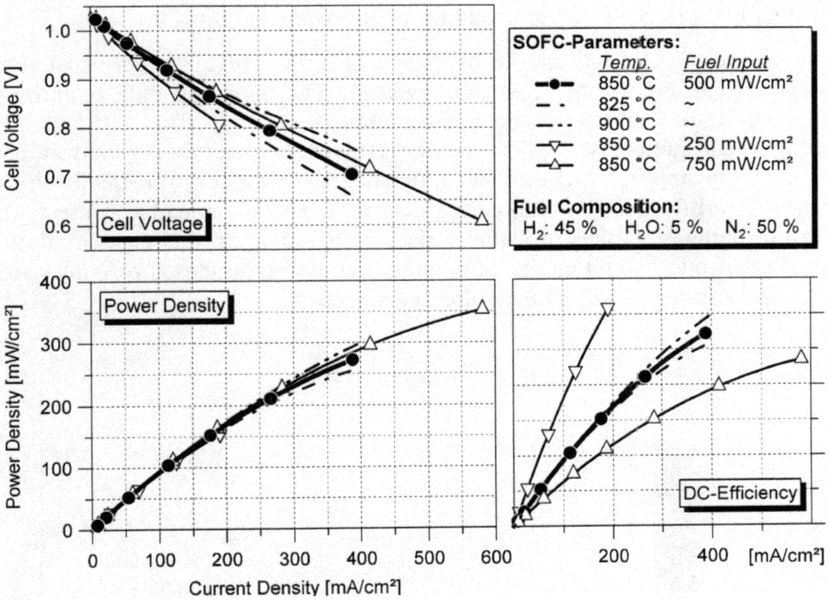

Figure 12.30. Typically performance data of a 3YSZ-SOFC (single cell)

12.3.6 Economics

Because fuel cell technology is still not ready for mass production, it is difficult to predict the economic efficiency of a fuel cell system. As described in the previous chapters, a high electrical efficiency is important. But there must also be a reduction in production costs since at present these are 2.5–20 times higher than the costs of comparable conventional systems. One reason for this discrepancy is because a market for such systems has not yet developed.

By improving production technologies the costs could be reduced as shown in Figure 12.31 for the Molten Carbonate Fuel Cell (MCFC) Systems. From 1996 to 2006 costs decreased from \$20,000/kW to \$4,900/kW for systems in a power range of P_{el}<1 MW; see Figure 12.31. For SOFC systems this will also happen during the next several years as increasing number of new systems are built.

Here are some of the applications where the usage of a fuel cell system could be cost effective:

- Grid interfacing combined with a wind turbine: the production of hydrogen is possible way to store the excess electrical energy produced by a wind turbine which cannot be supplied to the grid. Using a fuel cell system the hydrogen may be used to produce electrical energy when needed;
- Isolated networks: the fuel cell system could be used in isolated networks together with solar cells and wind turbines;
- Multiple niching: niche markets could be a useful mechanism for selling the first fuel cell products, for example auxiliary power units for yachts.

For stationary applications the average availability of a system is of great importance to determine the cost effectiveness. The operating data required to determine average availability is not yet available.
In addition, for stationary applications the efficiency compared to conventional systems is very important. If a higher efficiency and availability is demonstrated over the entire life cycle the fuel cell system may be economically efficient. The fuel cell technology's usage probability rises with increasing fuel costs. A further potential application could be the employment of currently unused resources such as bio-waste and residues from agricultural production.

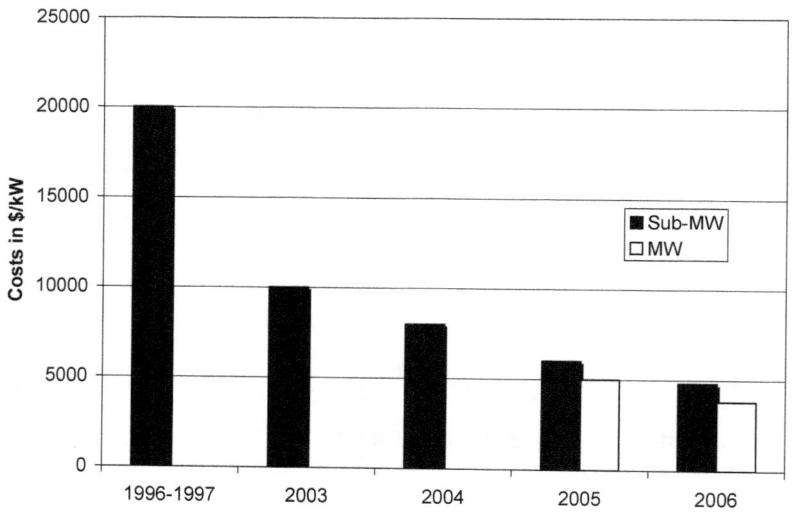

Figure 12.31. Decrease of the specific costs of an MCFC system

One critical point that has to be kept in mind is the energy required for the production of a fuel cell and the produced energy during the life cycle of the system. The lifetime is not only determined by the fuel cell itself but also by the so-called Balance of Plant (BoP) components such as fans, pumps, *etc.*

12.3.7 Future Development

Because of the wide power range in which fuel cells may be used, system concepts for different levels of electrical power output have been developed for different markets. The important future objectives can be summarized as cost reduction of used materials, production, *etc.* and increasing the lifetime of fuel cell systems including blowers, pumps, *etc.* Under laboratory scale conditions the fuel cell already demonstrated lifetimes of 40,000 h, but this must be verified in real world applications.

In Table 12.5 an overview of the different applications for fuel cells and the developments needed is given; see also [24]. With regard to grid integration the focus will be on stationary heat.

Table 12.5. Objectives for future developments of fuel cell systems

Electric power	application	Fuel cell type	Objectives
1 W	Rechargeable batteries (*e.g.*, in mobile phones)	PEMFC	- Efficient and cheap systems - Usage of liquid fuels - New production technologies for mass production
100 W	Mobile applications	PEMFC SOFC DMFC	- Cost reduction - Increasing lifetime - New production technologies for mass production
1 kW–10 kW	Stationary heat and power generation	PAFC PEMFC SOFC	- Cost reduction - Increasing lifetime - New production technologies for mass production
100 kW –1 MW	Stationary power generation	PAFC SOFC MCFC	- Cost reduction - Increasing lifetime - Increasing efficiency

PEMFC Proton Exchange Membrane Fuel Cell, DMFC Direct Methanol Fuel Cell, PAFC Phosphoric Acid Fuel Cell, AFC Alkaline Fuel Cell, MCFC Molten Carbonate Fuel Cell, SOFC Solid Oxide Fuel Cell

For the usage of renewable resources and to reduce CO_2 emissions there will be advantages of fuel cell systems in comparison with conventional power plants. By using hydrogen which is produced by solar and wind energy, fuel cell systems could be coupled with these power plants. Additionally agricultural biomass and biowaste could be used to produce electricity. The fuel cell technology may be an efficient way to stabilize the voltage in the grid in a future energy market which is based on renewable resources.

References

[1] MarketBuzz, (2007) Annual world solar photovoltaic industry

[2] Adamson KA, (2006) Small stationary survey. Fuel Cell Today Ltd., London, www.fuelcelltoday.com

[3] Adamson KA, (2007) Large stationary survey. Fuel Cell Today Ltd., London, www.fuelcelltoday.com

[4] Wagemann HG, Eschrich H, (1994) Basics of the photovoltaic energy conversion. (in German) Stuttgart, Teubner

[5] Deutsche Gesellschaft für Solarenergie, (2007) Planning and installing photovoltaic systems. Berlin, Geisel Druck

[6] Schulz D, Hanitsch R, (2000) Short-time simulation of solar inverters. 16th European Photovoltaic Solar Energy Conference, Glasgow:2430–2433

[7] Schulz D, (2002) Investigation of power quality parameters of grid-connected PV and wind energy systems. (in German). Dissertation thesis, TU Berlin

[8] Wilk H, Panhuber C, (1995) Power conditioners for grid interactive PV systems, what is the optimal size: 50W or 500kW? 13th European Photovoltaic Solar Energy Conference, Nice, France:1867–1870

[9] Woyte A, (2003) Design issues of photovoltaic systems and their grid interaction. Ph.D. thesis, Katholieke University Leuven

[10] Woyte A, (2002) Islanding and the risk of death by laughter. Photon International, vol.5, no.3:20–21

[11] Ropp M E, Begovic M, Rohatgi A, (2000) Determining the relative effectiveness of islanding prevention techniques using phase criteria and non-detection zones. IEEE Transactions on Energy Conversion, vol.15, no.3:290–296

[12] Hacker R, Thornycroft J, Knight J, (2000) Co-ordinated experimental research into PV power interaction with the supply network-phase 2. ETSU Report S/P2/00233/Rep2

[13] Schulz D, Hanitsch R, (2001) Islanding detection in Germany: current standards and development. 17th European Photovoltaic Solar Energy Conference and Exhibition, Munich

[14] Schulz D, Hanitsch R, (2002) Proposals for an international islanding detection standard. World Renewable Energy Congress VII, Cologne

[15] Schulz D, (2004) Power quality: theory, simulation, measurement and evaluation. (in German), Offenbach, VDE-Verlag

[16] Schulz D, Hanitsch R, Moutawakkil K, (2002) Power quality investigations of grid connected PV-plants. World Renewable Energy Congress VII, Cologne

[17] Schulz D, Lchamsuren B, Hanitsch R, (2001) Power quality improvements of solar systems: simultaneous inverter operation results in reduction of distortions. PCIM Power Quality Conference, Nuremberg:183–188

[18] Photon Consulting, (2006) Solar annual 2006. Solar Verlag

[19] O'Hayre R, Cha S, Colella W, Prinz FB, (2006) Fuel cell fundamentals. Wiley, New York

[20] VDEW/VDN, (2005) Generation units on the low voltage grid. (in German), Frankfurt/M., VWEW Energieverlag

[21] Fuel Cell Handbook 7th Edition, (2004) www.netl.doe.gov/technologies/

[22] VDI-Gesellschaft Energietechnik, (2007) VDI 4655 – reference load profiles of single and multi-family houses for the use of CHP systems. Beuth-Verlag, Berlin

[23] Larminie J, Dicks A, (2003) Fuel cell systems explained. 2nd Edition, Wiley, Chichester

[24] Sammes NM, (2004–2008) Journal of Fuel Cell Science and Technology. ASME Digital Library, www.asmedl.org/FuelCell

Index